Earth Surface Processes

Earth Surface Processes

Philip A. Allen

Formerly of the Department of Earth Sciences,
Oxford University;
currently at the Department of Geology,
Trinity College,
Dublin

**Blackwell
Science**

© 1997 by
Blackwell Science Ltd
Editorial Offices:
Osney Mead, Oxford OX2 0EL
25 John Street, London WC1N 2BL
23 Ainslie Place, Edinburgh EH3 6AJ
350 Main Street, Malden
 MA 02148 5018, USA
54 University Street, Carlton
 Victoria 3053, Australia

Other Editorial Offices:

Blackwell Wissenschafts-Verlag GmbH
Kurfürstendamm 57
10707 Berlin, Germany

Blackwell Science KK
MG Kodenmacho Building
7–10 Kodenmacho Nihombashi
Chuo-ku, Tokyo 104, Japan

First published 1997

Set by Setrite Typesetters, Hong Kong
Printed and bound in Great Britain
at the University Press, Cambridge

The Blackwell Science logo is a
trade mark of Blackwell Science Ltd,
registered at the United Kingdom
Trade Marks Registry

A catalogue record for this title
is available from the British Library

ISBN 0-632-03507-2

Library of Congress
Cataloging-in-publication Data

Allen, P.A.
    Earth surface processes/Philip A.
    Allen.
       p.   cm.
    Includes bibliographical references
    and index.
    ISBN 0-632-03507-2
    1. Earth.   2. Earth sciences.   3. Fluid
dynamics.   4. Sedimentation and
deposition.   5. Oceanography.
I. Title.
QB631.A386   1997
551.3—dc21   96-39095
                                    CIP

DISTRIBUTORS

Marston Book Services Ltd
PO Box 269
Abingdon
Oxon OX14 4YN
(*Orders*: Tel: 01235 465500
            Fax: 01235 465555)

USA
Blackwell Science, Inc.
350 Main Street
Malden, MA 02148 5018
(*Orders*: Tel: 800 759-6102
            617 388-8250
            Fax: 617 388-8255)

Canada
Copp Clark Professional
200 Adelaide St West, 3rd Floor
Toronto, Ontario M5H 1W7
(*Orders*: Tel: 416 597-1616
            800 815-9417
            Fax: 416 597-1617)

Australia
Blackwell Science Pty Ltd
54 University Street
Carlton, Victoria 3053
(*Orders*: Tel: 3 9347 0300
            Fax: 3 9347 5001)

Contents

Part two
Acting locally:
fluid and sediment dynamics

4 Some fluid mechanics 151

5 Sediment transport 179

6 Hyperconcentrated and mass flows 211

7 Jets, plumes and mixing at the coast 241

8 Tides and waves 267

9 Ocean currents and storms 307

10 Wind 341

Colour plate section falls between pp. 204 and 205

Preface

When astronauts in the Apollo mission looked down on Earth they saw the blue planet with its swirling weather systems. Speeding silently above the Earth, they passed over the brown wastes of the subtropical Sahara, and minutes later the peaks of the Himalaya shrouded in cloud. Not many minutes later the east coast of China disappeared from sight, the mighty Pacific Ocean was traversed and the Californian coast came into view. The whole Earth was circled in so short a time that its finiteness was scorched on the consciousness of those Apollo astronauts. And with its finiteness came a sense that this small planet permeated with water was our fragile home, strangely vulnerable. You can, it is said, see the Great Wall of China from space, but you cannot see political boundaries. Earth is a global system whose long evolution has resulted in a Garden of Eden potentially perfect for humankind. We do not live on a 'spaceship Earth'. Earth is the real thing, and for all purposes is the only natural environment which can serve as our home. We would do well to forget about spaceships and think about our own blue planet. We would do well to recognize our humanistic arrogance, sing our song of lament for Nature and work for the restoration of Eden.

> Sob, heavy world
> Sob as you spin
> Mantled in mist, remote from the happy.
> W.H. Auden (1907–73), *The Age of Anxiety*
> (1948), pt. 4, 'The Dirge'

Over the last century reductionism, that is, the method of understanding the whole by examining its parts, has served science and society well. The emergence of holistic disciplines such as Earth system science, however, signals the decline of the pre-eminence of reductionism. Holistic approaches to the Earth emphasize the linkages and feedbacks between its different components, and stress the connecting fluxes. They do not relegate reductionist approaches to redundancy, but rather draw on advances in relatively narrow sub-disciplines in building a multi-disciplinary picture of how complex systems work as a whole. This is the emphasis of *Earth Surface Processes*. It is important to appreciate the global, interactive models of the Earth's surface (Part one), but it is equally important to understand the physics under-pinning individual component processes (Part two), such as the settling of a grain through a fluid, the flow of ice in a valley glacier or ice sheet, or the instability of a slope. This then, is the broad rationale for the two-part subdivision of the book, the global and the local, which together give a dynamic and physically based appreciation of Earth surface processes.

It is a moot point as to what Earth surface processes include or exclude. My approach has been to concen-trate on the *physical* processes at the Earth's surface in the context of the global hydrological cycle, environ-mental change, and fluid and sediment transport. My subtitle is therefore something like 'Physical aspects of fluids and sediment in the environment'. I have only referred tangentially to associated problems in geochemistry, ecology, biogeography, meteorology and climatology. Nor has space permitted a more than cursory look at the impact of man on environmental change. All of these subjects could be claimed to be under the umbrella of Earth surface processes. Although I have been selective in emphasizing physical Earth surface processes, a glance at the contents list will reveal that I have been inclusive regarding my remit to involve both land surface and oceans. At a time when the coupling between oceans, atmosphere and land surface is increasingly focused upon, it seems to me essential to break down the disciplinary frontier at the water's edge. I therefore take the approach that physical Earth surface processes are a subset of the broader holistic enterprise of Earth system science. The material in this book draws on data and concepts found traditionally in physical geography, sedimentology, physical oceanography, mechanics and geology texts. It is hoped that as holistic approaches work their way through the curriculum in universities, this text will find a niche crossing these old interdisciplinary boundaries. I take my cue from Stuart Rojstaczer, who wrote in the *SEPM News* (Society for Sedimentary Geology) in November 1995:

> The skills necessary for geologists to excel in the field of environmental assessment are quite different from what is contained in the standard curriculum. The environmental geologist should be well acquainted with the Quaternary record, should understand quantitatively the development of sediment sequences and methods for their description, and should understand the physical and chemical state of fluids on the surface and in the shallow subsurface.

Earth Surface Processes is intended for advanced undergraduates taking courses in Earth science, environmental science and geography departments. It may also be useful for postgraduates who feel the need to brush up on some of the fundamentals underlying their more advanced studies. I have tried throughout the text to avoid presentation of results without an indication of how one arrives there. This means that for skilled mathematicians my development of theory will appear laboured. My experience as a teacher in Oxford, however, suggests that even here it is important to build logically the mathematical basis for a theory, clearly explaining the assumptions and boundary conditions. I prefer to avoid wherever possible presenting a mathematical result as if pulling it out of a hat. The practical exercises that accompany each chapter are designed to give the reader practice in using the quantitative aspects of the book. Course teachers could use these as they stand to accompany lectures, or they could adapt them by a careful choice of their own examples.

A word or two is required on my citation of references. I have tried to produce a text that is relatively uninterrupted by citations. However, the normal scholarly conventions require me to provide credits to the sources of data. Consequently, the 'References' section at the end of each chapter is not intended as a reading list, though some of the references found there will undoubtedly serve well as supplementary reading. Texts which I believe provide further background to the topics discussed in each chapter appear under 'Further Reading'.

My approach throughout is to illustrate physical principles of Earth surface processes by observation of the processes taking place on the Earth's surface today, backed up by an appropriate level of theory. This has involved a welcome sacrifice in the level of comprehensiveness and detail that so often afflicts Earth science and geography texts. The more inductive of readers can find a wealth of useful information in an arsenal of physical geography and 'environmental' books, including Goudie's *Environmental Change* (3rd edition, 1992, Clarendon Press) and *Human Impact on the Natural Environment* (1993, Blackwell Publishers); *Environmental Systems: an Introductory Text* (2nd edition, 1992, Chapman & Hall) by White *et al.*; *Geosystems: an Introduction to Physical Geography* (2nd edition, 1994, Macmillan) by Christopherson, and Summerfield's *Global Geomorphology* (1991, Longman) to name a few. I have found a number of books invaluable in their provision of physical insights into Earth surface problems. These include the seminal works of John R.L. Allen, such as *Principles of Physical Sedimentology* (1985, George Allen & Unwin), and of Gerry Middleton with co-workers, such as *Mechanics of Sediment Movement* (with John B. Southard, 1977, Society of Economic Paleontologists and Mineralogists) and *Mechanics in the Earth and Environmental Sciences* (with Peter R. Wilcock, 1994, Cambridge University Press). These books in particular have been deeply influential in forming my own approach to problems in the surface Earth sciences.

It is a great pleasure to thank a number of individuals for their help and criticism in preparing this text. Simon Rallison at Blackwell Science has been a constant source of encouragement and advice. I also thank a number of geologists and geographers who have read early drafts or near-completed chapters and freely gave their opinions; Pete Burgess (Caltech), Sanjeev Gupta (Edinburgh), Niels Hovius (Oxford), Jon Verlander (Oxford), Rachel McDonnell (Oxford), Hugh Sinclair (Birmingham). I am also grateful for the reviews of Mike Summerfield (Edinburgh), Andrew Goudie (Oxford), Mike Kirkby (Leeds) and Rudy Slingerland (Penn State). I have also benefited from discussions with a large number of friends and colleagues on certain aspects of the book, including Martin Brasier, Richard Corfield, Marc Audet, Philip England and Colin Stark in Oxford, Mike Leeder (Leeds), Peter Talling and Mike Stewart (Bristol). Finally, my wife Carolyn and children Lorna, Frances and Richard have supported me in a venture which had initially seemed rather straightforward and which turned into an over-consuming preoccupation. The result is dedicated to them.

Philip Allen

Part one
Thinking globally:
the global Earth surface system

1 Fundamentals of the Earth surface system

If God had consulted me before embarking on the Creation,
I would have suggested something simpler.

    Alfonso of Castille (15th century)

Chapter summary

The Earth surface system is dominated by the interaction of climate and topography, the energy for which comes from solar radiation and from the Earth's internal convection. Solar radiation drives a global hydrological cycle which stabilizes the climatic zones of the Earth and makes the planet habitable for humans. Processes deep within the interior of the Earth are responsible for the relative motion of lithospheric plates whose interaction generates much of the surface topography.

The solar radiation received by the Earth has a short wavelength which enables most of it to penetrate the atmosphere to reach the Earth's surface. However, the radiation emitted back from the Earth is of a longer wavelength, a substantial proportion of which is trapped within the atmosphere, a phenomenon known as the greenhouse effect. The greenhouse effect is sensitive to concentrations of certain gases in the atmosphere such as carbon dioxide.

The global climate system is mediated by the transfer of water between the reservoirs in the oceans, atmosphere and the upper part and surface of the lithosphere. The abundance of water on Earth and the coexistence of water in gaseous, liquid and solid states is crucial to the efficient stabilization of the climate system. This is a unique situation in the solar system. Each of the stores in the global hydrological cycle involves particular residence times of water molecules, and characteristic fluxes between them. The oceans have long residence times, and polar ice caps even longer. This buffers the climate system against rapid change, but also causes sudden perturbations to have

medium- to long-term effects that are not easily predicted.

Heat is transferred through hydrological processes by atmospheric motions, wind drift of the surface layer of the oceans, and by deeper, slow oceanic circulation. These atmospheric and oceanic systems work synergetically and interactively. The main driving forces in this complex global system are intense convection in the tropical atmosphere over the high sea surface temperatures of warm pools, in the polar ocean where cold, dense water sinks, and in the subtropical ocean where evaporation causes the sinking of saline surface waters.

The imbalances between precipitation and evapo-transpiration result in a surface runoff on the continents which is crucial for the denudation of the land surface and the transfer of particulate and dissolved matter into the oceans. Runoff reflects the interplay between topography and climate, with a rough latitudinal zonation disturbed by topographic effects and the influences of the distribution of land and sea.

The mass of the biosphere is minute within the Earth surface system yet its impact is profound. Chemicals pass through the pathways of the biosphere in biogeochemical cycles. The flux of nutrients such as nitrogen and phosphorus is critical in explaining the activity of the biosphere. The carbon cycle is driven by the forces of respiration and photosynthesis. Although the mass of carbon (in the form of CO_2) in the atmosphere is small compared to that of the ocean, it exchanges carbon very rapidly with ocean water, the soil and with terrestrial biota. Carbon dioxide is also a greenhouse gas, so plays a major feedback role in global temperatures and precipitation.

The topography of the Earth's surface is the result of the interaction of primary tectonic mechanisms and the modifying agents of erosion. Tectonic and erosive processes act at a variety of spatial scales. Topography is produced by the tectonic processes leading to heterogeneities in the lithosphere related to its density and/or thickness which result primarily from the horizontal motions of plates. Isostasy is the principle explaining how the Earth compensates for these heterogeneities. Topography is also produced by the response to processes involving viscous flow in the mantle, such as convection and mantle plume activity. This is termed dynamic topography.

Application of isostasy to sites of continental collision where the lithosphere is substantially thickened demonstrates that at certain wavelengths of topographic load, such as that of the Himalayan mountain chain, the lithosphere responds by a broad-wavelength flexure, causing sedimentary basins to form. At longer wavelengths of topographic load, such as the Tibetan plateau, a vertical, local, isostatic balance can be assumed. The topography at sites of extension can be explained by the isostatic effect of lithospheric thinning, or by the effect of the impingement on the base of the lithosphere of a hot plume of mantle material, as in East Africa. Part of the subsidence following extension can be modelled as similar to the cooling of newly formed oceanic lithosphere at mid-ocean ridges.

Dynamic topography may result from a number of mechanisms involving the dynamic effect of the mantle. Continental and oceanic surfaces are known to be uplifted over mantle plumes (hotspots) and their tracks. More speculatively, some broad sags may be related to mantle downwelling. Dynamic effects causing far-field tilting of continental plates may result from the onset and evolution of subduction of a cool plate beneath the overriding continent. Any signal produced by this mechanism is, however, difficult to decipher from other processes producing topography. Dynamic effects due to the insulation of supercontinental assemblies of plates in the geological past may be important in explaining long-term cycles of continental aggregation and supercontinent dispersal.

1.1 Introduction

The Earth's surface is constantly changing. The energy for this change comes from a number of sources. At the largest scale, the Earth has a planetary energy because of its position within a solar system powered by gravitational forces. It rotates on its axis as it moves around the Sun, the water in its oceans is affected by the gravitational pull of the Moon and to a lesser extent the Sun, and it is bathed in solar radiation. This solar radiation drives atmospheric circulation, controls the Earth's water budget and provides the energy for biological activity. These energy sources are all external. But the Earth's surface is also affected by processes from within. Heat generated in the interior of the Earth, largely from radioactive decay, drives a slow, deep convection. Thermal convection in turn is responsible for the very long-wavelength topography of the Earth's surface, and indirectly for the relative motion of a number of rigid plates across the Earth's surface. These plates, which comprise the relatively cool lithosphere, collide and override each other, forming ocean trenches, island arcs, mountains and plateaux. Where they separate, mid-ocean ridges, oceanic basins, continental margins and continental rifts are formed. Relative plate motion also generates earthquakes and volcanoes. This internal energy is therefore primarily responsible for the Earth's topography at various scales, providing potential energy for a host of denudational, transport and depositional processes.

The processes operating on the surface of the Earth (or in its seas) can be divided into those that are due to processes originating externally to the Earth (*exogenic*) and those caused by processes within the Earth (*endogenic*). Exogenic processes therefore include river, wind and glacial action on land, and tides, currents and waves in the ocean. Endogenic processes include volcanic activity and earthquakes, and horizontal and vertical motions of the Earth's surface caused by plate tectonics and mantle convection.

Solar radiation heats the atmosphere and surface of the Earth, resulting in a global average surface temperature of 288 K (15°C). Solar radiation drives the *hydrological cycle*, representing the continuous exchange of water in its solid, liquid and gaseous states between the atmosphere, oceans and land surface. The portion of the total incoming solar radiation reaching the Earth's surface as light supplies the energy for photosynthesis of plants and is therefore critical for the operation of the biosphere. The net excess of solar radiation in low latitudes and net deficit in higher latitudes results in a poleward energy transfer. This takes the form of a general atmospheric and oceanic circulation that controls and stabilizes the climatic zones on the Earth. The interaction of climate, topography and geology controls the weathering, transport and deposition of sediment on the Earth's surface (Chapter 3).

1.2 The Earth's energy balance

The primary energy source for the Earth is solar radiation, being 99.98% of all energy received (heat flow from the interior of the Earth accounts for 0.018% and tidal energy 0.002% [1]).

The energy radiated from a body such as the Sun or the Earth is proportional to its radiating temperature. The Sun's energy output includes radiation in the visible spectrum (c. 50%, 0.4–0.7 μm in wavelength), a great deal at longer wavelengths (nearly 50%, infrared, 0.7–1000 μm wavelength), and a small proportion of about 1% at short wavelengths (ultraviolet, 0.1–0.4 μm) (Fig. 1.1). The relation between the relative energy emitted and the wavelength is known as a *Planck curve*. Both the total energy and the wavelength of the maximum emission are determined by the radiating temperature. The total energy is given by the *Stefan–Boltzmann law*

$$E = \sigma T^4 \qquad (1.1)$$

where σ is the Stefan–Boltzmann constant ($5.5597 \times 10^{-8}\,\mathrm{W\,m^{-2}\,K^{-4}}$) and T is temperature. The peak emission is given by *Wien's law*, which states that the higher the radiating temperature, the shorter the wavelength of the maximum emission. Since the Sun has a surface temperature of 6000 K, the peak emission is $2897/T = 48$ μm, within the visible part of the spectrum. However, the Earth is much cooler than the Sun, with an average surface temperature of 288 K (15°C). Consequently, it radiates with a longer peak emission wavelength of about 10 μm—well into the infrared part of the spectrum. This marked difference between the incoming solar radiation and the outgoing radiation from Earth has a profound effect on the climate system.

The total energy emitted by the Sun is obviously not the same as the total solar radiation received by the Earth. The energy flux (in watts per square metre) received at the orbit of the Earth depends inversely on the distance between the two bodies. Consequently, if the Stefan–Boltzmann law (equation (1.1)) gives

$$E_{sun} = (5.559 \times 10^{-8})\,(6000)^4 = 7.20 \times 10^7\,\mathrm{W\,m^{-2}} \quad (1.2)$$

then the *solar constant* is

$$S = E_{sun}\left(\frac{R_{sun}}{R_{orbit}}\right)^2 = 1557\,\mathrm{W\,m^{-2}} \qquad (1.3)$$

where R_{sun} is the radius of the Sun, and R_{orbit} is the radius from the Sun to the orbit of the Earth. This gives a maximum envelope for the solar radiation received at the orbit of the Earth, making the assumption that the sun is a so-called *blackbody radiator*, or perfect emitter (Fig. 1.1). The energy received by the Earth now depends simply on the size of the Earth and the amount reflected back into space. If the radius of the Earth is R and the reflectivity of the Earth is A, otherwise known as *albedo* (typically about 0.3), the total solar energy received by Earth is

$$E_{in} = \pi R^2 S(1-A) \qquad (1.4)$$

The Earth can also be treated ideally as a blackbody radiator. The total energy emitted from the entire surface area of the spherical Earth is then

$$E_{out} = (4\pi R^2)\,(\sigma T^4) \qquad (1.5)$$

The energy balance between incoming solar radiation and outgoing radiation is therefore

$$\pi R^2 S(1-A) = 4\pi R^2\,(\sigma T^4)$$
$$\frac{S(1-A)}{4} = \sigma T^4 \qquad (1.6)$$

We can now substitute into equation (1.6) reasonable values for the solar constant ($1295\,\mathrm{W\,m^{-2}}$) and albedo (0.3) in order to calculate the average surface temperature of the Earth,

$$\frac{1295 \times 0.7}{4(5.5597 \times 10^{-8})} = 265\,\mathrm{K}$$

which is 23 K below the known average surface temperature of the planet. There must be another mechanism or set of mechanisms that are warming the Earth's surface. What has been neglected from the analysis is the blanketing effect of the Earth's atmosphere. We need to pause here and consider in a little more detail the fate of incoming solar radiation impinging on the outer surface of the atmosphere (Fig. 1.2).

Molecules which are common in the atmosphere, such as H_2O and CO_2 absorb radiation at particular wavelengths. Water vapour absorbs strongly at 12 μm, and CO_2 absorbs strongly at 15 μm. Ozone (O_3) and

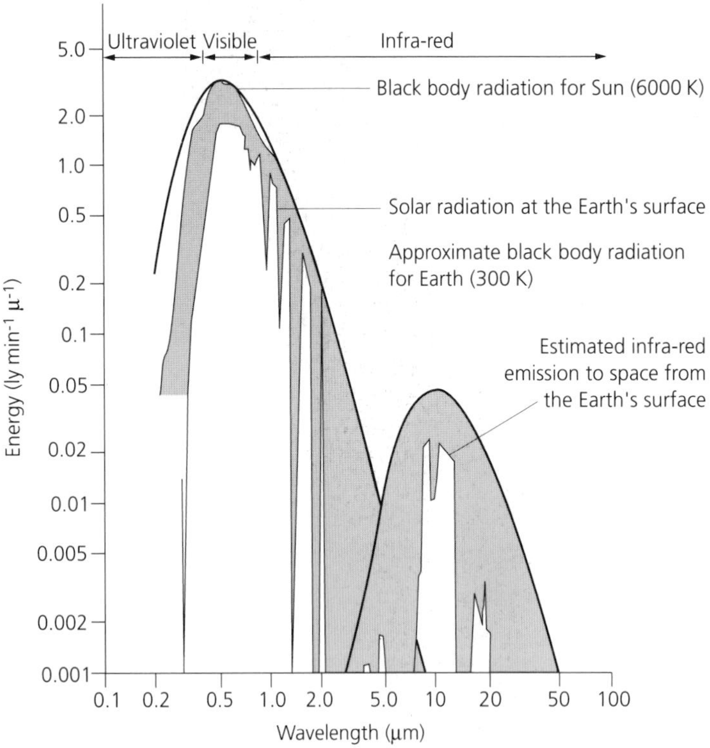

Energy (ly min^{-1} μ^{-1})

5.0 — Ultraviolet Visible | Infra-red

Black body radiation for Sun (6000 K)

Solar radiation at the Earth's surface

Approximate black body radiation for Earth (300 K)

Estimated infra-red emission to space from the Earth's surface

Wavelength (μm)

Fig. 1.1 Electromagnetic spectra for solar and terrestrial radiation (Planck curves) showing blackbody emittance as envelopes. After Sellers (1965) [2].

O_2 absorb radiation at shorter wavelengths. Much of the short-wavelength (<4 μm) incoming solar radiation therefore penetrates the atmosphere and reaches the ground as light. However, most of the shorter-wavelength radiation (ultraviolet, <0.4 μm) is absorbed in the upper atmosphere by O_3 and O_2 thereby protecting life on the surface of the Earth from its harmful effects. The bulk of the small amount of incoming longer-wavelength radiation (>4 μm) is absorbed by water vapour and CO_2 in the atmosphere. Consequently, only 51% of solar radiation reaches the Earth's surface. The roles of backscattering, absorption and reflection in the atmosphere are shown in Fig. 1.2.

The radiation absorbed in the atmosphere and received by the surface of the Earth is emitted back into the atmosphere as a long-wavelength (>4 μm, with a maximum in the infrared at 10 μm) radiation. This long-wavelength radiation is particularly susceptible to being absorbed by water vapour and CO_2 in the atmosphere. This property of allowing short-wavelength radiation from the Sun through the atmosphere to reach the Earth, but absorbing and retaining the outgoing long-wavelength radiation, is termed the *atmospheric greenhouse effect*. Apart

from radiation, the Earth also gives off a heat flux through conduction from its hot surface (sensible heat), and in the form of latent heat flux through the processes of evaporation and condensation (Section 1.3).

The energy balance can be rewritten to take into account the opacity to outgoing radiation or *emissivity*, e, of the Earth's atmosphere

$$e\sigma T^4 \frac{S}{4} = (1 - A) \tag{1.7}$$

This represents the fundamental relation between incoming radiation, reflection from the Earth's surface and absorbtion in the atmosphere that together control climate and climate change. It is a simplification for what is an extremely complex climate system, involving a number of important feedbacks. For example, an increase in temperature increases water vapour content in the atmosphere, driving a greenhouse effect and further warming. Another positive feedback effect from increased temperatures is the melting of snow and ice, decreasing the Earth's albedo and causing further warming.

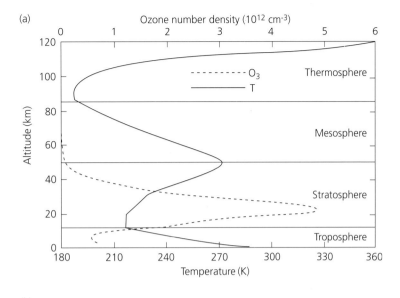

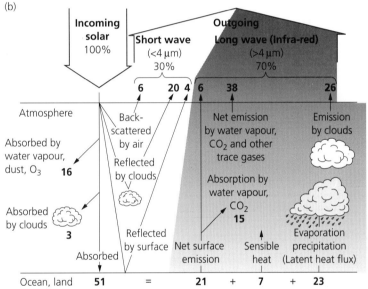

Fig. 1.2 The vertical stratification and heat budget of the atmosphere and Earth. (a) Vertical profile of temperature (solid line) and ozone density (dashed line). After Harrison *et al.* (1993) [3]. (b) Mean annual radiation and heat balance, with units assigned so that the total incoming radiation is 100. Data from the US Committee for the Global Atmospheric Research Program (1975).

The incoming radiation is thus balanced by the outgoing radiation. Over time, net gains in radiation, or net losses, should result in a global warming or cooling respectively. About 70% of the incoming solar radiation (the portion not reflected back into space) is used in interacting with the hydrological cycle.

1.3 The hydrological cycle
1.3.1 Role of the hydrological cycle in the global climate system
The hydrological cycle plays a crucial role in the climate system of the Earth. Water permeates all of the major components of the Earth's climate system—the ocean, the atmosphere, the lithosphere (or at least its upper part), the cryosphere (realm of snow and ice) and the biosphere. Any investigation of the global climate system must be based on a knowledge of hydrological processes. The abundance of water on Earth and its occurrence in multiple forms is unique in the solar system. The presence of water in the form of vapour, liquid and ice, and the relative ease with which water may transform from one phase to the other allows it to play a strongly stabilizing role in the Earth's climate.

Water stores and fluxes

Over 97% of water is stored in the oceans. Of the remainder, most is held in ice sheets and glaciers, some in groundwater, and minute percentages in lakes, the atmosphere, the soil and in rivers (Fig. 1.3). The transfers of water between these different stores or reservoirs have profound impact on Earth surface processes and require enormous inputs of energy. Water is transfered from the ocean to the atmosphere by evaporation and from the land by a combination of evaporation and transpiration from plants (*evapotranspiration*). A return flux of water is achieved by precipitation in the form of rain and snow. Over the continents the precipitation in general exceeds evapotranspiration, which leads to a flux of water from the continents to the oceans in the form of surface *runoff*. This runoff, although small in volume, is crucial in the physical evolution of the land surface, in the transfer of particulate sediment from erosional areas to depositional sites, and in the fluxing of chemical species in solution from the weathered land surface to the oceanic reservoir.

Since the surface of the Earth is dominated by the saline waters of the oceans, the human occupation of the land surface is dependent on the distillation of seawater to fresh water and its precipitation on the land. The relatively small percentage of fresh water in the hydrological cycle demands that it is continuously recycled. The fluxes of water between the different storages are large, a molecule of water having a characteristic residence time in each store (Table 1.1). For example, the turnover of water in the atmosphere is very rapid (with a residence time of only 8–10 days), the rivers only slightly less rapid (up to 2 weeks), whereas the oceans have long residence times of over 4000 years and water may remain in ice caps for tens of thousands of years [5]. The residence times of water resources are also of great relevance to the impacts of pollution, damage to ocean, lake and groundwater being much longer-lasting than to rivers. The extremely rapid turnover of the atmosphere requires the energy of a large proportion of the solar radiation received by the Earth. This can be appreciated by considering the

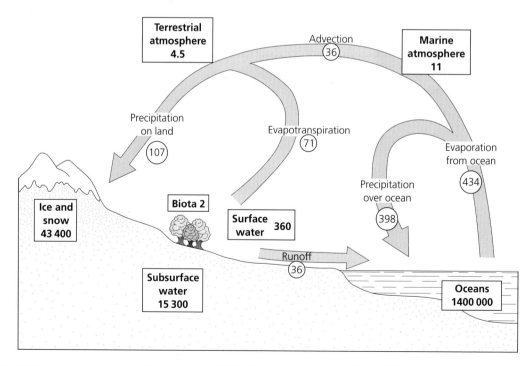

Fig. 1.3 The global water cycle, with its storages in 10^{15} kg (boxes) and fluxes in 10^{15} kg y^{-1} (figures circled). Water in the atmosphere constitutes 0.0001% of the total water in the hydrological cycle but is crucial for the efficient functioning of the system. Data from Chahine (1992) [4].

Table 1.1 Storages, fluxes and residence times in the global hydrological cycle. From Berner & Berner (1987) [1].

Storages	Fluxes
Atmosphere	Precipitation
Oceans and seas	Runoff slopes and channels
Lakes and reservoirs	Evaporation
Rivers	Horizontal vapour flux
Wetlands	Infiltration
Biological water	Percolation
Soil water	Groundwater flows
Groundwater	
Ice	

Residence times	
Atmosphere	8–10 days
Oceans and seas	4000 years +
Lakes and reservoirs	up to 2 weeks
Rivers	up to 2 weeks
Wetlands	years
Biological water	1 week
Soil water	2 weeks to 1 year
Groundwater	days to thousands of years
Ice	tens to thousands of years

behaviour of water vapour in the atmosphere. In its short residence time of 8–10 days, a molecule of water vapour is likely to have travelled an average distance of 1000 km.

Quantifying the global water balance is a formidable task requiring coordinated international efforts at gauging precipitation and runoff and estimating water volumes in all of the zones of the hydrological cycle. The fact that almost 70% of all fresh water is locked up in polar ice and in glaciers [6] has major implications for the effects of climate change, since large-scale warming or cooling causing shrinkage or expansion of the ice store will have profound effects on the entire hydrological cycle. Although some investigations have been made into the changes in the hydrological cycle over the last post-glacial period (approximately the last 10 000 years) (Chapter 2), remarkably little is known of the likely hydrological changes on a future warmer greenhouse Earth. This is now a major focus of research.

Why is the Earth unique in the solar system?

Among our solar system neighbours (Table 1.2), only

Table 1.2 Water in the solar system. Physical properties of the inner planets, their atmospheric compositions and disposition of water. Modified from Webster (1994) [7].

	Mercury	Venus	Earth	Mars
Specifications				
Planetary mass, 10^{23} kg	3.4	48.7	59.8	6.43
Planetary radius, km	2439	6049	6371	3390
Gravitation, m s^{-2}	3.8	8.9	9.8	3.7
Solar distance, 10^6 km	58	108	150	228
Solar irradiance, W m^{-2}	9200	2600	1393	596
State				
Mean surface temperature, K	442	700	288	210
Mean planetary albedo, %	6	71	33	17
Surface pressure, hPa	0	7900	1013	6
Atmospheric composition %				
CO_2	0	95	3×10^{-2}	>50
N_2, A	0	<5	79	<50
O_2	0	$<4 \times 10^{-3}$	21	1×10^{-1}
H_2O	0	1×10^{-2}	1	$\leq 1 \times 10^{-1}$
HCl	0	1×10^{-4}	0	0
HF	0	2×10^{-6}	0	0
CO	0	2×10^{-2}	1×10^{-5}	1×10^{-1}
Water disposition, kg				
Atmospheric mass	0	4.2×10^{20}	5.3×10^{18}	2.4×10^{10}
Liquid	0	0	1.4×10^{21}	0
Ice	0	0	4.3×10^{19}	1×10^{17}
Gas	0	4.2×10^{16}	1.6×10^{16}	2×10^{13}

the Earth has large reservoirs of liquid water and regions of ice over both continents and oceans [7]. The atmosphere contains water in all three forms: as water vapour comprising 1% of the entire atmospheric mass, and as suspended ice and water droplets in clouds. The bulk of the water on the Earth is, however, in liquid form. On Mars there is no known liquid water, though water vapour is an important constituent of the atmosphere and ice occurs in the polar ice cap and perhaps also as permafrost. On Venus water vapour exists in abundance in the atmosphere but there is no liquid water or ice, so that the total mass of water on the planet is far smaller than on the Earth. Without liquid water, the climate of Mars (and also of Mercury) is strongly coupled to changes in its radiation balance with outer space, giving rise to extremes in temperature between day and night. The large store of liquid water in the oceans dampens external fluctuations in supply of heat, releasing heat gradually and allowing a latitudinal transport of heat because of the liquid form of the water (Section 1.3.3).

The Clausius–Clapeyron relation

Why is it that water on the Earth exists close to its triple junction? This can be appreciated by considering the phase transitions of water as a function of temperature and water vapour partial pressure (Fig. 1.4a). The likely trajectories of the Earth, Venus and Mars [8] indicate clearly that the triple point where all three phases exist in equilibrium is very close to the Earth's present conditions, so that all three forms of water may coexist at virtually any point on the Earth's surface (with the exception perhaps of the polar regions during winter) (Fig. 1.4b).

The phase transition lines are nonlinear and can be written in the form of the *Clausius–Clapeyron equation*

$$\frac{\mathrm{d}\ln e_s}{\mathrm{d}T} = \frac{L_e, L_s}{R_v T^2} \tag{1.8}$$

where e_s is the saturation vapour pressure (in newtons per square metre) as a function of temperature T (Kelvin), L_e, L_s are the latent heats of evaporation $(2.5 \times 10^6\,\mathrm{J\,kg^{-1}})$ and sublimation (direct from solid to vapour) $(2.84 \times 10^6\,\mathrm{J\,kg^{-1}})$ respectively, and R_v is the specific gas constant for water vapour $(0.462\,\mathrm{J\,kg^{-1}\,K^{-1}})$. If it is assumed that the latent heats of evaporation and sublimation are constant with temperature, the Clausius–Clapeyron equation (1.8) can be solved to give the vapour pressure over liquid

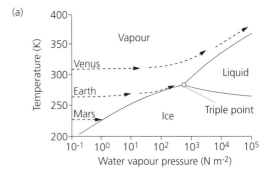

(a)

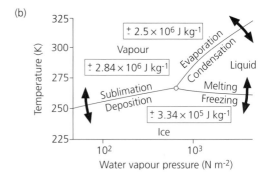

(b)

Fig. 1.4 (a) The phases of water as a function of water vapour pressure and absolute temperature. The phase transition curves between vapour, liquid and ice are given by the Clausius–Clapeyron equation (1.8). The dashed lines are hypothetical trajectories for the climate of Venus, the Earth and Mars through time. After Rasool & de Bergh (1970) [8]. (b) The water phase diagram close to the triple junction with the physical parameters pertaining to the Earth, with the latent energies for the phase transitions. After Webster (1994) [7].

water or solid ice as a function of temperature. Integrating the Clausius–Clapeyron equation gives

$$e_s(T) = e_s(T_0)\exp\left\{\frac{L_e, L_s}{R_v}\left(\frac{1}{T_0} - \frac{1}{T}\right)\right\} \tag{1.9}$$

where $e_s(T_0)$ is the saturation vapour pressure at temperature T_0. Inspection of (1.9) shows the saturation vapour pressure to depend exponentially on absolute temperature (in Kelvin). For example, the vapour pressure of the atmosphere in the tropics is more than an order of magnitude greater than that over the poles.

During the evolution of the Earth, outgassing of water vapour would have caused an increase in the vapour pressure with time, leading to greater absorbtion of outgoing radiation, creating an early greenhouse effect and leading to warming of the planetary surface. Surface warming in turn affects the phase transition equilibria because of the Clausius–Clapeyron effect. On Venus, the temperature–vapour pressure trajectory caused it to miss the vapour–liquid phase transition, leading to a 'runaway' greenhouse effect. On Mars, however, the trajectory intersected the sublimation–deposition transition, so that transfer is only possible between vapour and ice. This takes place, along with the deposition and sublimation of CO_2, at the Martian winter pole. The larger number of other thermodynamic possibilities on the Earth (Fig. 1.4b) explains the complexity and variability of the hydrological system.

We know from the Clausius–Clapeyron relation that saturation vapour pressure depends on temperature. It follows, therefore, that the global distribution of water in its three phases is determined by the global temperature structure. Before looking at this problem in further detail, it is important to stress the time-scales over which the various stores of water in the hydrosphere modulate climate. The ability of the ocean to mix vertically (see Chapter 9), rather than acting as a static pond, causes it to release and absorb heat on long time-scales. It takes of the order of hundreds to thousands of years for the global ocean to mix vertically throughout its entire depth. The modulation of climate by the oceans therefore involves the deep, slow circulation of its waters as well as the more rapid mixing at its surface (Section 1.3.3).

The Clausius–Clapeyron relation also fundamentally affects atmospheric dynamics. This is because of the following two main processes:
• Radiative absorption is a strong function of water vapour in the atmosphere (Section 1.2); this shows important geographical variations controlled principally by the Earth's temperature variations. The net radiative impact of water is a trade-off between the infrared absorption of water vapour in the atmosphere causing warming and the cooling caused by reflection of incoming solar radiation by the liquid water and ice in clouds. Consequently, the different phases of water affect climate in different ways.
• The global transfer of heat includes that due to the latent heat flux caused by evaporation at one locality and condensation elsewhere. The latent heat released by a saturated parcel of air depends on its initial temperature, so there are major variations between equator and poles determined by the Clausius–Clapeyron relationship. The latent heat released may be several times greater in the case of tropical air. This powers vigorous convection in low latitudes.

1.3.2 Global heat transfer

The circulation of the atmosphere and oceans is fundamentally caused by the fact that the amount of incoming solar radiation varies from a maximum at the equator to a minimum at the poles. This is caused by a number of factors:
• The angle of incidence of the Sun's rays changes from 90° at the equator to 0° at the poles. Less energy is therefore received at the poles because the energy is spread over a larger surface area at high latitudes.
• More reflection and absorption of incoming radiation takes place in high latitudes because of the greater thickness of atmosphere that must be penetrated.
• Variations in the amount of daylight are caused by the tilt of the Earth's axis relative to the plane of the Earth's orbit around the Sun, producing the seasons. The shortness of the daylight hours causes less annual radiation to be received per unit surface area in high-latitude than in low-latitude regions.

Since the long-wavelength radiation leaving the Earth does not vary greatly with latitude, radiation imbalances are set up, with surplus radiation between latitudes of about 50°N and 50°S and a deficit elsewhere. It is necessary to spread this heat imbalance over the surface of the Earth to prevent the tropics from getting increasingly warm and the poles increasingly cold. This heat transfer is accomplished through a strongly coupled circulation of the atmosphere and oceans.

1.3.3 Ocean–atmosphere interaction: driving mechanisms

The temperature gradient from equator to poles drives the atmospheric and oceanic circulations. The total heat transport is roughly equally balanced between ocean and atmosphere, but the types of flux of heat are somewhat different.
• Atmospheric motions are produced by heat fluxes at the atmosphere's lower boundary with land and ocean, and at its upper boundary by radiative cooling into outer space. The majority of heating takes place at the hot land and sea surface in the tropics, and in the low to middle troposphere of the tropics through the release of latent heat. This distribution forces a direct thermal circulation of the atmosphere shown in schematic form in Fig. 1.5.
• Ocean water motion is driven by wind stress at its upper surface, and horizontal density gradients arising from lateral variations in temperature, fresh water influx from precipitation, ice melting and river runoff from land. Whereas the atmospheric circulation is relatively efficient, the tendency for the upper tropical

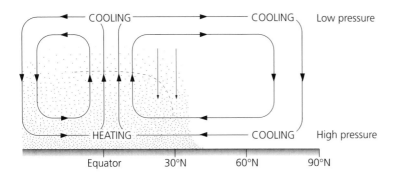

COOLING — COOLING — Low pressure

HEATING — COOLING — High pressure

Equator   30°N   60°N   90°N

Fig. 1.5 The rudimentary thermal circulation of the atmosphere driven by heating in low latitudes.

ocean to stratify causes the global ocean circulation to be slow and inefficient. Despite this, it appears to have a major control on long-term climate variation. We now examine these driving mechanisms in oceans and atmosphere in more detail.

Forcing of ocean circulation

The average temperature of the oceans is just 4°C and mean sea surface temperature 19°C. The mean ocean salinity is 35.5 parts per thousand, with mean sea surface salinity of 35.2 parts per thousand.

There is therefore a warmer and less saline layer of water at the surface of the ocean as a thin veneer over a deeper, colder and more saline body of water. This thin veneer is concentrated in the tropics. This correlates with a net excess of precipitation over evaporation over the warm tropical 'pools'. However, it is important to recognize that although the warm tropical oceans associated with vigorous atmospheric convection and resulting precipitation have a major impact on global climate, these warm (>28°C) waters constitute a minute part of the total water mass. Within the deeper water mass, the temperature and salinity distributions with depth are remarkably similar between the equatorial and subtropical regions. The main differences within the deeper water mass are found in higher latitudes where deep cold water is formed.

It is therefore a priority to understand the reasons for the thermal and salinity variations in the oceans. The emphasis here is on global patterns, but the mechanics of ocean circulation are discussed in some detail in Chapter 9.

In the open ocean, the net flux of fresh water is the balance between precipitation (P) and evaporation (E) (and at high latitudes by the balance between freezing and melting processes). The fresh water flux shows a latitudinal variation (Fig. 1.6):
• in the tropics and at high latitudes precipitation

exceeds evaporation, leading to a fresh, stable upper layer in the ocean;
• in the subtropics evaporation exceeds precipitation, causing an unstable surface saline layer.

The thermal effects and saline effects may oppose or reinforce each other (Fig. 1.7). In the subtropics, for example, dense saline surface water sinks and underflows the stable, warm equatorial pool created by the positive fresh water flux and high temperatures. On the other hand, in middle latitudes (20–60°) the thermal and haline circulations are opposed. In high latitudes summer melting of ice and solar heating generate a light surface layer; winter cooling and ice formation causes a dense surface layer to sink to great depths.

Ocean–atmosphere interaction: buoyancy The relation between water temperature, salinity and density is complex and nonlinear (Fig. 1.8). In warm water, a given density change $\Delta\rho$ is associated with a small temperature change ΔT_W and salinity change ΔS_W, where the subscript refers to 'warm' water. The same density change in cold water is associated with a similar salinity change ΔS_C but a much larger temperature change ΔT_C, where the subscript refers to 'cold' water. Since for the same $\Delta\rho$, $\Delta T_W < \Delta T_C$, whereas $\Delta S_W \approx \Delta S_C$, we can assume that circulation in the tropical ocean is forced by processes causing temperature changes, but that in high-latitude oceans the effects of salinity and temperature are comparable. There are therefore different regional responses to temperature and salinity variations [9].

This can be formalized by defining the *buoyancy* as the relative density of a parcel of ocean water compared to a neighbouring parcel

$$B = \alpha T - \beta S \qquad (1.10)$$

where α and β are the thermal and salinity expansion coefficients respectively, and T and S are temperature

(a)

Net heat flux (Wm⁻²)

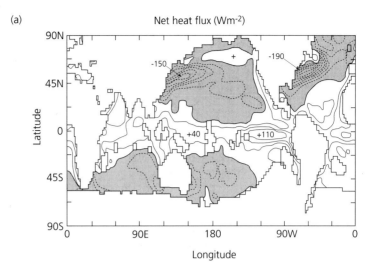

Fig. 1.6 (a) The annual mean net heat flux into the ocean (Q_T), showing regions of net heat loss (shaded) and net heat gain (blank). The tropical oceans have a net positive Q_T, the subtropics and high latitudes suffer moderate heat loss (negative Q_T), and the western parts of northern hemisphere ocean basins experience high net heat loss to the atmosphere (strongly negative Q_T). (b) Annual mean net fresh water flux into the ocean (F_W). Positive fluxes, where $P - E > 0$, are shown as shaded, and dominate the tropical regions where sea surface temperatures are highest. Negative fluxes, where $P - E < 0$, occur especially in the subtropics. Data are from Oberhuber (1988) in Webster (1994) [7].

(b)

Net fresh water flux (mm/month)

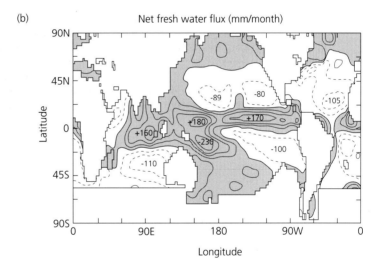

and salinity. If the ocean receives a total heat flux Q_T, and the flux of fresh water into the ocean is denoted by F_W (equal to precipitation minus evaporation, $F_W = P - E$), then the flux between the atmosphere and ocean can be thought of as a *buoyancy flux*

$$F_B = \alpha g Q_T + \beta g S(P - E) \tag{1.11}$$

(where g is acceleration due to gravity) written so as to make clear the thermal (first term on the right-hand side) and haline (second term) effects. Now it can be seen that if the fresh water flux is small compared to the heat flux, as in the tropics, buoyancy is driven by temperature variations, whereas in high-latitude oceans the melting and freezing of ice as well as temperature variations control buoyancy.

The global picture is of a small number of domains

Practical exercise 1.1: Regional sensitivity of water to changes in salinity and temperature

Water at the surface of the warm tropical pools is at about 29°C and has a salinity of 35 parts per thousand, whereas surface water in high-latitude regions has a temperature of about 7°C and salinity of 33 parts per thousand. What would be the effect on the density of these two different water masses of (a) a 2.5 parts per thousand increase in salinity; (b) a 5°C decrease in temperature?

Solution

Using Fig. 1.8, the warm tropical surface water has a
Continued on p. 14.

Practical exercise 1.1: *Continued*

density of 1022 kg m⁻³. The cool high-latitude water has a density of 1026 kg m⁻³. A 2.5 parts per thousand decrease in salinity causes a change in density of just less than 20 kg m⁻³ for both water masses. However, a 5°C temperature change causes a density change of less than 4 kg m⁻³ for the cold water compared to over 10 kg m⁻³ for the warm water. Warm water masses are therefore controlled by processes which influence temperature, whereas cold water masses are equally influenced by factors causing temperature and salinity changes. These regional sensitivities are controlled by the thermal expansion and salinity expansion coefficients for water. Although the thermal coefficient of expansion varies greatly with changes in water temperature between 2.5°C and 30°C (over a factor of 4 from 781×10^{-7} to $3413 \times 10^{-7} K^{-1}$), the salinity expansion coefficient is relatively constant over the same temperature range (8010×10^{-7} to $7490 \times 10^{-7} K^{-1}$).

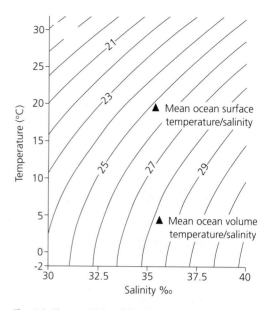

Fig. 1.8 The sensitivity of the density of seawater to changes in temperature and salinity. Density shown in kilograms per cubic metre in excess of 1000. Water masses in tropical and high-latitude regions have markedly different sensitivity to a given change in temperature or salinity. See Practical Exercise 1.1.

for the distribution of both fresh water flux and total heat flux. The heat flux, Q_T, is positive in equatorial–tropical oceans; negative in the subtropics

and high–latitude oceans, where there is moderate heat loss to the atmosphere; and strongly negative in the western parts of the northern hemisphere, where there is major heat loss. The fresh water flux, F_W, is positive in the tropical oceans, especially where the sea surface temperatures are highest; and negative in the subtropics.

With this modicum of theory we can now make more sense of the observed global pattern of atmospheric and oceanic circulation.

Observed oceanic and atmospheric circulation
Oceanic currents About a quarter of the heat surplus building up in low latitudes is carried by warm, wind-driven ocean currents such as the Gulf Stream, moving poleward from the region between 20°N and 20°S [2]. These warm currents heat the overlying atmosphere, moderating the climate of adjacent land masses. The water movement is caused by wind stress, the effects of rotation of the Earth, and by frictional interaction with continents (see Chapter 9). The circulation pattern that results is of *gyres* that flow clockwise in the northern hemisphere and anticlockwise in the southern (Fig. 1.9). Each gyre has

(a) **Ocean thermal and haline forcing**

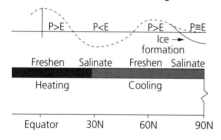

(b) **Ocean thermohaline circulation**

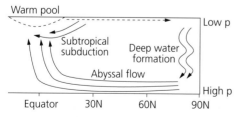

Fig. 1.7 Forcing mechanisms and circulation in the ocean. (a) Thermal and haline forcing. (b) Resulting thermohaline oceanic circulation (*P* = pressure). After Webster (1994) [7].

(a)

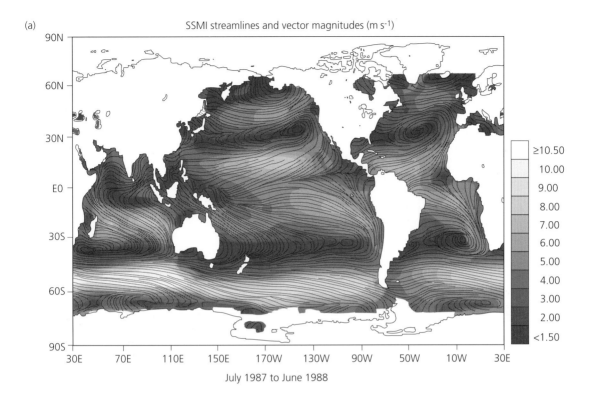

SSMI streamlines and vector magnitudes (m s⁻¹)

July 1987 to June 1988

(b)

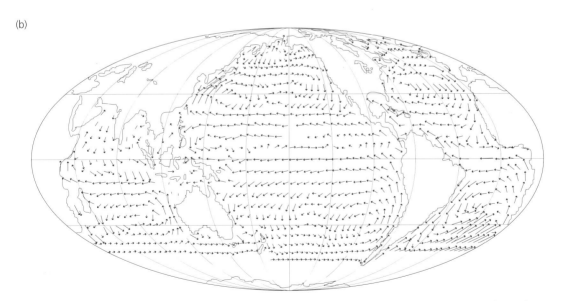

Fig. 1.9 (a) The ocean surface wind speeds (SSM/I data) as streamlines and vector magnitude coded by colour, from July 1987 to June 1988 (Atlas *et al*. (1993) [10]). (b) The mean total surface currents of the ocean derived from the average of historical ship drift observations (from Meehl (1982) [11]). The pattern is a resultant of geostrophic and wind-driven flows (see also Chapter 9).

a strong poleward current on the western side of the gyre and a weaker equatorward current on the east, a feature known as *western intensification*.

Whereas in the shallow ocean (the top few hundred metres) the circulation is driven by the wind, in the deep ocean the circulation is caused by density variations due to differences in temperature and salinity. This deep *thermohaline circulation* (Fig. 1.10), however, owes its origin to processes taking place at or near the surface, such as heating and cooling, evaporation, addition of fresh water, or abstraction of water as sea ice. At high latitudes in the Atlantic surface cooling and abstraction of sea ice (which leaves the seawater denser due to the concentration of salts not incorporated into the ice) causes the water to sink because of negative bouyancy. This cold, salty water then flows laterally as a deep oceanic circulation, diffusing slowly into surface layers and occasionally upwelling rapidly at continental edges. Where this upwelling of nutrient-rich water takes place high organic productivities result. The velocities of the deep thermohaline currents are very low, perhaps only 100 m per day, and the residence time of deep oceanic water is long, 200–500 years for the Atlantic and 1000–2000 years for the Pacific.

The actual pattern of thermohaline circulation (Fig. 1.10) is made complicated by:
• the finite size of the ocean basins;
• the interconnectedness of the different ocean basins;

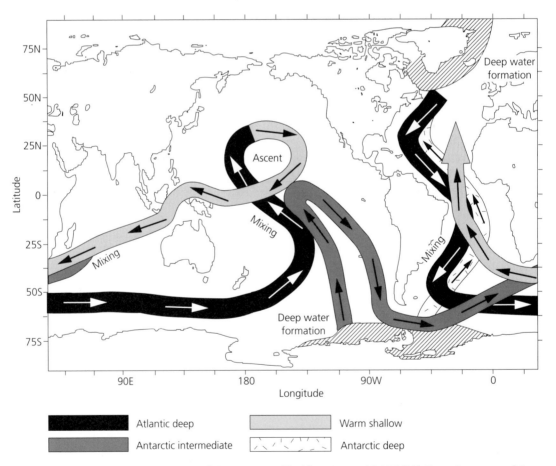

Fig. 1.10 The deep thermohaline circulation of the oceans, modified from Stommel (1958) [12]. The major sources of deep water at the present day in the north Atlantic and Weddell Sea are shown by hatching. The lack of deep water formation in the north Pacific may be due to the greater stability of the Pacific caused by the higher fresh water flux. The text describes the circulation patterns observed.

• the spatially and temporally varying forcing from the atmosphere.

The deep circulation of water through the ocean basins from regions of deep water generation in the north Atlantic and Weddell Sea area of the Antarctic Ocean was first demonstrated by Stommel in 1958. Deep water originating in the north Atlantic flows southwards along the western side of the Atlantic basin before turning east as a circumpolar current, then entering the Pacific basin. Removal of water from the Atlantic and build-up in the Pacific demands a return flow of some form. It has been suggested that the northward-moving deep water in the Pacific derived from the north Atlantic mixes with water derived from the Weddell Sea (Antarctic Intermediate Water) and leaves the Pacific basin through Drake Passage, between South America and Antarctica (Fig. 1.10). The remainder ascends in the northern Pacific and leaves the basin as a surface current through the Indonesian archipelago to the Indian Ocean. These two outlets re-enter the south Atlantic and flow northwards as a near-surface current towards the original site of the Atlantic Deep Water. There is therefore a complete circuit of oceanic circulation. The time for water to circulate in this global system is thought to be of the order of 10 000 years. Variations in the deep thermohaline circulation may be responsible for climate change on a similar time-scale, that is, the time-scale of the climatic fluctuations associated with the Pleistocene glaciations (Chapter 2). Much climate

change research is now focused on the dynamics of the deep thermohaline circulation.

The episodic massive release of icebergs into the north Atlantic Ocean during the last glaciation, known as *Heinrich events* [13], is thought to have strongly perturbed global climate through the effects on sea surface temperatures and ocean–atmosphere circulation. Further information is given in Chapter 2.

Atmospheric circulation The atmospheric circulation set up by the latitudinal radiation imbalances is responsible for over half of the heat transfer in the form of warm poleward-blowing winds and latent heat transfer. More information on the elementary physics of this global atmospheric circulation is given in Chapter 10.

We have seen that the simplest way in which the atmosphere might respond to the excess of heat in the tropics and deficit at the poles is in the establishment of a simple circulation of rising air at the equator, a high-level poleward motion, sinking at the poles as the air cools, and a return flow at low level to the equator (Fig. 1.5). This simple idealized pattern of two re-circulating cells, with a deflection of the winds by the rotation of the Earth (Coriolis force) (see also Chapters 9 and 10), was originally proposed in 1735 by George Hadley. The actual general circulation of the atmosphere (i.e. the mean annual winds) (Fig. 1.11) shows the Hadley circulation to be broken into several latitudinal zones. Instead of air rising at the

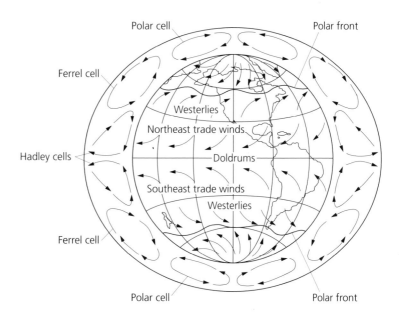

Fig. 1.11 The general circulation of the atmosphere showing the Hadley, Ferrel and Polar cells. Modified from Miller *et al.* (1983) [14].

equator and travelling all the way to the poles, it descends at about 30° latitude. The subtropical deserts are located under this descending dry air. It is dry for two main reasons: first, it has lost its moisture by condensation in the tropics; and second, it is descending and therefore warming, increasing its capacity to hold moisture. The air then travels equatorwards at low level as the trade winds. This low-latitude circulation is known as a *Hadley cell*. There is a region of low-level convergence of dry trade wind air, and moist equatorial air, known as the *intertropical convergence zone*. It is characterized by heavy rainfalls, and rainforests have developed under it, as in Amazonia and western equatorial Africa.

Air descending at about 30° latitude also flows polewards at low level as the westerlies until it meets cold polar air moving equatorwards at about 50°, forming the *polar front* of atmospheric instability. The instability generates storms and heavy precipitation. The air from low latitudes rises over the cold polar air, and returns to the subtropics at high level, completing the *Ferrel cell*. The polar air warms by mixing with the westerlies and by condensation at the polar front, rises, and flows back to the polar region where it cools and sinks. This is the *polar cell*. There are thus three major atmospheric cells per hemisphere.

In addition to the three latitudinal cells per hemisphere, high sea surface temperatures in the warm pools of the Indian and Pacific Oceans cause ascending air patterns, resulting in a series of east–west cells along the equator known as the *Walker circulation*. The maximum upward motion in the Walker circulation is associated with the highest sea surface temperatures, emphasizing the crucial link between ocean and atmosphere in determining global climate. The distribution and intensity of warm pools [15] rather than a latitudinally continuous zone of high sea surface temperatures is probably related to the distribution of the continental land masses. Air rises over the Indonesian region and descends over the eastern Pacific Ocean. It fluctuates in intensity, when it is termed the *southern oscillation*. When the circulation reverses, known as *El Niño events* (see also Section 9.6), major droughts may occur in the southern hemisphere, accompanied by wet weather in deserts.

At levels in the upper troposphere there are two channels of extremely high winds which travel as westerlies around the poles. These are known as the *jet streams*. One reaches as far equatorwards as the subtropics (the *subtropical jet stream*), and the other is restricted to higher latitudes (the *polar front jet stream*). These jet streams appear to delimit the circulatory cells, the poleward limit of the upper part of the Hadley cell being marked by the vigorous (<65 m s^{-1}) subtropical jet stream, and the weaker (<25 m s^{-1}) polar front jet stream being located in the zone of high meridional (longitudinal) pressure gradients at the tropospheric continuation of the polar front.

The distribution of pressure in the upper troposphere at high latitudes indicates a number of waves, whose motion facilitates the poleward transfer of heat. These *Rossby waves* appear to grow as disturbances in the jet stream, varying in their latitudinal position over weeks or months. Rossby waves may play an important role in the global circulation of heat in the atmosphere.

The *monsoon* is a large-scale circulation pattern which is asymmetrical about the equator. The forcing mechanism for the monsoonal circulation is the distribution of land and sea in the eastern hemisphere. The monsoonal circulation is characterized by extreme seasonality. In the winter Asia is typified by descending air. This reverses in southern Asia to a strong rising motion of air in the summer in response to intense heating, and is accompanied by heavy precipitation. Compensation for the rising of air over southern Asia causes a drawing in of air from the Middle East and north Africa (Fig. 1.12), where air descends, causing a strengthening of summer desert conditions. The vigorous upward motion forms part of a circuit involving a reverse flow in the upper troposphere.

The interannual variability of the monsoon can be correlated with reversals of the Walker cell and the occurrence of El Niño, causing drought over Indonesia and north Australia. Further discussion of monsoonal and Walker circulations is deferred to Section 10.2.3.

The tropical oceans and overlying atmosphere are extremely important in the global hydrological system. They are the sites of high sea surface temperatures and associated deep convection in the atmosphere. Consequently, the Walker cell and monsoonal circulations have a major impact on the global hydrological cycle.

Latent heat transfer Latent heat results from the change of state of water. It is given out when condensation takes place and consumed when evaporation takes place. Heat is transferred when latent heat stored in evaporated water (water vapour) is released elsewhere during condensation. Tropical air (between

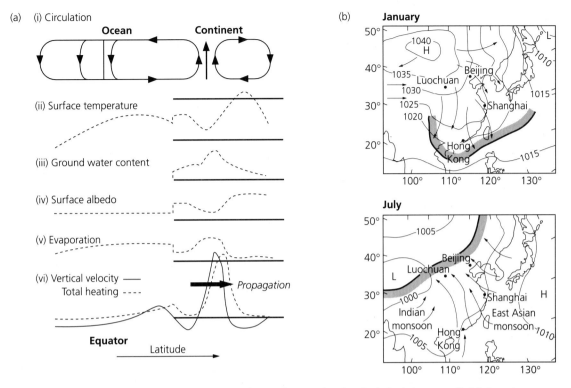

Fig. 1.12 The monsoonal circulation. (a) Interaction of atmospheric and surface hydrological systems. (i)–(vi) show the function of latitude during the poleward propagation of a monsoon convective event. After Webster (1994) [7]. (b) The winter and summer monsoon regimes affecting China, showing surface pressures (in millibars), dominant wind vectors (arrows) and polar front (January) and monsoonal front (July) (bold line). After Porter & Zhisheng (1995) [16].

10°N and 10°S) is heated both by latent heat released during condensation and by conduction from the warm land surface (sensible heat). Latent heat is consumed in the subtropical zone (15–30°N and S) where evaporation exceeds precipitation. It is released in mid-latitude zones (30–50°N and S) by condensation associated with storms. The tropical rainforests (see also Section 2.5) are extremely important in transferring latent heat high into the atmosphere. Very high rates of evaporation take place from the surfaces of rainforest plants, the water vapour produced condensing in large clouds extending to high altitudes, thereby releasing latent heat.

1.3.4 Summary: a global interactive model
The result of the heat transfers through oceanic and atmospheric circulations is the climatic zones of the Earth (Fig. 1.13). It is important to emphasize that

the ocean and atmospheric systems do not work in isolation but rather operate synergetically to maintain climatic stability. The global system can be viewed as a highly interactive entity operating at a number of different time-scales. This interactive system and its latitudinal structure are summarized in Fig. 1.14. Hydrological processes act as the linkages of the different interacting parts of the coupled atmosphere–ocean system, shown schematically in Fig. 1.15. The key elements of the interactive system are as follows.
- **Major convection** takes place:
  (i) in the tropical atmosphere (upward motion of moist air over high sea surface temperatures);
  (ii) in the polar ocean (downward motion of negatively buoyant water caused by radiative cooling in winter and abstraction of sea ice producing an increase in salinity);

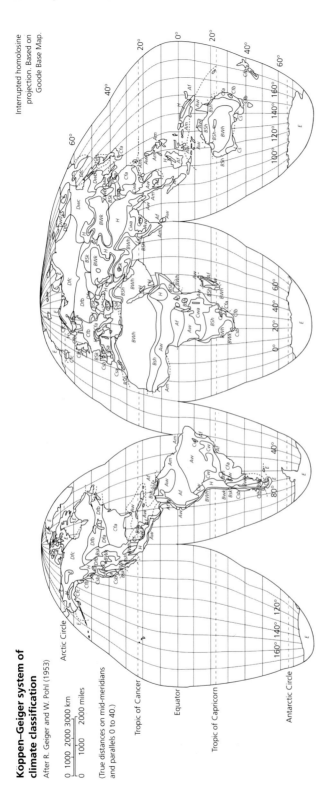

Köppen–Geiger system of climate classification

After R. Geiger and W. Pohl (1953)

0 1000 2000 3000 km
0 1000 2000 miles

(True distances on mid-meridians and parallels 0 to 40.)

Interrupted homolosine projection. Based on Goode Base Map.

Key to letter code designating climate regions:

First letter:

A C D Sufficient heat and precipitation for growth of high-trunked trees.

A Tropical climates. All monthly mean temperatures over 64.4°F (18°C).

B Dry climates. Boundaries determined by formula using mean annual temperature and mean annual precipitation.

C Warm temperature climates. Mean temperature of coldest month: 64.4°F (18°C) down to 26.6°F (-3°C).

D Snow climates. Warmest month mean over 50°F (10°C). month mean under 26.6°F (-3°C).

E Ice climates. Warmest month mean under 50°F (10°C).

Second letter:

S Steppe climate.

W Desert climate.

f Sufficient precipitation in all months.

m Rainforest despite a dry season. (i.e., monsoon cycle.)

s Dry season in summer of the respective hemisphere.

w Dry season in winter of the respective hemisphere.

Third letter:

a Warmest month mean over 71.6°F (22°C).

b Warmest month mean under 71.6°F (22°C). At least four months have means over 50°F (10°C).

c Fewer than four months with means over 50°F (10°C).

d Same as c, but coldest month mean under -36.4°F (-38°C).

h Dry and hot. Mean annual temperature over 64.4°F (18°C).

k Dry and cold. Mean annual temperature under 64.4°F (18°C).

H Highland climates.

Fig. 1.13 Climatic zones of the Earth, according to the Köppen–Geiger classification system. Originally developed by Vladimir Köppen (1846–1940), a German climatologist and botanist, it forms the basis for most morphoclimatic schemes such as that of Tricart & Cailleux (1972) [17].

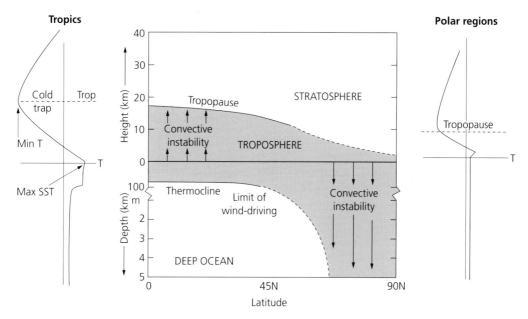

Fig. 1.14 The interactive zones of the atmosphere and ocean, encompassing the troposphere, upper ocean, and at high latitudes the entire ocean where deep water formation is involved. Interactive zones are those which operate at similar time-scales and thus directly affect each other. The stratosphere and deep ocean in lower latitudes operate at longer time-scales. After Webster (1994) [7].

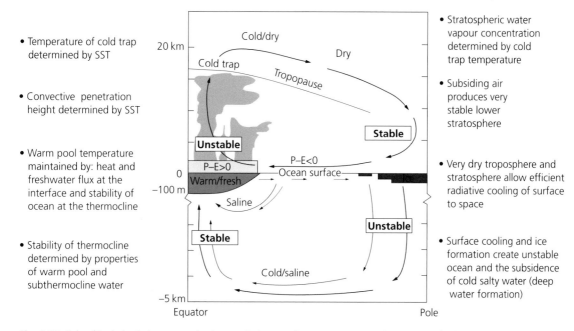

Fig. 1.15 Role of hydrological processes in the coupled atmosphere–ocean system. Deep atmospheric convection in the tropics produces a very cold tropopause which acts as a 'cold trap' for moist air passing through to the stratosphere. This ensures that the stratosphere remains dry, which in turn promotes the rapid cooling in polar regions, the formation of sea ice and of saline deep ocean water. The stable tropical ocean water ensures a surface warm pool which in turn drives atmospheric convection above. After Webster (1994) [7].

(iii) in the subtropical ocean intense evaporation at the surface causes sinking of saline waters.
- In contrast, **gravitational stability** characterizes:
  (i) the tropical ocean, with a warm upper layer overlying rising cold water;
  (ii) the polar atmosphere with its descending air flow.

Although this model does not take into account the complicated effects of the distribution of land and sea, it demonstrates the effectiveness of the linkages between the ocean and atmosphere circulations. The engine for the entire coupled system might be thought to lie in the warm pools of the tropical ocean.

1.3.5 Runoff
The surface water balance
We have seen that fresh water runoff plays a role in the coupled atmosphere–ocean system. It is also crucial in landscape development since most sediment transport on land is caused by runoff. The factors controlling denudation in the world's drainage basins, and the resulting sediment fluxes through and out of them, are discussed in Chapter 3. Although the quantity of water in the continental portion of the hydrological cycle is minute (Fig. 1.3), it has profound importance for the activities of humans and in sustaining natural ecosystems. Here, however, it is necessary to retain our global focus and to place runoff in its broad context in the hydrological cycle.

The runoff can be viewed in terms of its role in the surface water balance,

$$P = E + T + \Delta S + \Delta G + R \qquad (1.12)$$

where P is precipitation, E is evaporation, T is transpiration, ΔS is the change in the storage of water in the soil, ΔG is the change in the storage of water as groundwater, and R is the overland flow across the land surface as rills and gullies, streams and rivers (Fig. 1.16). This surface water balance differs between open water and land. In the former the balance is a simple relationship between the incident precipitation, the evaporation and the resultant change of volume of the water body. An excess of evaporation over precipitation results in a shrinkage of the water body, the situation common in evaporating pans. However, fresh water runoff from rivers may be important in both lakes and the open sea. In this case

$$P = E - \Delta F \qquad (1.13)$$

where ΔF is the inflow of water from land to ocean.

Globally, this relationship must balance, but there are considerable regional variations in rates of precipitation (Fig. 1.17), evaporation (Fig. 1.18) and fresh water inflow (Tables 1.3 and 1.4).

For a land surface the full surface water balance equation (1.12) is required. If we regard the soil and groundwater storages as not fluctuating significantly over long periods of time, the difference between the long-term precipitation and long-term evaporation is reflected in runoff. Since the global average depth of precipitation over land exceeds the global average depth of evapotranspiration, there is a surplus representing the surface runoff. Once again, there are important regional variations which will be examined below.

Runoff, together with its sediment load, is routed through river drainage systems or catchments (Chapter 3). The surface water balance equation shows that a number of components require measurement in order to assess the water balance of a catchment: streamflow, precipitation, evaporation, soil moisture and groundwater (Fig. 1.16). The relative importance of these components depends on the climatic, topographic and geological setting. The ratio of streamflow measured from the river and precipitation falling on its drainage basin is termed the *runoff coefficient*. It varies in both space and time between 0 and 100%. Evaporation losses resulting from evaporation *sensu stricto* plus infiltration losses are the difference between catchment precipitation and streamflow. Soil moisture and groundwater losses may be considerable in certain climates, particularly at times of drought in a humid-temperate climate when groundwater levels are normally high and soils relatively saturated. Soil moisture and groundwater therefore also have a major role to play in controlling runoff processes.

Global patterns of runoff
The annual evaporation from the oceans is equivalent to a layer of water 1.4 m thick over the ocean surface of the Earth. Most (90%) of this is returned to the oceans directly by precipitation on the oceanic surface. However, the remainder falls as precipitation on land, equivalent to an annual layer with average thickness of 1 m. Of this amount, about 35% is returned to the oceans in the form of runoff, the remainder being lost by evapotranspiration. The regional patterns in the relative importance of precipitation, evaporation and runoff, and the controls on these patterns, are briefly discussed below.

For rain or snow to form, there is a dual requirement of sufficient water vapour in the atmosphere and

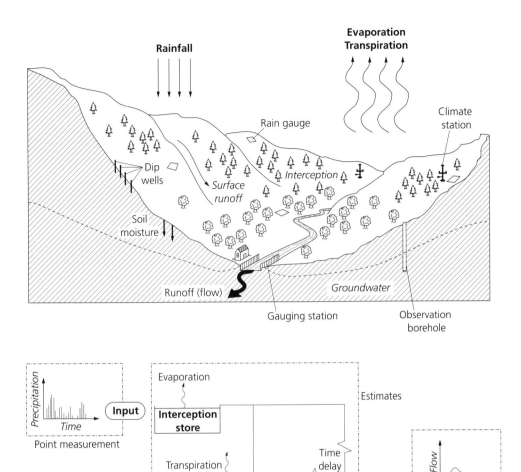

Fig. 1.16 The components of the hydrological cycle for a catchment. The individual storages (boxes) exist along a route from an input precipitation through to a river flow as an output from the catchment. After Newson (1994) [18].

rising air that can cause the vapour to cool, condense and cause precipitation. The geographical variation of mean annual precipitation (Fig. 1.17) shows it to be highest in tropical zones where humid air rises convectively through heating, and in regions where mountains force moisture-laden air to rise, so-called *orographic* precipitation. There is also a high amount of precipitation between 35° and 60°N and S where atmospheric instability associated with storms is common (Section 1.3.3).

Evaporation also varies greatly over the surface of the Earth (Fig. 1.18). High rates of evaporation require a large heat source (solar radiation), low moisture contents in the air, and a source of liquid water. Consequently, actual evaporation rates may be lower than expected in dry continental settings because of the lack of available water. The highest evaporation rates are found over subtropical oceans, from where currents such as the Gulf Stream transport warm water northwards into cooler, drier air conditions.

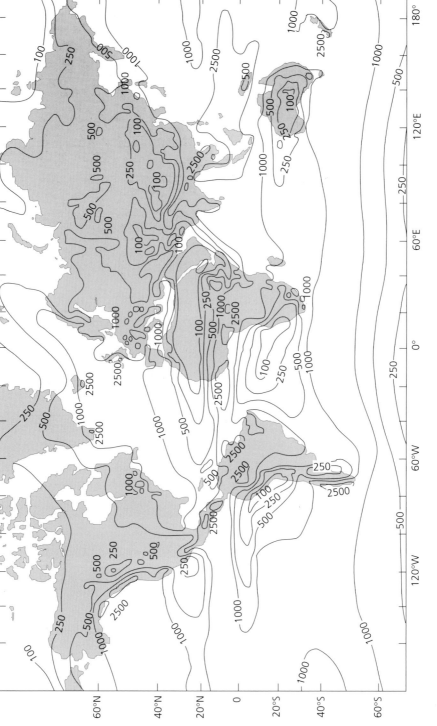

Fig. 1.17 The geographical variation of mean annual precipitation, in millimetres. After Lamb (1972) [19].

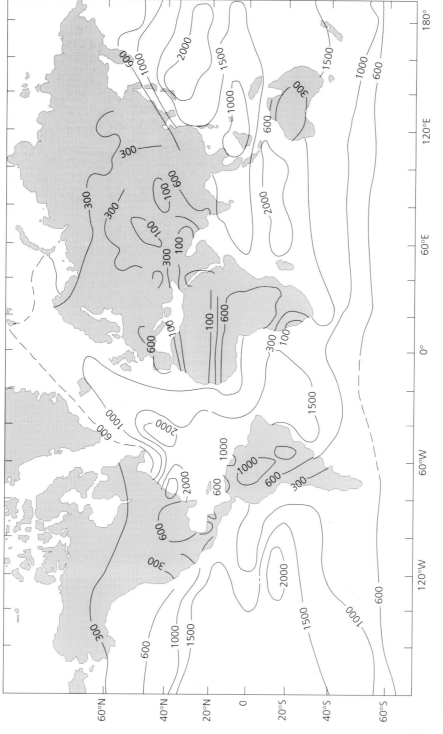

Fig. 1.18 Geographical variation of annual evaporation from ocean and evapotranspiration from land surface, in millimetres. After Barry (1970) [20].

Continent	Precipitation (mm)	Precipitation (km³)	Evaporation (mm)	Evaporation (km³)	Runoff (mm)	Runoff (km³)
Europe	790	8 290	507	5 320	283	2 970
Asia	740	32 200	416	18 100	324	14 100
Africa	740	22 300	587	17 700	153	4 600
North America	756	18 300	418	10 100	339	8 180
South America	1600	28 400	910	16 200	685	12 200
Australia and Oceania	791	7 080	511	4 570	280	2 510
Antarctica	165	2 310	0	0	165	2 310
Land as a whole	800	119 000	485	72 000	315	47 000
Areas of external runoff	924	110 000	529	63 000	395	47 000*
Areas of internal runoff	300	9 000	300	9 000	34	1 000†

Table 1.3 Water balance of the Earth's land surface in terms of precipitation, evaporation and runoff. From Shiklomanov (1993) [21].

* Including underground water not drained by rivers.
† Lost in the region through evaporation.

River	Average runoff (km³y⁻¹)	Area of basin (10³km²)	Length (km)	Continent
Amazon	6930	6915	6280	South America
Congo	1460	3820	4370	Africa
Ganges (with Brahmaputra)	1400	1730	3000	Asia
Yangzijiang	995	1800	5520	Asia
Orinoco	914	1000	2740	South America
Paraná	725	2970	4700	South America
Yenisei	610	2580	3490	Asia
Mississippi	580	3220	5985	North America
Lena	532	2490	4400	Asia
Mekong	510	810	4500	Asia
Irrawaddy	486	410	2300	Asia
St Lawrence	439	1290	3060	North America
Ob	395	2990	3650	Asia
Chutsyan	363	437	2130	Asia
Amur	355	1855	2820	Asia
Mackenzie	350	1800	4240	North America
Niger	320	2090	4160	Africa
Columbia	267	669	1950	North America
Magdalena	260	260	1530	South America
Volga	254	1360	3350	Europe
Indus	220	960	3180	Asia
Danube	214	817	2860	Europe
Salween	211	325	2820	Asia
Yukon	207	852	3000	North America
Nile	202	2870	6670	Africa

Table 1.4 Mean annual runoff of the world's largest rivers. From Shiklomanov (1993) [21].

The vast bulk of runoff (nearly 98%) enters the oceans directly, and only a small fraction is routed into closed interior basins such as the Caspian and Aral Seas in southern Asia. In such cases the entire runoff is evaporated. Substantial use by man of precious fresh water runoff in interior basins can result in fundamental changes to the water–salt balance of the water bodies into which the rivers flow. Catastrophic changes to the

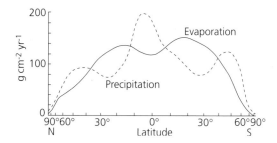

Fig. 1.19 Zonally averaged precipitation and evaporation as a function of latitude.

aquatic ecosystems and to the human societies that depend on them may result, as in the Aral Sea.

The water balance, measured in terms of the relative importance of precipitation, evaporation and runoff, varies according to latitudinal position or climatic zone (Table 1.3) [21]. The net precipitation, or precipitation minus evaporation (Fig. 1.19), is lowest in the subtropics (15–30°N and S) where the air is stable, and in the polar regions which have both stable air and very low moisture content because of the low temperatures. In some polar regions, the evaporation is so low that there is a positive net precipitation, allowing ice caps to form (Greenland, Antarctica). Very high percentages of precipitation are consumed by evapotranspiration in continents with large tropical and subtropical zones such as Africa (84%) and Australia (94%), lower percentages in continents with extensive temperate zones such as North America (62%) and Europe (57%), and a very low percentage in Antarctica (17%) with its very low evapotranspiration rates [22].

Table 1.3 shows that more than half of the global runoff occurs in Asia and South America (31% and 25% respectively), where it is concentrated in the equatorial zone. The Amazon River is outstanding in its average runoff of 6930 km^3 per year, more than 15% of the annual global runoff.

Chemistry of water in the global hydrological cycle

A comprehensive treatment of the chemistry of water in the different stores of the hydrological cycle is beyond the scope of this text. However, as attention in hydrological research is increasingly focused on problems at large spatial and long temporal scales, there is renewed interest in how chemical tracers can illuminate the dynamics of the global system. This is particularly true of the use of new methods of isotope geochemistry. The chemistry of water in the hydro-

sphere also determines nutrient supply and therefore impacts strongly on the biosphere (Section 1.4).

Precipitation falling over the land surface is modified in its chemical composition by reactions taking place with rock and soil, with plants and with decomposing organic matter. Land-based water is diluted or concentrated by additions of precipitation or losses through evaporation, but the most important factor determining the chemical composition of surface and soil water and the fresh water of lakes and rivers is the input from weathering of rock. This is dealt with in more detail in Chapter 3.

The global cycling of chemicals is strongly affected by the release of dissolved and particulate products from the weathering of rock and soil. In outline, waters from calcareous catchments contain high amounts of total dissolved solids (TDS), particularly Ca^{2+} and HCO_3^-, but small particulate loads. Waters draining varied igneous, metamorphic and sedimentary rock types have a correspondingly more varied chemical composition and higher particulate loads. This total flux from the land surface is partially responsible for the steady-state composition of the oceans.

The observed differences in the chemical composition of river water have been investigated in terms of both the chemistry of rainwater and the dissolved weathering products as a function of TDS [23]. The cationic ratio

$$\frac{Na^+}{Na^+ + Ca^{2+}}$$

and the anionic ratio

$$\frac{Cl^-}{Cl^- + HCO_3^-}$$

show actual water compositions to lie along two diverging arms (Fig. 1.20) controlled by three main mechanisms.

1 Control by precipitation: where the chemistry is dominated by rainwater rich in Na^{2+}, producing a cationic ratio of nearly 1, and low TDS since evaporated seawater falling as rain is dilute. Rivers draining areas of low relief with well-weathered bedrock and plentiful rainfall, such as the tropical rivers of Africa and South America, fall into this group. The low TDS values may also be affected by the large biomass of the lowland tropical forests effectively immobilizing chemicals from release into rivers.

2 Control by weathering: where the chemistry of river water is controlled by weathering reactions within their catchments. The cationic ratio very

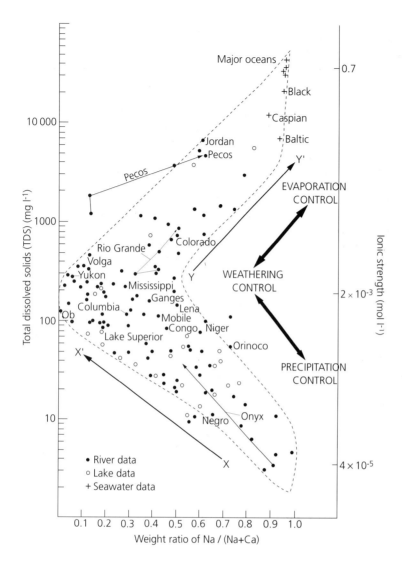

Fig. 1.20 Modified from Gibbs's (1970) [23] scheme for global water chemistry, wherein the total dissolved solids and ionic strength of surface waters are plotted against the cationic weight ratio of Na/(Na + Ca). The arrows connecting data points show the geochemical evolution of river waters from source downstream. Rivers plotting along the trend from X to X', such as the Mackenzie (Arctic Canada) and Ganges (southern Himalayas), occur in regions with highly active weathering processes. Rivers falling along the trend from Y to Y', such as the Jordan (Middle East), Rio Grande and Colorado (arid south-western North America) occur in areas experiencing high amounts of evaporation, and evolve towards the composition of seawater.

much depends on the type of rock being weathered, limestone bedrock producing the lowest ratio. TDS values are higher, reflecting the increased role of weathering (see Chapter 3). In the Amazon River basin, 85% of the solute load is derived from a relatively small area of weathering control in the Andes, whereas in the lowland part of the catchment the control is by precipitation.

3 Control by evaporation and crystallization: this is common in hot and arid climates where precipitation of calcite ($CaCO_3$) increases the cationic ratio, and evaporative losses of river water increase the TDS.

The chemistry of the river input to the ocean therefore varies from continent to continent in response to variations in the control of precipitation, weathering and evaporation. The total flux of dissolved matter into the ocean (Fig. 3.31) is determined by both the TDS and the discharge.

Naturally occurring isotopes in water can be used as tracers of the residence times, mixing ratios and fluxes of key elements of the hydrological cycle. They have an especially important role in assessing underground water masses of different isotopic composition and their mixing. The most commonly used isotopes are the stable isotopes deuterium (^2H), oxygen-18 (^{18}O) and carbon-13 (^{13}C). In addition, the radioisotopes tritium (^3H), carbon-14 (^{14}C) and radon-222 (^{222}Rn) are used.

Oxygen 18 and deuterium reside in the water molecule and are not affected greatly by reactions of water with rock, soil or vegetation. Since oxygen 18 has a different mass than oxygen 16, and deuterium likewise has a different mass than hydrogen, changes in phase of water from liquid to vapour to liquid to ice and back to liquid (representing evaporation, precipitation as ice and then melting) result in enrichment or depletion (fractionation) in one isotope compared to the other. Fractionation is a function of temperature. Consequently, winter precipitation is depleted in ^{18}O and depleted in ^{3}H compared to summer precipitation. Distinct differences also occur between the stable isotopic composition of precipitation at different latitudes and altitudes. Consequently, one can start to appreciate that stable isotopic compositions can be used to assess the contribution of melting of polar ice caps or of freshwater continental runoff, for example, to the oceans.

Other isotopes have been introduced into the hydrological cycle by humans, such as the large amounts of ^{3}H and ^{14}C discharged into the upper atmosphere by testing of thermonuclear weapons in the period 1957–64. Since the half-life of tritium is known (12.3 y), components of the hydrological cycle such as shallow groundwater can be 'tagged' and dated. Many other isotopes can be used.

1.4 Role of the biosphere

Although the relative mass of living (or once living) matter in the Earth surface system is very small, it has a profound effect on the upper veneer of the lithosphere, the hydrosphere and the inner part of the atmosphere, and mediates large parts of the hydrological cycle. The biosphere, the realm of life, is exceedingly complex in its components and its interactions. Although the role of the biosphere in Earth surface processes is referred to on numerous occasions in this book, the reader must look elsewhere for treatments beyond an outline of the global functioning of the biosphere dealt with here.

The basic functional organization of the biosphere is of solar radiation providing the energy for photosynthesis in the presence of chlorophyll in green plants, their growth and assimilation by herbivorous animals, and consumption by carnivorous animals. This organization represents a chemical energy flux. Energy is also transferred to a store of dead organic matter from the primary producers, herbivores and carnivores (Fig. 1.21). These layers within the functional organization of the biosphere are known as *trophic levels*. Any trophic level may experience a net gain or net loss of energy. For example, an increase in the biomass of a forest is indicative of net primary production. An increase in the number of grazing herbivores represents secondary production, also termed conversion.

The primary production of the land and ocean areas is estimated to be about $100 \times 10^{12} \, \text{kg y}^{-1}$ and $55 \times 10^{12} \, \text{kg y}^{-1}$ respectively (Fig. 1.22). This

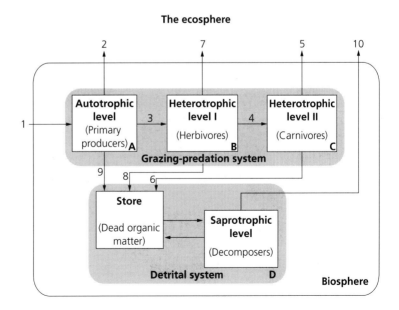

Fig. 1.21 Model of the trophic levels shown as compartments (A to D) of the biosphere, with energy transfers into and out of these compartments represented by arrows numbered from 1 to 10. After White *et al.* (1992) [24].

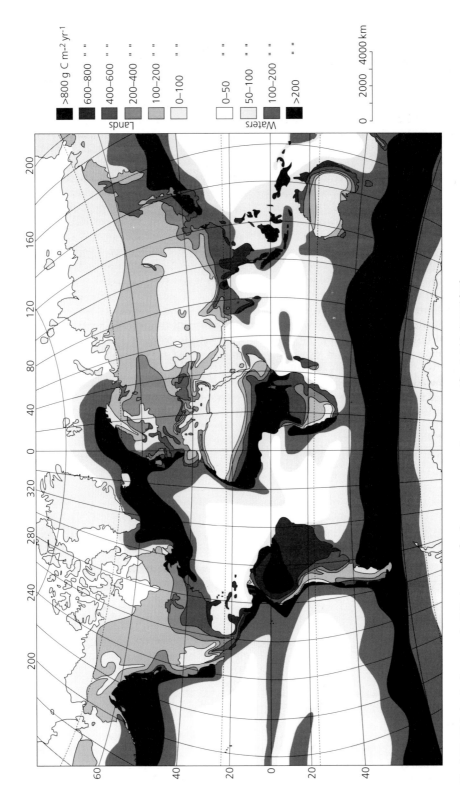

Fig. 1.22 Global net primary productivity in grams of carbon per square metre per year. After Reichle (1970) [25].

represents an extremely large quantity of energy fixed annually. The chemicals passing through the various pathways of the biosphere are involved in global geochemical cycles. Analysis of these cycles allows an estimation to be made of the fluxes involved, but a further significance is that the fluxes act as rate-limiting mechanisms. For example, the flux of nitrogen and phosphorus controls the rate of primary production of plankton and thereby of fish in the world's oceans. The flux involved in the weathering of rock controls the take-up by plant roots and therefore the production of grassland and forest. The pathways may be broken by long periods of residence of a chemical in a store, such as a deep sea sediment. This causes a slowing of the rate of geochemical cycling.

1.4.1 The carbon cycle

It is possible to study the geochemical cycles of a number of important elements such as phosphorus, nitrogen and sulphur. The approach taken here, however, is to focus briefly on the cycle driven by the opposing forces of respiration and photosynthesis involving carbon and oxygen. The main fluxes take place through the gaseous form of CO_2, but carbon is also fluxed through the activity of methane, CH_4. Methane is generated from wetlands and waterlogged

soils. It is present in the atmosphere at very low concentrations (1.7 ppm), with a residence time of about 10 years.

There are a number of stores of carbon in the global carbon cycle, with fluxes between the reservoirs (Fig. 1.23). By far the greatest reservoir of carbon is the deep ocean (34 Gt), together with its shallow surface waters (0.9 Gt). The atmosphere contains just 0.74 Gt, and similar amounts are stored in the terrestrial biomass (0.55 Gt) and soil (1.2 Gt). A substantial amount of carbon (5 Gt) is locked away in fossil fuels. However, the fluxes of carbon between the various reservoirs portray a rather different picture, because although the deep ocean is an immense reservoir, it does not exchange its carbon rapidly with surface waters, whereas the atmosphere exchanges freely and rapidly with the oceanic surface waters, soil and terrestrial biota (Fig. 1.23).

Carbon is transferred to and from the ocean surface and atmosphere by diffusion. It is transferred in the form of CO_2 from the land surface by a combination of weathering of rocks, burning of forest and grassland, volcanic emissions, burning of fossil fuels and other industrial outputs. Atmospheric CO_2 is consumed over the land surface by the photosynthesis of terrestrial biota.

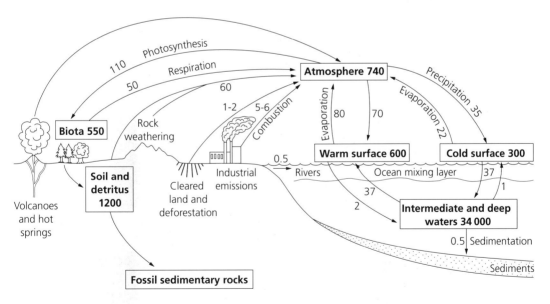

Reservoirs (boxes) 10^{15} g
Fluxes (arrows) 10^{15} g y^{-1}

Fig. 1.23 The global carbon cycle with estimates of storages (in gigatonnes, boxes) and fluxes (in gigatonnes per year, arrows). From various sources, including Berner & Berner (1987) [1] and McClain et al. (1993) [26].

The major source of carbon for the biological system on Earth is therefore from atmospheric CO_2. Despite its small quantity (0.03% by volume) it is extremely important in its role in the radiation balance, especially in absorbing the long-wavelength radiation from Earth (Section 1.2). It may therefore be called a greenhouse gas. The concentration of CO_2 in the atmosphere has been rising steadily over recorded periods (the last 120 years), due mainly to the anthropogenic release of the gas from the burning of fossil fuels (see also Chapter 2). Some of the additional CO_2 released from the burning of fossil fuels has stayed in the atmosphere, but 40% has been taken up in two reservoirs in the carbon cycle which exchange rapidly with the atmosphere on a human time-scale—the surface waters of the oceans, and the terrestrial biosphere plus soils.

Carbon is present in the oceans as bicarbonate and carbonate ions. When atmospheric CO_2 is added to ocean surface water, it is converted to bicarbonate. This exchange occurs rapidly. The surface waters of the ocean mix only slowly with the waters below the thermocline (depth of about 1000 m), so the exchange time of atmospheric CO_2 with the deep sea is very long. The activities of carbon-bearing organisms in the surface waters of the oceans, principally plankton, affect the coupling between atmosphere and ocean. Marine plankton fix in their tissues and skeletons a high proportion of the carbon flux. The size of the marine plankton biomass therefore affects the ability of the biomass to draw out of the carbon cycle (or sequester) the carbon present in the atmosphere.

Availability of important biolimiting nutrients such as nitrogen and phosphorus exerts a major control on the global carbon cycle through its effect on ocean productivity. Nutrient availability therefore has impact on ocean–atmosphere coupling and global climate change through the build-up or abstraction of the greenhouse gas CO_2.

It has been stated that the biosphere is 'hungry' for phosphorus and nitrogen [27], since the levels potentially used by the biosphere are far higher than average concentrations in the hydrosphere and lithosphere. Where light and temperature are optimal, the availability of these ions therefore limits primary production. Where P and N are plentiful, as in coastal and shelfal waters, primary productivity is also high.

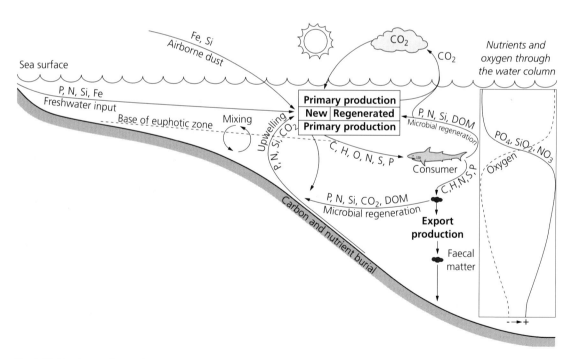

Fig. 1.24 Nutrient cycling in the ocean, showing the importance of the biological carbon 'pump'. Variations with depth of the biolimiting nutrients PO_4, SiO_2 and NO_3, and oxygen concentration are shown on the right. DOM is dissolved organic matter. After Brasier (1995) [27].

Nutrients are made available by runoff from the land or by mixing with deeper waters. Part of the organic matter involved in gross primary productivity is removed or 'exported' from surface waters by a biological 'pump' driven by grazing by zooplankton and fish, and sinking to the sea bed of faecal pellets (Fig. 1.24). On the sea floor the organic material may be used for further biological activity (deposit feeding, bacterial respiration) or locked into the sedimentary record as a black shale, phosphorite or biogenic chert, all of which are common under regions of high rates of export of productivity. Below the productive surface waters, nutrients and CO_2 are returned to the water column by microbial regeneration.

The most likely way of bringing about large changes in atmospheric CO_2 is, therefore, by varying the rate of carbon fixation by primary productivity in the ocean under the influence of biolimiting nutrients. Zones of upwelling may be crucial in this process, since their high productivities cause a removal of CO_2 from the atmosphere, and carbon burial in sea bed sediments. The conditions leading to upwelling, such as the intensity of wind stress (see Chapter 9), are important elements in the larger picture of global climate change through the impact on nutrient supply, productivity and sequestering of atmospheric CO_2.

There is also a rapid exchange of atmospheric CO_2 with the terrestrial biosphere. Increased storage in the terrestrial biosphere, however, is counteracted by the effects of *deforestation* which leads to an addition of CO_2 in the atmosphere (see also Section 2.5). The problem of net loss or net uptake by the biosphere is presently unsettled, but the flux is thought to be much smaller than the uptake of CO_2 by the oceans.

The increase in CO_2 (and other greenhouse gases) in the atmosphere has climatic and hydrological effects. Climatic greenhouse models predict that a doubling of the CO_2 content of the atmosphere would result in a worldwide increase in the mean annual temperature of between 1.5 and 3°C [28]. As we have previously seen (Section 1.2), some of these processes act in a positive feedback.

1.5 Topography and bathymetry

The topography of the Earth's surface and the bathymetry of its oceans (Plate 1.1, facing p. 204) are determined by the interplay of the primary 'tectonic' mechanisms and the modifying exogenic forces causing denudation and deposition. Clearly, topography can be viewed at a range of scales. With our global perspective, the topography at small scales, such as that of an individual hillslope, is not of great importance. At the medium scale, topography may be dominated by endogenic processes causing, for example, the formation of volcanic edifices, or fault scarps at the Earth's surface. At a still larger scale, however, the primary mechanisms for the topography of continents and the bathymetry of ocean basins are related to the horizontal and vertical motions of the lithospheric plates. Plate tectonic theory proposes that the lithospheric plates are sufficiently rigid to move over the surface of the Earth as reasonably coherent entities, their relative motion causing deformation to be concentrated at their edges.

However, plate convergence may result in thickening of the lithosphere very far from plate boundaries, suggesting that the continental plates may act at large time and spatial scales more like viscous sheets undergoing a continuum of deformation. Whether concentrated at plate boundaries or widely distributed into plate interiors, the crustal and lithospheric thickness changes or lateral variations in density caused by the relative motion of plates result in the major features of the Earth's topography. The driving forces for the relative motion of the lithospheric plates are not fully understood, but are probably dominated by the negative bouyancy of cold oceanic lithosphere being subducted into the Earth's interior. Such deep subduction is part of a slow and very large-scale convection of the mantle. Other large-scale topography appears not to be related to relative plate motion *per se*, but to the direct interaction between the underlying mantle and the overlying plates. The uplift of the oceanic and continental lithosphere over hotspots is an example. The mechanisms for the support of topography can therefore be broadly classified as:

- **isostatic**—the response to density and thickness variations in the lithosphere. Such variations result primarily from the horizontal motions of plates.
- **dynamic**—the direct response to processes involving flow in the viscous mantle. Large-scale three-dimensional variations in mantle density are most likely caused by whole-mantle thermal convection. Smaller-scale heterogeneity is probably caused by mantle plume activity.

1.5.1 The shape of the Earth

The Earth is not a perfect sphere. If it were a perfect sphere, the gravitational acceleration would be the same at every point on its surface, but it is known that this is not the case. Instead, when the effects of rotation are taken into account, gravity observations show the Earth to be approximated by an *oblate spheroid*, the shape generated by rotating an ellipse around its shorter axis. There is a bulge at the equator and a flattening at the poles. The oblate spheroid is the theoretical gravitational *equipotential surface*. However, at a smaller, subglobal scale, the Earth shows important deviations from the oblate spheroid. The actual equipotential surface for the Earth, which can be thought of as the level to which still water accumulates in the ocean, is the so-called *geoid*. The geoidal surface can be viewed as the ultimate base level for terrestrial denudation.

Lateral variations in the density distribution within the Earth give rise to deviations of the geoid height from the reference oblate spheroid; these are called *geoid height anomalies*. At long wavelength, the geoidal surface shows some impressive departures from the reference oblate spheroid, with a deep (−80 m) geoid low in the Indian Ocean and a geoid high (+60 m) over the western Pacific (Plate 1.2, facing p. 204). Geoid height anomalies are accompanied by gravity anomalies, troughs in the geoid being associated with negative gravity anomalies (mass deficit at depth) and peaks in the geoid being associated with positive gravity anomalies (mass excess). Geoid height anomalies can therefore be used to calculate the variation in density with depth.

1.5.2 Isostatic topography
Hypotheses of Airy and Pratt

Isostasy is a principle requiring that the rigid surface layer of the Earth, the lithosphere, is able to 'float' on a fluid substratum, the aesthenosphere. It is assumed that below a certain depth, the *depth of compensation*, all the pressures are equal and hydrostatic (the same in all directions). The excess mass represented by the topography of a mountain is therefore compensated for by a mass deficiency beneath it at a depth above the compensation depth. Conversely, the mass deficit of a deep trough at the surface may be compensated for by a dense mass at depth above the compensation depth. Two competing hypotheses were proposed in the mid-1850s. The *Airy hypothesis* involves compensation by variations in the thickness of the upper layer, rather like blocks of wood floating in a tank of water (Fig. 1.25a) so that

$$h_{root} = \frac{h_{mt}\rho_1}{\rho_2 - \rho_1} \tag{1.14}$$

where ρ_1 and ρ_2 are the densities of the upper and lower layers respectively, h_{root} and h_{mt} are the depth of the root below the level of the base of the upper layer associated with a surface at sea level, and the

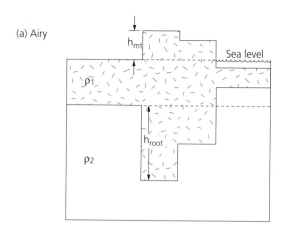
(a) Airy

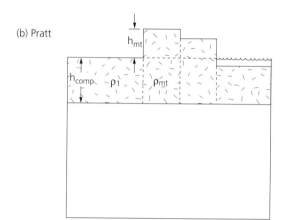
(b) Pratt

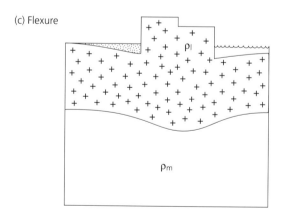
(c) Flexure

Fig. 1.25 Isostasy. Contrasting (a) Airy, (b) Pratt and (c) flexural models. The notation is explained in the text.

height of the mountain above sea level, respectively. The upper layer is customarily taken as the crust and the lower layer as the mantle lithosphere. Since the aesthenosphere is weak and can tolerate only small lateral pressure gradients, the depth of compensation must lie at relatively shallow depths within or at the base of the lithosphere.

In the *Pratt hypothesis* the base of the upper layer is assumed to be constant, isostatic equilibrium being achieved by density variations within the upper layer (Fig. 1.25b). In this hypothesis mountains are not compensated by possessing a root, but by being underlain by light material compared to adjacent regions. In such a situation

$$\rho_{mt} = \rho_1 \left(\frac{h_{comp}}{h_{mt} + h_{comp}} \right) \tag{1.15}$$

where h_{comp} is the depth of compensation (below sea level) and ρ_{mt} is the density of material underlying the mountain. In the Pratt hypothesis oceans are underlain by denser material than adjacent regions.

Flexural isostasy

In both the Airy and Pratt hypotheses the pressures at the depth of compensation under imaginary, narrow, vertical columns of rock are constant. Compensation can be thought of as local. However, there are instances where the upper part of the lithosphere appears to behave elastically by a process of long-wavelength flexure (Fig. 1.25c). A mountain range or an oceanic seamount chain acts as a load, causing

a regional flexure, the wavelength of which depends on the rheological properties of the plate, specifically its *flexural rigidity* (or elastic thickness). The characteristic response of an elastic lithosphere to an applied load system is of a maximum deflection under the centre of gravity of the load, the deflection decreasing laterally and passing into a region of relative uplift, the *flexural forebulge*. The sea floor bathymetry around oceanic islands such as the Emperor-Hawaiian seamount chain clearly shows the expression of the flexural moat and flanking bulge.

The way in which the lithosphere compensates flexurally for the excess mass of a mountain belt (or any other load) is determined by the rigidity of the plate, but also by the wavelength of the load (Fig. 1.26). Consider a sinusoidally varying load of wavelength λ at the Earth's surface with a maximum amplitude h_0 which produces a sinusoidal deflection beneath it of the same wavelength and amplitude w_0. If the flexural rigidity of the elastic lithosphere is D, then

$$w_0 = \frac{h_0}{\dfrac{\rho_m}{\rho_\ell} - 1 + \dfrac{D}{\rho_\ell g} \left(\dfrac{2\pi}{\lambda} \right)^4} \tag{1.16}$$

where ρ_m is the density of the mantle displaced by the downward deflection of the elastic lithosphere, ρ_ℓ is the density of the material comprising the load, and $2\pi/\lambda$ is termed the *wavenumber*. If the wavelength of the load is short, the denominator becomes very large, indicating that the deflection is very small compared

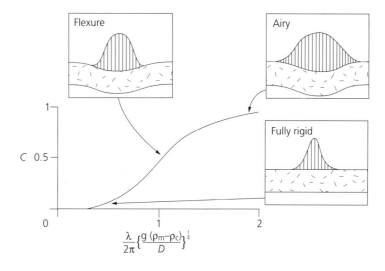

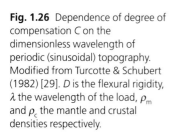

Fig. 1.26 Dependence of degree of compensation C on the dimensionless wavelength of periodic (sinusoidal) topography. Modified from Turcotte & Schubert (1982) [29]. D is the flexural rigidity, λ the wavelength of the load, ρ_m and ρ_c the mantle and crustal densities respectively.

to the load. The lithosphere appears to behave very rigidly to loads under these circumstances. However, if the wavelength of the load is large, the equation for the maximum deflection simplifies to

$$w_0 = \frac{\rho_\ell h_0}{(\rho_m - \rho_\ell)} \qquad (1.17)$$

which is recognizable as the solution for Airy (local) isostasy. At long wavelengths, therefore, the lithosphere appears to have little flexural strength. We can define the *compensation* as the ratio of the actual deflection to the maximum (Airy) deflection. It is illustrated in Fig. 1.26.

In summary, loads of very long wavelength are highly compensated, the lithosphere appearing in such cases to be very weak in its ability to support loads. Consequently, the load represented by the water in the Pacific Ocean is compensated locally by the underlying oceanic lithosphere. Loads of small wavelength, however, may be uncompensated, the lithosphere appearing to be extremely rigid in supporting the load. At intermediate wavelengths the lithosphere deforms by flexure under the load. Flexural depressions and bulges are commonplace in the ocean, either at the trenches of subduction zones or the moats surrounding seamounts, and on the continents as foreland basins adjacent to mountain belts.

Apart from mountain belts, some of the greatest topography on the surface of the Earth is associated with continental rifting (e.g. E Africa) and with continental margins (e.g. SW Africa) (Plate 1.4). Both of these settings involve stretching, and therefore thinning, of the lithosphere. It is therefore important to note that considerable topography may be associated with both mountain building at zones of collision and extension at sites of rifting or continental margin development.

Flexural uplift of rift flank and ocean margins Studies of the elastic thickness (flexural rigidity) of regions undergoing active extension such as East Africa suggest that the lithosphere has a finite rigidity rather than behaving according to local (Airy) isostasy. The unloading of the lithosphere by extension along major rift-bounding crustal faults may therefore cause flexural uplift (Fig. 1.27). This may explain some rift flank and continental margin uplifts, such as the South African Atlantic margin.

The wavelength of the rift flank uplifts should depend on the flexural rigidity of the lithosphere

Practical exericse 1.2: Flexural isostasy

1 The Alps of western Europe have a width of about 150 km and are responsible for flexing down the underlying lithosphere. Assuming the density of the load to be 2800 kg m^{-3}, the density of mantle to be 3200 kg m^{-3}, and flexural rigidity 10^{23} N m, what is the degree of compensation of a sinusoidal load represented by the Alpine mountain belt?
2 The topographic load represented by the Tibetan plateau has an average width of 1400 km. Assuming the same density terms as above, and a flexural rigidity of 5×10^{23} N m, what is the degree of compensation for the Tibetan load?

Solution
1 The wavelength of the Alpine load is approximately twice the width, $\lambda = 300$ km, giving a wavenumber of $2\pi/\lambda = 1.925 \times 10^{-5}$. Consequently, the degree of compensation is $C = 0.17$. This suggests that the lithosphere responds flexurally to the Alpine load.
2 The wavelength of the Tibetan load is approximately 2800 km, giving a wavenumber of $2\pi/\lambda = 2.24 \times 10^{-6}$. The degree of compensation is $C = 0.99$, indicating that the Tibetan Plateau can be treated as being in Airy isostasy.

being unloaded, the amplitude depending on the magnitude of the unloading. In the case of small grabens in Tibet, flexural isostatic compensation is thought to be facilitated by flow in a viscous lower crust, giving a short-wavelength rift margin uplift, rather than in the aesthenosphere, which would produce the longer-wavelength feature recognized in the flank uplifts of the Baikal Rift (Siberia) and Shanxi Rift (China) (Fig. 1.27). If so, it would lend support to the idea that the upper crust and mantle are capable of decoupling along a low-viscosity lower crust.

Isostasy in zones of convergence
A vivid illustration of the topographic effects of mantle and lithospheric processes is provided by the site of collision of the Indian 'indenter' and the continent of Eurasia (Fig. 1.28). The date of collision of India and Eurasia is thought to be about 50 Ma, with approximately 2900 (±900) km of convergence since that time being accommodated by the thickening of crust both in the Himalaya and the Eurasian area beyond. This thickening may result from the underthrusting of the intact Indian plate beneath the

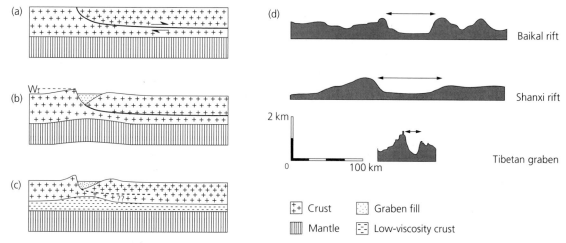

Fig. 1.27 Effect of the depth of the low-viscosity layer on the wavelength of rift flank uplift. (a), (b) and (c) illustrate basic concepts. (a) Extension along fault soling in lower crust creates graben filled with sediment. Regional compensation for the negative load of the graben (compared to normal thickness crust) produces margin uplifts of height w. In (b) the crustal viscosity is high and compensation takes place in the aesthenosphere, giving a long flexural wavelength. In (c) the crust has a low viscosity, compensation takes place in the crust itself and the wavelength is short. (d) Comparison of topographic profiles shows the Baikal (Siberia) and Shanxi (China) rifts to have long-wavelength flank uplifts, whereas the graben in the Tibetan plateau have very short-wavelength uplifts. After Masek *et al.* (1994) [30].

frontal edge of Eurasia, or from widespread north–south shortening of Eurasia's crust. The Tibetan plateau, at about 5.5 km average elevation, is the isostatic result of this crustal thickening.

Where convergence takes place over broad zones, it is likely that the thickening does not take place by the simple thrusting of one plate beneath the other, but by a wholesale thickening of the entire lithosphere (Fig. 1.29). In such a case, the cold, mantle lithosphere would also contribute to a thickened root. Since this root of mantle lithosphere is denser than the aesthenosphere at the same depth, it possesses negative bouyancy. An isostatic balance can be carried out as follows [31].

Consider crust of thickness h_c and density ρ_c overlying a mantle which increases in temperature linearly from the base of the crust to the top of the aesthenosphere at depth h_0 with a gradient γ. The temperature of the aesthenosphere is about 1300°C, corresponding to a density of 3200 kg m^{-3} (decompressed to surface pressures). The density of the mantle lithosphere decreases with depth below the crust because of the progressive warming at a rate determined by the coefficient of thermal expansion α,

$$\rho_\ell = \rho_a\{1 + \alpha T_\ell\} \tag{1.18}$$

where ρ_a is the density of the aesthenosphere, ρ_ℓ is the density of the mantle lithosphere and T_ℓ is its temperature. Hence at a depth y

$$\rho_\ell(y) = \rho_a\{1 + \alpha\gamma(h_0 - y)\} \tag{1.19}$$

Now if we instantaneously thicken this lithosphere by a factor f, we can isostatically balance the rock column of the thickened lithosphere with the initial lithosphere down to a depth equal to the depth of the thickened lithosphere. After thickening the new crustal thickness becomes $h_c f$, the new lithospheric thickness becomes $h_0 f$, the new temperature gradient becomes γ/f. However, this is accompanied by an elevation change of the overlying mountain or plateau Δh in response to the increased lithospheric root. If h is the thickness of the increased root of mantle lithosphere, then

$$\Delta h = h_0(f-1) - h \tag{1.20}$$

Performing an isostatic balance down to a depth $h_0 + h$,

$$\Delta h = (f-1)\left\{\frac{(\rho_a - \rho_c)h_c}{\rho_a} - \frac{1}{2}\alpha\gamma(h_0 - h_c)^2\right\} \tag{1.21}$$

where the first term in the curly bracket refers to the

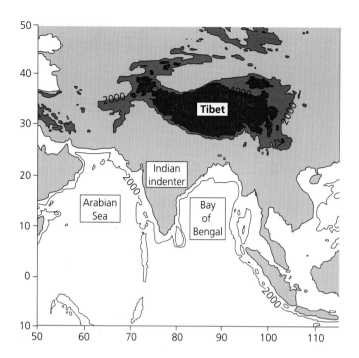

Fig. 1.28 Location of the collision zone of the Indian indenter with Asia and the resulting topography. Bathymetric contour at −2 km and topographic contours at 2 km and 4 km are shown. After Molnar *et al.* (1993) [31].

effect of crustal thickness changes (Airy isostasy), and the second gives the effect of the thickened root of mantle lithosphere.

The downward penetration of a cold lithospheric root into the aesthenosphere causes lateral temperature gradients in the aesthenospheric mantle which may cause the development of convective instabilities. Consequently, the deep lithospheric root may be swept away by detaching from overlying stronger lithosphere, and be replaced by hotter, more buoyant material. This should have the isostatic response of surface uplift. The removal of the cold lithospheric root should result in a surface elevation change (uplift) equivalent to but with opposite sign to the second term in equation (1.19), i.e. *c.* 1.7 km in the practical exercise.

This explains the present surface topography of Tibet. The time-scale of the uplift due to convective removal of the lithospheric root is thought to be relatively short compared to the time-scale of the initial thickening (20–40 My). Removal is thought to have taken place during the late Miocene (*c.* 6–9 Ma),

Practical exercise 1.3: The elevation of Tibet

The average elevation of the Tibetan plateau today is 5.5 km. The crust appears to have been thickened by a factor of 2. Assuming that the lithosphere has thickened by the same factor, what is the predicted elevation change of the Tibetan plateau resulting from the thickening? If only crustal thickening takes place, what is the prediction from Airy isostasy? Use the following parameter values:

$h_c = 35 \, \text{km}$

$h_0 = 135 \, \text{km}$

$\rho_c = 2800 \, \text{kg m}^{-3}$

$\rho_a = 3200 \, \text{kg m}^{-3}$

$\alpha = 3.5 \times 10^{-5} \, °\text{C}^{-1}$

$\gamma = 15 \, °\text{C km}^{-1}$

Solution

The elevation change caused by thickening the entire lithosphere by a factor of 2 is 3.6 km. This is made of the components due to Airy isostasy (5.3 km) and due to the temperature differences in the mantle lithosphere (−1.7 km). Thickening of the Asian lithosphere by a factor of 2 should have resulted in a plateau with a lower elevation than observed today. The prediction from Airy isostasy, based on crustal thickening only, is much closer to the observed elevation. This is discussed further below.

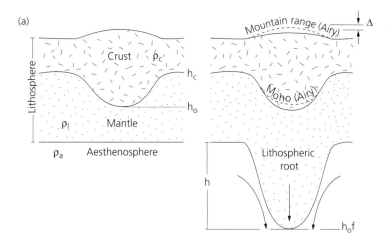

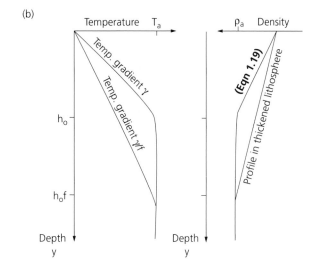

Fig. 1.29 Consequences of lithospheric thickening under Tibet, after Molnar *et al.* (1993) [31]. (a) Cross-sections to illustrate the difference between crustal thickening (left) and thickening of the entire lithosphere (right). The excess mass in the thickened lithospheric root pulls the overlying crust down and lowers the topography of the mountain belt. (b) Profiles of temperature (left) and density (right) in normal and thickened lithosphere.

and is synchronous with the onset of the Asian monsoon [31].

A similar but not identical mechanism has been invoked to explain the uplift of the Colorado plateau, western USA. The mechanism envisaged has been termed *continental delamination* to emphasize that the entire mantle lithosphere peels off from the overlying crust and sinks into the aesthenosphere [32].

Isostasy at zones of extension

The active seismicity, presence of large normal faults at the surface, high surface heat flows, generally alkaline volcanic activity, elevated Moho indicated by deep seismic investigations, and negative Bouguer gravity anomalies suggest that continental rifts are sites of lithospheric stretching. The following two models satisfy a wide range of geological and geophysical observations:

1 'Active', 'mantle-activated' or 'open system' rifting caused by thermal activity in the mantle, and characterized by large volumes of extrusive rocks. Active rifts are probably related to hotspot activity in the mantle, which leads to lithospheric heating, thinning, the release of melts by decompression and rift formation.

2 'Passive', 'lithosphere-activated' or 'closed-system' rifting, where the upwelling of aesthenosphere is a passive response to, not a cause of, mechanical stretching of the lithosphere. Passive rifts have little or no volcanism.

These two models have somewhat different dynamics, active rifting being dealt with in Section 1.5.3. Passive rifting presents the simplest scenario since we can ignore any temperature and density variations in the aesthenosphere (Fig. 1.30). Let us assume that before stretching the crust has a thickness h_c, and lithosphere thickness h_ℓ. If the crust and lithosphere are stretched uniformly by a stretch factor β, the new crustal thickness becomes h_c/β and the new lithospheric thickness becomes h_ℓ/β. There is an elevation change Δh at the surface of the Earth above the region of stretching which depends on the amount of stretching and on the ratio h_c/h_ℓ. Let the average crustal, mantle lithosphere and aesthenosphere densities be ρ_c, ρ_m and ρ_a respectively, and the density of the material infilling the space caused by the elevation change be ρ_i. The pressure under the lithospheric column before rifting is therefore

$$h_c\rho_c g + (h_\ell - h_c)\rho_m g \qquad (1.22)$$

Because the acceleration due to gravity is common to all lithospheric columns we can ignore it in the following analysis. After stretching, the pressure at the depth of the original lithospheric thickness (ignoring g) is

$$\Delta h\rho_i + \left(\frac{h_c}{\beta}\right)\rho_c + \left(\frac{h_\ell - h_c}{\beta}\right)\rho_m + \left(h_\ell - \frac{h_\ell}{\beta} - \Delta h\right)\rho_a \quad (1.23)$$

Equating the two columns before and after rifting and regrouping the terms,

$$\Delta h = \frac{(1-1/\beta)}{(\rho_a - \rho_i)}\left\{h_\ell \rho_a - h_c\rho_c - (h_\ell - h_c)\rho_m\right\} \qquad (1.24)$$

It only remains to specify the density structure in the region of stretching. Let us assume that the crust can be given a single value of ρ_c that remains constant, but that the mantle lithosphere changes in density as a function of its temperature, from the aesthenospheric temperature at its base, with a geothermal gradient given by $\gamma_0\beta$, where γ_0 is the original linear geothermal gradient. The average temperature of the mantle lithosphere is therefore

$$\rho_m = \rho_a\left\{1 + \alpha\left(T_a - \gamma_0\frac{h_\ell - h_c}{2}\right)\right\} \qquad (1.25)$$

where T_a is the aesthenospheric temperature. It will be

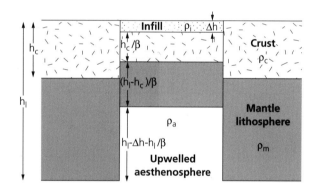

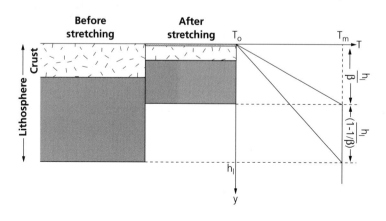

Fig. 1.30 Notation and thermal consequences of simple uniform stretching of the lithosphere.

seen in the practical example below that the effect of the temperature increase on the density of the mantle lithosphere is minimal and can in most cases be neglected.

Practical exercise 1.4: Subsidence in rifts

Calculate the elevation change in a region of passive rifting for the following parameter values:

$\beta = 2$

$h_c = 35\,\text{km}$

$h_\ell = 125\,\text{km}$

$\rho_c = 2800\,\text{kg m}^{-3}$

$\rho_a = 3200\,\text{kg m}^{-3}$

$T_a = 1333°\text{C}$

$\alpha = 3.5 \times 10^{-5}°\text{C}^{-1}$

$\gamma_0 = 30°\text{C km}^{-1}$

Solution

The elevation change is 3.1 km subsidence for a water filled rift basin, or 4.9 km for a rift basin filled to the brim with sediment of average bulk density 1800 kg m^{-3}.

Unless the initial crustal thickness is very small compared with the lithospheric thickness, lithospheric stretching should therefore be accompanied by large amounts of subsidence. Yet we know that active rift zones such as the East African rift are located on elevated topographic domes. The simplest way to account for this is to increase the heat flow from the mantle. In the case of active rifting, we may increase the aesthenospheric temperature by 100–200°C compared to its normal temperature, by placing a hot plume head under the lithosphere (Section 1.5.3).

Thermal subsidence following stretching Stretching of the lithosphere results in an elevation of the aesthenosphere to shallower depths determined by the stretch factor. Consequently, the entire lithospheric column is heated (Fig. 1.31). Once the source of the stretching ceases, the isotherms in the lithosphere relax to their pre-stretching position. This results in thermal contraction of the lithosphere and subsidence. Thermal subsidence may characterize a post-rift stage of continental rifts where the rift zone and margins sag after rifting is aborted, or the history of continental margins where rifting has been successful in completely splitting the continent (passive margins).

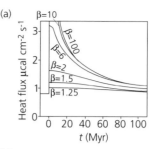

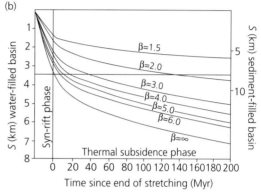

Fig. 1.31 The heat flow and thermal subsidence following stretching as a function of the stretch factor β. (a) The magnitude of the stretching-related heat flux is dependent on the stretch factor, but at long periods after the cessation of stretching, the heat fluxes are very similar irrespective of the amount of stretching. (b) The rate of thermal subsidence decreases exponentially with time. The curves shown refer to a lithosphere of thickness 125 km and crust of thickness 31.2 km. After McKenzie (1978) [33].

The form of the thermal subsidence phase of sedimentary basins is similar to that of the cooling of the oceanic lithosphere at mid-ocean ridges (Section 1.5.5). However, we should expect the stretch factor β to determine the heat added to the lithosphere during stretching, and therefore the heat lost after stretching. The form of the thermal subsidence over time must also be affected by the thermal material properties of the lithosphere.

The thermal subsidence as a function of time t is given by

$$S(t) \approx E \frac{\beta}{\pi}\left(\sin\frac{\pi}{\beta}\right)\left(1 - e^{-t/\tau}\right) \qquad \textbf{(1.26)}$$

where

$$E = \frac{4h_\ell \rho_a \alpha T_a}{\pi^2(\rho_a - \rho_i)}$$

and τ is the thermal time constant of the lithosphere

$$\tau = \frac{h_\ell^2}{\pi^2 \kappa} \qquad (1.27)$$

where κ is the *thermal diffusivity* (in metres squared per second).

We can now calculate the thermal subsidence history for a rift by taking the parameter values from Practical Exercise 1.4, and adding the thermal diffusivity $\kappa = 10^{-6}\,m^2\,s^{-1}$. Assuming the basin to be filled with water ($\rho_1 = 1000\,kg\,m^{-3}$), the thermal subsidence follows an exponential curve (Fig. 1.31). After one lithospheric time constant, about 50 My, the thermal subsidence is 1.383 km. Failed rifts and continental margin basins are commonly fully or partially filled with sediment during the thermal subsidence phase. These sediment loads drive further subsidence which amplifies the thermal contraction signal. As a result, some continental margins such as the US eastern seaboard have up to 16 km of preserved sediment. Failed or aborted rifts such as the North Sea in NW Europe contain 3–4 km of post-rift sediment overlying the fault-bounded syn-rift sediments. Using the example above where $\beta = 2$, for a basin filled with sediment of density $1800\,kg\,m^{-3}$, the thermal subsidence after 50 My is 2.173 km.

1.5.3 The bathymetry of the ocean floor

Inspection of a bathymetric chart of the ocean shows immediately that the ocean gradually deepens away from the mid-ocean ridges. It is generally accepted that the bathymetry of the sea floor is controlled at first order by the thermal contraction of oceanic lithosphere formed at the mid-ocean ridge system.

When new ocean floor is formed at mid-ocean ridges it cools as it moves away from the site of spreading. We can treat the cooling of the new oceanic crust by the cold seawater as instantaneous (Fig. 1.32). The distance over which the sudden cooling is felt after a time t is determined by the thermal diffusivity κ of the ocean crust. This thermal diffusion distance must be equal to $\sqrt{\kappa t}$. The cooling new oceanic material forms the oceanic lithosphere as a

cold thermal boundary layer above the easily deformed aesthenospheric mantle. The base of the new oceanic lithosphere can therefore be defined by a

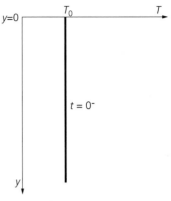

Just before instantaneous heating

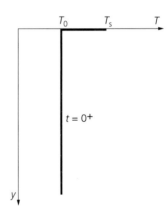

Just after instantaneous heating

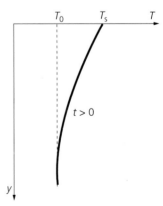

At time t after instantaneous heating

Fig. 1.32 *Right.* Heating of a half-space by an instantaneous increase of temperature from T_0 to a new surface temperature T_s. This might apply to the temperature field following the injection of a dyke into cool country rocks. A similar solution can be applied to the sudden cooling of hot mantle rock injected at a mid-ocean ridge.

characteristic isotherm (1330°C, or about 1600 K), and its thickness is clearly a function of its age. If the spreading rate is u and the horizontal distance from the ridge crest is x, the age is x/u, and the thermal diffusion distance $\sqrt{\kappa x/u}$.

Introducing a dimensionless temperature ratio

$$\theta = \frac{T - T_0}{T_s - T_0} \tag{1.28}$$

where T_0 is the initial temperature (the aesthenospheric temperature), T_s is the constant temperature of the space into which the ocean lithosphere is emplaced (the seawater temperature), and T is the temperature at time t, the solution for the temperature is then

$$\theta = \mathrm{erfc}\left(\frac{x}{2\sqrt{\kappa t}}\right) \tag{1.29}$$

where erfc is the complementary error function.

The cooled oceanic material forms a thermal boundary layer. It is a largely arbitrary choice of the definition of the boundary layer, but if we define it as the thickness to where $\theta = 0.1$, it is found that the thermal boundary layer is 2.32 times the thermal diffusion distance $\sqrt{\kappa t}$ or $\sqrt{(\kappa x/u)}$. The thickness of the oceanic lithosphere at an age of 50 My, taking the thermal diffusivity as $10^{-6}\,\mathrm{m^2\,s^{-1}}$, is therefore approximately 92 km. The oceanic lithosphere should increase parabolically in thickness with age, and therefore also with distance from the ridge crest.

By cooling, the oceanic lithosphere contracts, increases its density and exerts a higher stress on the underlying mantle. An isostatic balance shows that the sea floor must therefore subside. Any lithospheric columns through the oceanic lithosphere can be balanced isostatically to give the depth of the ocean floor as a function of distance from the ridge crest, or time (Fig. 1.33).

The observed heat flows in the ocean, and the observed bathymetry, are in general agreement with a model of instantaneous cooling of new oceanic lithosphere and its loss of heat through time by conduction resulting in subsidence. There is therefore a simple relationship between the bathymetry of the ocean floor and its thermal age, of the form

$$h = h_{\mathrm{ridgecrest}} + A\sqrt{t} \tag{1.30}$$

where h is the depth of the ocean floor, $h_{\mathrm{ridgecrest}}$ is commonly in the region of 2.5 km, A is a coefficient and t is the age of the oceanic lithosphere. Since the age of the oceanic lithosphere is directly related to the

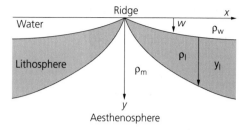

At any point, mass per unit area $= \int_0 y_l\, \rho_l dy + w\rho_w$
At ridge crest, mass per unit area at depth
$(w + y_l) = \rho_m(w + y_l)$
For equilibrium, $w(\rho_w - \rho_m) + \int_0 y_l\,(\rho_l - \rho_m)\,dy = 0$

Fig. 1.33 The principle of isostasy requires the ocean to deepen with age to offset the effects of thermal contraction of the oceanic lithosphere. The water depth below the level of the ridge crest is $w = h - h_{\mathrm{ridgecrest}}$, the thickness of the oceanic lithosphere is y_ℓ, and ρ_m, ρ_w and ρ_ℓ are the mantle, water and lithospheric densities, respectively.

distance from the ridge crest by the spreading rate, the oceanic bathymetry increases gradually away from the sites of spreading. An example from the North Atlantic is given in Figure 1.33. The isostatic balance will of course be affected by the presence of marine sediments. However, in the deep sea the cover of pelagic sediments is generally only a thin veneer.

1.5.4 Dynamic topography

Dynamic topography can be defined [34] as the deflection of the solid interface of the Earth which results from buoyancy forces (due to lateral density variations) in the viscous mantle; these buoyancy forces are thought to originate primarily from thermal convection. That dynamic topography exists in the present day can be supported from two main lines of evidence.

1 The observational evidence that very long-wavelength variations in dynamic topography exist is provided primarily by an analysis of global topography and bathymetry [35]. The observed topography is first corrected for shallow-density variations in the lithosphere, including the effects of the cooling of oceanic lithosphere as a function of age (Section 1.5.3). The residual dynamic topography has an amplitude of about 300 m, and a very long-wavelength spatial pattern (dominated by degree 2, which means that two wavelengths occupy the circumference of the Earth), similar to that of the geoid. The degree 2 pattern in the geoid and dynamic

topography shows highs located over Africa and the western Pacific, and lows located over the eastern Pacific and Antarctic–Indian Ocean region (Plate 1.4). It is most likely that this very long-wavelength topographic signal results from anomalies in the deep mantle.

2 Measurements of the velocities of seismic body waves travelling through the mantle show a three-dimensional picture of relatively fast- and slow-velocity regions. These small but significant velocity anomalies are interpreted in terms of density variations caused by temperature differences due to a slow convection of mantle material. These density variations can in turn be used to generate a model of the geoid. Computer modelling of convection in the upper mantle suggests that there should be positive geoid height anomalies over the hot rising limbs of convection cells. These anomalies arise because although there is a density deficit associated with the hot rising limb, it is more than counteracted by the dynamic upward-acting force due to positive buoyancy. The seismic anomalies of the deep mantle are capable of reproducing the observed geoid, and predict large-scale (degree 2–6) variation in dynamic topography with an amplitude of 1 km.

Although the amplitudes of dynamic topography outlined above are considerably smaller than the isostatic topography of mountain ranges caused by crustal thickness changes, dynamic topography is of large wavelength, and is therefore an important control on global topography.

Plumes, hot spots and hot spot tracks

Two processes acting away from plate margins that are not readily explainable by plate tectonics are the chains of volcanic oceanic islands thought to be linked to hotspots, and large, rapidly erupted igneous provinces typified by flood basalts on land and basaltic plateaux under the sea. Both oceanic island chains and continental flood basalt provinces contain basalts whose geochemistry indicates that they come from the melting of mantle elevated above the normal temperature of the aesthenosphere. They can be explained by the rising of hot plumes from the hot thermal boundary layer at the core–mantle boundary. The lithospheric plates, on the other hand, are the cold thermal boundary layer at the surface of the Earth. Both plumes and plates are therefore associated with a particular mode of mantle convection.

Fluid dynamical work suggests that low-viscosity plumes may initiate from within the Earth and ascend as a spherical pocket of fluid (plume head) fed by a pipe-like conduit (plume tail) continuously supplying bouyant, hot material to the head region. It is likely that enlargement of the plume occurs by the entrainment of material heated by the ascent of the plume from the core–mantle boundary. The flood basalt provinces have been interpreted as originating through the melting of the heads of newly started plumes, whereas the oceanic island chains represent the trails of the relatively long-lived plume tails as the plate migrates over the mantle [36].

Fluid dynamical experiments suggest that the diameter of a new plume head varies according to the volume flux and temperature excess of the source material provided to the plume head, and the thermal and viscosity properties of the lower mantle into which the plume starts to ascend (Fig. 1.34a). The plume head then grows by entrainment as it ascends through the mantle, so that the plume head diameter grows as a function of the distance travelled (Fig. 1.34b). By the time the plume head has penetrated into the upper mantle it should have cooled to only 100–150°C above the ambient temperature.

Upon nearing the surface of the Earth, the plume head spreads out into a disc of hot material with positive buoyancy. This produces surface uplift which in some cases may lead to extension of the lithosphere and the formation of 'active' rifts. The scaling of laboratory experiments suggest that the timing and magnitude of the surface uplift depend strongly on the viscosity of the upper mantle. The results from laboratory experiments in which a plume head is sourced from the core–mantle boundary with a buoyancy flux of $3 \times 10^4 \, \mathrm{N \, s^{-1}}$ and a source temperature in excess of 300°C, lower mantle dynamic viscosity of $10^{22} \, \mathrm{Pa \, s}$ (kinematic viscosity of $2.5 \times 10^{18} \, \mathrm{m^2 \, s^{-1}}$) ascending into an upper mantle with a viscosity of $3 \times 10^{20} \, \mathrm{Pa \, s}$, is shown in Fig. 1.35 [37, 38]. The surface is initially weakly uplifted while the plume head is entirely within the lower mantle (–25 My). When the plume head enters the low-viscosity zone of the upper mantle (–3 My) the surface uplift takes place rapidly, reaching a maximum elevation of 600 m after further ascent to just beneath the lithosphere. At this stage the plume head has a diameter of 1300 km. Two factors may increase the maximum elevation of the topographic dome: (i) penetration of the hot plume into the cold lithosphere; (ii) a volume increase caused by melting. The release of large amounts of basalts by melting of the plume can only take place once the plume head has reached shallow depths. It should be noticed that this is only possible several millions of years after the maximum surface uplift. The development of smaller-scale gravitational instabilities over the cap of the plume as cold, dense material is squeezed between the ascending plume head and the Earth's surface

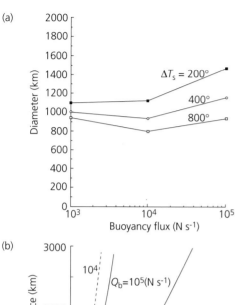

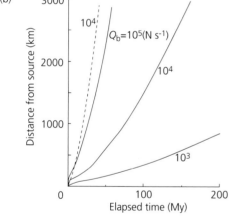

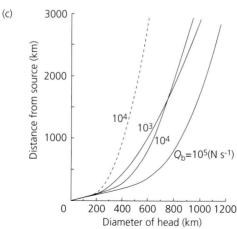

may facilitate the last-stage ascent of the plume and produce the high surface uplifts seen today in locations such as East Africa (Fig. 1.36) and the large outpourings of basalts in the geological past such as the Siberian and Deccan (India) Traps.

Two sorts of uplift pattern and igneous activity should result from plume activity, depending on whether the lithosphere migrates over a plume head or plume tail. Plume head provinces should be areally extensive and equant (1500–2500 km across) whereas plume tail provinces should be narrow (<300 km wide) and linear, that is, they should be hotspot tracks.

Hotspots are at their most obvious in the oceans [39], where they are associated both with volcanism and strongly elevated bathymetry. The Hawaiian Islands, Iceland and Canaries are all excellent examples (Table 1.5). The recognition of hotspots on the continents is far more problematic. However, the highspots associated with alkaline volcanism in north central Africa, such as the Hoggar and Tibesti domes, may be related to activity in the underlying mantle. Doming over hot plumes has also been interpreted as responsible for particular centrifugal palaeodrainage patterns, as well as for the eruption of vast piles of basalts [40].

If there are upwellings beneath continents producing hotspots, we should also expect there to be downwellings, which would be cold. Downwellings should therefore produce dynamic subsidence of the continental surface. The Congo basin may be situated over one such downwelling. It is adjacent to the East African dome which is thought to overlie a plume head (Fig. 1.36).

1.5.5 Continental hypsometries

A convenient way of visualizing the broad morphological characteristics of the surface of the Earth is

Fig. 1.34 *Left.* Evolution of a plume arising from the core–mantle boundary. (a) The predicted diameter of a starting plume from the core–mantle boundary for a mantle viscosity of 10^{22} Pa s, typical of the modern lower mantle, for a range of source temperature anomalies ΔT_s in degrees Celsius. (b), (c). The height of (b) rise and (c) plume head diameter for $\Delta T_s = 400°C$ in a mantle of viscosity 10^{22} Pa s (solid lines) and viscosity 10^{21} Pa s, appropriate for the modern upper mantle and Archaean lower mantle (dashed lines). Plumes with low-bouyancy fluxes (<10^4 N s^{-1}) are likely to degenerate into a sequence of diapirs before reaching the surface after a slow ascent from the core–mantle boundary. Consequently, there is a lower bound to the buoyancy of plumes affecting surface uplift. After Griffiths & Campbell (1990) [37].

(a)

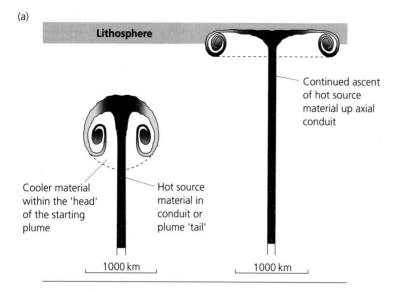

(b)

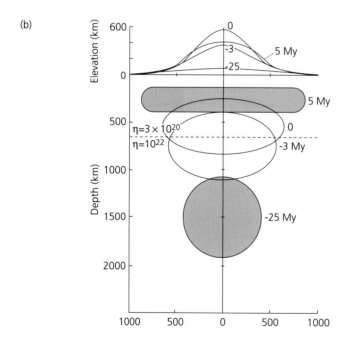

Fig. 1.35 (a) Cartoon of the ascent of a plume through the mantle and its mushrooming into a disc beneath the lithosphere, with darker shading indicating higher temperatures. After Hill (199) [41], v99, pp. 66–78, with kind permission from Elsevier Science-NL, Sara Burgerhartstraat 25, 1055 KV Amsterdam, The Netherlands. (b) The dimensions of a starting plume together with the predicted surface uplift, based on laboratory experiments. During the lateral spreading of the head the input of material from the source is discontinued, simulating the carrying away of the head from the source region by plate motion. After Griffiths & Campbell (1991) [37].

to plot the cumulative frequency distribution of elevation, or *hypsometric curve*. This reveals the proportion of the surface above a given height (Fig. 1.37). The hypsometric curve shows two important features:

1 the major change in topographic slope occurs at the edge of the continental shelf in present-day water depths of about 200 m;

2 only a very small proportion (less than 10%) of the Earth's surface is at elevations of greater than 1 km. When the hypsometric curves for the continents are compared it is seen that Africa is unusual in having a relatively large percentage of its surface at high elevations. This is entirely accounted for by the high topographic elevation of eastern (East African Rift System) and southern Africa. East Africa is thought to

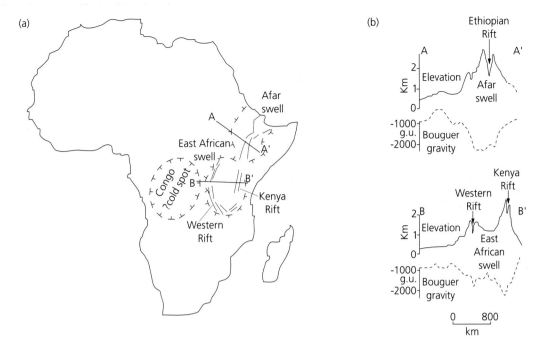

Fig. 1.36 The topographic swells of Africa (Afar and East African (Kenyan) domes) are due to uplift over hotspots. Both are rifted and are the sites of extensive volcanism. The Congo basin has been proposed to overlie a mantle coldspot (Hartley & Allen (1993) [42]. To right: topographic and Bouguer gravity profiles of Afar and East African swells. After Ebinger *et al.* (1989) [43].

Table 1.5 Buoyancy fluxes of the world's hotspots. From Sleep (1990) [44].

Hotspot	Flux* (Mg s^{-1})	Reliability	Flux† (Mg s^{-1})	Hotspot	Flux* (Mg s^{-1})	Reliability	Flux† (Mg s^{-1})
Afar	1.2	good		Juan Fernandez	1.6	poor	1.7
Australia, East	0.9	fair		Kerguelen	0.5	poor	0.2
Azores	1.1	fair		Louisville	0.9	poor	3.0
Baja	0.3	poor		Macdonald	3.3	fair	3.9
Bermuda	1.1	good	1.5	Marqueses	3.3	fair	4.6
Bouvet	0.4	fair		Martin	0.5	poor	0.8
Bowie	0.3	poor	0.8	Meteor	0.5	poor	0.4
Canary	1.0	fair		Pitcam	3.3	fair	1.7
Cape Verde	1.6	good	0.5	Réunion	1.9	good	0.9
Caroline	1.6	poor		St Helena	0.5	poor	0.3
Crozet	0.5	good		Samoa	1.6	poor	
Discovery	0.5	poor	0.4	San Felix	1.6	poor	2.3
Easter	3.3	fair		Tahiti	3.3	fair	5.8
Fernando	0.5	poor	0.9	Tasman, Central	0.9	poor	
Galapagos	1.0	fair		Tasman, East	0.9	poor	
Great Meteor	0.5	poor	0.4	Tristan	1.7	fair	0.5
Hawaii	8.7	good	6.2	Yellowstone	1.5	fair	
Hoggar	0.9	fair	0.4	Sum	54.9		
Iceland	1.4	good					
Juan de Fuca	0.3	fair					

* Sleep (1990). [44]
† Davies [1988]. [45]

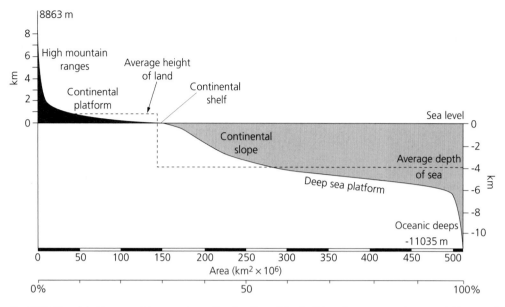

Fig. 1.37 The global hypsometric curve shows the elevation of the Earth's surface expressed as cumulative area (in millions of cubic kilometres) and as a percentage of the total surface area above a given height. The present-day land surface accounts for 29% of the Earth's surface area, but the break in slope on the hypsometric curve corresponding to the edge of the continental shelf gives a figure of 37% for the true spatial extent of the continents.

be situated over a mantle plume head. The present high elevation of southern Africa has been attributed to (i) the presence of underplated igneous bodies following plume activity in the Mesozoic, or (ii) the presence of an underlying present-day or recently active hotspot.

Hypsometry is strongly influenced by dynamic topography (Section 1.5.4). The continental hypsometry also determines the extent to which continents are flooded by certain absolute sea level changes. Consequently, an analysis of the extent of continental flooding in the geological past may reveal past variations in continental hypsometry. The simplest pattern to emerge is of two main cycles through the Phanerozoic (the last 540 My), with elevated sea levels particularly in the Ordovician and Cretaceous, and low sea levels particularly in the Cambrian and Permo-Triassic. Some authors link this broad pattern to the aggregation (inflation and exposure) and dispersal (continental flooding) of supercontinents. However, separating the effects of absolute (eustatic) sea level change from those of variations in isostatic or dynamic topography is a difficult task. Since most of the observations on continental flooding were made on stable continental interiors which have not suffered a great deal of

tectonics leading to lithospheric thickness changes, the topographic contribution is most likely due to dynamic topography. By comparing different continents and deciphering the common eustatic signal, it can be seen, for example, that Africa has undergone broad uplift in the Tertiary compared to other continents. This may explain its present anomalous hypsometry.

Further reading

P.A. Allen & J.R. Allen (1990) *Basin Analysis: Principles & Applications.* Blackwell Scientific Publications, Oxford.

E.K. Berner & R.A. Berner (1987) *The Global Water Cycle: Geochemistry and Environment.* Prentice Hall, Englewood Cliffs, NJ.

D.M. Moore (1982) *Green Planet: the Story of Plant Life on Earth.* Cambridge University Press, Cambridge.

J.M. Wallace & P.V. Hobbs (1977) *Atmospheric Science: An Introductory Survey.* Academic Press, San Diego, CA.

References

1 E.K. Berner & R.A. Berner (1987) *The Global Water Cycle: Geochemistry and Environment.* Prentice Hall, Englewood Cliffs, NJ, Table 2.3, p. 20.

2 W.D. Sellers (1965) *Physical Climatology.* University of Chicago Press, Chicago.

3 E.F. Harrison, P. Minnis, B.R. Barkstom & G.G.

Gibson (1993) Radiation budget at the top of the atmosphere. In: *Atlas of Satellite Observations Related to Global Change* (eds R.J. Gurney, J.L. Foster & C.L. Parkinson). Cambridge University Press, Cambridge, pp. 19–38.

4 M. Chahine (1992) The hydrological cycle and its influence on climate. *Nature* **359**, 373–80.

5 R.C. Ward (1975) *Principles of Hydrology*, 2nd edn. McGraw-Hill, Maidenhead.

6 R. Keller (1984) The world's fresh water: yesterday, today and tomorrow. *Applied Geography and Development*, **24**, 7–23.

7 P.J. Webster (1994) The role of hydrological processes in ocean–atmosphere interaction. *Reviews of Geophysics* **32**, 427–76.

8 S.I. Rasool & C. de Bergh (1970) The runaway greenhouse and the accumulation of CO_2 in the Venus atmosphere. *Nature* **266**, 1037–9.

9 A. Gill (1982) *Atmosphere–Ocean Dynamics*. International Geophysical Series **30**, San Diego, CA.

10 R. Atlas, R.N. Hoffman & S.C. Bloom (1993) Surface wind velocity over the oceans. In: *Atlas of Satellite Observations Related to Global Change* (eds R.J. Gurney, J.L. Foster & C.L. Parkinson). Cambridge University Press, Cambridge, pp. 129–40.

11 G.A. Meehl (1982) Characteristics of surface current flow from a global ocean current data set. *Journal of Physical Oceanography* **12**, 538–55.

12 H. Stommel (1958) The abyssal circulation. *Deep Sea Research* **5**, 80–2.

13 H. Heinrich (1988) Origin and consequences of cyclic ice rafting in the northeast Atlantic Ocean during the past 130 000 years. *Quaternary Research* **29**, 142–52.

14 A. Miller *et al.* (1983), *Elements of Meteorology*, 4th edn. Charles E. Merrill.

15 E.G. Njoku & O.B. Brown (1993) Sea surface temperature. In: *Atlas of Satellite Observations Related to Global Change* (eds R.J. Gurney, J.L. Foster & C.L. Parkinson). Cambridge University Press, Cambridge, pp. 237–50.

16 S.C. Porter & A. Zhisheng (1995) Correlation between climate events in the North Atlantic and China during the last glaciation. *Nature* **375**, 305–8.

17 J. Tricart & A. Cailleux (1972) *Introduction to Climatic Geomorphology* (translated by C.J. Kiewiet de Jonge). Longman, London.

18 M. Newson (1994) *Hydrology and the River Environment*. Oxford University Press, Oxford.

19 H.H. Lamb (1972) *Climate: Present, Past and Future. Vol. 1: Fundamentals and Climate Now*. Methuen, London.

20 R.G. Barry (1970) A framework for climatological research with particular reference to scale concepts. *Transactions of the Institute of British Geographers* **49**, 61–70.

21 I.A. Shiklomanov (1993) World fresh water resources. In: *Water in Crisis* (ed. P.H. Gleick). Oxford University Press, Oxford, pp. 13–24.

22 A. Baumgartner & E. Reichel (1975) *The World Water Balance*. Elsevier, Amsterdam.

23 R.J. Gibbs (1970) Mechanisms controlling world water chemistry. *Science* **170**, 1088–90.

24 I.D. White, D.N. Mottershead & S.J. Harrison (1992) *Environmental Systems*, 2nd edn. Chapman & Hall, London.

25 D.E. Reichle (1970) *Analysis of Temperate Forest Ecosystems*. Springer-Verlag, Heidelberg.

26 C.R. McClain, G. Feldman & W. Esaias (1993) Oceanic biological productivity. In: *Atlas of Satellite Observations Related to Global Change* (eds R.J. Gurney, J.L. Foster & C.L. Parkinson), Cambridge University Press, Cambridge, pp. 251–63.

27 M.D. Brasier (1995) Fossil indicators of nutrient levels. 1: Eutrophication and climate change. In: *Marine Palaeoenvironmental Analysis from Fossils* (eds D.W.J. Bosence & P.A. Allison), Geological Society Special Publication 83. Blackwell Science, Oxford, pp. 113–32.

28 J.T. Houghton, G.J. Jenkins & J.J. Ephraums (eds) (1990) *Climate Change: The IPCC Scientific Assessment*. Cambridge University Press, Cambridge.

29 D.L. Turcotte & G. Schubert (1982) *Geodynamics*. Wiley, New York.

30 G.J. Masek, B.L. Isacks, E.J. Fielding & J. Browaeys (1994) Rift flank uplift in Tibet: evidence for a viscous lower crust. *Tectonics* **13**, 659–67.

31 P. Molnar, P.C. England & J. Martinod (1993) Mantle dynamics, uplift of the Tibetan plateau, and the Indian monsoon. *Reviews of Geophysics* **31**, 357–96.

32 P. Bird (1979) Continental delamination and the Colorado plateau. *Journal of Geophysical Research* **84**, 7561–71.

33 D.P. McKenzie (1978) Some remarks on the development of sedimentary basins. *Earth & Planetary Science Letters* **40**, 25–32.

34 M. Gurnis (1991) Continental flooding and mantle-lithosphere dynamics. In *Glacial Isostasy, Sea Level and Mantle Rheology* (ed. R. Sabadini *et al.*). Kluwer Academic Publishers, Dordrecht, p. 446.

35 A. Cazenave, A. Souriau & K. Dominh (1989) Global coupling of the Earth surface topography with hotspots, geoid and mantle heterogeneity. *Nature* **340**, 54–7.

36 W.J. Morgan (1981) Hotspot tracks and the opening of the Atlantic and Indian oceans. In: *The Sea* (ed. C. Emiliani). Wiley, New York, vol. 7, pp. 443–87.

37 R.W. Griffiths & I.H. Campbell (1990) Interaction of mantle plume heads with the Earth's surface and onset of small scale convection. *Journal of Geophysical Research* **96**, 18 295–310.

38 R.W. Griffiths, M. Gurnis & G. Eitelberg (1989) Holographic measurements of surface topography in

laboratory models of surface hotspots. *Geophysical Journal* **96**, 477–95.

39 N.H. Sleep (1990) Monteregian hotspot track: a long-lived mantle plume. *Journal of Geophysical Research* **95**, 21 983–90.

40 K.G. Cox (1989) The role of mantle plumes in the development of continental drainage patterns. *Nature* **342**, 873–7.

41 R.I. Hill (1991) Starting plumes and continental break-up. *Earth & Planetary Science Letters* **104**, 398–416.

42 R.W. Hartley & P.A. Allen (1993) Interior cratonic basins of Africa: relation to continental break-up and role of mantle convection. *Basin Research* **6**, 9–114.

43 C.J. Ebinger *et al.* (1989) Effective elastic plate thickness beneath the East African and Afar plateaus and dynamic compensation of the uplifts. *Journal of Geophysical Research* **94**, 2883–2901.

44 N.H. Sleep (1990) Hotspots and mantle plumes: some phenomenology. *Journal of Geophysical Research* **95**, 6715–36.

45 G.F. Davies (1988) Ocean bathymetry and mantle convection, 1, large-scale flow and hotspots. *Journal of Geophysical Research*, **93**, 10 467–80.

2 Environmental change: past, present and future

Let the great world spin for ever down the ringing grooves of change.

Alfred, Lord Tennyson (1809–92), *Locksley Hall* [1842], l. 182

Must helpless man, in ignorance sedate,

Roll darkling down the torrent of his fate?

Samuel Johnson (1709–84), *The Vanity of Human Wishes* [1749], l. 345

Chapter summary

An enormous amount of information on environmental change is stored in the relict landscapes and sedimentary deposits of the Quaternary period, representing the last 1.65 million years of Earth history. The timescale of environmental change recognized in the Quaternary is high-frequency, dominated by the glacial and interglacial cycles with frequency of the order of 100 ky. The start of the Quaternary in the Pleistocene epoch cannot, however, be placed unequivocally at the onset of global climatic deterioration. The picture is more of a progressive cooling through the Tertiary, and ice is known to have existed in Antarctica since at least the Oligocene (26 Ma).

The extent of northern hemisphere ice sheets and periglacial fringes has been accurately delineated. The breakthrough since the first descriptions of the four glacial phases of the Alpine region of Europe by Penck and Brückner in 1909 has been the recognition of a far larger number of climatic oscillations during the Quaternary. These findings come from a number of powerful methods, including the analysis of the oxygen isotopic compositions of marine organisms. These isotopic compositions reflect the process of isotopic fractionation during the abstraction of water into ice, and seawater palaeotemperatures. The pattern of 17 glacial cycles in the Quaternary evaluated from the marine oxygen isotopic record has been corroborated from many other lines of evidence, such as the isotopic and mineralogical composition of ice cores from Greenland and Antarctica, the record of aeolian dust deposition in the deep sea, and the distribution and stratigraphy of loess.

The global dynamics of environmental change during the Quaternary can be investigated by detecting environmental signals in different latitudes on land and under the sea. One result from a myriad of findings is the relationship between glacial maxima and aridity in low latitudes. There is now convincing evidence that the major glacial maxima were associated with massive extensions of the subtropical deserts, but that in other latitudes, such as that of the Great Basin of the USA, glacial maxima are linked with increased hydrological activity in so-called pluvial phases when Pleistocene lakes were much expanded. The geographical variation of the behaviour of lakes or the extent of deserts is seen to reflect the complex hydrological response to climate change. The marked geographical variations recognized in the Quaternary are also a feature of the predictions from general circulation models which show both 'winners' and 'losers' during global climate change.

The environmental changes since the last glacial phase bear the unmistakable imprint of humans. The increased time resolution and the use of historical records allow the impact of individual events, such as large volcanic eruptions, to be assessed. We have just emerged from a period of global cooling caused by the eruption of Mount Pinatubo in the Philippines in 1991.

A set of forcing mechanisms can be identified for climate change relating to changes in the amount of solar radiation received by the Earth, changes in the composition of the atmosphere, and changes to the surface of the Earth. The most convincing explanation for the glacial cycles of the Pleistocene derives from the work of the Serbian physicist Milankovitch, who suggested that variations in the radiative energy received from the Sun were caused by orbital variations of the Earth. Such variations operate at a range of frequencies comprising 21 ky, 41 ky and 100 ky. Milankovitch band orbital variations act as a pacemaker for climate change in an Earth surface system with complex internal dynamics and built-in lag times.

Changes in the concentration of greenhouse gases such as CO_2, tectonic factors causing changes in the distribution of land and sea and the elevation of mountains, and the radiative properties of the land surface can all be identified as factors influencing past climates and act as components in global climate models.

Sea level has fluctuated strongly in the geological past at a range of frequencies. In the Quaternary, high-frequency changes in sea level mimic the marine oxygen isotopic record, suggesting that sea level responds closely to the abstraction of water in continental ice caps, a concept known as glacio-eustasy. More local relative sea level changes associated with glaciation include the isostatic rebound following melting of glacial loads.

The current epoch of the Quaternary, the Holocene, is our very own mass extinction event. The impact of humankind on the environment is profound. The causes and consequences of global warming are focused upon. It is now official that global warming due to anthropogenic emissions of greenhouse gases is taking place. In addition, changes to important ecosystems such as the destruction of tropical rainforests pose important questions in a complex Earth surface system containing dangerous positive feedbacks. The increased vulnerability to natural hazards recognized in the last decade does not bode well for the future.

2.1 Introduction: environmental change

The fact that the Earth's surface is subject to change gives its study a strongly historical perspective. A knowledge of the recent climatic and physical evolution of the Earth's surface is important for at least two main reasons. First, the environmental changes that have taken place in the last couple of million years are so great, and their rate of change so rapid, that present-day landscapes are not necessarily in equilibrium with currently acting processes. To understand currently operating systems it is therefore necessary to know something of their history. Second, the extremes of climate in the geologically recent past, and the recorded geomorphic responses, afford an invaluable insight into the way in which the global environment may respond to future perturbations, whether forced by human or natural causes. An enormous amount of data on environmental change is stored in the relict landscapes and sedimentary deposits of the Quaternary period, that is, the last 1.65 million years, encompassing the Pleistocene glaciations and interglacials and the Holocene warming. This period spans a large part of the emergence, and eventually dominance, of humankind. We shall see how profound has been the human impact on the environment, and speculate on the likely changes to the Earth surface system in the future caused by this impact.

Delving more deeply into the Earth's geological record gives us glimpses of processes that operate at time-scales too long to be revealed by study of the Quaternary, or which have produced extremes of palaeogeography or climate that were not witnessed in the last couple of millions of years. The geological past gives us a record of mass extinctions, for example in the Triassic and end-Cretaceous, that allow us to place a yardstick against the biodiversity of the present—our very own mass extinction event. The geologic record provides evidence of assemblies of continental plates into supercontinents such as Pangaea. Such an assembly had profound impact on global climate and the workings of the hydrological cycle. The geological record also gives us reliable evidence of long-term global sea levels, at times elevated at hundreds of metres above present levels, as in the mid-Cretaceous. All of these geological events have relevance to the testing of actualistic models of Earth surface processes, such as the general circulation models for climate, and thereby enable the more reliable forecasting of future environmental change.

We saw in Chapter 1 that the Earth surface system represents a highly dynamic and complex system with many positive and negative feedbacks whose component processes operate at particular spatial and temporal scales. Looking into the past may be the only way of appreciating the impact of the rare but high-impact event. Different Earth surface processes vary considerably in their magnitude and their frequency of occurrence, these characteristics themselves varying from one topographic and climatic setting to another. River discharge, for example, tends to be equable in temperate, low-relief regions, but short-lived and intense in semi-arid settings with high local relief. High-magnitude events tend to be rare, but may have profound climatic, geomorphic and stratigraphic effects.

The emphasis on the effects of the rare, high-magnitude event is embodied in the term *catastrophism*. Unfortunately, the overemphasis on uniformitarianism since its first formal usage by James Hutton and Charles Lyell has clouded the recognition of the extreme importance of the rare event. The importance of the high-magnitude, low-frequency event can be demonstrated using the following example. The enormous Spokane flood following the catastrophic drainage of glacial Lake Missoula in the northwest USA between 13 000 and 18 000 BP is well documented, discharges reaching over $20 \times 10^6 \, m^3 \, s^{-1}$ and flow velocities in some channels reaching $30 \, m \, s^{-1}$. Furthermore, it is thought that at least five further major floods have occurred in the same vicinity during the Quaternary. The Pleistocene Lake Bonneville flood (15 000 BP), for example, rivals the Spokane flood. Stream powers are thought to have been in the region of $75 000 \, N \, m^{-1} \, s^{-1}$. By way of comparison, recent floods on the Mississippi and Amazon Rivers had powers of just $12 \, N \, m^{-1} \, s^{-1}$.

There is now a vast amount of data on the mechanics and rates of short-term processes acting on the Earth's surface or under the sea. The problem is the application of these short-time-scale data to longer-term change. The superb coverage of the Earth's surface by satellite imaging offers unprecedented opportunities to monitor environmental change, but once again the time-frame is short. Aerial photography only extends back to the 1940s in Britain and in parts of Europe and the USA. Topographic maps provide a longer time coverage, but the errors in earlier surveys may be greater than the amount of change. It is more customary, therefore, to substitute space for time in studying long-term change in geomorphic problems. This methodology relies on the recognition of temporal stages in development, William Davis's scheme of river evolution (youthful, mature and old) being a well-known example published in 1902. The method must be used with caution and restricted to where there is clear evidence of sequential (though not necessarily continuous) development, linked by a causal mechanism. The concept is formalized in the so-called *ergodic principle*, which proposes (from a statistical viewpoint) that sampling a distribution of a variable in space is equivalent to sampling the distribution in time. This principle is used routinely (and sometimes blindly) by

Earth scientists, for example in the reconstruction of different environmental elements in a palaeogeography (i.e. spatially) by using their occurrence in a preserved succession of sedimentary rocks (i.e. temporally), when it is called Walther's law. The dangers are considerable, since the statistical distribution of preserved sediments in time is clearly very different to the statistical distribution of sediments in space, a strong bias being introduced by the effects of selective preservation.

Long-term change can also be studied through *physical models*. Such models include laboratory experiments, for example, studying the mechanics of sediment transport under flowing water, or the development of erosional rills on a slope, which are then applied to the natural world. These laboratory experiments are extremely informative, but usually suffer from the problem of their small spatial and temporal scale of observation. Some of these problems can be lessened by the use of a dimensionless analysis (e.g. Reynolds–Froude scaling in laboratory flumes: see Chapter 4). Other physical models (analogue models) involve the replacement of the natural materials with other materials that enable the processes of interest to be better visualized and measured. An example is the use of clay to study rock or ice deformation. Numerical models also enable predictions to be made of long-term change, either through the use of a set of known relations between parameters (deterministic) or through the incorporation of random or probabilistic components (stochastic). The advantage of numerical models is that, in addition to simulating the real world, they are capable of increasing our understanding of the fundamental processes at work and their relative contribution to an observed product (*sensitivity analysis*). The climate general circulation models (GCMs) are examples of three-dimensional, time-dependent numerical models which are based on the fundamental equations governing the motions of the atmosphere and oceans.

2.1.1 Significance of the Quaternary

The Quaternary is the most recent period of the Cenozoic, and contains two epochs, the Pleistocene and the Holocene (Fig. 2.1a), though it could be argued that the Pleistocene, a time of glacial advances and retreats, is still in progress and that the Holocene does not merit the status of a separate epoch. It should be immediately recognized that from the geological perspective the time-scale of environmental change discerned in the Quaternary record is very high-frequency. These rapid changes should be set

(a)

ERA	Sub-era/period/sub-period/Epoch			Age (A) (Ma BP)	Age (B) (Ma BP)
Cenozoic	Quaternary	Holocene		0.01	0.01
		Pleistocene		1.64	1.6
	Tertiary — Neogene	Pliocene		5.2	5.3
		Miocene	Late	14.2	15.8
			Early	23.3	23.7
	Tertiary — Paleogene	Oligocene		35.4	36.6
		Eocene		56.5	57.8
		Paleocene		65.0	66.4
Mezozoic	Cretaceous	Late		97.0	97.5
		Early		145.6	144
	Jurassic	Late		157.1	163
		Middle		178.0	187
		Early		208.0	208
	Triassic			245.0	245
Palaeozoic	Permian			290.0	286
	Carboniferous			362.5	360
	Devonian			408.5	408
	Silurian			439.0	438
	Ordovician			510.0	505
	Cambrian			570.0	570
	Precambrian				

(b)

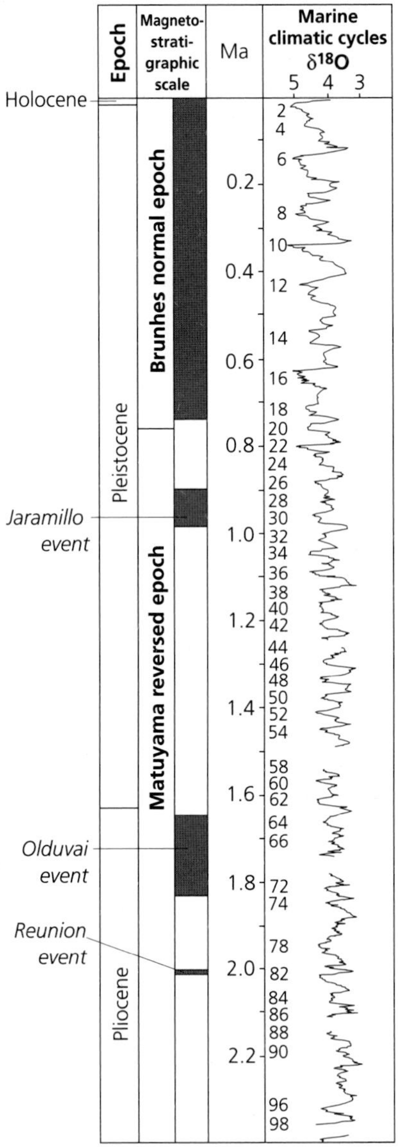

Fig. 2.1 (a) The geological time-scale with absolute ages according to (A) Harland *et al.* (1990) [1], and (B) Geological Society of America (1983) [2]. (b) Magnetic reversal stratigraphic scale for the late Pliocene to Holocene, with climatic cycles derived from marine isotopic records (Section 2.2.2). After Harland *et al.* (1990) [1].

against a background of the lower-frequency changes which have affected the Earth's surface and which can only be detected from a study of the longer-term geological record. The Earth has clearly undergone a nested set of climatic cycles of different frequency (Fig. 2.2). The sophistication of the climatic reconstructions in the Quaternary is in large part due to the availability of techniques which are either inappropriate or hazardous to use on older geological deposits. The elucidation of Quaternary environmental changes has benefited from the study of varves, tree rings, zoological and botanical evidence, especially pollen, ^{14}C and other isotopic dating techniques, as well as the recently developed techniques of analysis of the amino acid composition of bone, thermoluminescence and electron spin resonance, all of which provide information on dating. Quaternary deposits can also be calibrated against a detailed magnetic reversal global standard (Fig. 2.1b). This has enabled Quaternary events to be correlated between widely separated regions, and, importantly, the record from deep sea cores to be compared with terrestrial deposits. The reader is referred to other texts describing and applying these techniques [4].

The start of the Pleistocene can be defined from two lines of evidence. On faunal evidence, the Pleistocene

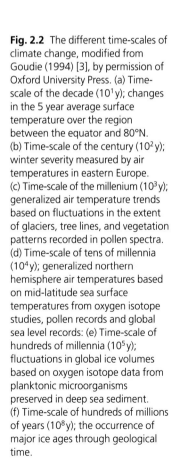

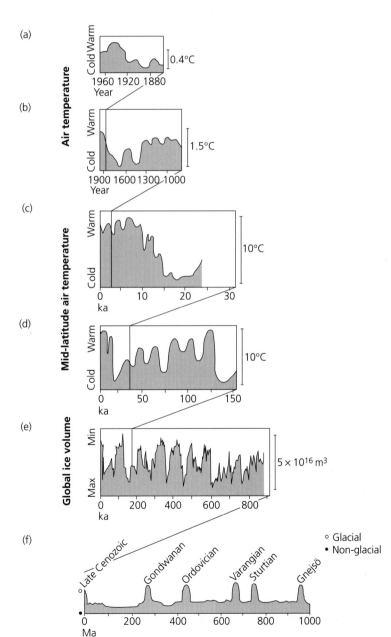

Fig. 2.2 The different time-scales of climate change, modified from Goudie (1994) [3], by permission of Oxford University Press. (a) Time-scale of the decade (10^1 y); changes in the 5 year average surface temperature over the region between the equator and 80°N. (b) Time-scale of the century (10^2 y); winter severity measured by air temperatures in eastern Europe. (c) Time-scale of the millenium (10^3 y); generalized air temperature trends based on fluctuations in the extent of glaciers, tree lines, and vegetation patterns recorded in pollen spectra. (d) Time-scale of tens of millennia (10^4 y); generalized northern hemisphere air temperatures based on mid-latitude sea surface temperatures from oxygen isotope studies, pollen records and global sea level records: (e) Time-scale of hundreds of millennia (10^5 y); fluctuations in global ice volumes based on oxygen isotope data from planktonic microorganisms preserved in deep sea sediment. (f) Time-scale of hundreds of millions of years (10^8 y); the occurrence of major ice ages through geological time.

is viewed as the time of the first appearance of many modern genera such as elephant, camel and horse. On this basis, the start of the Pleistocene is taken as 1.65 Ma, which coincides with a major magnetic reversal (top of the Olduvai event). Alternatively, the boundary can be defined on climatic evidence as the start of glaciation, or of sudden temperature decline. This latter approach is problematic for two reasons.

1 The transition to cooler climates appears to have taken place in a series of pulses over a period of time stretching back into the Tertiary. Glaciation is thought to have started in Antarctica in the Oligocene, based on the occurrence of ice-rafted debris in sea floor sediments of this age (26 Ma). In addition, buried ice in the Beacon valley area of East Antarctica is covered by a till containing relict ice wedges which themselves contain volcanic ash thought to have been deposited by direct airfall into an open crack. The ashes can be dated ($^{40}Ar/^{39}Ar$) to 7.9–8.1 Ma, implying that the till is Miocene or older in age. The oxygen isotopic composition of the buried ice suggests that it formed under conditions similar to those of today, which indicates that glacier outlets for the East Antarctic ice sheet have existed for at least the last 8 million years. This argues strongly for an early onset of glaciation in the south polar region, and also for the long-term stability of the East Antarctic ice sheet [5]. Away from the polar regions, there is evidence of ice-rafted debris at 4–5 Ma in the north Pacific and Arctic Oceans. Glaciation at lower latitudes was well established by 2.5–3 Ma (Section 2.2.1).

2 Isotopic studies (see also Section 2.2.2) show there to have been a *progressive* cooling of ocean waters through the Cenozoic from the Eocene to the present at all latitudes (Fig. 2.3) rather than a distinct cooling to have taken place at the onset of the Quaternary. For these reasons, the Pleistocene loses its distinctiveness as a glacial epoch.

The onset of mid-latitude glaciation appears to have taken place at about the same time as the development of high aridity in north Africa. From about 2.5 Ma the large tropical inland lakes of Africa began to dry out and the African continent suffered considerable vegetational change. We will see in a later section how glacial advances and retreats correlate with environmental changes in lower latitudes (Section 2.2.6).

2.2 Environmental change associated with glaciation: the record of the Pleistocene

Perhaps the most fundamental attribute of the Pleistocene is that it was not an interval of uniform

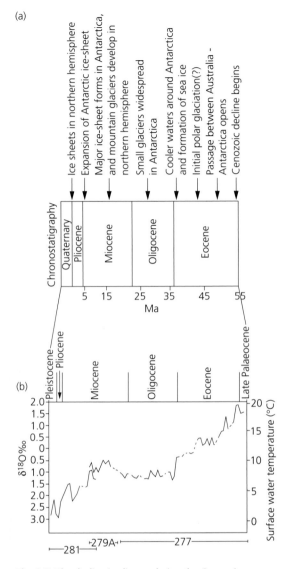

Fig. 2.3 The decline in climate during the Cenozoic indicated by the isotopic composition of the shells of planktonic foraminifera at three sub-Antarctic sites 277, 279 and 281, and interpreted surface water palaeotemperatures. Data from Kennett & Shackleton (1975) [6].

cold, but one of cold phases (glacials and stadials) and warmer phases (interglacials and interstadials) (Fig. 2.4). Pioneering studies on land, particularly in the Alpine regions of Europe, notably by Penck and Brückner in their three-volume work *Die Alpen im Eiszeitalter* (1909), established a model for the Pleistocene characterized by four great glaciations,

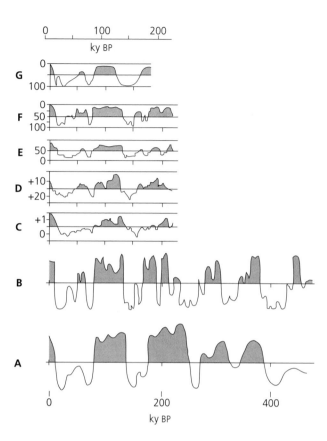

Fig. 2.4 Various indices of climatic fluctuations of the last 400 000 years based on information from deep sea cores. (A) Generalized curve for the equatorial Atlantic based on the ratio of cold- to warm-water foraminifera. (B) Generalized curve for the NE Atlantic (Polar Front region) based on the ratio of cold to warm foraminifera. (C), (D) Oxygen isotope curves for core P6304-9 at 15°N in the Caribbean and core 280A at 40°N in the Atlantic respectively. (E), (F) Curves for percentage coccolith carbonate and percentage polar foraminiferal fauna respectively in core V-23-83 from 51°N in the Atlantic. (G) Curve based on the percentage by weight of glacial detritus in the sand fraction of core RE5-36 at 50°N in the Atlantic. Data sources in Goudie (1994) [3], Fig. 1.7; by permission of Oxford University Press.

the Würm, Riss, Mindel and Günz, separated by mild interglacials. It is unsurprising that a scheme based on the record in the European Alps should now be at variance with the Pleistocene sequences recorded elsewhere, which demonstrate there to have been far more glacial cycles than previously appreciated.

2.2.1 The northern hemisphere ice sheets and fringes

Geomorphic evidence shows that at their maximum extent ice sheets and glaciers covered nearly $50 \times 10^6 km^2$, more than three times their present extent ($15 \times 10^6 km^2$). Most of the expansion in glaciated area involved the great northern ice sheets of North America (Laurentide) and north-west Europe (Scandinavian) (Fig. 2.5). Beyond the limits of glaciation *sensu stricto* lay vast fringes of open tundra characterized by permanently frozen ground conditions (permafrost) (Fig. 2.6). Today, the southern limit of permafrost in northern Europe coincides with the −5 or −6°C mean annual isotherm. A number of features characterize areas with permafrost [7], such

as polygonal ice wedges, so that the former extent of permafrost conditions can be reconstructed. This shows a marked displacement southwards (for northern hemisphere ice sheets) of permafrost conditions during Pleistocene glaciations. Only the Iberian and Italian peninsulas and south-west France were unaffected by permafrost in western Europe. If the present-day isotherm is a valid limit to apply to the past, this southerly limit of permafrost in Europe represents a temperature change of as much as 15–20°C. Most areas that are temperate today were tundra, cold semidesert, or occupied by lakes fed with glacial meltwater at the time of the last glacial maximum (see also Section 2.2.6) (Fig. 2.6).

The temperature changes associated with past glaciations can also be estimated by looking at the former extent of faunal and floral provinces, and geomorphological evidence such as the altitudes of cirque floors, which today cluster at about the 0°C summer isotherm. These results can then be compared with the palaeotemperatures derived from the marine oxygen isotope record (Section 2.2.2). The results are

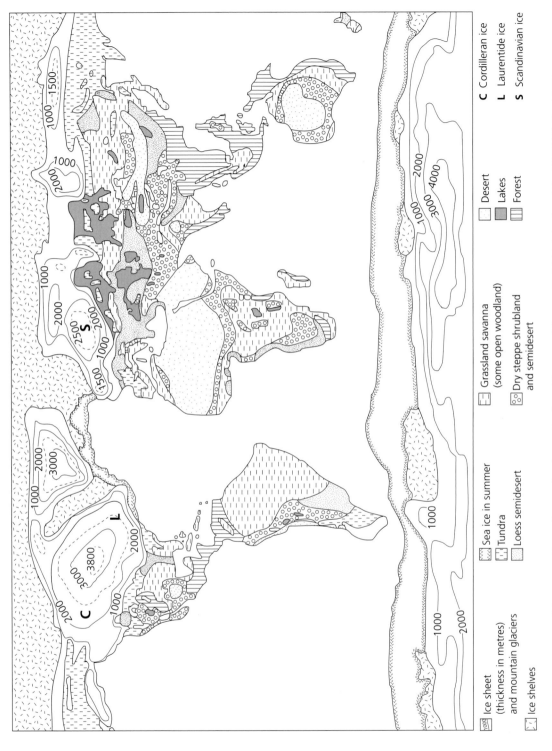

Fig. 2.5 Reconstruction of the extent of ice and vegetational zones for the coldest part of the Last Glacial (18 ka). After Selby (1985) [8], Fig. 16.29.

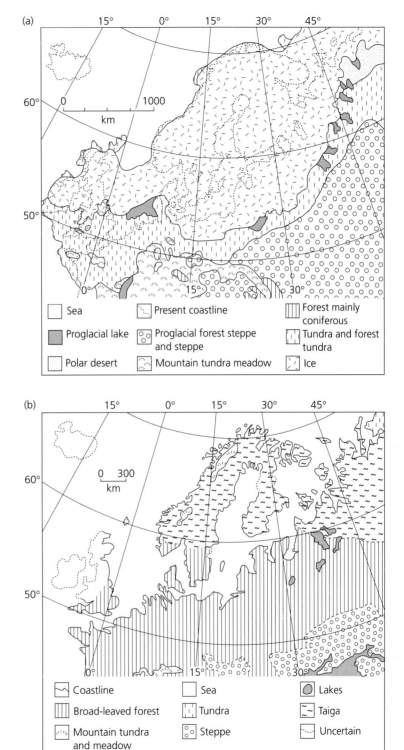

Fig. 2.6 Comparison of the vegetational and geomorphological zones of NW Europe between (a) the Last Glacial and (b) the Last Interglacial. After Gerasimov (1969) in Goudie (1992) [3], Figs 2.15 and 2.22.

in broad agreement that temperatures during glacial periods were depressed by 5–15°C, with values varying between the method chosen and between geographical area.

The climate of the interglacials, judged from the vegetational types, appears to have been similar to that of today, with tundra being replaced by forest in the temperate zone of the northern hemisphere (Fig. 2.6). The vegetational succession, as climate first improves into an interglacial and then worsens as a new glacial is approached, has been elaborated by a number of authors. However, the response time of vegetation may well have been different than the response time of glaciers and landforms to a given climatic change. Vegetation probably has a very rapid response to climatic change, with estimated advance rates of trees following the last glaciation of 200 km per thousand years. Ice caps, in contrast, would be very slow to respond. The retreat of the Greenland ice cap at about 9500 BP at the maximum rate of deglaciation is measured at 3 km per hundred years.

The timing of the most recent deglaciation varies from location to location. Indeed, in some low-latitude parts of the Earth the glacial activity appears to have been out of phase with global variations in temperature. However, if one considers the great ice sheets of the northern hemisphere, rapid deglaciation started approximately 14 000 years ago. This is supported by the isotopic composition of cores from the Gulf of Mexico, which show a major influx of fresh (isotopically light) water (see Section 2.2.2), presumably derived from the melting Laurentide ice sheet via the Mississippi River system, between 15 000 and 10 000 BP.

2.2.2 The marine stable isotope record

Study of cores from the uppermost sediment layers of the ocean has revolutionized thinking on Pleistocene environmental change (Fig. 2.7). Although the material has been subjected to a number of scientific techniques, the most informative has been the analysis of the oxygen isotopic composition of the shells of calcareous microfossils known as foraminifera.

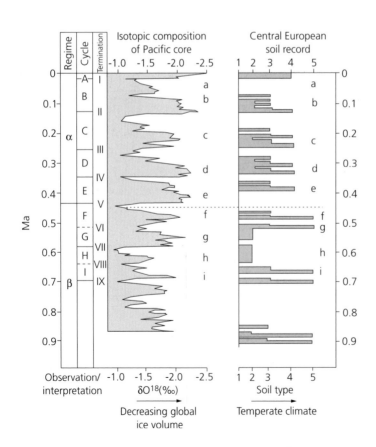

Fig. 2.7 Comparison of isotopic composition of planktonic foraminifera from Pacific core V28-238 and the soil types in loess from the Brno area of the Czech Republic over the last 900 ky. Modified from Goudie (1994) [3], Fig. 2.5, by permission of Oxford University Press.

The $^{18}O/^{16}O$ ratio in the tests of foraminifera is controlled by (i) the temperature-dependent fractionation of the isotopes from seawater into the calcite skeleton of the micro-organism, and (ii) the isotopic composition of the seawater. The oxygen isotopic composition of the seawater is determined by the extent of the preferential incorporation of the lighter isotope ^{16}O in the ice crystals of ice sheets, thereby affecting the $^{18}O/^{16}O$ ratio in the global ocean. Abstraction of ^{16}O in ice leads to a concentration of the heavier isotope ^{18}O in the ocean. During glacial and interglacial cycles fluctuations in the extent of the Antarctic and Greenland ice sheets should cause the oxygen isotopic composition of seawater to vary from relatively positive (isotopically heavy) during glacials to relatively negative (isotopically light) during interglacials. The recorded $^{18}O/^{16}O$ ratio in the calcite of an ocean-dwelling organism is therefore made of two contributions, one reflecting the temperature of the water in which the organism lived, and another the oxygen isotopic composition of the seawater reflecting the amount of global ice. Fortunately, these two effects are in phase, so excursions of the $^{18}O/^{16}O$ ratio in shells can be used for fine-scale correlation between deep sea cores. The quantitative use of the $^{18}O/^{16}O$ ratio in shells, however, must rely on the ability to discriminate between the two contributing effects. Furthermore, it is necessary in all investigations to be able to eliminate the effects of post-burial changes to the oxygen isotopic composition of shell calcite, and care must be taken in using foraminiferal species which are known to have no metabolic effect on the temperature-dependent fractionation of the isotopes.

Oxygen isotopic compositions are measured relative to a global standard known as Standard Mean Ocean Water (SMOW). The oxygen isotopic composition of a sample can then be expressed as per mille differences relative to SMOW:

$$\delta^{18}O = 10^3 \left[\frac{\left(^{18}O/^{16}O\right)_{sample} - \left(^{18}O/^{16}O\right)_{SMOW}}{\left(^{18}O/^{16}O\right)_{SMOW}} \right] \quad (2.1)$$

Positive values of $\delta^{18}O$ indicate enrichment of the sample in the heavier isotope, negative values indicate depletion. A different standard is generally used for oxygen contained in carbonate. This is the isotopic composition of a marine Upper Cretaceous fossil (Belemnite) from the PeeDee Formation of South Carolina that is, PeeDee Belemite (PDB):

$$\delta^{18}O_{SMOW} = 1.030\,86\,\delta^{18}O_{PDB} + 30.86 \quad (2.2)$$

The palaeotemperature can be found from the isotopic ratio of the carbonate as long as the isotopic composition of the seawater can be estimated. Assuming a fixed amount of water in the ocean-ice sheet part of the hydrosphere, global (eustatic) sea levels can be used to estimate the amount of water locked up as ice in ice sheets, and thereby the $\delta^{18}O$ of the seawater. Variations through Pleistocene glacial–interglacial cycles are thought to result in differences in the $\delta^{18}O$ of the seawater of about 1.2 per mille. Note that this non-temperature-related variation is a high proportion of the total $\delta^{18}O$ signal in the calcite of foraminifera.

In the deep ocean, temperatures are likely to have fluctuated little during glacial–interglacial cycles, so the $\delta^{18}O$ of benthic foraminifera is likely to represent changes in $\delta^{18}O$ of seawater due to ice volume changes. Likewise, in the warm subtropical pools, maximum temperatures are self-limited by evaporation rates. Foraminifera living in these warm surface waters will also have oxygen isotopic compositions of shell calcite that vary principally due to changes in $\delta^{18}O$ of seawater caused by ice volume fluctuations. However, foraminifera living elsewhere in near-surface waters are likely to preserve a decipherable palaeotemperature record in their shell calcite. A further caveat is that on a smaller spatial and temporal scale, the oxygen isotopic composition of seawater is affected by the ratio of evaporation to precipitation. This is likely to be of greatest importance in the ocean's surface layers characterized by large fresh water fluxes (Chapter 1).

The palaeotemperature obtained from the oxygen isotopic ratio of carbonate can be found from

$$T_{water} = 16.9 - 4.4\,(\delta^{18}O_{calcite} - \delta^{18}O_{water}) \quad (2.3)$$
$$+ 0.1\,(\delta^{18}O_{calcite} - \delta^{18}O_{water})^2$$

where the temperature T is in degrees Celsius, and the $\delta^{18}O$ values are measured relative to SMOW for water and the PDB standard for calcite [9]. In the Palaeocene, when the Earth was ice-free, the value of $\delta^{18}O_{water}$ is estimated to be 1.2 per mille. During the Pleistocene, it should vary according to the extent of glaciation.

One particular core from the equatorial Pacific (V28-238) has been chosen as a standard. It shows eight glacial cycles in the last 700 ky, the time of the last interval of normal magnetic polarity (the Brunhes), each cycle having secondary fluctuations (Fig. 2.8). The eight major cycles do not vary a great deal in amplitude, wavelength or shape over the last

Practical exercise 2.1: Sea surface palaeotemperatures

The $\delta^{18}O$ values from foraminifera preserved in the top 10 m of a core obtained from the sea floor of the Caribbean (core V12-122), spanning the last 450 ky, range between 0 per mille and –2 per mille [5] (Fig. 2.8). Assuming that the foraminifera lived at the sea surface, what temperature range of surface water do these isotopic ratios represent?

At the glacial maxima, assume that the oxygen isotopic composition of seawater was 0 per mille. Take the value of $\delta^{18}O$ for seawater during interglacials to be 0.4 per mille.

Solution

Using equation (2.3), the palaeotemperature for the glacial phases is 19°C. During the interglacials it is 28°C. This temperature range is therefore approximately 4.5°C for 1 per mille variation in the recorded $\delta^{18}O$ in shell calcite.

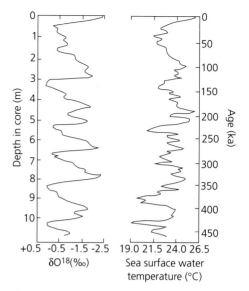

Fig. 2.8 Isotopic record of Caribbean core V12-22, derived from planktonic foraminifera inhabiting near-surface waters, and the interpreted sea surface water palaeotemperatures. Data from Imbrie *et al.* (1973) [10].

700 ky, and the warm peaks are all at about the same level. This suggests that present-day temperatures and ice volumes are similar to the peak interglacials of the past. The typical duration of a full glacial cycle is of the order of 100 ky, whereas interglacials only last an average of 10 ky. Our present climate therefore must be seen as being unusual in terms of the last 700 ky.

Study of older deep sea cores suggests that glaciations have been taking place over the last 3.2 My, with no less than 17 glacial cycles in the last 1.65 My, constituting the duration of the Pleistocene.

2.2.3 Information from ice cores

Polar ice also contains data reflecting past climatic change. Ice cores provide information on atmospheric gases and aerosols in the past, and also serve as a proxy for palaeotemperatures (Fig. 2.9). This latter property of allowing calculation of past temperatures is possible because of the link of the isotopic composition of water ($^{18}O/^{16}O$ and D/H ratios) and the precipitation temperature at cloud level. Analysis by mass spectrometry of the meltwater from ice samples collected from the upper part of an ice core reveals seasonal variations. In deeper, older parts of the core, however, these seasonal variations are lost and only longer-term variations are found. This enables climatic changes over several hundred thousand years to be investigated.

Ice cores also contain air bubbles which are sealed in the ice when the compacted snow (*firn*) transforms to ice (see also Chapter 11). The bubbles of air can be analysed for the common greenhouse gases such as carbon dioxide and methane. Although care must be taken in the recognition of the age difference between the air bubble and the ice sealing it (which may be as large as 3000 years), the air bubbles provide direct samples of previous atmospheres. The chemical analysis of impurities found in the ice also yields palaeoenvironmental information [11]. These impurities may be contributed by sea spray salt, marine biogenic activity (gaseous production from phytoplankton which oxidizes in the atmosphere), volcanic eruptions, biomass burning, lightning (followed by poleward transport), extraterrestrial fallout and airborne aerosols.

The impact of major volcanic eruptions is particularly marked in ice cores. However, the atmospheric effects of most volcanic eruptions are local. Volcanic ash from nearby volcanoes may be preserved as sedimentary layers within the ice. More distant volcanic events may be recognized in the sulphate signal due to the emission of acidic gases such as sulphur dioxide. The eruption of Tambora in 1816, for example, is recorded as a spike in ice recovered

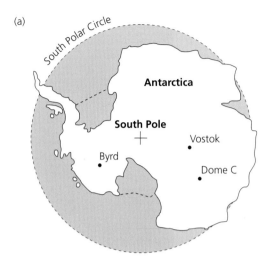

(a)

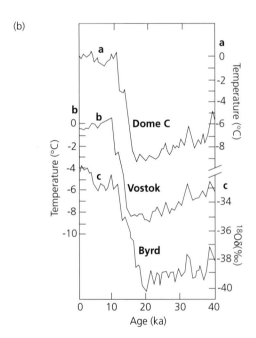

(b)

Fig. 2.9 Stable isotope curves and palaeotemperatures from three localities in the Antarctic ice sheet. (a) Location of Vostok, Byrd Station and Dome C. (b) Stable isotope record for the last 40 ky, showing the change at the last glacial–interglacial transition. After Delmas (1992) [11].

from both Antarctica and Greenland (Fig. 2.10). Other historical eruptions are recorded unequally in the ice cores from the two polar regions, or are recorded in only one. Cataclysmic eruptions which emit huge volumes of fine dust into the stratosphere

may also cause deposition of fine dust over polar regions as a direct volcanic acid fallout. These very fine particles can be detected by continuous electro-conductivity measurements (ECM) of the ice cores. However, there is not at present a clear correlation between the ECM record of past volcanic activity and the isotope record of past climate.

The ice cores from Vostok Station Antarctica

The ice cores recovered from Vostok Station (location shown in Fig. 2.9a) provide a palaeoclimatic record over a full glacial–interglacial–glacial cycle and allow some of the principles described above to be elaborated [12]. The first cores were recovered in the 1970s by Soviet scientists, allowing preliminary studies of the oxygen isotope composition of the ice. In the 1980s and continuing to the present (1995) deeper holes have been cored to depths of as great as 2755 m (in 1993).

The palaeoclimatic record can be extracted from ice cores because of the fractionation that takes place between an air mass and the precipitation from it, leading to a depletion of the isotopes deuterium and oxygen-18 in the precipitation. It is known that in polar regions this fractionation leads to a linear relationship between the average isotopic content of snow and the mean annual temperature. For example, a cooling of 1°C in the air where the precipitation is formed results in a decrease in deuterium of 9 parts per thousand. Consequently, by mesuring the deuterium content of the ice we gain a direct record of changing air temperatures through time, but it is also necessary to have an accurate chronology of the ice if these data are to be of use in palaeoclimatic studies. Establishing this chronology is far from simple. One method is to model two critical processes: one is the ice flow rate into the Vostok region; the other is the accumulation rate which, because it depends on the saturation vapour pressure, may be estimated from the palaeotemperature record. This modelling method can be tested and validated by the use of cosmogenic isotopes such as beryllium-10. Since the deposition of this isotope is thought to be constant over time, and its half-life is known, measurements of ^{10}Be in Vostok ice cores allows the age of the ice to be dated. In this way, the age of the ice at 2456 m in core 4G, for example, is calculated to be 220 ± 20 ky.

Some of the results are given in Fig. 2.11. The deuterium measurements show elevated values (about −440 per mille relative to SMOW) in the youngest ice corresponding to the last *c.* 10 ky, and a second period of elevated deuterium values corresponding to

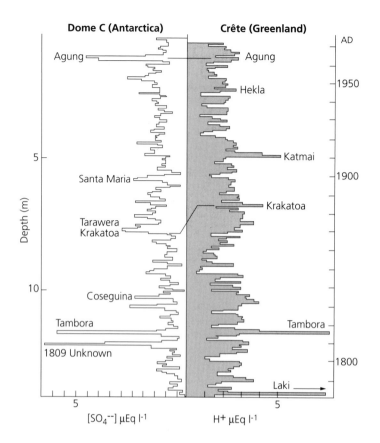

Fig. 2.10 Comparison of volcanic horizons in Greenland and Antarctic snow over the last 210 years. Acidity profile from Crête (Greenland) derived from electroconductivity measurements. Profile from Dome C (Antarctica) based on sulphate measurements is chronologically less accurate, thereby explaining the slight mismatch between the two profiles. After Delmas (1992) [11].

an age of *c.* 120–130 ky. These short warm periods separate longer periods with more negative deuterium values (*c.* 460–480 per mille relative to SMOW) corresponding to colder periods. The temperature changes between these glacial and interglacial periods is of the order of 6°C. Within the last ice age (Fig. 2.11) it is possible to recognize three minima separated by three warmer periods. These are the stadials and interstadials. Further confidence in the Vostok results comes from correlation of the ice record with the oxygen isotope record obtained from the calcite in benthonic foraminifera from the southern Pacific Ocean (Fig. 2.11c). Temperature variations obtained from the ^{18}O record correlate strongly with the variations obtained from the deuterium in ice. Furthermore, there is a good corelation between the concentration of dust in the Vostok core and the mass accumulation rates of wind-blown dust in deep sea cores recovered from the Indian Ocean (Section 2.2.5). There seems little doubt, therefore, as to the validity and importance of the ice core information for large-scale, perhaps global, palaeoclimatic reconstructions.

Two other aspects of the data acquired from the Vostok ice cores deserve mention. First, the occurrence of wind-blown terrestrial aerosols indicates much higher dust concentrations during full glacial periods than during interglacials. This suggests that glacial periods were characterized by more extensive deserts and/or more efficient dispersal of dust towards the poles in the atmospheric circulation (Section 2.2.5). Second, the air trapped in the Vostok ice provides a direct record of variations in the atmospheric concentrations of CO_2 and CH_4 over a complete glacial–interglacial cycle. Both greenhouse gases increase in concentration during interglacial periods. This topic deserves a certain amplification.

The levels of CO_2 and CH_4 have increased since pre-industrial times by 40% and 100% respectively (Section 2.5.1). It is known that CO_2 and CH_4 concentrations correlate with the Vostok isotopic temperature record, but there are lags in the greenhouse gas–temperature system. For example, at the end of the last interglacial, the CO_2 decrease significantly lagged behind the Antarctic cooling, while during the warmings between the glacials and

Depth (m) in Vostok ice core

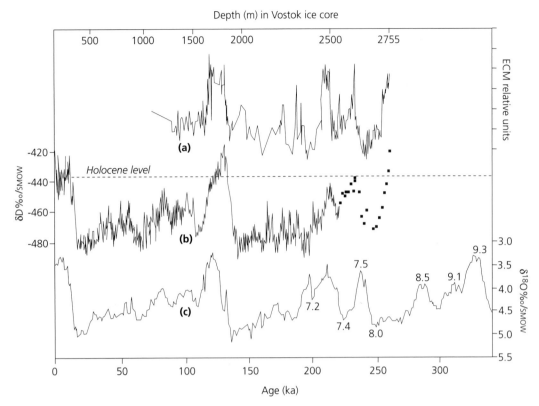

Fig. 2.11 Variations of electrical conductivity (ECM, curve (a) shown in arbitrary units with an inverted scale) and deuterium content (curve (b)) from Vostok ice cores, and for comparison, the oxygen isotopic record of benthonic foraminifera from Pacific Ocean deep sea core V19-30 (curve (c)). After Vostok Project Members (1995) [13].

following interglacials the CO_2 increase and Antarctic palaeotemperatures appear to vary in concert. Information from the northern ice sheets suggests that the increase in trace gas contents during deglaciation precedes the onset of most melting of northern ice sheets by several thousand years. There appears to be a positive feedback between the greenhouse gas concentrations and climate change. That is, the forcing mechanism for climate change, which is almost certainly due to orbital processes (Section 2.4), is amplified by changes in greenhouse gas concentration.

2.2.4 Wind-blown dust on land: loess
Loess is a terrestrial wind-blown silt deposit (10–50 µm sized particles) consisting chiefly of quartz, felspar, mica, clay minerals and carbonate grains in varying proportions [14]. When unweathered it is typically homogeneous and highly porous, and commonly buff in colour. Wind-blown loess may be reworked by running water and hillslope processes

and accumulate in valley bottoms. *Weathered loess* is modified by weathering, soil formation and diagenesis. It contains more clay-grade material than unweathered loess but is commonly decalcified. Precipitation-driven pedogenesis of loess produces very fine-grained magnetic minerals which strongly influence the magnetic susceptibility signal of the loess. Thick successions of loess show alternations of unweathered and weathered loess which correlate closely with the global marine oxygen isotope record.

The thickest and most extensive loess blankets are found in China, central Asia (former Soviet Union), the Ukraine, central and western Europe, the Great Plains of the USA and Argentina (Fig. 2.12). Many loess deposits, such as those of the Mississippi Valley in the USA, occur downwind of major river valleys which served as dust sources from glacial outwash during colder periods in the Quaternary, and decrease in thickness downwind. In addition, loess deposits typically fine downwind, so that slowly deposited distal loess is considerably richer in clay minerals and more weathered than typical loess deposited close to source. Where a dust transport path crosses different climatic zones, for example from an arid zone to a sub-humid zone, the distal loess may be pedogenetically altered and be particularly rich in clays. This then becomes an extremely useful criterion by which to infer past climate change from a vertical succession of loess deposits of varying character, a classic use of the ergodic principle (Section 2.1). Breaks in deposition also give weathering an opportunity to modify the loess deposits.

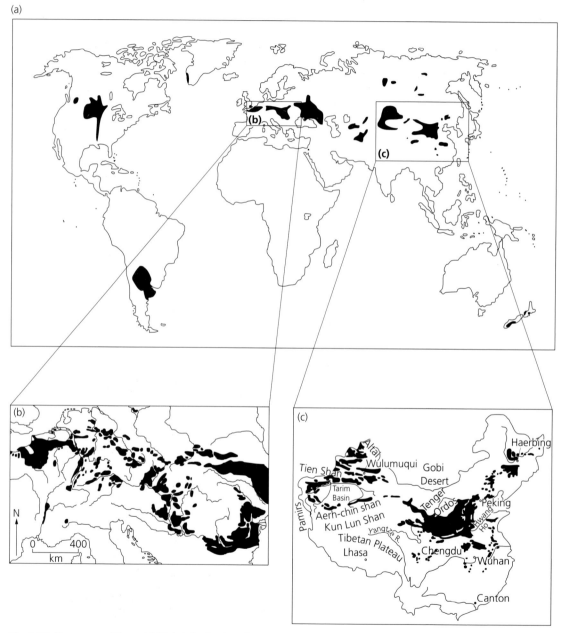

Fig. 2.12 Occurrence of loess. (a) Global distribution of main deposits. Distribution of loess in (b) Europe and (c) China. After Pye (1987) [14].

The prerequisites for the formation of extensive loess deposits are the existence of bare land surfaces composed of poorly sorted sediment with a high proportion of silt, and winds to transport the finer fraction away. There is overwhelming evidence that mid-latitude loess formation is associated with the glacial periods of the Pleistocene, when outwash plains fed by glacial meltwaters were deflated by winds blowing outwards from continental high-pressure systems. In Europe, there is good evidence that during interglacial and interstadial periods, ice sheets re-treated, dust deflation of glacial outwash plains ceased and previously deposited loess was altered in soil zones in relatively warm and humid conditions. In China, however, loess is thought to have been derived from a desert rather than glacial source and to have been transported by winds blowing from the cold, high-pressure system of central Asia.

The Quaternary history of loess deposition gives a striking confirmation of the palaeoclimatic changes indicated by the marine stable isotope record. In central Europe (Austria, the Czech Republic and Slovakia) [15] wind-blown loess is interbedded with hillwash sediment and forest soils which formed in interglacials and interstadials. At least 17 periods of wind-blown loess deposition are known within the last 1.7 My, with eight depositional cycles of wind-blown loess to palaeosol recorded in the Bruhnes palaeomagnetic epoch representing the last 700 ky (Fig. 2.13). The loess cycles are separated by horizons representing rapid environmental change which can be correlated with marked increases in global ice volume derived from the marine oxygen isotope record. In Austria, the oldest loess is older than the Matayuma–Gauss boundary at 2.48 Ma.

In China, a very thick loess succession is found at Luochuan (Shanxi province). At Luochuan the lowermost loess pre-dates the Olduvai event (1.67–1.87 Ma) and probably began to accumulate at about 2.4 Ma, in close agreement with the central European date. Within the Bruhnes palaeomagnetic epoch there are eight loess cycles. In combination with the central European data, therefore, the onset of northern hemisphere glaciation can be taken as about 2.4 Ma,

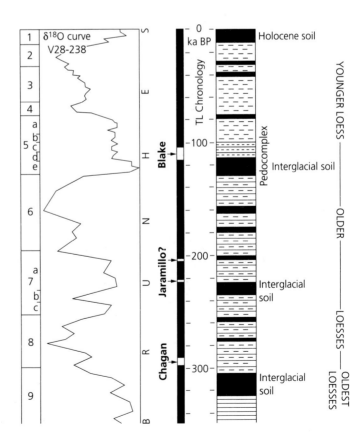

Fig. 2.13 The loess stratigraphy of southern Poland correlated with the oxygen isotope curve from core V28-238 and magnetic reversal stratigraphy. The warm palaeotemperatures from the isotopic record correlate with well-developed soils in the loess stratigraphy. After Pye (1987) [14].

with a frequency of glacial stages of about 100 ky in the Bruhnes interval. The grain size variation in the youngest Pleistocene loess at Luochuan (Malan loess) shows a high-frequency, high-amplitude pattern which correlates extremely well with the timing of Heinrich events in the north Atlantic (Fig. 2.14). These events relate to massive iceberg discharge from ice sheets surrounding the north Atlantic at times of exceptional atmospheric cooling. This is a particularly important discovery because it links a cooling of sea surface temperatures in the north Atlantic with atmospheric circulation in Asia during the last glacial period. This linkage may have been through the zonal westerly winds which may have been enhanced in intensity during cold intervals due to stronger hemispherical temperature gradients (see Section 2.2.5).

In the former Soviet republics of central Asia such as Uzbekistan and Tajikistan the loess record is more variable. The results support a general northern hemisphere cooling in the region of 2 million years ago and the cyclicity of environmental changes in the

last 700 ky (Bruhnes), but also provide information on additional mechanisms for cooling, such as the tectonic uplift of the Pamirs and Tien Shan. This would have affected river incision and sediment available for deflation by the wind as well as the extent of glaciation. The uplift of a plateau such as Tibet would also profoundly influence climate (Section 1.5.2).

Loess accumulation rates appear to have been at a maximum (0.5–3 mm y^{-1}) during the last glacial maximum.

2.2.5 Wind-blown dust in the deep sea

Mixed in with the biogenic sediment of the ocean floor is a wind-blown component of dust. It is now known that wind-blown (aeolian) dust accounts for up to half of the mineral component of pelagic sediments in general, and essentially all of the mineral component of the non-biogenic deep sea sediments known as pelagic clays. By unravelling the flux of airborne particulate matter and the intensity of the winds in the past obtained by analysis of the grain size

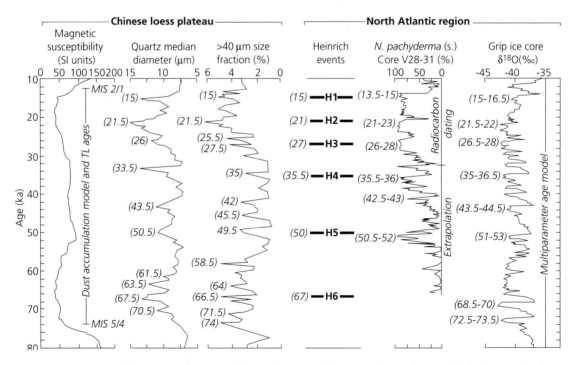

Fig. 2.14 Comparison of the Luochuan loess record (magnetic susceptibility and grain size characteristics) with events in the north Atlantic region (Heinrich events, abundance of *Neogloboquadrina pachyderma* (s.) and oxygen isotope variations from the Greenland ice sheet (GRIP core). Figures in brackets are ages of peaks rounded to the nearest 0.5 ky. After Porter & Zhisheng (1995) [16].

distribution of the wind-blown component, it is possible to draw conclusions about the extent of aridity in the recent geological past, and to link changes in aridity to better-understood variations in ice sheet volumes. There has consequently been considerable interest in analysing the wind-blown dust found in deep sea cores [17].

What is the present-day relationship between dust transport and climate? It has been known since the late 1930s that the ocean floor of the Atlantic downwind of the Sahara contains wind-blown dust. The transport paths of dust storms are now well delineated by atmospheric scientists, and it is generally accepted that dust from north Africa dominates the subtropical Atlantic north of the equator, and dust from Asia dominates the entire Pacific basin north of the intertropical convergence zone (ITCZ). All the major sources of dust and all the significant aeolian deposits in the deep sea are found in the northern hemisphere. The amount of particulate matter transported by dust storms is directly linked to the climate of the source region, commonly with a variation on the El Niño (year to several years) to decade time-scale. Dryness in the source region is positively correlated with deposition in the deep sea. This relationship between intensity and frequency of dust storms and mean annual precipitation has been called the *dust potential* of a continental surface [18]. The peak in dust potential is found in arid (rather than hyperarid) climates.

The size of the grains transported by wind is an indication of the energy of the transporting wind, the larger grains being transported by the stronger winds (see also Chapter 10). The vast bulk of the mineral material in pelagic sediments is in the grain size range finer than 6–8 μm, i.e. in the clay and very fine silt range. These grain sizes, which constitute the 'background aerosol', are capable of being transported thousands of kilometres in the wind, but coarser grains (coarse silt reaching up to sand size) of over about 50 μm are incapable of being transported far from their source region. Such grains are found immediately downwind of the deserts of Asia and the Sahara.

It is a misunderstanding to presume that the *amount* of wind-blown dust delivered to distal locations is related to the strength of the zonal tropospheric winds such as the trades or westerlies. The present-day picture is of maximum far-field aeolian transport in the western Atlantic from the Sahara and in the northern Pacific from Asia during the spring, at times of decreased rather than increased zonal wind activity. Furthermore, dust fluxes and grain sizes are not correlated in samples obtained from a number of deep ocean sites. Previous confusion about this may have been caused by the erroneous identification of hemipelagic material in cores as aeolian. Hemipelagic material is derived from the continents and is advected in plumes and dilute suspensions far into the ocean. Hemipelagic fluxes are greater at times of high runoff from the continents. They may dominate deep sea deposition within hundreds of kilometres, and perhaps up to 1000 km, of the coast.

The total airborne flux from the continents to the ocean is a small proportion (about 5%) of the flux from run-off (Chapter 3). However, it is very widely distributed. The average present-day mass accumulation rate of wind-blown dust on the ocean floor is 200–250 mg cm^{-2}ky^{-1} or (2–2.5 kg m^{-2}ky^{-1}). This present-day value can be compared with mass accumulation rates at times of different climate in the Quaternary. The present-day pattern (representative of accumulation in the Holocene) of deposition of wind-blown dust is shown in Fig. 2.15. These maps show that there are only three areas that supply substantial amounts (in excess of 1000 mg cm^{-2}ky^{-1}) of wind-blown dust to the adjacent ocean floor: the deserts of central and western China and Mongolia; the Sahara and Sahel; and Arabia and the Horn of Africa.

Most high-resolution studies of palaeoclimate during the Quaternary rely on the establishment of an oxygen isotope record as a proxy for changes in global ice volumes. This then serves also as a chronological standard. The mass accumulation rate of wind-blown quartz has a variability which spectral analysis always shows to be concentrated at a period of about 100 ky, usually also at 41 ky and commonly at 19–23 ky. These periods correspond to those predicted by orbital forcing (Section 2.4). The coherence of the aeolian dust records and the oxygen isotope record suggests relationships between glaciation, aridity in source regions and aeolian deposition in the deep sea. North Pacific, Arabian Sea and central Atlantic cores all show a conclusive correlation between high mass accumulation rates and increased aridity during glacial periods.

Longer-term records from Deep Sea Drilling Project and Ocean Drilling Programme cores show that although the central and north Atlantic Ocean have been receiving wind-blown particles from Africa and to a lesser extent North America throughout their history, there is a marked increase in dust flux at about

(a)

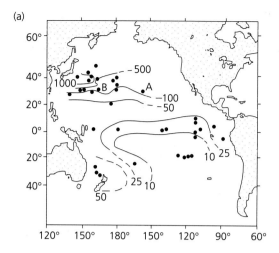

(b)

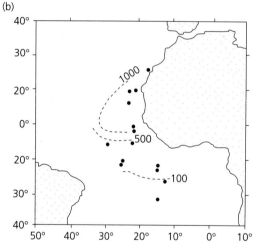

(c)

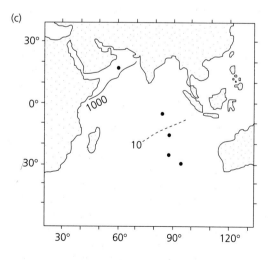

2.4 Ma which has been linked to the onset of northern hemisphere glaciation.

The best long-term record so far established comes from the north Pacific (core LL44-GPC3) (Fig. 2.16a). This core shows that mass accumulation rates of aeolian dust were very low from 80 to 25 Ma, at which point they doubled, steadily increased through the Miocene–Pliocene and then increased by about an order of magnitude at 3 Ma. This is thought to coincide with the drying of central Asia and eventually the onset of northern hemisphere glaciation. Other north Pacific cores show an almost identical trend in the mass accumulation rates of aeolian material (Fig. 2.16b).

In the southern hemisphere, there are no signals that correlate with the onset of northern hemisphere glaciation at 2.4 Ma. Instead, cores taken east of Australia indicate an increase in the aridity of the Australian source region at about 13 Ma. In the subtropical south Pacific an increase in trade wind activity is suggested by a marked increase in the grain size of wind-blown grains at 9 Ma, several millions of years before the expansion of ice volume at 3.5 Ma indicated by the oxygen isotope record. This may imply that the 13.5 Ma abstraction of water into continental ice did not result in major changes to the pole to equator temperature gradients and therefore trade wind intensity.

The highlights and implications of the deep sea dust record can be summarized as follows.

• The major palaeoclimatic event of the last 20–30 million years is the onset of northern hemisphere glaciation in the late Pliocene at about 2.5 Ma, but the record of this event is entirely found within the northern hemisphere. The ocean dust records fundamentally reflect profound aridity in the Asian and African source regions.

• Since the southern hemisphere records show no evidence of the 2.5 Ma palaeoclimatic event, we can infer a *hemispherical asymmetry* in responding to climate change [19]. In other words, the reduction

Fig. 2.15 *Left.* Mass accumulation rates of aeolian dust expressed as an aeolian flux in milligrams per square centimetre per thousand years. Values are usually obtained from the uppermost Quaternary sample available, and occasionally represent averaged rates through the Holocene. (a) Pacific Ocean, A is core LL44-GPC3, B is DSDP core 576; (b) central Atlantic; (c) Indian Ocean. After Rea (1994) [17].

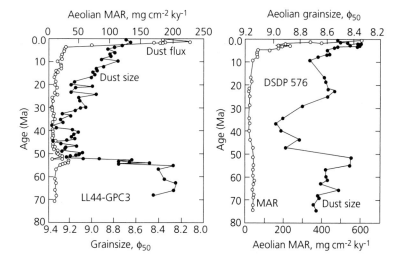

Fig. 2.16 Long-term record of dust deposition in the deep sea in terms of mass accumulation rate and grain size. (a) Core LL44-GPC3; (b) DSDP core 576 in North Pacific. Both are located on Fig. 2.15. After Rea (1994) [17].

in temperature leading to Arctic glaciation at 2.5 Ma was not mirrored by similar temperature changes in an already cold and glaciated Antarctica, nor in an increase in intensity of southern hemisphere tropospheric zonal winds.

• The extent of hemispherical asymmetry may be reflected in the position of the ITCZ, being offset far into the warmer hemisphere because of the smaller temperature gradients and more sluggish circulation in that hemisphere. Movement of the ITCZ in turn offsets the climatic zones of tropical rainfall, which act as barriers to the transport of airborne dust, and subtropical aridity where erosion by wind is enhanced.

• During the Oligocene and early Miocene when Antarctica was glaciated (from *c.* 35 Ma) but the northern hemisphere was warm, the ITCZ should have been displaced far to the north of its present latitude (perhaps at 15–20°N). This may be reflected in the reduction in dust grain sizes and change in mineralogy found in cores in the central north Pacific. The ITCZ should have moved back much closer to the equator after the onset of northern hemisphere glaciation at 2.5 Ma, which equalized temperature gradients and circulation intensity.

2.2.6 Geomorphic change in low latitudes

One of the most interesting problems in the Pleistocene is how environmental changes in the subtropics and tropics may be correlated with the ice advances and retreats in higher latitudes. Successful linkage of the two areas allows a much better appreciation of global responses to climatic change. How did the Earth's vegetational and climatic zones respond

to the glacial advances and retreats at high latitudes? This has been partly answered by looking at the record of deposition of wind-blown dust in the deep sea (Section 2.2.5). However, there is also direct geomorphic evidence for environmental change on land.

The environmental changes that have taken place at low latitudes during the Pleistocene can be divided into two categories: those due to increased aridity; and those due to increased hydrological activity, the so-called *pluvial* phases.

Geomorphic evidence of increased aridity

The former extent of desert areas during the Pleistocene can be assessed by studying fossilized dune fields. These can often be recognized on satellite or aerial photographs from their characteristic texture of linear (longitudinal), transverse and more complex dune crestlines. Ground observations can confirm that the dune systems are relict rather than active from the presence of deep weathering and soil development, stabilization by vegetation, or dissection by runoff. Some old dune fields are now partially flooded with lakes. Surprisingly small increases in rainfall can profoundly affect the possibilities for stabilization by vegetation, but a figure of 100–300 mm is generally regarded as the mean annual precipitation required to inhibit active aeolian dune migration.

The picture that emerges is that desert areas were much more extensive between 30°S and 30°N during the last glacial maximum (*c.* 18 000 BP) than today [20] (Fig. 2.17). The approximately fivefold expansion of the desert areas during the last glacial maximum appears to have been accompanied by a marked

(a) **Today**

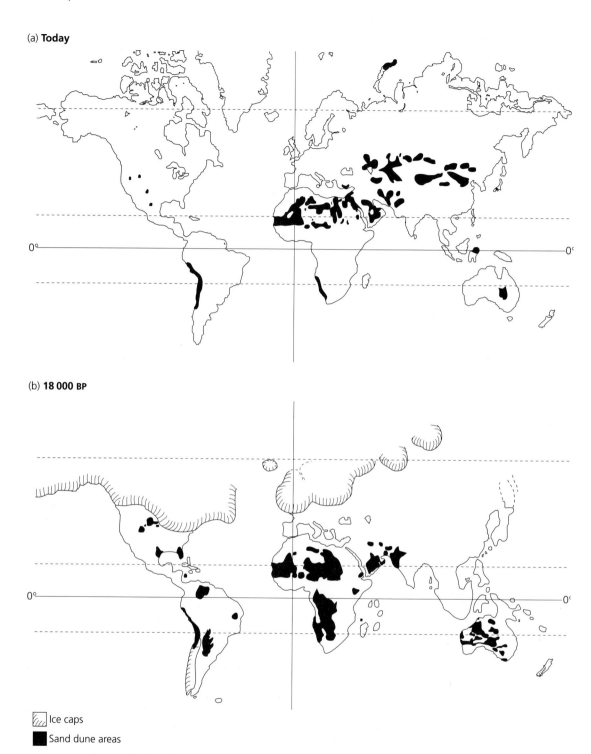

(b) **18 000 BP**

◲ Ice caps

■ Sand dune areas

Fig. 2.17 Comparison of deserts (a) today (after Goudie (1994) [3], by permission of the Oxford University Press) and (b) at the last glacial maximum (after Sarnthein (1978) [20]).

shrinkage of the areas occupied by tropical rainforests and savannahs.

In areas such as Rajasthan, northern India, aeolian dunes are weathered, pedogenically altered (build-up of carbonate) and deeply gullied. The dunes are overlain by small tools used by mesolithic humans, showing that very little aeolian deposition has taken place since abandonment. The extent of fossil dunes can therefore be mapped out in detail [21], and includes areas that now have mean annual rainfalls as high as 750–900 mm.

In southern Africa relict dunes cover large areas of Botswana, Angola, Zimbabwe and Zambia, far beyond the present-day limits of the Kalahari desert. Pleistocene sand dunes may also underlie parts of the Congo equatorial rainforest [22]. The savannah and forest of west Africa, where present-day rainfalls may be as high as 1000 mm per year, is underlain by fossil extensions of the Sahara. In the Lake Chad region old dune fields have been flooded by an expansion of the lake.

Large expanses of fossil dunes have also been studied on the North American continent, principally the High Plains (including the Nebraska and Dakota sandhills), in South America (including the Orinoco basin, where the Llanos grassland was once an active aeolian dune field, and southern Brazil), and in Australia. In Australia there was a major phase of dune construction between 25 000 and 13 000 BP, but some examples may date back to pre-Pleistocene times, and even to the end of the Eocene (34–37 Ma).

The shift in rainfall isohyets associated with glaciation varies from place to place. In Sudan the two phases of dune activity recognized there indicate southward shifts of the rainfall and wind belts of 450 km and 200 km. Fossil dunes extended as far south as 10°N, crossing the Nile, which is therefore assumed to have dried up at the time of the increased aridity. In western Australia, the rainfall isohyets may have moved equatorwards by as much as 900 km.

Increased hydrological activity: pluvial phases

Periods of increased hydrological activity can be recognized from trends in the water levels of lakes at particular times in the past. The widespread nature and consistent chronology of the trends suggest that the lake level fluctuations were driven by climatic change rather than having an origin in local factors. In contrast to the expanded arid environments previously described, which were hostile for the habitation of early humans, the fringes of lakes were favoured sites of occupation and have left behind a rich archaeological record.

The greatest concentration of pluvial lakes probably occurs in the Great Basin of the western USA, where late Pliocene and Pleistocene extensional faulting caused the ponding of over 100 Pleistocene lakes (Fig. 2.18). Lake Bonneville was the largest, reaching at its maximum a size similar to that of present-day Lake Michigan. These Pleistocene lakes recurrently drained by catastrophic flooding along spillways, the Lake Missoula flood being a widely cited example (see Section 2.1). Most lakes had their highest water levels during the period between 24 000 and 14 000 BP, and in the period between 10 000 and 5000 BP were at their lowest, and commonly dry [23]. The time of lake expansion therefore correlates well with the time of the last glacial maximum in the western USA (Fig. 2.19).

During the Last Glacial the Caspian and Aral seas were united in a lake larger than any other known $(1.1 \times 10^6 \text{km}^2)$. It is thought to have been fed by glacial meltwaters from the Volga, Ural and Amu Darya rivers. In the Middle East, the Dead Sea Rift was occupied by the 220 km-long Lake Lisan between 18 000 and 12 000 BP, that is, the time of increased aridity in nearby East Africa. The lake occupying the Konya basin in Turkey also reached its greatest extent at 23 000–17 000 BP.

Expanded pluvial lakes occur extensively in north central Africa (Lake Chad), north-east Africa, east Africa and southern Africa in a variety of tectonic settings. Improved dating (principally ^{14}C), however, shows that almost all of the east African lakes attained their highest levels at 8000–9000 BP, which is the early post-glacial period (early Holocene) rather than the period of glacial maximum. Lakes were at low stands at the end of the Pleistocene (14 000–13 000 BP). The dating of previous events in the Pleistocene is less certain, but there was most likely a wet phase in north Africa between about 20 000 and 40 000 BP. This wetter period is reflected in evidence from Australian lakes which show high water levels from about 40 000 to 25 000 BP, and an arid phase from 25 000 BP, peaking at about the time of the last glacial maximum. By about 13 000 BP, the aeolian dunes had become stabilized by vegetation as a wetter period took over.

Although it has long been believed that pluvial phases correspond to glacials, it is now clear that hydrological activity has varied from place to place in response to glacial cycles. In addition, we have already seen some of the compelling evidence provided by wind-blown dust in deep sea cores that increased mass accumulation rates of airborne dust are correlated

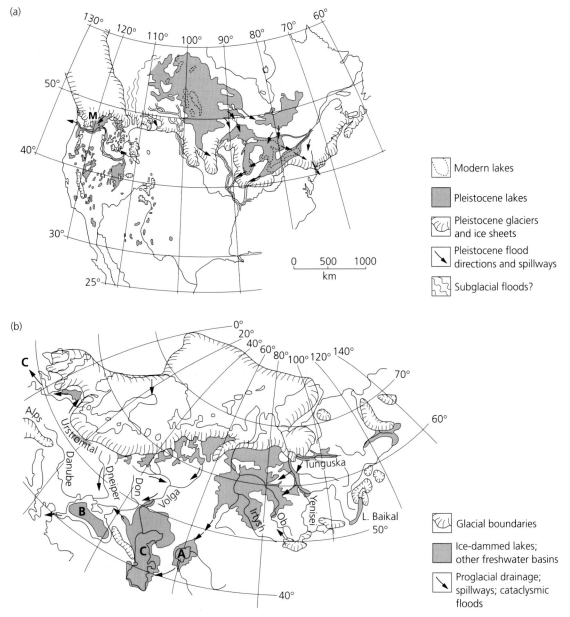

Fig. 2.18 Late Pleistocene lakes and cataclysmic flood spillways in (a) North America and (b) Eurasia. Specific glacial spillways are shown for the Lake Missoula outburst (M) and for the Pleistocene predecessors of the Aral Sea (A), Caspian Sea (C) and Black Sea (B). After Baker (1996) [24].

with cold glacial periods. In addition, mineralogical and compositional changes through time (e.g. feldspar content, kaolin–quartz ratio, pollen type, biogenic content) are thought to reflect changes in terrestrial weathering and vegetation type, and thereby changes in humidity and temperature (Chapter 3). Pollen

analysis of deep sea cores shows that humid tropical rainforest expanded considerably in Africa during the Last Interglacial (*c.* 124 000 BP), and studies of younger deposits indicate that drier, colder vegetation occupied east Africa during the last glacial maximum.

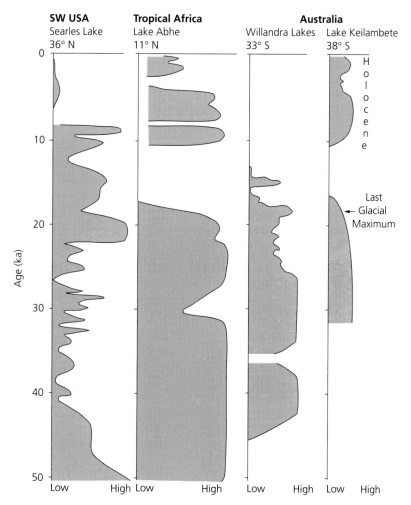

Fig. 2.19 Late Quaternary lake levels in four basins across three continents. Pluvial Searles Lake in California reached its highest elevations (between 10 and 20 ka) when Lake Abhé in tropical Africa and the Australian lakes were very low. After Goudie (1994) [3], by permission of Oxford University Press.

The different responses of lakes to climatic forcing in different regions is discussed in the following paragraphs.

Relationship of temperature variations and hydrological activity

To understand the geographical variations in the behaviour of lakes and the extent of deserts, we need to consider them in the light of the hydrological response to global climate change. A positive water balance is a prerequisite for the existence of lakes. Fluctuations in lake level therefore fundamentally depend on the balance between, on the one hand, precipitation on the lake and in the catchment, and, on the other hand, evaporation and transpiration. We know that the present pattern is of a positive water balance in the tropics, a deficit in the subtropics, and a positive balance again in most of the high latitudes (Chapter 1). With respect to the late Cenozoic and Quaternary, there are two factors that may have affected rates of precipitation: global temperature and topography.

Global warming should increase the global rate of precipitation, global cooling should reduce it. This is an outcome of the Clausius–Clapeyron relationship, since evaporation rates should increase with temperature, which then results in increased precipitation. Climate models for the mid-Cretaceous indicated a more than 20% increase in precipitation caused by a global average temperature increase of 10°C.

High topography affects precipitation by causing prevailing winds to rise, cool and cause precipitation on the windward side of the topographic barrier, and a rain shadow in the lee. The uplift of topographic obstacles also affects monsoonal air circulation and therefore precipitation patterns. The uplift of topographic obstacles such as the Tibetan plateau may therefore cause regional changes in precipitation patterns.

A global drop in temperature in a glacial phase should reduce evapotranspiration in a number of ways:
• evaporation rates are reduced by the lower air temperatures;
• transpiration rates are reduced by the smaller biomass;
• lower air temperatures reduce the air's capacity to hold water vapour through the Clausius–Clapeyron relationship (Section 1.3.1).
Elevated lake levels during glacial phases may therefore have resulted not from higher precipitation rates but from reduced evapotranspiration. This explanation has been applied to some Australian and south-west USA pluvial lakes, and may also explain the history of Lake Lisan in the Dead Sea Rift. In contrast, during the early Holocene, when lake levels were elevated in large parts of Africa, the high lake levels more likely reflect increased runoff associated with higher precipitation as global air temperatures rose.

However, it is most important to recognize that the predictions of general circulation models suggest that during climate changes there are both 'winners' and 'losers' according to geographical location. These sorts of geographical variation have been described from the African continent over the last 18ky, that is, since the Last Glacial Maximum [25]. At the time of the Last Glacial Maximum the southern European Mediterranean area was typified by dry conditions, the northern Sahara had high lake levels, and tropical Africa had low lake levels. All of these areas experienced pluvial conditions during the early interglacial at 8000–9000 BP. Clearly each continent, or fraction of a continent, responded in its own particular way, and correlations between and within continents need to be made with extreme care and with the very best dating available.

2.3 Post-glacial changes up to the present day

2.3.1 Climatic changes in the Holocene

There is an increasing amount of data available on environmental change as one proceeds to the present day. Such data provide information on very high-frequency change. The Holocene is increasingly recognized as an epoch of considerable climatic change within the context of an interglacial phase.

Insect remains and vegetation types indicate that the time of transition from the maximum of the Last Glacial to the Holocene was marked by fluctuations in climate, with minor stadials and interstadials. A number of lines of evidence (such as insect remains in Britain, Greenland ice core isotopes and dust concentrations) suggest that the cool conditions of the last stadial (the Younger Dryas) ended rather abruptly at about 10 000 BP. The large vegetational changes that followed, as, in Europe, former tundra and grasslands were transformed into forest, had profound implications for the development of human societies, marking the change from palaeolithic to mesolithic cultures. In Africa, former grasslands were desiccated, causing humans to migrate to wetter regions or develop new methods of cultivation and animal husbandry. The climatic change coincides with the eradication (at *c.* 11 000 BP) of many of the world's large terrestrial mammals from North and South America and northern Europe. Many authors believe this to have been caused primarily by the hunting activities of humans rather than climate change *per se*, but a combination of factors, with complex feedbacks, may have been responsible.

We can investigate the environmental changes in northern latitudes and, as in the Pleistocene, correlate them with changes in lower latitudes. In addition, during the Holocene, the impact of humans on environmental change can be discerned, particularly on vegetational changes by forest clearance, and more recently in accelerating soil deterioration (Chapter 3).

In both northern Europe and North America the vegetational changes through the Holocene suggest an expansion of forested areas following rapid deglaciation, followed by a small retreat at about 5000 BP (Atlantic Post-Glacial zone of Europe) due to cooling, and a further retreat at about 2500 BP due to the occurrence to cool, wet, oceanic conditions. The fluctuation of the boundary between prairie (grassland) and forest in northern USA during the Holocene has also been derived from the pollen record. In the Saharan region, radiocarbon dating of material in lake sediments and archaeological evidence (including rock paintings of savannah fauna and flora) suggest that there were a number of phases in the Holocene that were wetter than at present. Studies of pollen from lake sediments in the eastern Sahara indicate that savannah existed until 6000 BP, after which there was a trend towards aridity. Other vegetation studies suggest that the boundary between the southern Sahara and the Sahel moved about 1000 km northwards during deglaciation to a position close to the Tropic of Cancer, until about 6000 BP when it reversed its migration about 600 km to its present position.

The period between the rapid temperature rise at the start of the Holocene and a later deterioration is sometimes called the climatic optimum. This period of elevated temperatures, which were higher by 1–3°C than today's, varies from place to place and according to dating method. The end of the optimum may be at 5000–6000 BP in some localities and as young as about 2500 BP in others. Following the optimum, temperatures declined until a minor warming in early medieval times (AD 750–1300), which was followed by a significant cooling known as the Little Ice Age or *neoglaciation*. During the Little Ice Age glaciers advanced further than at any time since the Pleistocene. The timing of this climatic deterioration once again varies according to location—as early as the mid-twelfth century in some places, but more commonly with a maximum between the mid-fourteenth and mid-nineteenth centuries. The ending of the Little Ice Age can be recognized in the warming of the first half of the twentieth century in meteorological records (Fig.

2.20). Some areas experienced a slight cooling mid-century, but all areas have been subject to a warming in the last couple of decades (Section 2.5). This has been accompanied by changes in rainfall patterns.

In Europe, the former Soviet Union, and the USA the warming in the second half of the twentieth century has been accompanied by increases in rainfall, particularly in the winter. The reverse is true for North Africa and the Middle East, which have experienced less rainfall over the same period, especially in the summer. There has been an increase in the frequency of hurricanes in low latitudes (e.g. the south-eastern USA, Japan, Australia and the Indian Ocean). These observations are capable of interpretation in different ways. What seems clear is that the variability of temperature and rainfall is very large, and complex in its spatial distribution.

In high latitudes/altitudes, mountain glaciers have been in overall retreat since the ending of the Little Ice Age. Land left exposed after the retreat of glaciers and ice caps has become colonized by a succession of vegetation. The glacial landscape is out of equilibrium with current climatic conditions, which has importance for the occurrence of hazards such landslides, debris flows and rock falls.

2.3.2 Effects of volcanic activity
The volcanic forcing of climate has attracted much

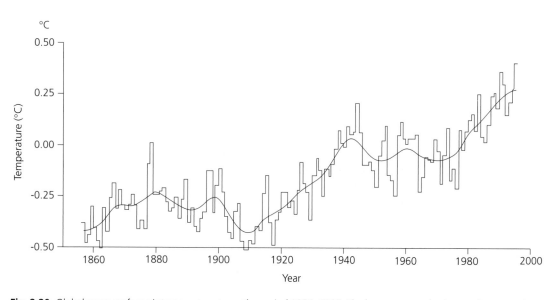

Fig. 2.20 Global mean surface air temperatures over the period 1856–1995. The bars are annual values as departures from the mean temperature for the period 1961–90. The smooth curves shows the filtered annual values. After issue no. 19 of *Tiempo*, published by The International Institute for Environment and Development and the University of East Anglia, UK.

recent attention. Volcanic activity can affect climate through outgassing (the greenhouse gas CO_2 and the ozone-destroying Cl_2 being examples), but more important at shorter time-scales is the emission of large quantities of dust. Dust concentrations can be evaluated by the attenuation of solar radiation through the atmosphere; volcanic dust can be differentiated from terrestrial dust and that originating from human industrial activities.

Much of the fine dust from volcanoes never reaches the stratosphere and is instead concentrated at lower levels in the troposphere. There have been a number of major eruptions in recent historical times that have resulted in important atmospheric effects. The eruption of Krakatoa in 1883, in the Straits of Sunda between Java and Sumatra, culminated in the explosive disintegration of a number of volcanic peaks and

the formation of calderas. Clouds of ash were sent 80 km into the atmosphere, turning day to night for 48 hours 200 km downwind of the craters. Dust initially occupied the belt of tropical easterlies, but within 3 months had spread to the northern hemisphere atmosphere (Fig. 2.21a), reducing the solar radiation received at the Earth's surface by as much as 10% three years after the event. Yet Krakatoa is not exceptional. More ancient explosions in the Mediterranean area, such as those of Santorini in 1628 BC and Vesuvius in AD 79, are thought to have produced even more atmospheric dust than Krakatoa. The Tambora eruption in Indonesia in 1815 and the Cosequina eruption in Nicaragua in 1835 both released four times as much dust as Krakatoa, and were part of a phase of strong global volcanic activity in the eighteenth and early nineteenth centuries. This

(a)

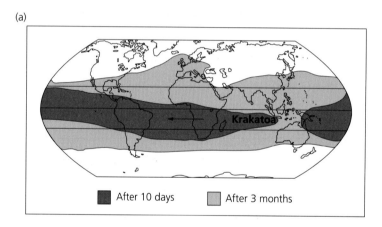

(b)

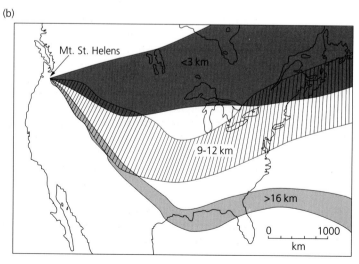

Fig. 2.21 Atmospheric loadings of volcanogenic dust. (a) Distribution of stratospheric dust following the eruption of Krakatoa on 27 August 1883; (b) distribution of tropospheric dust at three altitudes following the eruption of Mount St Helens in 1981. After Bryant (1991) [26].

corresponds with the reduced surface temperatures of the Little Ice Age. The dimming of the Sun and stars after the Tambora eruption lasted for 3–4 years (Figure 2.22).

The 1980 eruption of Mount St Helens involved a sideways blast and most of the ash settled relatively close (<700 km) to the mountain. Some dust reached the top of the troposphere (Fig. 2.21b), however, encircling the globe at about 9–12 km altitude within 17 days. Although surface temperatures immediately downwind of the eruption were reduced by 0.5°C for a few weeks, the effects on global climate were negligible. The discharge of large volumes of dust into the atmosphere from Mount Pinatubo in the Philippines in 1991, however, has had a recognizable cooling effect on global temperatures for several years following the eruption.

The volcanic forcing of short-lived climatic cooling can be tested by an investigation of tree rings. Tree rings have extremely high precision in their dating, being annual and absolute rather than relative. During anomalously cool summers tree wood growth is restricted. Analysis of tree rings across different continents since 1600 shows a strong correlation of anomalously cool summers and the dates of known large volcanic eruptions [28, 29].

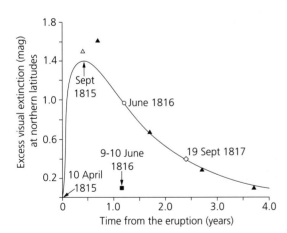

Fig. 2.22 The dimming of the Sun and stars for northern latitude sites following the Tambora eruption of 1815, expressed by different methods. 1816 was known as the 'year without a summer'. From Stothers (1984) [27].

2.4 Causes of past climate change
2.4.1 The forcing mechanisms of climate change
If we know the forcing factors for climate change, we have the possibility of building models useful for deciphering past climates as well as predicting future climate change. These forcing factors fall into three groups, though there are sometimes links between them:

1 changes in the amount and distribution of solar radiation received at the top of the Earth's atmosphere;
2 changes in the composition of the atmosphere affecting emissivity;
3 changes in the nature of the surface of the Earth, caused by:
    (i) tectonic factors and
    (ii) sea level changes.

Astronomical factors
Changes in solar radiation may be caused by variations in the Sun's luminosity, by orbital fluctuations of the Earth, or by galactic mechanisms such as the passage of clouds of interstellar matter.

The temperature of the Sun and its *luminosity* vary with the burning of hydrogen to helium. It is estimated that there has been a 30–40% increase in solar luminosity since the formation of the sun 4.5 billion years ago [30]. This is a particularly important conclusion for those interpreting the Earth's ancient history. Disturbances of nuclear equilibrium in the burning reactions in the sun may result in variations in the luminosity on time-scales of millions to tens of millions of years.

Changes in the inclination of the Earth's orbit relative to the plane of the solar system (a plane which passes through the orbit of Jupiter) may affect the amount of extraterrestrial material entering the Earth's upper atmosphere. Such changes in inclination have a periodicity of about 100 ky [31] and have been proposed to explain the glacial cycles of the Pleistocene.

However, the rapid changes in the Pleistocene from glacial to interglacial are most commonly explained as an effect of changing radiative energy received from the Sun caused by orbital variations. The Serbian physicist Milankovich proposed that there were three different mechanisms responsible for variations in solar radiation received at the top of the Earth's atmosphere (Fig. 2.23).
• The Earth's orbit around the Sun is an ellipse, not a circle. The shape of the ellipse varies with a period of about 100 ky. This variation is therefore due to the *eccentricity* of the Earth's orbit.

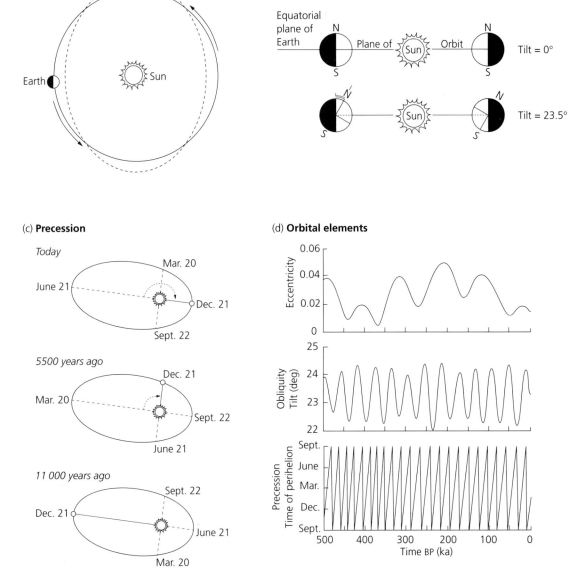

Fig. 2.23 Orbital mechanisms in the Milankovich band: (a) eccentricity; (b) obliquity; (c) precession; and (d) the associated frequencies.

- The Earth wobbles in its spin about its axis of rotation. The axis of rotation itself varies in inclination with a period of about 21 ky, the so-called cycle of *precession*.
- The axis of rotation leans with respect to the plane of its orbit around the Sun, but the angle of lean varies between 22.1° and 24.5°. This variation of lean of 2.4° occurs with a period of about 41 ky. It is known as the *obliquity* of the ecliptic.

The calculated variation in solar radiation at the top of the atmosphere can be calculated as a function of latitude and season for the last 500 ky. This shows the

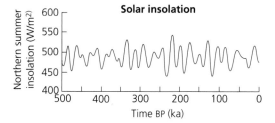

Fig. 2.24 Variations in solar input caused by orbital variations. After Berger (1978) [33].

northern hemisphere values to fluctuate by up to nearly 10% either side of the mean [32] (Fig. 2.24). Estimates have also been made of the variations in solar radiation going back to 2500 Ma. But the greatest interest has been aroused in the linking of these orbital forcing mechanisms to the growth and shrinkage of ice sheets in the Quaternary. The close match between the calculated solar radiation variations and the records of palaeotemperature have led to the conclusion that forcing factors of Milankovich type are responsible for the periodic climate changes seen in the Quaternary, and by analogy for the more ancient geological past. The 100 ky eccentricity cycle appears to be particularly well correlated with the glacial–interglacial variations of the last 700 ky. It is important to recognize, however, that the different periods and amplitudes of the three Milankovich components interfere to produce a potentially complex resultant signal. In addition, there are lags built into the response of the Earth surface system to changes in the Milankovich forcing. It is perhaps better to see the Milankovich orbital variations as providing a pacemaker for climate change, driving changes in a complex climate system with built-in lag times due to internal dynamics.

How variations in solar energy are converted into global variations in climate depends on the ways in which heat is transferred over the Earth's surface (Chapter 1). Since this is achieved by ocean and atmospheric circulation, which are in turn affected by the distribution of continental and oceanic areas, we should expect Milankovich mechanisms to have varied in their efficacy in the geological past.

Atmospheric factors

We have already seen that CO_2 is a greenhouse gas. Long-term variations in CO_2 before the marked increase due to anthropogenic emissions are likely to have been caused by variations in the rate of volcanic outgassing and utilization in weathering, since weathering of silicate rocks removes CO_2 from the atmosphere (Chapter 3). A number of different methods have all concluded that atmospheric CO_2 levels were considerably higher 100 million years ago in the Cretaceous than in the recent past. We can therefore view the mid-Cretaceous period as that of a greenhouse Earth. Climate experiments for the mid-Cretaceous have been carried out with CO_2 values at four times their present concentration [32]. Increased CO_2 results in warming at all latitudes, with a global average temperature increase of 3.6–5.5°C. Since high levels of CO_2 should cause the tropical oceans to overheat, ocean–atmosphere heat transport must have been highly efficient in redistributing this heat to polar regions.

The injection of particulate matter into the atmosphere scatters or absorbs a fraction of the incident solar radiation causing cooling, but this may be offset by the screening of outgoing infrared radiation. The result of this trade-off depends on the amount, size, composition and distribution of the particles. Injection of dust into the stratosphere ensures a long residence time and a maximum climatic effect. Single eruptions are unlikely to have a climatic effect of geological interest. For geologically recognizable climate changes volcanic activity needs to be intense and long-lived. This may have been the case with the extensive volcanic activity during the Triassic and at the Cretaceous–Tertiary boundary. High dust loads might also result from the collision of an asteroid with the Earth, a theory favoured by many for the major environmental changes and mass extinctions found at the Cretaceous–Tertiary boundary.

Tectonic factors

The position of the continents, which has varied throughout geological time because of plate tectonics, may affect climate through a number of mechanisms.
• How much land is in each latitudinal zone affects the climate system through its affects on albedo. The proportion of land at high latitudes, or *polar continentality*, may be particularly important because of the very high albedo of snow-covered surfaces.
• Continents affect the flow of ocean currents through the presence of ocean gateways and ocean barriers. For example, there is a close correlation between the opening of the Drake Passage between Antarctica and South America, and the separation of Australia from Antarctica with the development of the modern deep ocean circulation. Similarly, the closing of the gateway through the area now occupied by the

isthmus of Panama is likely to have had important effects on deep ocean circulation and therefore global heat transport.

• The distribution and elevation of mountains affects climate through the now familiar effects on albedo, and their impact on atmospheric circulation patterns such as the triggering of the monsoon by the uplift of Tibet. Large topographic obstacles may even affect waves in the jet stream through their effect on surface frictional drag which thereby affects general circulation patterns, and their effect on the infrared opacity of the atmosphere, increasing outgoing infrared radiation and thereby causing cooling.

• major changes in vegetation type affect climate through variations in albedo, rates or evapotranspiration and frictional drag.

The effects of tectonics can be illustrated by the supercontinental assembly of Permo-Triassic Pangaea, with a single world ocean (Panthalassa). This represents an entirely different configuration of land and sea than is seen today (Fig. 2.25). The geological record shows extremes of climate in the Permian and Triassic, such as extensive Permian ice sheets, and warm, arid Triassic environments. The Pangaean assembly of a supercontinent therefore offers a good test of continental configuration as a climate forcing factor. The climate models present consistent predictions of continental aridity and extreme climatic seasonality.

• The temperatures of continental interiors depend on the advection of heat, which is controlled by the size of the continent, and the solar radiation, which is controlled by latitude. Seasonal variations should increase with continental size and latitude. This is seen in the cold polar flora for the late Permian in Gondwanaland at an approximate palaeolatitude of 50°S. Since ice sheets develop by extension from land, a key factor in explaining the extensive glaciation in the Permo–Carboniferous is the presence of a sizeable continental landmass in polar latitudes.

• Supercontinental geometry also has a major effect on the seasonal shift of atmospheric pressure patterns. Today, strong high-pressure systems dominate the

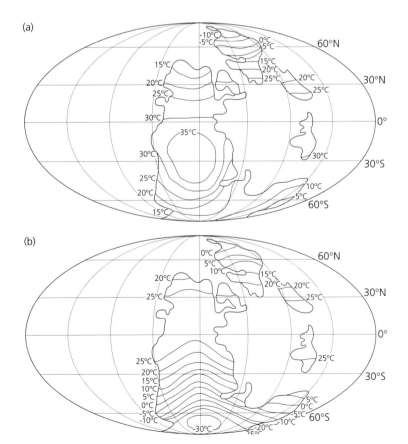

Fig. 2.25 Pangaean supercontinental assembly at 255 Ma (Kazanian) with summer (a) and winter (b) temperatures from an energy balance model. Areas below 0°C and above 30°C shaded. After Crowley *et al.* (1989) with permission of *Geology* **17**, the Geological Society of America, Boulder, Colorado USA, [34].

continental areas in winter, and low pressure dominates in summer. This effect is likely to have been exaggerated during the supercontinental assembly of Pangaea.

- Precipitation today is associated with the rising limb of the Hadley cell in low latitudes, and with atmospheric instability in mid-latitudes. In contrast, aridity is found under the descending limb of the Hadley cell in the subtropics, in orographic rain shadow areas and in continental interiors characterized by high pressure because of differential heating. Palaeoclimatic models for Pangaea indicate summer monsoons on either side of Tethys, but elsewhere extensive aridity within the tropics and subtropics with negative moisture balances (precipitation minus evaporation).

Global sea levels

Sea level is linked to global climate change through changes in atmosphere–surface coupling through the heat balance and thermal inertia, changes in ocean circulation, and changes in ocean–atmosphere chemical composition.

- Different surface types have different balances between incoming radiation, and outgoing heat losses through sensible heat and evapotranspiration. Take three different surfaces (Fig. 2.26). If continental flooding replaces subtropical desert with subtropical ocean, there is a fall in the albedo and increased evaporation. Apart from the effect on atmospheric temperatures through the albedo effect (a 0.01 increase in albedo causing a 1°C increase in global temperature) this also increases cloudiness and precipitation. The heat capacity also controls the rate at which heat is released from different surfaces in a transient temperature field, that is, the *thermal inertia*. The oceans mix in their surface layers relatively slowly, which dampens seasonal temperature contrasts. Bare land surfaces, however, are prone to very rapid response to temperature fluctuations, promoting high seasonality.

- Sea level change may also affect oceanic circulation by its effects on ocean gateways and through the bathymetric control on current directions. Thus, important heat-transporting flows such as the Gulf Stream may be deflected, weakened or strengthened by sea level change.

- Sea level change may affect ocean and atmospheric chemical composition by its effects on the land surface area undergoing erosion, which has impact on the carbon cycle, and specifically the CO_2 concentration of the atmosphere (see Chapter 3). Sea level change may affect oceanic phytoplankton activity, which also

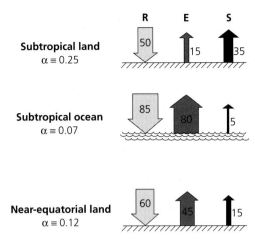

R = Net solar and IR radiation (Wm^{-2})
E = Evaporation / transpiration (Wm^{-2})
S = Sensible heat (Wm^{-2})
α = Surface albedo

Fig. 2.26 Effects of three different surfaces on heat balance: subtropical land, subtropical ocean and near-equatorial land. After Barron & Moore (1994) [32].

is important in mediating the carbon cycle, but in addition may result in variations in the fluxes of compounds such as gaseous dimethylsulphides which oxidize rapidly in the atmosphere to SO_2.

When the causes of climate change in the Quaternary are considered it is important to recognize that there is good evidence of major glaciations affecting the Earth's surface in previous geological times (Fig. 2.2). The record of previous ice ages includes extensive morainic material called *tillites*, *dropstones* due to the fall of blocks drifted by ice into fine-grained sediments on the sea floor, glacially striated bedrock and a number of other features indicative of glacial and periglacial action (Chapter 11). It is clear that there have been ice ages deep into the geological past, recurring with a frequency of about 150 My. There are most likely two different sets of mechanisms to explain, on the one hand, the long-term climatic changes outlined above, including the gradual climatic deterioration from the Eocene to the Quaternary, and, on the other hand, the short-term oscillations of period 100 ky typical of the Pleistocene which most satisfactorily are accounted for by Milankovich mechanisms. It is known that the lithospheric plates have undergone cycles of aggregation into supercontinents, followed by dispersal. The key question of how the presence of aggregated rather than dispersed land areas may have affected global climate has been addressed in relation to the supercontinental assembly of Pangaea.

The various factors outlined above do not operate in isolation. It is instructive to finish this section by considering the general pattern of climate change during the Tertiary, since this gives some insights into the many factors involved in determining global heat transfer. During an ice-free Eocene (at *c.* 50 Ma), the Pacific Ocean was substantially wider than today (and the Atlantic Ocean narrower), and North and South America were widely separated by a sea. South America was connected to Antarctica. Heat transfer from the equator to the poles by ocean currents and winds is thought to have been highly efficient, causing a very wide tropical climatic belt and temperate climates within the polar circle. During the Oligocene, the opening of a passage (Drake Passage) between Antarctica and South America allowed the establishment of a circum-Antarctic current which may have prevented oceanic gyres from warming the Antarctic continent, causing it to become glaciated. Once Antarctica had become substantially glaciated it may have fed the oceans with cold bottom waters. These bottom waters currently upwell along the western sides of continents, causing marked aridity on adjacent land areas. For example, the instigation of the Benguela Current off south-western Africa at about 10 Ma is thought to have been important in causing the development of the Namibian desert, and on occasions the extension of dry conditions northwards into equatorial Africa. Thus, there is a causal link between cooling and aridity. Since the Oligocene, the joining of South America and North America has reduced the very wide longitudinal sweep of tropical ocean, which may have led to their cooling. The widening of the north Atlantic may have allowed the northward penetration of the warm North Atlantic Drift and the development of storm tracks carrying snow to the Greenland and northern European areas. The formation of topographic barriers in the Alpine system of Europe and southern Asia has restricted the inflow of moist air into central Asia. By the early Pliocene air temperatures may have dropped sufficiently for subtropical high-pressure systems to expand, forcing the intertropical convergence zone southwards in Africa. This would have brought dry conditions to the southern Sahara, and wet conditions in central Africa and the Guinea coast of west Africa.

2.4.2 Sea level change

The major environmental changes of the geological past have been accompanied by variations in *absolute sea level*. The record of variations during the Quaternary is particularly rich. There are considerable concerns about future sea level rise resulting from global warming (Section 2.5). One of the problems of deciphering the absolute sea level variations that have taken place in the past is that observations of sea level change are most often made with respect to shorelines. Movement of shorelines, however, is a response to *relative sea level* rise or fall. Relative sea level change results from the interplay of absolute sea level change, tectonic uplift or subsidence of the land, and rate of sediment delivery to the coast. The key criterion in distinguishing between absolute (eustatic) sea level change and that due to local or regional factors is the presence or absence of global synchroneity of the sea level change. In looking at the geological record, the ability to answer this question depends to a large extent on the quality of the data providing age dating. The mechanisms causing long-term sea level change and the evidence for it are presented in other texts listed in the Further Reading section at the end of this chapter. Here, the mechanisms for short-term, high-frequency change are focused on.

Sea level change in the Quaternary

Relative sea level change can be inferred from a number of pieces of evidence.

1 Raised beaches allow the relative sea level change to be evaluated if they can be dated. This dating has recently been improved by the application of thermoluminescence techniques.

2 Uplifted coral reef terraces in the tropics gives a high-precision record of relative sea level change. Tropical areas also have the advantage of not being affected by isostatic rebound from the removal of ice masses.

3 The oceanic oxygen isotope record from deep sea cores provides an estimate of global ice volumes if the effects of temperature variations can be removed (Section 2.2.2). From ice volume changes can be calculated the absolute sea level fluctuation, other factors being constant.

4 High-resolution imaging of the subsurface by, for example, seismic reflection techniques and ground-penetrating radar can be used to infer relative sea level fluctuations from the geometries of depositional units in the shallow subsurface.

The record of sea level change in the last 700 ky of the Quaternary can be reconstructed by calibrating the oxygen isotope record with other evidence such as flights of uplifted coral terraces (Fig. 2.27). Although the match of highstands from isotope evidence and

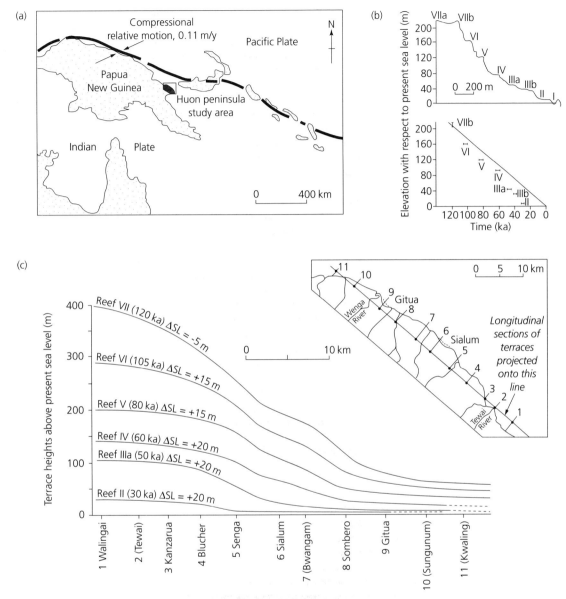

Fig. 2.27 Uplifted coral reef terraces, Huon Peninsula, Papua New Guinea. (a) Location of Huon Peninsula in a region of tectonic uplift due to compressional relative motion of the Pacific and Indian plates; (b) The pattern of sea level change over the last 120 ky on the assumption that the tectonic uplift rate is linear (solid line). (c) Longitudinal profiles of reef terraces along surveyed traverses labelled 1–11. The terrace heights have been corrected for variations in sea level (ΔSL) as indicated. After J. Chappell (1974) [35] and Chappell (1987) [36].

from tropical uplifted shorelines and coral reefs is less than perfect, there is a reasonable correlation in the last 400 ky [35, 36]. Of interest is the number of high-stand terraces preserved at various localities: 12 in the last 730 ky at Taranaki, New Zealand; 10 at Atauro,

Indonesia; 19 at Huon Peninsula, Papua New Guinea; 15 in South Australia. Allowing for the vagaries of preservation, and the existence of local factors, this is in broad agreement with the number of interglacials indicated from the marine isotopic record. However, it is more difficult to estimate the amplitude of sea level change through the Quaternary. At Huon Peninsula, the deviation of the height of uplifted marine terraces from a line representing uniform uplift of the peninsula over the period of generation of terraces (120 ky) indicates relative sea level changes of up to about 50 m (Fig. 2.27). The sea level variations interpreted from the oxygen isotope record and tropical coral terraces corelate well with the high-resolution late Quaternary stratigraphy of certain coastal regions such the Louisiana coast of the Gulf of Mexico. Here and elsewhere the entire continental shelf was emergent and incised by rivers during the maximum lowstand, when sea level may have been as much as 180 m below the present level.

Mechanisms for short-term change

Abstraction of water from the oceans to form land-based ice sheets should produce a glacio-eustatic sea level fall. Melting should result in a glacio-eustatic rise. The loading of the lithosphere by an ice sheet and its unloading during melting result in a flexural depression under the mass of the ice sheet and a flexural rebound during melting. These are known as *glacio-isostatic* adjustments (Fig. 2.28).

Glacial loading and postglacial unloading are able to affect the elevation of the land surface and therefore relative sea level because the mantle has a viscosity such that it deforms relatively easily when acted upon by a force system. When the loading or unloading process is very slow, as in the case of mountain

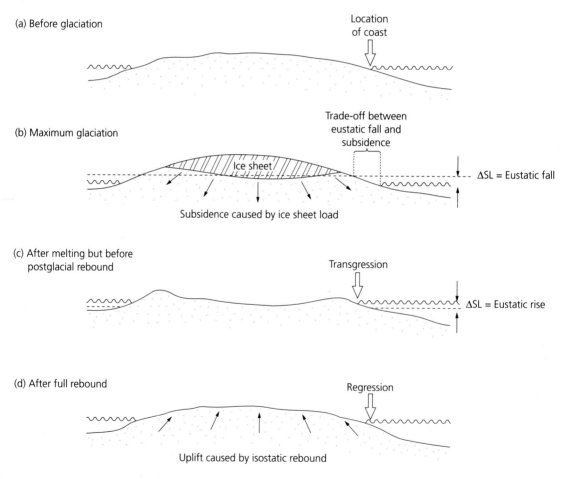

Fig. 2.28 Effects of flexural subsidence and uplift and eustatic sea level change during ice sheet growth and melting. The movement of the shoreline is a function of the time behaviour of eustatic change in response to ice sheet formation and melting traded off against the time behaviour of flexural subsidence and rebound.

building, the mantle is constantly able to maintain an isostatic balance (Chapter 1). However, when the loading process is very rapid, such as during the growth and decay of ice sheets, the mantle adjusts itself dynamically to the changing surface conditions. There is therefore a characteristic response time of the mantle determined essentially by its viscosity. Under the surface load a pressure builds up which causes horizontal pressure gradients. This may induce flow in the mantle. The same is true of unloading. As the mantle flows away from the region of excess pressure, the elevation at the surface changes exponentially with time. If the displacement at any time t is w, and w_0 is the initial displacement of the surface, and if the initial displacement is small compared to the wavelength, then

$$w = w_0 \exp(-t/\tau) \tag{2.4}$$

where τ is the characteristic time of the exponential relaxation of the initial displacement which is given by

$$\tau = 4\pi\mu/\rho g\lambda = 4\pi v/g\lambda \tag{2.5}$$

where μ and ρ are the viscosity and density of the mantle, v is the kinematic viscosity, and λ is the wavelength of the initial displacement. The relaxation time for postglacial rebound is of the order of thousands of years.

The sea level change at any point as a result of the melting of ice sheets is therefore a resultant of the postglacial rebound causing relative uplift, and the addition of water to the world ocean causing relative sea level rise. Close to ice sheets the postglacial uplift commonly dominates, whereas elsewhere ice sheet melting is responsible for coastal flooding. These different responses can be seen in the coastal evolution of Maine, north-east USA, which was close to the great Laurentide ice sheet, compared to the Gulf of Mexico coast of the USA, which was unaffected by postglacial rebound.

Smaller ice sheets are compensated by flow in the mantle at a shallower depth. Since the aesthenosphere is very weak, the relaxation time constants for smaller ice sheets may be shorter. The uplifted beaches around the Swedish coast (Fig. 2.30), for example, suggest relaxation times of 4400 years and mantle viscosities of about 1×10^{21} Pa s.

Melting of the entire volume of today's ice sheets (principally Antarctica and Greenland; ice shelves are of no great consequence since they merely displace their own mass of water), taking into account the depression of the ocean floor by the added mass of

Practical exercise 2.2: Postglacial rebound

It is estimated that the Laurentian ice sheet over the Hudson Bay area of Canada had a thickness of 4 km and a wavelength of about 5000 km prior to its melting at about 10 ka (Fig. 2.4). Assume that ice with a density of 800 kg m^{-3} displaced mantle with a density of 3300 kg m^{-3}.

1 Assuming local isostatic equilibrium (no flexural rigidity), what was the maximum deflection of the land surface under the ice sheet?

2 The Hudson Bay area has a negative free air gravity anomaly (0.3 mm s^{-2}) that is assumed to be due to unfinished rebound following melting at 10 ka. The anomaly suggests that the present land surface is still deflected downwards by 0.22 km relative to its fully unloaded elevation. Calculate the postglacial relaxation time. What is the dynamic viscosity of the mantle?

Solution

1 We perform an isostatic balance of a column far from the ice sheet and a column through the centre of the ice sheet (Fig. 2.29). The depth of compensation is at the depth of the crust plus the maximum deflection.

$$\rho_c h_c + \rho_i h_i = \rho_c h_c + \rho_m w_0 \tag{2.6}$$

where ρ_m, ρ_c and ρ_i are the densities of mantle, crust and ice respectively, and h_c, h_i and w_0 are the thickness of the crust, maximum height of the ice sheet and the maximum deflection of the land surface under the ice sheet respectively. Substituting the correct parameter values, $w_0 = 970$ m.

2 The relaxation time constant can be found from equation (2.4) by letting $w_0 = 970$ m, $w = 220$ m and $t = 10^4$ y. The solution is $\tau = 6739$ y. Conse-quently, from equation (2.7) the viscosity of the mantle is $\mu = 2.7 \times 10^{21}$ Pa s.

water in the ocean basins, should result in an absolute sea level rise of about 50 m. The estimated sea level fall resulting from the locking up of water in ice sheets during the last glacial maximum, the extent of which can be estimated from Quaternary geology, is about 75 m. The glaciation at 120–190 ka is thought to have been more extensive than the last glacial. In such a case, the total height of sea level change associated

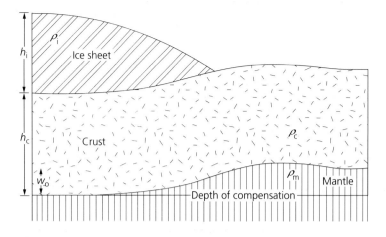

Fig. 2.29 Set-up for isostatic compensation of ice sheet in Practical Exercise 2.2.

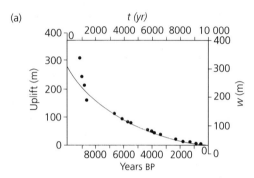

(a)

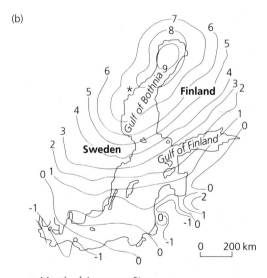

(b)

* Mouth of Angerman River

Fig. 2.30 (a) Post glacial uplift of the Baltic coast, Sweden, over the last 10 ky, compared with the exponential model in equation (2.4). (b) Isobases of present-day uplift of land surfaces in Sweden and Finland due to glacial unloading in millimetres per year. After Eronen (1983) [37].

with complete melting of Pleistocene-scale ice sheets is well in excess of 100 m and perhaps as much as 150 m. These magnitudes of sea level change are quite likely to have exposed to subaerial erosion many of the present-day continental shelves. This possibility is supported, for example, by the river drainage patterns underlying the recent sediments of the English Channel on the NW European continental shelf, and the unequivocal evidence of emergence of the US Gulf of Mexico coast revealed from seismic stratigraphy.

The added effect of thermal expansion of the oceanic water during glacio-eustatic highstands is likely to have been small compared to the sea level change resulting from the melting of continental ice sheets. Ocean water at a constant salinity (of 35 parts per thousand) which changes in *average* temperature (not sea surface temperature) by 5°C, from 5°C to 10°C, would be associated with a density change of less than $1 \, kg \, m^{-3}$. If we assume that the average depth of the ocean is 3770 m and the area of the oceans (excluding the continental shelves) is $3.1 \times 10^{14} \, m^2$, the associated fractional volume change in the ocean of 6.8×10^{-4} would produce a sea level rise of about 2.5 m due to the thermal expansion effects.

It is therefore widely accepted that land-based ice sheet growth and decay are potent and effective mechanisms to explain rapid and high-amplitude sea level variation.

2.5 Human impact

Throughout geological time the natural environment, and Earth surface processes, have been constantly changing. Our investigation of the dramatic changes in the Quaternary serves well as a benchmark of the dynamism of the Earth's surface environment. The Earth has suffered major revolutions in its ecosystems, such as the faunal and floral extinctions across the Cretaceous–Tertiary boundary about 65 million years ago. But there can be no doubt that the effects of humankind, and the rapidity of these anthropogenic changes, are enormous. Although conclusive evidence is difficult to obtain, it is generally thought that we are living at a time of alarming loss of animal and plant species, and there are acute concerns about the future irreversible climate change which is already under way. The biosphere, the realm of life which interacts pervasively with the environment, exercises important and critical controls on the climate system (Chapter 1) and on the sediment routing systems bringing about geomorphological change (Chapter 3).

During the last billion years the Earth has undergone profound changes in its climate and life systems, but it has never become completely uninhabitable. One might contrast Venus, which probably had watery oceans, but where the greenhouse effect has run away, leading to a boiling off of the oceans and the creation of surface temperatures of 470°C (Section 1.3.1). On Earth, life systems have a number of negative and positive feedbacks into the physical climate system. Some of these serve to buffer the system against excessive change. Others have the following dangerously escalating effects.

• **A negative feedback:** in warm periods the precipitation of calcium carbonate as limestone, the proliferation of vegetation, and the formation of coal and oil from the remains of organic tissue take CO_2 out of the atmosphere, reducing the greenhouse effect, and therefore reducing temperatures. In cold periods vegetation is reduced, so less CO_2 is taken from the air, increasing the greenhouse effect and increasing temperatures.

• **A positive feedback:** changes in cloud cover, water and snow cover, and proportion of vegetation types, affect the reflectivity of the Earth's surface (albedo), which in turn controls its surface temperature. Increased low-level cloudiness, snow cover and glossy leaved vegetation should increase reflected energy, causing cooling. Ice cover therefore promotes further cooling through its albedo effect. Water bodies absorb radiation, so decrease reflected energy losses. Plankton in the ocean may change its colour and thereby alter its reflectivity.

We know the CO_2 and CH_4 contents of air during the last glaciation from the analysis of bubbles in ice cores. The CO_2 level was about 190 parts per million by volume (ppmv) just prior to the start of deglaciation (13 500 BP), and the CH_4 level about 350 parts per billion by volume (ppbv). Within a very short period of time, a few centuries, the CO_2 level of the air increased to 260 ppmv, and the CH_4 level doubled to

650 ppbv. The effect of these increases in greenhouse gases was to flip the Earth out of profound glaciation. Although we may not know *why* the CO_2 and CH_4 contents increased so dramatically, we do know that the increases of 70 ppmv and 300 ppbv respectively were enough to cause a major climatic reorientation of the Earth's surface. In the last century the activities of humankind have contributed substantially to further increases of 70 ppmv CO_2 and 1000 ppbv CH_4. It seems inconceivable that this will not have strong effects on the Earth's climatic system.

How have humans brought about this change? Our impact lies principally in two areas: consumption (burning) of fossil energy sources (coal, oil, natural gas); and changes in the Earth's vegetation, especially the massive clearance of the boreal and tropical forests.

Although CO_2 is produced naturally through volcanic emissions, humankind has now exceeded natural production through the activities associated with industrialization and with the clearing and burning of forests. Over the last 30 years or so there has been an accelerating build-up of CO_2 in the atmosphere (Fig. 2.31), paralleling the rate of increase of the world's population. This means that the annual production of CO_2 is now far outstripping the annual consumption through plant growth and dissolution in ocean water, to the tune of 11 billion tonnes of CO_2, or 3 million tonnes of carbon.

Carbon dioxide dissolved in the immense oceanic water reservoir is used by plankton during photosynthesis. Their shells fall through the ocean water column, dissolve and thereby transfer carbon to deep ocean water. In shallower water depths, their shells may reach the sea floor and be buried by overlying sediment, sealing the carbon from interaction in the ocean–atmosphere system. The ocean therefore has considerable potential to take up a great deal of the additional CO_2 in the atmosphere, but the mechanism of sedimentation works on a long time-scale of thousands of years. On land, most plants grow vigorously in a CO_2-rich atmosphere, fixing carbon in their tissue. In the short term, CO_2 is incorporated in new plant growth in the spring and released during decay in the autumn. Some carbon is temporarily stored in organic matter in soil, in tree trunks, and over longer time periods in fossil carbon in the form of peat, coal, gas and oil. The residence time of carbon in the ocean is much longer than in forests.

Although these sinks are powerful ways of removing CO_2 from the atmosphere, there are opposing processes at work. For example, a warming of ocean water affects CO_2 solubility, so that CO_2 is emitted back to

(a)

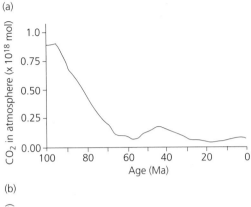

(b)

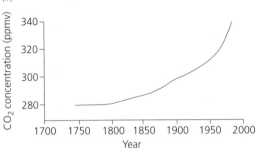

(c)

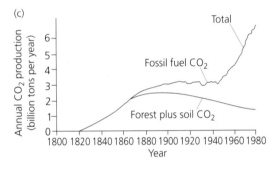

Fig. 2.31 Carbon dioxide levels at geological and historical time-scales. (a) Decline in CO_2 since the greenhouse world of the Cretaceous; (b) build-up of CO_2 in the atmosphere over the last 300 years; (c) breakdown of emissions into burning of fossil fuels, and those resulting from deforestation. After Nisbet (1991) [38].

the atmosphere. The uncertainty in predicting anthropogenic emissions in the future, and of the trade-offs in the ocean–land–atmosphere carbon system, makes the estimation of future atmospheric CO_2 levels difficult, but it is agreed that concentrations are bound to increase so markedly as to take us into new and uncharted territory in terms of climate.

Methane concentrations are very rapidly increasing (by 1–1.5% per year) in the atmosphere. Although

its residence time in the atmosphere is very short (7–10 years), it is a powerful greenhouse gas. It also contributes to ozone production in the troposphere (that is, unwanted low-level ozone, which has its own greenhouse effect). Methane comes from the burning of fossil fuels, from the melting of permafrost and also from wetlands, peat bogs, paddy fields, landfills and from cattle.

There are a number of other important trace gases which have implications for atmospheric chemistry, such as carbon monoxide, nitrogen oxides, chlorofluorocarbons (which have a severe effect in depleting protective high-level ozone), ammonia, and the hydroxyl radical (which has been described as the 'policeman' of the atmosphere since it scavenges various substances and makes them less harmful by oxidizing them). The reader is referred to other texts listed in the Further Reading section at the end of this chapter.

2.5.1 Global warming

The accurate prediction of future environmental change depends on our ability to understand the present. Computer models which successfully simulate the present-day distribution of rainfall and temperature, and their distribution at times in the Pleistocene on a strongly climatically perturbed Earth, may be used to forecast the results of global warming (Fig. 2.32). It is sometimes said that global climate models should be used but not believed. This is the crux of the problem. The Earth's climate system is exceedingly complex and nonlinear; it contains many thresholds and feedbacks; and it is extremely sensitive to tiny variations in initial conditions.

But is there evidence that the Earth is actually warming? The problem is that any evidence for warming is set against a background of considerable natural variation. The fall in temperatures in the early 1990s resulting from the eruption of Mount Pinatubo in the Philippines is an example of an important variation introduced by a single event. Although there are difficulties in changes in the measurement techniques used, or in the location of measuring stations, there is growing agreement that the Earth is indeed getting warmer. On a statistical basis, the temperatures over the last couple of decades have been unusually warm. More convincing evidence comes from the temperature of permafrost in the Arctic obtained from drill-holes. The whole Arctic seems to have warmed by several degrees Celsius over the last 50 years. Climate models predict that the first place where signs of planetary warming will be

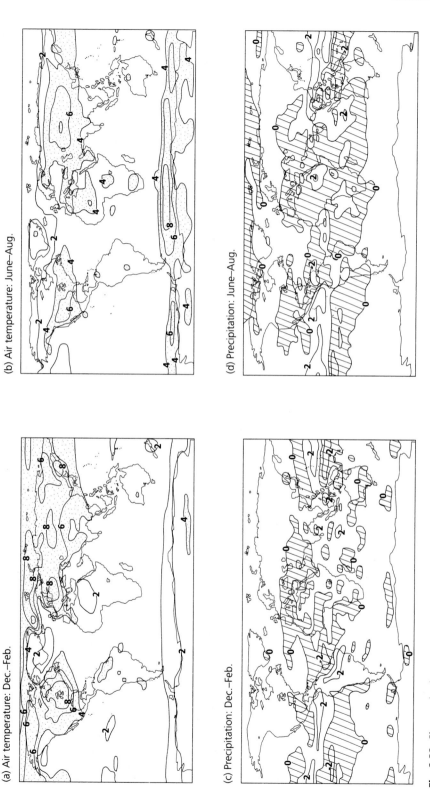

(a) Air temperature: Dec.–Feb.

(b) Air temperature: June–Aug.

(c) Precipitation: Dec.–Feb.

(d) Precipitation: June–Aug.

Fig. 2.32 Changes in air temperature and precipitation in winter and summer for a world with twice the present CO_2 levels, based on the United Kingdom Meteorological Office model. After Houghton et al. (1990) [39]. Patterned areas are those with temperature changes of greater than 4°C and with net precipitation losses.

detected is the Arctic. So it's official: we have global warming.

Effects in the ocean

The oceans are a potentially enormous reservoir of heat: they are capable of delaying atmospheric warming, but they would also delay recovery by slowly releasing heat in the event of atmospheric cooling. Warming of the oceans will have a number of effects.

• Warming causes thermal expansion, and therefore sea level rise. This is augmented by melting of polar ice. The actual sea level rise predicted for certain atmospheric changes in the future is speculative, partly because increased temperatures may cause more snowfall in Antarctica, lowering sea level rather than raising it. Many commentators believe that a 1 m rise may result from a doubling of atmospheric CO_2. This may be achieved at some point in the twenty-first century. The Intergovernmental Panel on Climate Change [39] suggested a rise of about 6 cm per decade.

• Warming may substantially change oceanic circulation patterns, since these patterns are dependent on the contrast in density between cold, dense, polar waters and warmer, less dense waters elsewhere. It is likely that warming would lead to an intensification of the Gulf Stream, causing an increased polar transfer of heat in the northern hemisphere.

• Warming may increase storminess because of the increased heat fluxes from equator to poles, and increased globally averaged rainfall (remember there are always winners and losers), but there are likely to be major geographical variations, as can be appreciated from the record of environmental change during the Quaternary.

Vegetational changes on land

Most plants should, within limits, grow more luxuriantly in a CO_2-rich atmosphere. This should increase crop yields and forest growth. But growth of vegetation also depends on rainfall and availability of nutrients, so vegetational changes will once again be subject to large geographical variation.

It has already been stated that the biosphere exerts a critical control on the physical environment, and there is perhaps no more important element in the biosphere than the rainforests. To appreciate the importance of all terrestrial vegetation, and especially the tropical rainforests, we must place the problem in the context of the carbon and hydrological cycles (Chapter 1).

In the dense Amazonian rainforest only about a quarter of the rain that falls ends up as runoff in rivers, the remainder being evaporated or transpired from leaves. The rainforest therefore returns to the atmosphere three-quarters of the water received by it. This is important for a number of reasons:

• The rainforest creates its own rainfall by recycling water vapour—if the rainforest were cut down, the amount of rain falling on the cleared land would be considerably less than had fallen over the rainforest.

• Clearing of the rainforest increases runoff (and soil erosion: see Chapter 3), so that the proportion of water available for evapotranspiration back into the atmosphere is reduced.

• Transpiration releases latent heat, helping to keep the forest cool—if the rainforest were cut down, latent heat would not be released, causing excessive surface heating.

• Water vapour transpired from the forest is taken high into the troposphere, releasing latent heat by condensation—clearance of the rainforest would result in only the heating of the very low-level atmosphere by sensible heat, a much less efficient process in mixing and distributing heat.

In short, the rainforests act like an efficient air-conditioning system, keeping the equatorial regions of the Earth cool, and helping to warm higher latitudes. Although the rainforests occupy only 3.4% of the Earth's surface (this figure is shrinking yearly) their primary productivity represents 35% of the total for all terrestrial ecosystems. The possible climatic effect of deforestation in Amazonia on the climate of Amazonia itself is illustrated in Fig. 2.33. The predicted surface temperature increase in Amazonia is accompanied by predicted decreases in rainfall in Amazonia, northern South America (especially Venezuela) and north-east Brazil. The change in atmospheric circulation would most likely have a significant effect on the climate of USA.

The rainforests are under threat from two directions: first, the direct activities of humans in the form of felling; and second, indirectly though the effects of global warming. What evidence do we have of the response of the rainforest to these environmental stresses?

In almost all the Earth's major ecosystems there has been change resulting directly or indirectly from human activity. The sorts of change that have taken place can be illustrated by the Amazonian rainforest [40].

Although Amazonia is being rapidly deforested, the causes of deforestation vary. Rondônia is an area of Amazonia with rich soils and was recognized as having agricultural potential. This led to a land rush in 1974, and the population of about 10 000 native indians was rapidly transformed into a population of over a

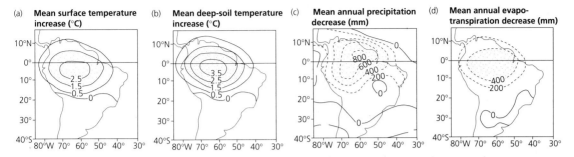

Fig. 2.33 Results of a computer simulation of the likely effects of the total deforestation of Amazonia. After Shukla *et al.* (1990) [41].

million. The ancient forests of Rondônia, with their phenomenal biodiversity, have been cleared at an accelerating rate. After a decade of clearance, Rondônia had already lost over 10% of its forest. In other less fertile areas of Amazonia vast tracts of forest have been felled, burned and cleared to be replaced by cattle pastures for a few years before the soil deteriorates and the area passes to poor quality, unproductive grassland. Other areas have been deforested as a result of petroleum and mineral exploration, particularly gold, tin and iron ore, and subsequent development.

Similar stories of ecosystem destruction can be told of many of the world's rainforests. Particularly poignant is the deforestation of Madagascar, an island, like Australia, with a unique fauna including the lemur and giant flightless birds. Madagascar has lost over 80% of its forest over the last few decades for timber exports and local fuel and to make way for cattle and agriculture. Soil erosion has increased dramatically, as satellite images of plumes of red sediment streaming from its rivers show.

What is the global scale of the problem? Recent estimates are that roughly half of the Earth's tropical rainforest has been destroyed, and that about 2% is being removed or severely damaged each year. At current rates of deforestation, we are facing their effective eradication within one human generation. The remarkable biological diversity of the tropical rainforests is increasingly being impoverished as species after species becomes extinct. The solution to the problem of deforestation is complex and essentially economic and political; it is beyond the scope of this text to discuss these aspects. The importance to students of Earth surface processes is that tropical rainforests perform a crucial role in the Earth's air-conditioning system, and therefore have impact on the hydrological and carbon cycles. On a smaller scale, the forests play a vital role in

the hydrology of catchments (drainage basins), and therefore in their sediment yield (Chapter 3) and sediment transport mechanics.

In Amazonia, the biggest threat to the ecosystems is directly from humans, not indirectly through climate change. In more marginal environments, however, such as parts of Asia and Africa, global warming and its attendant climatic shifts pose serious problems. The prediction of future environmental changes is largely crystal ball gazing, and much will depend on how human activities respond to the climatic changes.

2.5.2 Natural hazards and global climate change

There are two major issues driving approaches to environmentally sustainable development. One is climate change caused by human activities, the other is the reduction of human and economic losses caused by natural hazards. The two issues are related. Patterns of hazard occurrence and of natural disasters may be a manifestation of global climate change. This is the subject of interest in this concluding section.

Although the topic is complex and controversial, there are indications that the frequency and intensity of natural hazards are increasing in the second half of the twentieth century [42]. But if this trend is real, it is most likely that it is the vulnerability of human communities that has increased rather than major changes in the hazards themselves. Put differently, the problems of global change do not arise from a global temperature rise *per se*, but from a rapidly increasing population concentrated in hazard-prone areas. The International Decade for Natural Disaster Reduction has established targets to be achieved by the year 2000. These are national assessments of risk, plans for mitigation, and access to warning systems and rapid dissemination of warnings. Many of these goals rely not only on the transfer of appropriate technology and knowledge, but also on the development of education and public information programmes.

The International Panel on Climate Change (IPCC), which first reported in 1990, predicts a global temperature rise of between 0.2 and 0.5°C per decade, and a sea level rise by the end of the twenty-first century of about two-thirds of a metre (with an uncertainty of one-third of a metre). As we have seen, these figures imply strong regional variations in climate, and possibilities of affecting oceanic and atmospheric circulation patterns such as the El Niño. From the point of view of natural hazards, the most serious concern is the likely effects on the frequency and intensity of tropical cyclones. At present, the evidence is conflicting: elevated sea surface temperatures should generate more cyclones, but warming of the troposphere may enhance atmospheric stability and dampen cyclone genesis. What is certain is that low-lying coastal zones, and, critically, small Pacific Ocean island states, will be increasingly vulnerable to storm surge hazard under elevated sea levels, as well as problems due to coastal erosion, submergence and salination of groundwater.

There are serious and probably under-documented implications for glacier-related hazards posed by global warming. Warming in the last 100–150 years since the Little Ice Age has resulted in considerable glacier retreat from mountainous areas. This has been associated with sudden downslope movements of ice detached from the glacier terminus (glacier avalanches), increased land instability, particularly of steep valley sides carved by ice action and of loose, morainic material left from the last glacial advance, and catastrophic glacial lake bursts (*Jökulhlaups*). Glacial retreat and thinning also affect the hydrology of rivers fed by glaciers, and thereby have implications for water resource management in mountainous regions. Glaciers act as a storage for meltwater in the early melt season, buffering the release of water to rivers and spreading the peak spring discharge over a longer time period. Glacier retreat therefore increases spring flood danger and reduces late summer flow.

There are important potential effects of global warming on the incidence of drought. The most commonly used tool for predicting the initiation of drought under different future levels of atmospheric CO_2 is the general circulation model (Fig. 2.32). Reduced precipitation, or the distribution of precipitation, propagates through the terrestrial part of the hydrological cycle as reduced runoff, reduced soil moisture, reduced streamflow and lowered groundwater levels. General circulation model predictions of soil moisture changes in a world with doubled CO_2 levels show extensive and major effects (up to 50–60% reductions) in mid-continental areas of both North America and Eurasia. However, the uncertainties are large. We have previously seen in the Amazon case study that deforestation has an important positive feedback into desertification. Shifted wind and precipitation patterns may also result in new areas being prone to soil erosion and wind-blown dust hazards (Chapter 10).

Further reading

A. Goudie (1992) *Environmental Change*, 3rd edn. Clarendon Press, Oxford.

A. Goudie (1993) *The Human Impact on the Natural Environment*, 4th edn. Blackwell Publishers, Oxford.

E.G. Nisbet (1991) *Leaving Eden: to Protect and Manage the Earth*. Cambridge University Press, Cambridge.

K.T. Pickering and L.A. Owen (1994) *An Introduction to Global Environmental Issues*. Routledge, London.

References

1 W.B. Harland, R.L. Armstrong, A.V. Cox, L.E. Craig, A.G. Smith & D.G. Smith (1990) *A Geologic Time Scale 1989*. Cambridge University Press, Cambridge, 263pp.

2 Geological Society of America (1983) *Decade of North American Geology Geologic Time Scale*. Geological Society of America, Boulder, CO.

3 A. Goudie (1994) *Environmental Change*, 3rd edn. Clarendon Press, Oxford.

4 J.J. Lowe & M.J.C. Walker (1984) *Reconstructing Quaternary Environments*. Longman, London.

5 D.E. Sugden, D.R. Marchant & N. Potter (1995) Preservation of Miocene glacier ice in East Antarctica. *Nature* **376**, 412–14.

6 J.P. Kennett & N.J. Shackleton (1975) Laurentide ice-sheet meltwater recorded in Gulf of Mexico deep sea cores. *Science* **188**, 147–50.

7 C. Embleton & C.A.M. King (1975) *Periglacial Geomorphology*. Edward Arnold, London.

8 M.J. Selby (1985) *Earth's Changing Surface*, Oxford University Press, Oxford.

9 R.M. Corfield (1995) An introduction to the techniques, limitations and landmarks of carbonate oxygen isotope palaeothermometry. In: *Marine Palaeoenvironmental Analysis from Fossils* (eds D.W.J. Bosence & P.A. Allison), Geological Society of London Special Publication 83. Blackwell Science, Oxford, pp. 27–42.

10 J. Imbrie, J. van Donk & N.G. Kipp (1973) Paleoclimatic investigation of a late Pleistocene Caribbean deep sea core: comparison of isotopic and faunal methods. *Quaternary Research* **3**, 10–38.

11 R.J. Delmas (1992) Environmental information from ice cores. *Reviews of Geophysics*, **30**, 1–21.

12 J. Jouzel (1994) Ice cores north and south. *Nature* **372**, 612.

13 Vostok Project Members (1995) *EOS (Transactions of the American Geophysical Union)* **76** (17), Fig. 2.

14 K. Pye (1987) *Aeolian Dust and Dust Deposits.* Academic Press, 334pp.

15 J. Fink & G.J. Kukla (1977) Pleistocene climates of central Europe: at least seventeen interglacials after the Olduvai event. *Quaternary Research* 7, 363–71.

16 S.C. Porter & A. Zhisheng (1995) Correlation between climate events in the north Atlantic and China during the last glaciation. *Nature* **375**, 305–8.

17 D.K. Rea (1994) The paleoclimatic record provided by eolian deposition in the deep sea: the geologic history of wind. *Reviews of Geophysics* **32**, 159–95.

18 K. Pye (1989) Processes of fine particle formation, dust source regions, and climatic changes. In: *Palaeoclimatology and Palaeometeorology: Modern and Past Patterns of Global Atmospheric Transport* (eds M. Leinen & M. Sarnthein). Kluwer Academic, Norwell, MA, pp. 3–30.

19 H. Flohn (1981) Ahemispheric circulation asymmetry during late Tertiary. *Geologische Rundschau* **70**, 725–736.

20 M. Sarnthein (1978) Sand deserts during Glacial Maximum and climatic optimum. *Nature* **272**, 43–6.

21 A.S. Goudie, B. Allchin & K.T.M. Hedge (1973) The former extensions of the Great Indian Sand Desert. *Geographical Journal* **139**, 243–57.

22 A.T. Grove (1969) Landforms and climatic change on the south side of the Sahara. *Geographical Journal* **135**, 192–212.

23 G.I. Smith and F.A. Street-Perrott (1983) Pluvial lakes of the western United States. In: *Late Quaternary Environments of the United States* (ed. S.C. Porter). 190–212pp.

24 V.R. Baker (1996) Megafloods and glaciation. In: *Global Changes in Postglacial Times: Quaternary and Permo-Carboniferous* (ed. I.P. Martini). Oxford University Press, Oxford.

25 R.A. Street and A.T. Grove (1976) Environmental and climatic implications of late Quaternary lake level fluctuations in Africa. *Nature* **261**, 385–90.

26 E.A. Bryant (1991) *Natural Hazards.* Cambridge University Press, Cambridge.

27 R.B. Stothers (1984) The great Tambora eruption in 1815 and its aftermath. *Science* **224**, 1191–8.

28 R.S. Bradley (1988) The explosive volcanic eruption signal in northern hemisphere continental temperature records. *Climatic Change* **12**, 221–43.

29 H. Lamb (1971) Volcanic activity and climate. *Palaeogeography, Palaeoclimatology, Palaeoecology* **10**, 203–20.

30 D. Gough (1977) Theoretical predictions of variations in solar output. In: *the Solar Output and its Variations* (ed. O. White). Association University Press, Boulder, CO, pp. 451–74.

31 R.A. Muller & G.J. MacDonald (1995) Glacial cycles and orbital inclination. *Nature* **377**, 107.

32 E.J. Barron & G.T. Moore (1994) *Climate Model Application in Palaeoenvironmental Analysis*, Short Course Notes 33. SEPM (Society for Sedimentary Geology), Tulsa, OK. [A useful overview.]

33 A. Berger (1978) Longer-term variations of daily insolation and Quaternary climate change. *Journal of Atmospheric Sciences* **35**, 232–6.

34 T.J. Crowley, W.T. Hyde & D.A. Short (1989) Seasonal cycle variations on the supercontinent of Pangaea. *Geology* **17**, 457–60.

35 J. Chappell (1974) Geology of coral terraces, Huon Peninsula, New Guinea: a study of Quaternary tectonic movements and sea level changes. *Bulletin of the Geological Society of America* **85**, 553–70.

36 J. Chappell (1987) Late Quaternary sea level changes in the Australian region. In: *Sea Level Changes* (ed. M.J. Tooley & I. Shennan). Institute of British Geographers Special Publication 20. Blackwell Scientific Publication, Oxford, Fig. 10.11.

37 M. Eronen (1983) Late Weichselian and Holocene shore displacement in Finland. In: D.E. Smith & A.G. Dawson (eds.) *Shorelines and Isostasy.* London, Academic Press, pp. 183–207.

38 E.G. Nisbet (1991) *Leaving Eden*, Cambridge University Press, Fig. 3.2, p. 56.

39 J.T. Houghton, G.J. Jenkins & J.J. Ephraums (eds) (1990) *Climate Change: the IPCC Scientific Assessment.* Cambridge University Press, Cambridge.

40 G.T. Prance & T.E. Lovejoy (1985) *Amazonia.* Pergamon, Oxford. [A masterly work on the Amazonian ecosystem.]

41 J. Shukla, C. Nobre & P. Sellers (1990) Amazon deforestation and climate change. *Science* **247**, 1322–5.

42 R. Bras (ed.) (1992) *The World at Risk: Natural Hazards and Climate Change*, AIP Conference Proceedings 277. American Institute of Physics, Cambridge, MA.

3 Liberation and flux of sediment

Unfortunately, the [fractal] dimension of topography . . . is the same as that of
numerous other surfaces, for instance the surface of a bowl of porridge,
with significantly different underlying dynamics.

P.O. Koons [1]

Chapter summary

The weathering and erosion of the Earth's surface are not only responsible for its landforms and soils; they also play a vital role in the functioning of the entire Earth surface system, acting as a dynamic component in global geochemical cycling of particulate and dissolved species. Runoff from the continents is focused in cells of activity known as sediment routing systems. The patterns and rates of denudation are strongly coupled to tectonics and climate.

Weathering processes are interrelated but for simplicity are divided into physical weathering, involving mechanical processes and physical change, and chemical weathering, involving reactions leading to chemical breakdown and the formation of new minerals. Rates of weathering are strongly affected by the presence of water, biological activity and temperature acting upon bedrocks of certain composition and mineral reaction kinetics.

The starting material for chemical weathering is the silicate minerals comprising the bulk of the rocks of the Earth's crust and its sedimentary cover. Silicates are made of building blocks of SiO_4 tetrahedra, with various metal cations occupying sites in the silicate lattice. The style of network of silicon–oxygen ions, expressed by the degree of polymerization, controls the ease of weathering, and gives rise to the Goldich series of reactivity. Chemical weathering may involve a number of processes including solution, oxidation and reduction, hydration, acid hydrolysis, cation exchange and chelation.

The regolith is the weathered mantle overlying pristine bedrock. Its thickness is determined by the

rate of bedrock weathering at its base, and the rate of removal by erosion at its surface. Since weathering reactions decrease in intensity downwards through the regolith, efficient removal by erosion is a prerequisite for maintaining high rates of bedrock weathering. The regolith typically contains a suite of clay minerals caused by the breakdown of silicate rocks. The clay minerals present are a sensitive record of the parent rock mineralogy, extent of leaching and climate. The soil is the upper portion of the regolith. Soil taxonomy reflects the varied processes of leaching of constituents from upper horizons (eluviation) and accumulation in lower horizons (illuviation). Apart from their supreme importance in land use, soils and the biological activity they support play a crucial role in the global climate system by their consumption of CO_2 by weathering reactions and metabolism.

A sediment routing system is the entire network between eroding hillslopes and depositional sites for particulate material, in contrast to the catchment which involves only the subaerial compartment. The concept of the sediment routing system is therefore valuable in considering mass balances and budgets. The Indus sediment routing system starting in the Himalayas and ending in the deep water Indus fan of the Indian Ocean is illustrated as an example. The erosional 'engine' of the sediment routeing system is the hillslope system, dominated by soil creep, overland flow, gullying, debris flows, landslides and rockfalls. Hillslope erosion feeds sediment and dissolved load to fluvial systems which transport the material away. The rate of river incision therefore sets a boundary condition for surrounding hillslopes by controlling hillslope gradients. The landscape has a characteristic response time to a perturbation depending on the physical processes involved. Different landscapes therefore have different sensitivities to forcing mechanisms such as tectonic uplift or climate change.

A modelling approach suggests that at slope angles below a critical value, hillslopes evolve in form according to a diffusive law which relates the local flux to the topographic gradient. At slope angles above critical, landsliding dominates. High channel incision rates push the hillslope system above this threshold slope value.

Sediment and solute transport rates in rivers give information on the denudation rate of the drainage basin. Mass may be transported as bedload, suspended load, and as dissolved load. Information on bedload sediment transport rates is sparse. A large database is available on suspended loads of the world's major rivers, which demonstrates that the highest sediment yields are in an arc extending from northern China and Japan, around SE Asia (including New Zealand) to the western Himalayas. This arc is characterized by rapidly uplifting mountain belts, steep topography, high rates of precipitation and high temperatures. Most of the solute load is made up of four components—the bicarbonate, sulphate and calcium ions and silica. Rivers can be categorized according to their main source of solute load into those dominated by atmospheric precipitation, by rock and soil-derived solutes, and those dominated by evaporation and subsequent precipitation. The global average of solute load is roughly one-fifth of the global average of suspended load. The ratio between sediment yield and solute yield varies a great deal, but in general is high in mountainous terrains and low in lowlands.

There have been a number of attempts to quantify denudation rates in relation to environmental factors. In addition, it is now clear that the rate of tectonic uplift is an important consideration. Tectonic uplift, irrespective of topography, keeps regoliths replenished with bedrock, thereby promoting high rates of weathering and denudation.

A number of coupled tectonic–erosion models are emerging from a variety of settings, including the vigorous systems of collisional mountain belts such as the Southern Alps of New Zealand, the tilted fault block terrains of extensional provinces such as the Basin and Range of western North America, and uplifted continental margins such as those of South Africa and eastern Australia. These models rely on a resolution of the correct spatial and temporal scales of erosion and tectonic uplift and their interplay.

Denudation and sediment yield are very sensitive to changes in land use caused by man, notably through deforestation, agricultural practices, and engineering projects such as dam construction. Deforestation and desertification illustrate the important links between vegetation, soil deterioration and sediment yield.

3.1 Introduction

There are few more dramatic illustrations of the dynamic nature of the Earth's surface than the relentless denudation of the continental surface and the transport of particulate and dissolved matter through routeing systems to the coast and, eventually, the deep ocean. The weathering and erosion of the Earth's surface is not only responsible for the creation of our landscapes but also plays a crucial role in the functioning of the entire Earth surface system. Long-term denudation rates affect global geochemical cycling, including the carbon budget. Denudation therefore has important climatic feedbacks. The flux of material over long time-spans from net erosional zones to net depositional zones leads to the formation of new sedimentary rocks. The sedimentary basins in which these sediments are deposited represent vast repositories of geological data on Earth history, and contain a high proportion of the planet's natural resources, particularly hydrocarbons. A knowledge of the ways in which these denudational systems operate is important in understanding the evolution of landscape, the supply and filling of sedimentary basins, the mitigation of natural hazards, and the consequences for the Earth surface system of global climate change.

It is increasingly recognized that there is a strong coupling, at medium to large spatial scales, between tectonics and erosion. Tectonics, in general, creates the regional topographic surfaces upon which weathering and erosion act. Such surfaces are commonly the result of the uplift of rocks due to lithospheric shortening and thickening, the situation prevalent in collisional mountain belts (Section 1.5.2). However, significant topography may result from tectonic uplift in other settings, such as the linear uplifts along the passive margins of continental plates, and the domal uplifts associated with hotspots (Section 1.5.4). For a surface that stays at a constant elevation, denudation implies that there is a flux of rock into the zone of weathering. This exhumation flux is necessary to maintain high rates of denudation of the continental surface.

The removal of weathering products is achieved initially through a set of geomorphological processes dominated by unchannellized hillslope erosion and mass flows (landslides, debris flows), the details of which are presented in Chapter 6. However, the volume of material derived from these upland processes is eventually disposed of within a different geomorphological subsystem by the action of rivers. Rivers provide an efficient means of routeing water, particulate sediment and dissolved (solute) load from topographically high to topographically low regions. The aim of this chapter is to describe and explain the global patterns of weathering and erosion, with an emphasis on the information obtained from the particulate and solute loads of rivers.

3.2 Weathering and soils

Weathering is the decay and disintegration of rock *in situ* at the Earth's surface. All igneous and metamorphic rocks are formed at temperatures and pressures higher than those at the Earth's surface, and consolidated sedimentary rocks have generally been buried to depths of kilometres in sedimentary basins. Rocks exposed at the Earth's surface at atmospheric temperatures and pressures, and in the presence of water and biological activity, are consequently unstable. That this is the case is well demonstrated by the sheer abundance of clay minerals in the ocean. These fine-grained minerals result from the breakdown of a wide variety of rocks during weathering.

Weathering processes are complex, varied and interrelated, but for simplicity can be divided into: *physical weathering*, involving only mechanical processes causing physical change; and *chemical weathering*, involving chemical reactions leading to the breakdown of minerals and the formation of new ones.

The rate of weathering is strongly influenced by the presence of water. This is for a number of reasons. First, water is a powerful solvent. Second, water easily dissociates into hydrogen (H^+) and hydroxyl (OH^-) ions. Water may therefore act as an acid or alkaline substance. But the influence of water is of such profound importance because it enters pores and cracks in bedrock, and the interconnected void spaces

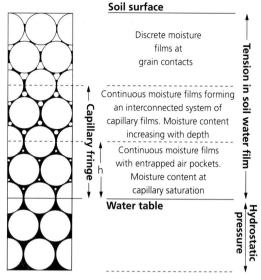

h=height of equivalent rise in a capillary tube

Fig. 3.1 The zones of moisture in regolith, showing the extent of the capillary fringe. After Carson 1969 [12].

of soils, to form continuous or discrete moisture films around grains. Water moves up and down in the weathered zone, the *regolith*, by downward percolation under gravity, and upwards by *capillary suction*. Capillary forces are inversely proportional to the diameter of the conduits or pores transmitting the water. As a result, moisture is drawn up from the water table into the overlying soil or regolith in a capillary fringe in which grains are surrounded by a continuous moisture film (Fig. 3.1). The capillary fringe varies in thickness from a few millimetres in sand-grade regolith, to many metres in clay-grade soils or regolith. Only in the regolith very near the surface are grains free from the effects of an enveloping and reactive moisture film.

3.2.1 Mechanical weathering

Physical weathering processes can be subdivided into those involving a volumetric change in the rock mass, and those involving volumetric changes caused by the introduction of material, commonly water but also salts, into pores, void spaces and fissures in a rock mass. Into the first category fall exfoliation, insolation weathering and hydration weathering. In the latter are freeze–thaw processes and salt weathering.

Exfoliation is the spalling off of sheets of rock along lines of weakness. The weaknesses may be caused by the presence of joints produced during unroofing.

Exhumation of a rock mass to lower pressures results in expansion due to stress release, causing joints to form. Other weaknesses may be related to the crystallographic or tectonic fabric of the rock. Lateral expansion of steep valley walls may also promote exfoliation, as in Half Dome in Yosemite Valley, USA. Rocks may also weather mechanically by *granular disintegration* into individual crystals.

Volume changes caused by temperature fluctuations may result in *insolation weathering*. The volume change accompanying a temperature change is given by the volumetric coefficient of thermal expansion. This coefficient varies according to rock type, with common rocks having values of $1.6–3.0 \times 10^{-5} K^{-1}$ (gabbro 1.6, granite 2.4, limestone 2.4, sandstone 3.0). A temperature increase of 50°C in a granite therefore only results in a minute 0.12% volume change. Large temperature variations, such as those between day and night in desert regions (diurnal changes of 50°C are not uncommon), at the surface of a rock are conducted to or from the rock interior. This sets up stresses which may overcome the yield strength of the rock, but the rate of conduction through rocks is slow. Rinds of chemically weathered material may undergo a volume increase compared to pristine rock, which causes the rind to spall off physically. Since the chemical changes generally involve hydration, especially in clays known as smectites (Section 3.2.2), this process is known as *hydration weathering*.

A very common mechanism of mechanical weathering in Arctic and mountain environments is the shattering of rock by the pressures exerted by ice growth, known as *frost weathering*. The volume change caused by the freezing of water is a 9% expansion. Experimental studies suggest that rapid freezing and rock moisture content are of critical importance. Saturated rocks are particularly prone to this type of mechanical weathering because pressures caused by ice growth cannot be dissipated by the displacement of water into air-filled cavities. Frost weathering is more intense in environments experiencing frequent freeze–thaw cycles than in polar regions suffering seasonal rather than diurnal variations. Frost action associated with ice lenses is also responsible for heave in unconsolidated soils, and may give rise to many kinds of patterned ground in regions with very cold winters.

In arid and semi-arid environments the precipitation and expansion of salt crystals in rock pores and cracks may cause physical breakdown. This is known as *salt weathering*. Concentration of ions in surface and soil water by high rates of evaporation leads to growth of crystals of a number of evaporite minerals, such as gypsum (hydrated calcium sulphate), mirabilite (hydrated sodium sulphate) and rock salt (sodium chloride). Minerals may also expand, once precipitated, by heating and by hydration. The latter process is well known in the transformation of the anhydrous calcium sulphate *anhydrite* to the hydrous form *gypsum*. Similar considerations apply to salt weathering as to frost weathering in terms of the need to overcome rock strength by the growth or expansion of salt crystals in void spaces. Where sediments containing saline solutions are unconsolidated, however, the sediments may be displaced by heave, giving rise to patterned ground characterized by polygonal salt wedges and upthrust ridges (*teepee structures*). The salts may be derived from sea spray blown inland, from precipitation, or as a product of chemical weathering. In arid climates, ephemeral lakes and ponds may become desiccated, leaving a layer of salt which may be subsequently deflated by the wind and used in salt weathering of bare rock surfaces. Salt weathering commonly causes granular disintegration.

3.2.2 Chemical weathering
The starting material: silicate minerals
The land surface consists essentially of a basement of continental igneous and metamorphic rocks, and a veneer, which is locally absent, of sedimentary rocks, comprising 80% of the total land surface. This sedimentary cover comprises carbonates (20%), sandstones (15%) and evaporites (5%) but is dominated by fine-grained mudrocks composed of clay minerals and quartz (60%). Weathering and erosion of the igneous and metamorphic basement and sedimentary cover results in a flux of material *via* river systems to depositional sites in the ocean. It is therefore perhaps unsurprising that the average compositions of continental rocks, sediment suspended in rivers, and sediment found on the sea floor are very similar (Table 3.1).

Most continental rocks are composed of silicates such as feldspars and quartz which have as their building blocks tetrahedra of SiO_4, with one silicon cation situated at the centre of a tetrahedron of oxygen anions (Fig. 3.2). It is important to pause to consider why SiO_4 tetrahedra are so important in crustal and Earth surface materials.

Although in some molecules individual atoms bond together by sharing electrons, many crystalline inorganic materials bond together by donating and accepting electrons from atom to atom. This is known as *ionic bonding*. An example is NaCl, which is formed when chlorine gains one electron and sodium loses

	Average upper continental crust (wt%)	Average sedimentary mudrock (wt%)	Average suspended load (rivers) (wt%)
SiO_2	66.0	62.8	61.0
TiO_2	0.5	1.0	1.1
Al_2O_3	15.2	18.9	21.7
FeO	4.5	6.5	7.6
MgO	2.2	2.2	2.1
CaO	4.2	1.3	2.3
Na_2O	3.9	1.2	0.9
K_2O	3.4	3.7	2.7
Total	99.9	99.9	99.4

Table 3.1 Average composition of upper crustal continental rocks, mudstones derived by weathering from upper crustal continental rocks, and suspended load of rivers (average of Amazon, Congo, Ganges, Garronne and Mekong). Data from Taylor & McLennan (1985) [3].

(a)

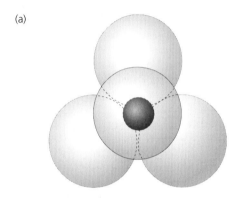

(b)

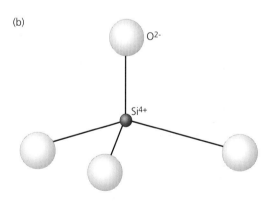

Fig. 3.2 Structure of the SiO_4 tetrahedron (a), with the bond lengths much exaggerated in (b) to clarify structure.

one electron so as to achieve stable atomic structures, producing a compound which is electrically neutral. It is common for ionic bonding to involve the acceptance and donation of one, two or three electrons, but atoms requiring transfers of more than three electrons generally bond by sharing of electrons in the outer shell—that is, by *covalent bonding*. Gases, water and organic compounds are typically formed by covalent bonds.

In ionically bonded crystals the densest possible packing arrangement of anions, represented by equal-sized spheres, is of stacks of planar layers involving mutual contact between six spheres (hexagonal symmetry). In such an arrangement there are characteristic gaps between neighbouring spheres. If we joined the centres of spheres by straight lines, there would be two possible geometries: the first would produce a tetrahedron involving four adjacent spheres; and the second would produce an octahedron involving six adjacent spheres (Fig. 3.3). Now cations in ionic crystals occupy these tetrahedral or octahedral sites, and the type of cation that will fit into these sites is determined by the ratio of the ionic radii of the cation and anion:

$$r_{cation}/r_{anion} \tag{3.1}$$

which must be 0.414 for an exact fit in an octahedral site giving the strongest possible bond. In real crystals, the ionic radius ratio is somewhat less or more than 0.414, but this weakens the compound. If the ratio of the radii is smaller, the bond length is too long, causing the structure to collapse into a more tightly packed geometry. If the ratio is larger, the packing of anions is expanded because of the presence of a large cation in the octahedral site. When the cation is sufficiently large (at a ratio of 0.732), it may eventually touch eight equidistant neighbours, producing a second stable configuration with an optimum bond length.

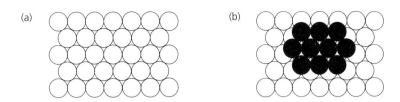

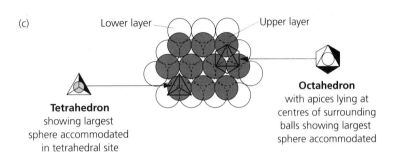

Fig. 3.3 Tetrahedral and octahedral coordination in silicates. (a) Spheres in a planar layer with hexagonal symmetry. (b) A second layer of spheres stacked on the layer in (a) so that each sphere in the upper layer fits into the depression between three spheres in the lower layer. (c) Lines drawn between centres of spheres define polyhedra — either tetrahedra or octahedra. After Andrews *et al.* (1995) [4].

This theory can now be directly applied to silicates where the layered stacks of spheres comprise oxygen anions. Which cations fit within these layers of oxygen anions depends on their relative ionic radii and the geometry of coordination (Table 3.2). It can be seen that tetrahedral coordination with oxygen is predicted for silicon Si^{4+} because, being a small cation, it fits well into the small tetrahedral sites. On the other hand,

Table 3.2 Coordination of different cations with oxygen on the basis of the radius ratio. From Raiswell *et al.* (1980) [5].

Critical radius ratio	Predicted coordination	Ion	Radius ratio, $r_{cation}/r_{O^{2-}}$	Commonly observed coordination numbers
	3	C^{4+}	0.16	3
	3	B^{3+}	0.16	3, 4
0.225				
	4	Be^{2+}	0.25	4
	4	Si^{4+}	0.30	4
	4	Al^{3+}	0.36	4. 6
0.414				
	6	Fe^{3+}	0.46	6
	6	Mg^{2+}	0.47	6
	6	Li^+	0.49	6
	6	Fe^{2+}	0.53	6
	6	Na^+	0.69	6, 8
	6	Ca^{2+}	0.71	6, 8
0.732				
	8	Sr^{2+}	0.80	8
	8	K^+	0.95	8–12
	8	Ba^{2+}	0.96	8–12
1.000				
	12	Cs^+	1.19	12

octahedral coordination (6 adjacent spheres) with oxygen is predicted for a number of common metal cations such as Fe^{3+} and Fe^{2+}, Mg^{2+}, Na^+ and Ca^{2+}. Other ions too big to occupy octahedral sites, such as Sr^{2+} and K^+, require crystals with more open packing arrangements such as cubic.

When Si^{4+} combines with four oxygen (O^{2-}) neighbours, there is a net charge of -4. In other words, only half of the bonding capacity of the four oxygens is satisfied by the silicon, leaving additional bonding capacity unsatisfied. The remaining bonds can be used by attracting other elements which have ionic bonding with oxygen, such as Mg^{2+}. Alternatively, each oxygen anion may combine with two silicon ions in covalent bonds rather than with one Si^{4+} ion. This formation of networks of silicon and oxygen is called *polymerization*, and silicates are normally classified according to the degree to which it occurs (Fig. 3.4). The order of increasing polymerization is as follows.

• *Monomer silicates* are made of isolated SiO_4 tetrahedra bonded to metal cations. Consequently, each oxygen anion is bonded to just one Si^{4+} cation. This is expressed as having four non-bridging oxygens. Minerals of this type, such as olivine and garnet, are known as *orthosilicates*, with the general formula SiO_4.

• *Chain silicates* consist of chains of linked SiO_4 tetrahedra in which each tetrahedron shares two of its oxygens. There are therefore two non-bridging oxygens per tetrahedron, producing a Si:O ratio of 1:3. Chain silicates, such as the *pyroxene* group, have the general formula SiO_3.

• *Double chain silicates* are formed by cross-linking of single silicate chains so that alternate tetrahedra share an oxygen with the neighbouring chain. For every four tetrahedra, two share two oxygens and two share three oxygens. Consequently, the structure has 1.5 non-bridging oxygens, giving a Si:O ratio of 4:11 and a general formula of Si_4O_{11}. Double chain silicates include the *amphibole* group.

• *Sheet silicates* involve the cross-linking of chains into sheets whereby every tetrahedron shares three oxygens with neighbouring tetrahedra. This produces a structure with only one non-bridging oxygen, giving a Si:O ratio of 4:10 and a general formula Si_4O_{10}. Sheet silicates include the *mica* group and *clay minerals*.

• *Framework silicates* are composed of a three-dimensional network where every oxygen is shared between two tetrahedra. There are consequently no non-bridging oxygens, giving a Si:O ratio of 1:2, and a very simple general formula of SiO_2. This is the

(a) **SiO_4 tetrahedron**

Apex up   Apex down

(b) **Simplified structure for monomer silicate—olivine**

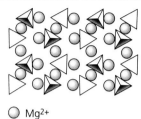

○ Mg^{2+}

(c) **Chain silicate—e.g. pyroxene**

End view

(d) **Double chain silicate—e.g. amphibole**

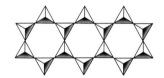

(e) **Sheet silicate (hexagonal rings)—e.g. mica**

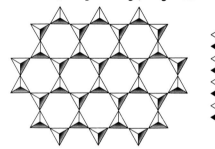

Fig. 3.4 Increasing degree of polymerization of adjacent tetrahedra in silicates. After Gill (1989) [6].

formula for quartz. Aluminium Al^{3+} is a small ion which fits well into tetrahedral sites and may substitute for Si^{4+}. Consequently, many framework silicates occur as aluminosilicate minerals, such as the very commonly occurring *feldspar* group. The charge imbalance caused by substituting the trivalent Al^{3+}

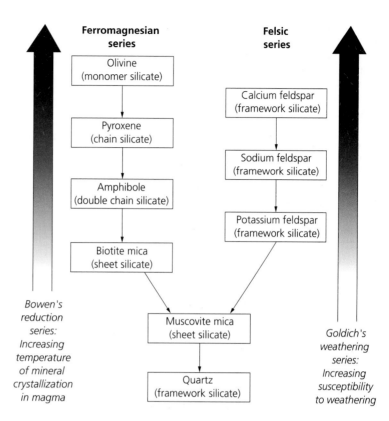

Fig. 3.5 The Goldich series of ease of weathering as the inverse of Bowen's reaction series. Minerals formed at high temperatures dominated by ionic bonding are more susceptible to weathering.

for the tetravalent Si^{4+} is neutralized by the presence of other cations. For example, K^+ may be incorporated for every substituted tetrahedral Al^{3+} ion, giving $KAlSi_3O_8$, known as *orthoclase*, which is a member of the feldspar group.

This structural organization of the silicate minerals is of great importance in controlling the reactivity of the mineral and therefore its behaviour during weathering. This is expressed in the Goldich series [7] (Fig. 3.5), which effectively mirrors the well-known Bowen's reaction series.

Processes of chemical weathering

There are a number of processes important in chemical weathering, though combinations of these processes commonly occur together.

Solution This involves the action of water as a solvent, breaking down the ionic bonds. The tendency for a mineral to dissolve in water is expressed by its *equilibrium solubility*. It is affected by the temperature and pH of the local environment (Fig. 3.6). For example, quartz has a low solubility below a pH of 10, and is highly soluble in very alkaline waters above this

value. Alumina (Al_2O_3) is only soluble in conditions seldom found in nature, below a pH of 4 and above a pH of 9. Consequently, alumina accumulates as a residue during weathering, whereas silica may be slowly leached. Calcium carbonate ($CaCO_3$), in contrast, has a steadily decreasing solubility in alkaline waters. However, the low solubility of $CaCO_3$ in pure water is rarely applicable in the natural environment, because dissolved CO_2 in water causes $CaCO_3$ to be replaced by calcium bicarbonate $Ca(HCO_3)_2$, which is more highly soluble (see 'Acid Hydrolysis' below).

The reaction rate during dissolution slows as the water in close proximity to the grain surface becomes saturated with reactants. Consequently, where minerals are highly soluble, the rate of throughput of unsaturated water is also an important parameter in explaining rates of dissolution. A measure of the throughflow of solute compared with the rate of diffusion down a concentration gradient is expressed by a dimensionless parameter known as the *Peclet number*,

$$Pe = \frac{\bar{u}L}{\kappa} \tag{3.2}$$

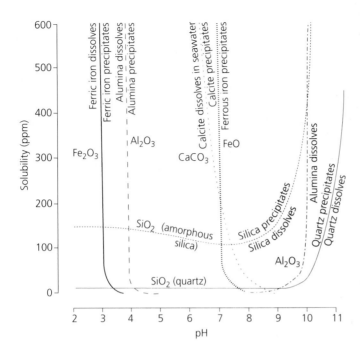

Fig. 3.6 Equilibrium solubility (in parts per million) of crystalline quartz, amorphous silica, alumina, ferrous and ferric iron, and calcium carbonate as a function of pH. Modified from Summerfield (1991) [8]

where $\bar{u}$ is the mean velocity (m s^{-1}) of the flow, known as the *Darcy velocity* for flow through a porous medium, κ is the diffusivity (m^{2}s^{-1}) and L is the length scale for the transmitting medium, such as aquifer or regolith thickness.

Oxidation and reduction These involve the gain or loss of charge by the addition (reduction) or loss (oxidation) of negatively charged electrons. The oxygen dissolved in water is the most common oxidizing agent. Oxidation results in the formation of oxides and hydroxides, such as those of iron. An important example is the oxidation of ferrous iron to ferric iron

$$4FeO + O_2 \rightarrow 2Fe_2O_3 \qquad (3.3)$$

Sulphides such as pyrite (FeS$_2$) are common in sedimentary rocks such as black shales, in mineral veins and in coal deposits. Spoil heaps of waste products from mining operations may be particularly rich in sulphides which are oxidized by the atmosphere or groundwater. Oxidation under aerobic conditions produces sulphuric acid:

$$2FeS_{2(s)} + 7\tfrac{1}{2}O_{2(g)} + 7H_2O_{(l)} \rightarrow 2Fe(OH)_{3(s)} \\ + 4H_2SO_{4(aq)} \qquad (3.4)$$

Consequently, the drainage from old mine workings is commonly highly acidic (pH as low as 1 or 2). Oxidation is catalysed by the activity of bacteria (see

below), commonly resulting in an iron oxide, goethite (FeOOH), which can be seen as a distinctive yellow-orange crust.

The sulphuric acid may also react with calcium carbonate, for example from fossils, to form gypsum (CaSO$_4$.2H$_2$O), accompanied by a major increase in volume. Pyritous shales may therefore have white efflorescences of gypsum on their surfaces due to oxidative weathering.

Anaerobic conditions may cause the reduction of ferric iron to ferrous. Since ferrous oxides are more soluble, they may be mobilized and removed in drainage. This explains why waters draining anaerobic bogs are stained brown by the presence of reduced iron oxides.

The oxidation of organic matter in soils is caused by bacteria, producing CO$_2$ and therefore generating acidity. Carbon dioxide levels in soils may be many times the amount expected from equilibrium with atmospheric CO$_2$ concentrations. The acidity generated is then used in the acid hydrolysis of silicate minerals (see below).

The tendency for oxidation or reduction to take place is indicated by the *redox potential* (Eh), measured in millivolts. Abundant oxygen dissolved in pore waters leads to strongly oxidizing conditions (positive Eh), but in soils where oxygen has been depleted by bacterial activity and decomposition, reducing conditions (negative Eh) may occur. The redox potential also varies with pH, as shown in Fig. 3.7.

Hydration This is a process whereby minerals absorb water into their crystal structures. This absorption of water may make the lattice more porous, and therefore more susceptible to weathering. A common weathering reaction involving hydration is the transformation of the iron oxide hematite to the hydrated iron hydroxide limonite:

$$2Fe_2O_3 + 3H_2O \leftrightarrow 2Fe_2O_3.3H_2O \qquad (3.5)$$

Acid hydrolysis This is the reaction of a mineral with acidic weathering agents, where the acidity may be derived from a variety of sources, such as:
• the dissociation of atmospheric CO_2 in rainwater, giving carbonic acid (H_2CO_3);

• the dissociation of CO_2 in soil zones, again producing carbonic acid (the carbon dioxide is produced predominantly by the respiration of plant roots and the bacterial decomposition of plants);
• sulphur dioxide emitted naturally from volcanoes and anthropogenically from industry giving carbonic acid and sulphuric acid.
The hydrolysis involving CO_2 is commonly termed *carbonation*.

Hydrolysis involves the replacement of metal cations in the crystal lattice such as K^+, Na^+, Ca^{2+} and Mg^{2+} by the hydrogen or hydroxyl ions of water. The released cations combine with further hydroxyl ions, commonly to form clay minerals. An example of a possible weathering reaction involving hydrolysis is the weathering of albite (plagioclase feldspar, $NaAlSi_3O_8$), a mineral abundant in granites, to the clay mineral *kaolinite*, $Al_4Si_4O_{10}(OH)_8$, with the

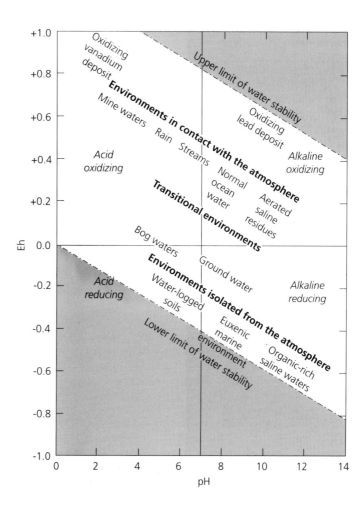

Fig. 3.7 The Eh–pH conditions found in nature, with the common minerals forming under various Eh–pH conditions. Data from Bass Becking *et al.* (1964) [9].

release of silica and sodium and hydroxyl ions in solution:

$$4NaAlSi_3O_8 + 6H_2O$$
$$\rightarrow Al_4Si_4O_{10}(OH)_8 + 8SiO_2 + 4Na^+ + 4OH^- \quad \textbf{(3.6)}$$

The production of hydroxyl ions in solution suggests that pore waters should be alkaline, but they are commonly neutral or slightly acid, which indicates that pure hydrolysis is not the dominant process at work. It is more likely that the abundance of soil CO_2 favours carbonation.

Carbon dioxide dissolves in water forming a weak carbonic acid which may dissociate into H^+ and bicarbonate (HCO_3^-) ions. The hydrolysis reaction for albite can now be rewritten in a more realistic form to take account of the production of bicarbonate ions:

$$4NaAlSi_3O_8 + 6H_2O + 4CO_2$$
$$\rightarrow 8SiO_2 + Al_4Si_4O_{10}(OH)_8 + 4Na^+ + 4HCO_3^- \quad \textbf{(3.7)}$$

A very similar reaction accounts for the breakdown of the potassium-rich orthoclase feldspar to the common clay mineral illite:

$$6KAlSi_3O_8 + 4H_2O + 4CO_2$$
orthoclase
$$\rightarrow K_2Al_4(Si_6Al_2O_{20})(OH_4) +$$
illite
$$12SiO_2 + 4K^+ + 4HCO_3^- \quad \textbf{(3.8)}$$
silica

with the removal of silica and potassium and bicarbonate ions in solution. The silica may be removed as a silicic acid in humid environments where water is plentiful, leaching is intense and pHs are very acid because of the activity of plants. This results in the formation of a kaolinitic residue.

Carbonation dominates the weathering of limestones. Calcite dissolves in the weak carbonic acid produced by the reaction of water and carbon dioxide. The weathering reaction

$$CaCO_{3(s)} + H_2CO_{3(aq)} \Leftrightarrow Ca_{(aq)}^{2+} + 2HCO_{3(aq)}^- \quad \textbf{(3.9)}$$

is dependent on the amount of CO_2 available. Adding CO_2 causes the formation of more carbonic acid (the reaction runs forward to the right) which dissolves more limestone. Reducing the CO_2, on the other hand, causes the reverse reaction (running to the left) and the precipitation of limestone. The forward reaction is responsible for the dissolution of limestone in karst terrains. The reverse reaction is responsible for the familiar stalactites and stalagmites in caves and travertines at springs caused by the outgassing of CO_2 into the atmosphere from groundwater. The precipitation of calcium carbonate in soil zones as

nodules and layers gives rise to a hardened soil type known as calcrete (Section 3.2.3).

Cation exchange This is the exchange of one cation for another, and is particularly common in clay minerals which have loosely bonded cations on their surfaces. Clay minerals may therefore change from one type to another rather easily (see below).

Chelation This refers to the mobilization of metal cations by organic compounds produced directly by secretion by organisms, or through microbial decomposition of humus in soils. Chelation may release cations such as Fe^{3+} and Al^{3+} that are almost insoluble under normal Eh–pH conditions. Bacteria may be responsible for the oxidation and concentration of iron and manganese in weathering films known as *desert varnish* or *rock varnish*.

The regolith and its constituents

The weathered mantle overlying pristine bedrock, the *regolith*, varies greatly in thickness according to the trade-off between two rate effects: the rate of bedrock weathering, which is strongly controlled by climatic factors; and the rate of removal by denudation, which is determined by climatic, tectonic and topographic factors. Regoliths may be particularly thick (over 100 m) in localities with humid, warm climates and subdued topography. Regolith growth is, however, self-limiting, since the development of a thick weathered mantle decreases permeabilities and reduces the throughput of water (and therefore Peclet numbers) to the weathering front at the contact between unaltered bedrock and regolith. Rates of chemical weathering of bedrock then become minimal. Chemical changes under these conditions of low bedrock weathering are restricted to intense leaching of the soil zone. Mountain slopes in humid climates rarely reach their maximum or potential regolith thicknesses, despite high rates of chemical weathering at the weathering front. This is because the regolith is removed during slope processes (overland flow, saturated throughflow, rill and gully development, soil creep, debris flows and landslides). There is therefore a profound difference between the regolith thickness, which for optimal development reflects landscape stability, and the rate of denudation, which is an indicator of active tectonics generating topography.

The mineralogical composition of regolith is determined not only by the type and intensity of chemical weathering processes, controlled essentially by climatic factors, but also by the parent bedrock.

There are considerable differences in the way basalts and granites weather under the same climate. Most studies are focused on the weathering of granites under tropical and subtropical conditions.

Clay minerals The chemical weathering of bedrock results in the selective breakdown of pre-existing minerals, and in the growth of secondary minerals. The reason why susceptibility to weathering and temperature of crystallization are closely related is the strength of the bonding between oxygen and cations in the mineral structures. The most common secondary minerals are clay minerals, sheet silicates with complex layered structures of silica, alumina and various metal cations, sometimes with layers of water. In addition, a number of oxides and hydroxides commonly occur as secondary minerals. Table 3.3 gives commonly occurring secondary minerals.

Clay minerals composed of layered silicates are built of tetrahedra or octahedra of O^{2-} ions, sometimes in combination with OH^- ions, with a centrally located Si^{4+} ion. Differences among the layered silicates depend on whether tetrahedra or octahedra are present, the way in which these structural units are linked into the general structure of the mineral (as a chain, as a sheet or in a ring), and the cations co-ordinated with the oxygen or hydroxyl ions. Al^{3+}, Na^+,

K^+, Ca^{2+} and Fe^{2+} are commonly introduced into the mineral structure to form the common layered silicate clay minerals (Table 3.3).

The main classes of layered silicate minerals are defined by the way in which tetrahedral and octahedral sheets are arranged. A combination of one tetrahedral sheet and one octahedral sheet produces a 1:1 crystal lattice (kaolinite and halloysite), a combination of two tetrahedral and one octahedral sheets is known as a 2:1 crystal lattice (illite, vermiculite, montmorillonite), and a combination of double sheets of tetrahedra and octahedra is a 2:2 crystal lattice (the chlorite group, not found in weathering zones). Ionic substitution is common in 2:1 minerals, but not in 1:1 structures.

The thickness of one layer (comprising two or more sheets) plus the interlayer space may be expanded by the introduction of water or organic molecules between the layers. Clay minerals where this happens are said to be in a swollen or expanded state, and are termed *expandable clays*. Montmorillonite is the best example.

The crystalline or amorphous oxides and hydroxides of iron and aluminium are known as *sesquioxides* and occur in particular abundance in strongly leached soils in the humid tropics.

The clay minerals present in a weathering profile are sensitive indicators of the parent bedrock, hydrology

Table 3.3 The commonly occurring secondary minerals.

Mineral group	Composition	Common varieties
Clay minerals (layered silicates)		
Kaolinite group	Hydrous aluminium silicate	Kaolinite
Illite group	Hydrous potassium silicate	
Smectite group	Complex hydrous magnesium aluminium silicate	Montmorillonite
Chlorite group	Hydrous silicate of aluminium, iron and magnesium	
Vermiculite group	Hydrous magnesium silicate	
Palygorskite and sepiolite group	Hydrous magnesium silicate	
Mixed-layer group	Hydrous silicate	Illite–smectite mixed layers
Oxides and hydroxides (sesquioxides)		
Oxides of silicon	Silicon dioxide	Quartz, amorphous silica, opaline silica
Hydroxides of aluminium	Hydrous alumina	Gibbsite, boehmite
Oxides and hydroxides of iron	Iron oxide and hydrous iron oxide	Hematite (oxide) and goethite (hydroxide)
Oxide of titanium	Titanium dioxide	Anatase

and geochemical environment during chemical weathering. A primary factor is the extent of leaching, which strips minerals of their metal cations and eventually of their silicon and iron, leading to a stable aluminium-rich residue (gibbsite). Large throughputs of water are required to cause advanced stages of leaching. Where leaching is less intense, the cations released by bedrock breakdown promote the formation of cation-bearing clay minerals such as those of the illite and smectite groups. However, the extent of leaching varies not only between climatic zones, but also within a single weathering profile. The flux of water generally decreases with depth because of reduced permeability caused by compaction of the regolith, infiltration of fines, and reduction in biological activity. Consequently, there are commonly variations in the clay minerals present with depth. Residual minerals such as kaolinite and gibbsite are found at the top of weathering profiles, whereas smectite and illite are found in deeper zones where leaching is less intense. The evolution of the regolith over time may cause a progressive development of different clay minerals as, for example, regolith material is first subjected to low leaching intensities close to the weathering front, and then, as the weathering front descends into the bedrock and as surficial erosion takes place, higher leaching intensities characteristic of near-surface zones occur.

The effect of parent rock is also important. Bedrocks lacking aluminosilicates, such as sandstones, will be dominated by aluminium-rich oxides, hydroxides and silicates, such as boehmite, gibbsite, hematite, goethite and kaolinite.

Summary: controls on the rate of chemical weathering reactions and global patterns

The previous discussion has highlighted the role of the biosphere in chemical weathering, primarily in the acidification of waters by the CO_2 released by biological activity in soils. However, other climatic factors affect the rate of chemical weathering reactions in their effect on temperature and availability of water, and underlying bedrock composition controls the minerals available for weathering and the chemistry of pore waters. The reaction kinetics of individual minerals must also have a role in determining rates of chemical weathering. These factors operate at different scales in a complex system with many feedbacks. Consequently there are grave risks in overgeneralization. But at a very large scale, there appears to be a rough latitudinal zonation, as illustrated in Figs 3.8 and 3.9. Very high rates of chemical weathering always

result from a favourable combination of the factors summarized briefly below.

- **Organic activity in soils**: the presence of soil organic matter and its decomposition increases CO_2 concentrations in waters moving through the zone of weathering. The organic matter in soils increases retention of water in soils and, together with the fine mineral matter present, an open structure with high surface area for attack by acidic waters. The release of organic acids and the supply of carbon dioxide to soils from the decomposition of litter vary according to ecosystem, with very high values in tropical rainforests and minimal values in the semi-arid and arid subtropics. Biological activity is thought to enhance weathering rates compared to abiotic situations by orders of magnitude.

- **Climatic effects—temperature and availability of water**: the energy supplied by heat speeds up chemical reactions, causing a doubling of the reaction rate for a 10°C increase in temperature. There should therefore be important latitudinal variations in chemical weathering rates between the poles and the equator due to this effect. However, climatic variations are responsible for marked variations in rainfall as well as temperature. These in turn determine the density and type of natural vegetation and the level of biological activity in the soil. Since water is a solvent, the enclosure of grains in a capillary fringe promotes rapid chemical weathering. Dry subtropical deserts, therefore, despite being hot, have low rates of chemical weathering. In the humid tropics, on the other hand, high amounts of flushing of soil zones by water and high temperatures promote the leaching out of the soluble components.

- **Kinetics of mineral reactions**: the chemical weathering of minerals requires the pore waters to be undersaturated with respect to the mineral being weathered. Once saturation is approached, high flow rates of water will not cause further reaction. The weathering rate of relatively insoluble minerals such as silicates and carbonates is determined by the rate at which ions can be detached from crystal surfaces rather than by the rate of removal of the solute. The weathering of highly soluble minerals such as evaporites, where ions detach rapidly from crystal surfaces, however, is determined by the rate of flushing of the saturated solution that builds up adjacent to the crystal surfaces. The rate-limiting factor in this case is the rate of water flow rather than the mineral reaction kinetics.

- **Bedrock composition**: the type of parent rock determines the stability of the mineral components to

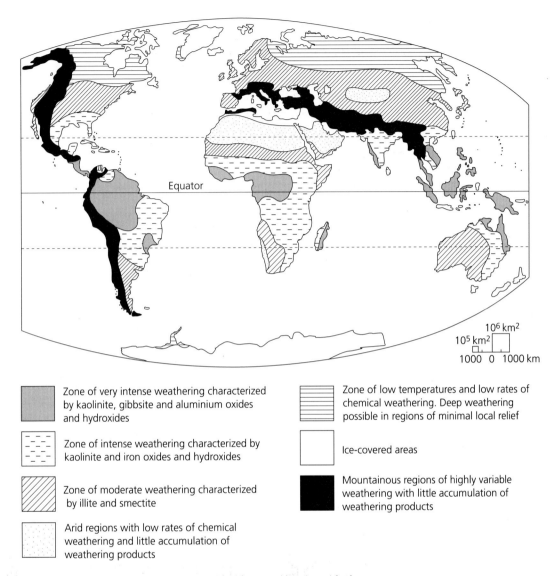

Fig. 3.8 The global pattern of weathering. Modified from Strakhov (1967) [10].

Legend:

- Zone of very intense weathering characterized by kaolinite, gibbsite and aluminium oxides and hydroxides
- Zone of intense weathering characterized by kaolinite and iron oxides and hydroxides
- Zone of moderate weathering characterized by illite and smectite
- Arid regions with low rates of chemical weathering and little accumulation of weathering products
- Zone of low temperatures and low rates of chemical weathering. Deep weathering possible in regions of minimal local relief
- Ice-covered areas
- Mountainous regions of highly variable weathering with little accumulation of weathering products

weathering through their degree of polymerization. Monomer silicates have weak ionic bonds between the metal cations and oxygens which are easily broken, so minerals such as olivine are very easily weathered. At the other end of the range, framework silicates are robust in their bonding, so quartz is extremely resistant to chemical weathering. This is reflected in the Goldich series (Fig. 3.5). Parent material also affects chemical weathering by its effects on regolith permeability, which strongly influences rates of chemical change.

- **Topography**: this is important in influencing the rate of loss of regolith by denudation, which serves con-tinually to invigorate the chemical weathering process. It also controls the flow rate of water through the regolith. Flow rates are high for a well-drained slope, and very low for a flat, poorly drained landscape.

- **Time**: this is required for chemical changes to take place, and weathering profiles are seldom completely in equilibrium with surface conditions because of the rapidity of climatic change. In some regions, such as central Australia, deep weathering profiles have been inherited from an earlier more humid climatic phase.

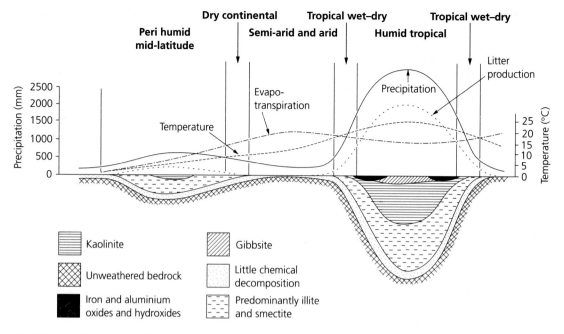

Fig. 3.9 Latitudinal zonation of regolith thickness and type. Modified from Strakhov (1967) [10].

Karstification

The term *karst* describes a landscape exhibiting a number of features caused by dissolution of limestone, such as a lack of surface drainage, a subterranean system of caves and fissures, and a thin and discontinuous soil. The development of karst landscapes, *karstification*, is favoured by the presence of thick, mechanically strong, jointed limestones in the subsurface, a humid climate, and sufficient relief for rainwater to travel through to a deep water table.

A host of small-scale solutional forms known as *karren* are found at the surface of karst terrains. Larger features include closed-surface depressions known as *dolines*, some of which originate by the collapse of the roof of a cave, some by enhanced solution at vulnerable points in the limestone fabric such as the intersection of faults, and some at the sink holes of streams where they disappear into the underground drainage system. There are many other characteristic solutional landforms associated with different stages of karst development.

3.2.3 Soils

A very large part of the Earth's landscape is covered by soil. Interaction of the land surface with the hydrosphere and atmosphere must therefore take place through the medium of soils. Soils have enormous

importance in agriculture and food production, economics and thus also politics, but this human dimension is not discussed here. The physics, chemistry, biology and geography of soils is a vast subject that can only be touched upon in this text. The interested reader can turn to a large number of books on soils, some of which are listed in the Further Reading section at the end of this chapter. What is of interest here is the role of soils, in the coupling between atmosphere and the terrestrial surface, the information they contain on environmental change, and the problems caused by their deterioration and loss (Section 3.6).

Basic taxonomy of soils

As well as having particular physical, chemical and biological properties, soils are characterized by a vertical structure made up of layers, or soil horizons. These soil horizons are the building blocks of soil classification. Near the surface organic matter is concentrated which transforms into humus, partly through the decomposing activities of micro-organisms. At the base of the soil profile is consolidated bedrock or unconsolidated substrate. The soil horizons in between are characterized by a number of weathering processes bringing about mineralogical changes. In general, weathering causes a leaching or physical transport of constituents from upper horizons (*eluviation*) and accumulation in lower horizons (*illuviation*).

Soils are commonly classified by: their colour, which reflects their chemical composition and organic content; texture, which is due to the physical size and sorting of particles in the soil; structure, which may be granular, platy, blocky or prismatic according to the arrangement of the clusters of soil particles known as *peds*; consistency, which refers to the cohesion in soil, so that soils may be sticky, plastic, friable or cemented; porosity and permeability, which reflects the fractional volume and interconnectedness of voids filled with water or air, often produced by the activity of roots and tunnelling animals, or by ploughing; and moisture level, ranging from dry for large portions of the year, to almost permanently wet.

Russian soil scientists such as V.V. Dukachaev first recognized in the second half of the nineteenth century that soils were a response to linked climatic, physical and biological processes and the passage of time. Since then a number of soil classification systems have been devised, such as that of the US Soil Conservation Service [11]. This exhaustive and comprehensive manual identifies 11 general soil orders which can be mapped world-wide (Fig. 3.10).

1 Oxisols: found in hot and humid tropical regions and characterized by high amounts of leaching leading to *laterization*.

2 Aridisols: of the hot, dry deserts in which little alteration of parent material takes place, but where illuviation of carbonates and evaporites may take place. Carbonate illuviation leads to the hard crusts known as calcretes.

3 Mollisols: of the subhumid and semi-arid grasslands, known as *chernozems*, are rich in organic material with well-developed horizonation.

4 Alfisols: are grey-brown moderately weathered soils of humid temperate forests.

5 Ultisols: are red-yellow highly weathered soils of subtropical forests.

6 Spodosols: otherwise known as *podzols*, are the soils of northern conifer and cool humid forests, with a highly leached horizon and lower illuviated horizon rich in Fe/Al clays.

7 Entisols: are recent soils which have no developed profile and many inherited properties from the parent rock; they are found in all climates, but particularly in the tundra.

8 Inceptisols: are embryonic in development and are found in some humid regions and under subarctic forests.

9 Vertisols: or *tropical black clays*, are characterized by expandable clays in the tropics and subtropics, which cause large cracks to develop during dry periods.

10 Histosols: such as are found in peatlands and bogs have in excess of 20% organic matter.

11 Andisols: form in areas affected by frequent volcanic activity providing ash and volcanic glass for chemical weathering.

The typical processes and horizonation during laterization in oxisols, calcification in aridisols and mollisols, and podzolization are illustrated in Fig. 3.11.

Role of soils and silicate weathering in the Earth surface system

The consumption of CO_2 by the weathering reactions in soils should lower the partial pressure of CO_2 in the atmosphere. We know that atmospheric CO_2 concentrations have an important feedback on global climate through their greenhouse effect. Consequently, an increase in consumption of CO_2 in weathering should cause cooling.

It is clearly important to be able to quantify the strength or weakness of the negative feedback between increased temperature and CO_2 consumption. The weathering of silicate minerals is responsible for over half of the consumption of carbonic acid, the product of the reaction of CO_2 with water, in continental weathering [12]. Consequently, the weathering reaction rates of, for example, feldspars as a function of temperature need to be accurately determined. The natural weathering rates, however, are affected by a large number of factors of which temperature is only one. By comparing two drainage basins which are identical in underlying geology, vegetation and aspect, but of different elevation and therefore temperature, it is possible to better constrain the natural weathering rates of feldspars [13]. These results show that there is a strong negative silicate weathering mediated feedback between atmospheric CO_2 and temperature. This would modulate the wide temperature variations predicted by some global carbon cycle models [14].

Before the colonization of the land surface by vascular plants at about 400 Ma (late Silurian to early Devonian) it might be imagined that chemical weathering of the land surface was essentially abiotic. Since we know that biological activity greatly enhances chemical weathering, lower global rates of chemical weathering prior to 400 Ma may have been responsible for atmospheric warming. The multiple occurrence of major ice ages in the Proterozoic, Cambrian and Ordovician casts some doubt on this

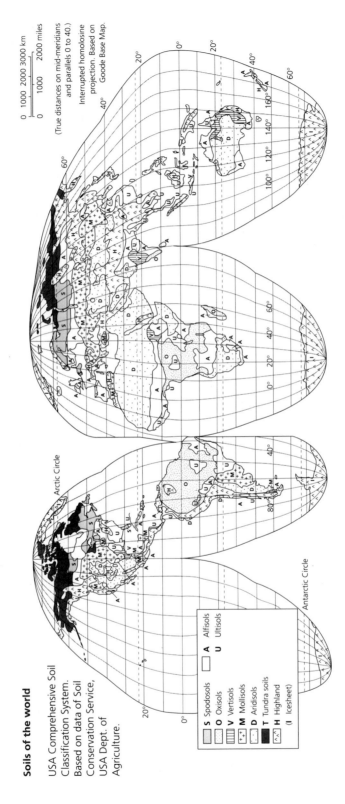

Soils of the world

USA Comprehensive Soil
Classification System.
Based on data of Soil
Conservation Service,
USA Dept. of
Agriculture.

S	Spodosols	**A**	Affisols
O	Oxisols	**U**	Ultisols
V	Vertisols		
M	Mollisols		
D	Aridisols		
T	Tundra soils		
H	Highland		
	(**I** Icesheet)		

Arctic Circle

Antarctic Circle

0 1000 2000 3000 km
0 1000 2000 miles

(True distances on mid-meridians
and parallels 0 to 40.)
Interrupted homolosine
projection. Based on
Goode Base Map.

Fig. 3.10 Distribution of major soil orders. Adapted from US Soil Conservation Service (1975) [11].

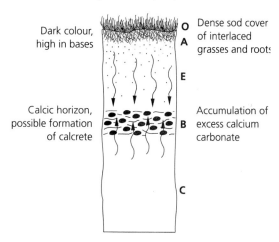

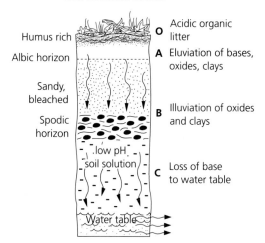

Fig. 3.11 Laterization, podzolization and calcification in soils. Laterization is common in highly leached, humid tropical and subtropical climates. Calcification typifies climates where evapotranspiration exceeds precipitation, producing aridisols and mollisols. Podzolization takes place in cool and moist climates.

proposal. It is possible that despite the absence of vascular plants the land surface before 400 Ma may have been occupied by primitive algae, bacteria and lichens which consumed soil CO_2, kept chemical weathering rates high and promoted global cooling.

It has been proposed that periods of mountain building are associated with increased chemical weathering of silicates, which has the effect of drawing down atmospheric CO_2 [15, 16]. The increase is thought to result from increased rainfall from a strengthened monsoon on the newly erected mountains and the accelerated erosion of the steeper terrain. Deep sea cores indicate a lowering of the calcium carbonate compensation depth during the late Cenozoic (c. 5–10 Ma). This has been attributed to the rapid uplift of the Himalayas causing an increased flux of calcium to the ocean by transport as solute load in rivers. If the weathering of silicates also showed a marked increase in the late Cenozoic, a large increase in the CO_2 sequestered from the atmosphere may have resulted, driving global cooling. The effect may have been strong enough to have triggered the northern hemisphere glaciation.

Use of soils as a record of environmental change

Soils are a sensitive indicator of climate change. Soil profiles preserved over long periods of time but which are not being actively developed are known as *palaeosols*. Certain characters of a soil are more likely to be preserved than others. The zones of illuviation are particularly liable to be fossilized. These are termed *duricrusts* (Fig. 3.12). Through their mechanical durability they commonly form distinct and extensive surfaces in the landscape, often capping hillslopes. A number of duricrusts occur with distinctive mineralogies. The most important components are oxides and hydroxides of iron and aluminium, as well as silica, calcium carbonate and gypsum.

Duricrusts rich in aluminium and iron are termed *alcretes* and *ferricretes*, respectively, and form in the humid to subhumid tropics. Weathering materials rich in aluminium and iron are commonly called *laterites*. Relative enrichment in aluminium is favoured by high rainfalls and intense leaching causing the loss of more mobile components. Enrichment at levels high enough for commercial extraction results in *bauxites*. Laterization may be promoted by deforestation and soil erosion following removal of natural vegetation.

Duricrusts rich in silica, *silcretes*, are found in both humid and arid tropical environments. Calcium carbonate concentration in *calcretes* is found especially in semi-arid climatic zones with rainfall between 200 mm and 600 mm. Gypsum duricrusts, or *gypcretes*, are confined to very arid regions with annual precipitation below 250 mm. Silcretes, calcretes and gypcretes form through absolute concentration, rather than relative enrichment, in silica, calcium carbonate and gypsum. These minerals may be sourced from the weathering of bedrock, wind-blown detritus, groundwater, plant residues, rainfall and floods. Material may be transferred to the site of precipitation in solution by upward rise through the capillary fringe, or as solids by downward percolation. An appropriate geochemical environment, especially its pH, is essential for the solution and precipitation of the various cements. The alkalinity is important in explaining the occurrence of silcretes, since we know that SiO_2 has a very low solubility up to a pH of about 9. Silica may be dissolved in local areas of high pH, such as alkaline lakes in arid and semi-arid regions, and precipitated in weathering zones where the pH is lower. Silcretes may also form in the acidic conditions of humid tropical environments with high organic productivities by the mobilization of aluminium, thereby leaving a siliceous residuum. This goes some way to explaining the occurrence of silcretes in diverse climatic settings (Fig. 3.12).

Duricrusts take considerable times to form, and once formed, they are long-lived features of the terrestrial landscape. After formation, climate may change. This makes the precise linkage of climatic and other factors with the more resistant types of duricrust such as alcretes, ferricretes and silcretes difficult.

3.3 Sediment routing systems

The drainage basin is the natural unit within which many geomorphological processes operate. But the drainage basin is, in general, not a closed system for either water, dissolved load or particulate sediment transport. We must view the drainage basin as a topographic unit acting as a physical process–response system but with outputs, or fluxes, of material to the coast and beyond. A more satisfactory way of viewing the erosion, transport and deposition of sediment on the Earth's surface is the *sediment routing system*, sometimes referred to as a *denudation cell*. A typical sediment routing system consists of: a source region dominated by denudational processes such as hillslope erosion and river incision; a transportational zone dominated by sediment transfer, typically a river network; and a depositional region, dominated by the eventual deposition of the sediment supply, which may be a subaerial floodplain, a shallow marine environment such as a delta or continental shelf, or commonly the lower part of the continental slope and the abyssal plains of the deep sea. These zones are

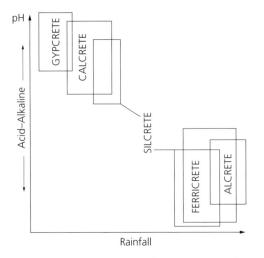

Fig. 3.12 Occurrence of duricrusts according to climate and pH. After Summerfield (1991) [8].

characterized by a dominance, but not exclusivity, of either erosion, transport or deposition. For example, river systems are in general part of the transportational zone, but may contain segments dominated by erosion and deposition. The advantage of considering entire sediment routeing systems is that they are, except for the very finest particles and the dissolved load, essentially closed systems for calculating sediment volumes. We can think of the erosional, transpor-

tational and depositional zones as distinct process subsystems. The fluxes of sediment through the routeing system are dependent not only on the processes operating within the subsystems, but also on the linking processes between the subsystems. This is particularly true of the transfer of sediment from the erosional zone to the transportational zone. The sediment transport through the river network will reflect the supply of material to it from upland denudation.

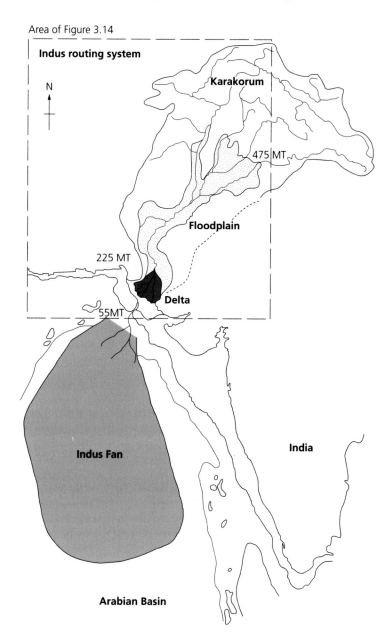

Fig. 3.13 The Indus sediment routing system, with sediment fluxes in megatonnes per year. After Milliman *et al.* (1994) [17].

The concept of the sediment routing system can be illustrated by the Indus system of northern India and Pakistan (Fig. 3.13).

3.3.1 The Indus sediment routing system

The Indus River is one of the world's largest rivers in terms of its drainage area and discharge. It drains the arid and semi-arid regions of the western Himalayas, including the high Karakorum. The Indus system receives tributaries from Afghanistan (Kabul River), from major antecedent rivers crossing the Himalayan chain (Sutlej and main Indus trunk stream), and from the frontal Himalayan ranges (Chenab, Jhelum, Ravi). On leaving the mountain range near Attock, the river flows across its floodplain for about 1000 km before reaching its delta near the head of the Arabian Sea. The delta feeds an enormous submarine fan 1500 km long and 960 km wide at its maximum. In summary, the entire sediment routing system has dimensions of the order of 4400 km for the length of the Indus River and Indus Fan combined, a drainage basin (catchment) area of nearly $1\,000\,000\,km^2$ for the subaerial Indus system and a very similar figure of $1\,100\,000\,km^2$ for the submarine fan (Fig. 3.13).

The upland parts of the Indus catchment are heavily glaciated (Chapter 11) and rugged in relief. Large areas are covered with loose moraine and outwash left from previous glacial advances. On active glaciers, huge amounts of eroded material accumulate from mass movements and snow avalanches. The summer heat melts the ice rapidly, especially in the deep valleys, producing great thicknesses of loose debris on the glacier surfaces and moraine downvalley. This morainic material is then easily transported by flood meltwaters. In addition, the very steep slopes are subject to failure, producing soil creep, landslides, debris flows and rock avalanches at a range of scales. Many major landslides have taken place in the middle and upper Indus, some unequivocally associated with earthquakes. Other mass movements are triggered by undercutting by rivers, or by saturation during periods of rain or snowmelt. All of these factors contribute to the very high sediment loads of the Indus river system.

The water and sediment discharges of the Indus system vary considerably between gauging stations. However, in general, discharges are lowest from November to mid-spring when water is locked up in snow and ice at high altitudes and little precipitation falls over the drainage basin. Discharges then increase due to snowmelt, and reach a maximum in July at the peak of the monsoonal rainy season. Some of the sediment load from the mountainous reaches of the Indus system rivers is deposited as the Indus valley flattens and broadens on leaving the Himalayan chain close to Darband (Fig. 3.14). This is predominantly sand-sized material.

Sediment loads increase as major tributaries add their load to the Indus. Some of this sediment is deposited as the Indus flows across its wide floodplain, and by the most downriver gauging station at Sehwan, some 300 km from the coast, 20–50% of the total sediment load has been deposited. This illustrates very well that the river system is an agent of transportation, but also one of selective deposition of the coarser grain sizes. This pattern has been severely disrupted by the activities of humans since the early 1950s, principally the various engineering activities along the Indus. This has resulted in a fall in discharge through the system as a whole, with dramatic effects immediately downstream from major dams. It is estimated that about 250 Mt of sediment, mostly silt and clay, reached the estuary annually prior to the phase of dam construction. By 1974–75 this had fallen to about 100 Mt per annum, and it is believed that present rates are negligible [17].

The present Indus delta occupies over 3000 km², but the presently active delta has shrunk to a tenth of this area as a result of diminished sediment discharges. Sedimentation in the northern Arabian Sea has been dominated by terrigenous input from the Indus River, though a wind-blown contribution from the deserts of Africa and Arabia is discernible. Much of the terrigenous supply is transported to the Indus Fan via the Indus submarine canyon. The submarine canyon, 15 km wide at its head, 170 km long and with an average depth of 800 m, efficiently routes sediment to the deep sea by deeply incising the shelf to a point just 6.5 km offshore from the main Indus River distributary. It is believed that the principal method of transport is by muddy turbidity currents. Loss of river-derived sediment by longshore transport is thought to be minimal.

Knowledge of the geological evolution of the Indus sediment routing system allows something to be said about average denudation rates, though there are many uncertainties. Mineralogical data from deep sea cores suggest that the Indus system came into existence in the Oligocene, presumably roughly at the same time as uplift in the Himalayas consequent upon the indentation of the Indian plate with Asia. In the submarine fan depocentre located closest to the delta, over 11 km of sediment has been imaged from seismic reflection surveys [18, 19]. Most of this sediment is thought to be late Oligocene and younger in age.

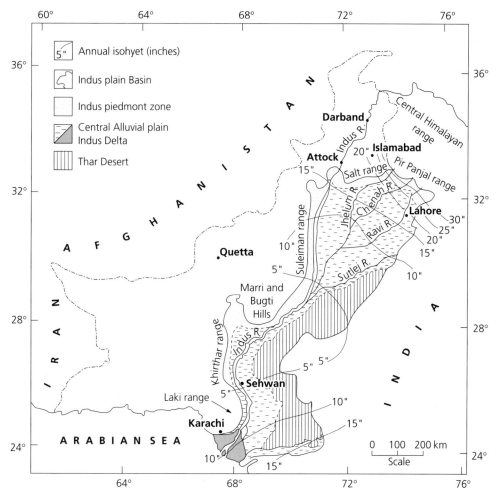

Fig. 3.14 Geography of the subaerial part of the Indus sediment routing system.

Since the submarine area of the Indus Fan is comparable to the subaerial area of the Indus catchment, these large sediment thicknesses in the depositional part of the sediment routing system give a vivid illustration of the high sediment fluxes from the rapidly exhuming High Himalayas, Pamir and adjacent ranges.

3.3.2 Modelling the erosional engine of the sediment routing system

Landscapes are created by a range of erosional and depositional processes acting upon surfaces that have in most cases been interactively created with tectonics. In Chapter 1 the processes leading to the creation of topography were examined, and isostasy was seen to be a key concept in linking topography and tectonics.

This section outlines some of the important concepts useful in understanding the erosion of the continental surface at regional to global scales. The process systems involved can be divided into two groups, between which there is a dynamic linkage: the *hillslope system* which provides erosional material by soil creep, saturated throughflow and overland flow, gullying, debris flows, landslides and rockfalls, and the *fluvial system*, which through its river channel network transports the material away and incises into the hillslopes. The headward incision of rivers therefore sets a boundary condition for the surrounding hillslopes.

The landscape is also characterized by particular time-scales of response—what might be called *landscape sensitivity*. Hillslope systems respond rapidly, channel longitudinal profiles less rapidly, and the planview channel pattern most slowly. In other words, once 'etched' into the land surface, a channel network is relatively constant, whereas hillslopes are susceptible to rapid change. This is a not unfamiliar concept, since we are quite used to thinking of antecedent drainage which cuts across tectonically produced topographic obstacles that post-date the channel network. In contrast, hillslopes can be assumed to respond instantly to geomorphic change, producing a steady-state relief.

This section focuses on the hillslope system, the erosional engine for the sediment routing system. Although the transport of sediment and dissolved load through fluvial systems is an important component of the sediment routing system, it is controlled by the boundary condition imposed by the hillslope system. This is sufficient justification for considering the processes operating in the hillslope system in some detail. The measurements of sediment and dissolved loads in the fluvial system are dealt with in Section 3.4.

Hillslope processes

The denudation of hillslopes takes place by a variety of processes, ranging from the relatively slow and gradual dislodgement of soil particles by the splashing of raindrops, to the catastrophic, high-magnitude but low-frequency failures of the slope by landsliding or rock avalanching.

Soil erosion by water Soil erosion is accomplished by the detachment of loose particles by raindrop impact (*rainsplash*) and their subsequent removal by runoff in the form of shallow sheets, or in rills and gullies. The collection of large amounts of field data on erosion types and rates, in combination with the measurement of variables assumed to be important in the soil erosion process, allows general equations for soil loss to be generated. An example is the well-known universal soil loss equation (equation (3.12)). An alternative approach is to develop models based on the physics of the processes thought to be taking place.

Soil erosion is controlled by a large number of variables which relate to climate, especially rainfall characteristics; topography, particularly slope; vegetation; soil characteristics; and practices of land use. The processes by which rainfall causes soil erosion are first outlined; the influence of non-climatic factors is then briefly discussed.

Raindrop erosion The initial detachment of particles from soil by the kinetic energy of rain impact is of crucial importance in explaining soil erosion [20]. The kinetic energy of rain must be related to raindrop mass and size, size distribution, raindrop terminal velocity and rainfall intensity. The total kinetic energy of rain over a certain land surface also depends on the frequency and duration of rainstorms. If the kinetic energy can be calculated it can be compared with observations of soil detachment and transportation to obtain values of *erosivity*. There are a number of different relations between erosivity and rainfall characteristics which may be more or less appropriate according to local conditions. On sloping ground rainsplash results in a downslope transport of detached particles.

Runoff erosion Runoff erosion can be described from the point of view of overland flow [21] or saturated throughflow [22].

When the rainfall exceeds the *infiltration capacity* of the soil, water begins to accumulate in and on the soil, eventually leading to surface runoff. This is known as a positive supply rate. The process is complicated by the fact that the infiltration capacity varies during the course of a rainstorm, and between different soils and vegetational cover. Vegetation intercepts falling raindrops as well as increasing infiltration

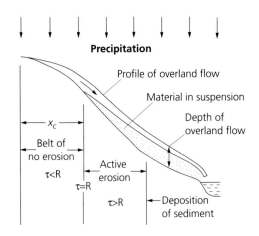

τ: eroding stress
R: shear resistance of soil surface
x_c: critical distance of overland flow

Fig. 3.15 Overland flow occurs when the infiltration capacity of the soil is overreached. Erosion takes place after a critical distance of overland flow x_c. After Horton (1945) [21].

capacity compared to barren soil. As surface runoff flows downslope as an overland flow, after a critical distance it generates enough shear stress on the land surface to erode soil (Fig. 3.15). It is only after this critical distance that gully erosion can occur; before it, rainsplash erosion dominates. It can be shown that the slope of the land surface and its roughness and the runoff intensity (supply rate) control the critical distance. Once flowing and eroding, the runoff takes place in rills or gullies. Concentration of the flow in rills and channels increases its flow depth and velocity, leading to enhanced erosion.

The saturated throughflow model stresses the importance of downslope movement of water through the saturated upper soil horizon (Fig. 3.16). Throughflow rates are slow compared with overland flow and determined by the Darcy equation for flow through porous media. To calculate the throughflow velocity, therefore, it is necessary to know something of the range of permeabilities of soils in the drainage basin.

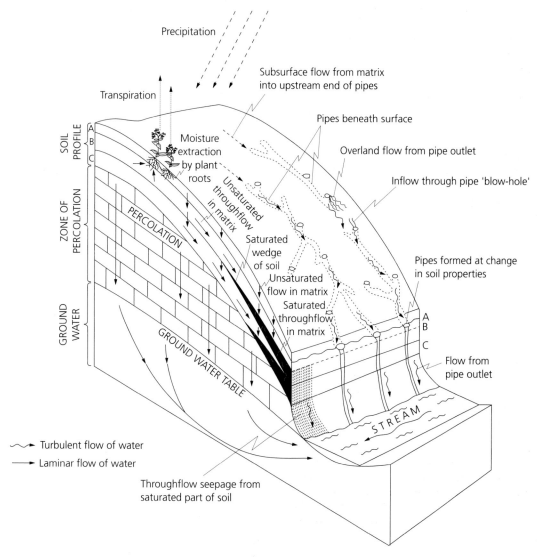

Fig. 3.16 Types of subsurface flow on a hillslope to emphasize the flow of water within the unsaturated and saturated matrix of soil and its focusing along pipe-like channels developed along interfaces in the soil profile. After Atkinson [23].

The factors controlling throughflow are slope, soil characteristics, rainfall intensity and distance downslope. In contrast to overland flow, throughflow is favoured by proximity of streams, concave slopes and hollows, and thin and poorly permeable soils, all of which promote early saturation of the soil. Throughflow is not determined by distance from the hillslope crest, as is the case with overland flow models. The greatest flux of throughflow occurs on the steepest gradient, so gullies should be located just below the maximum slope.

It is likely that the two models are ideal abstractions of nature. Both may operate together to varying degree, with overland flow dominating poorly vegetated semiarid slopes and throughflow more humid settings.

Much of the erosion of hillslopes is caused by the development of rills—short-lived and discontinuous shallow channels focusing runoff. Rills commonly develop into longer-lived, deeper gullies which grow by headward retreat, bank collapse and basal scouring. Gullies may develop preferentially where the surface is disrupted by, for example, walls and ditches, or off-road vehicle tracks.

Factors affecting soil loss These may be categorized into four types.

1 Topography: it is clear from the discussion of the overland flow and saturated throughflow models that a number of topographic or spatial variables affect raindrop and runoff erosion. These parameters include land surface slope, length of hillslope undergoing runoff, and surface roughness. Surface slope and slope length are commonly combined in a topographic parameter and related directly to the rate of erosion per unit area.

2 Vegetation: cover decreases soil erosion in a number of different ways. The leaves of vegetation intercept falling raindrops and thereby reduce their kinetic energy on impact with the ground surface. Vegetation decreases the water available for runoff by improving the infiltration capacity of the soil through the effect of the increased permeability caused by root systems, and by using water in metabolism and evapotranspiration. Plants also increase the surface roughness, which reduces the velocity of overland flow, and decrease the erodibility of soils by the binding action of roots. Different vegetation types have relatively different effects on these inhibiting processes, depending on canopy characteristics, root density, timing of growth and so on. Cutting of forest and ploughing of grassland are likely to have severe effects on soil erosion (Section 3.6.2).

3 Soil characteristics: the erodibility of soil depends on a number of technical properties of the soil that affect its susceptibility to detachment and transport by rainsplash and runoff. The properties include the dispersion (a measure of the amount of fine-grained material in the soil), its strength (the shear resistance of the soil to movement) and its transmissibility or permeability. These properties are controlled by a soil's physical and chemical make-up, soil moisture and organic content, density, texture and fabric (Section 3.2.3). The tendency for the soil to form stable aggregates resistant to erosion is of fundamental importance. The field measurement of soil loss under controlled rainfall conditions can allow the importance of the different technical properties of the soil to be evaluated.

4 Land-use practices: the impact of land-use practices on soil loss and soil conservation is a huge area that cannot be treated in any detail here. The subject is touched upon in Section 3.6, and suggestions for further reading are provided. Soil conservation practices involve making use of the protective effect of vegetation cover in reducing the impact of rainfall as well as improving soil characteristics to withstand rainsplash and runoff erosion. In addition, erosion-control measures can be used to restrict soil loss on slopes, such as terracing, contour planting and 'strip cropping' involving the planting, usually in rotation, of different crops or grasses in strips of land parallel to contours. Land-use practices such as these can be factored into soil loss models or formulations.

The universal soil loss equation The soil loss predicted from a knowledge of the parameters outlined above is given by the widely used universal soil loss equation, which has the form

$$A = R \times K \times LS \times C \times P \qquad \text{(3.10)}$$

where A is the average annual soil loss, R incorporates the factors attributable to the rainfall regime, K likewise incorporates a number of factors describing the erodibility of the soil, LS is a topographic factor obtained from the product of hillslope length and gradient, C is a factor related to crop management, and P is a factor to account for supporting erosion-control practices. The equation, despite its name, is not universal in its applicability, but instead is appropriate for cropped fields and small construction areas subject to sheet and rill erosion. The equation may be modified to better suit particular settings, for example, areas where gullying is important, in which case a factor to account for the erosion of open-channel flows would need to be incorporated. More seriously, the equation does not account for any feedbacks or nonlinear reactions between the component factors. The equation therefore does not provide any great physical insights into the soil erosion process.

Slope instability The sudden downslope movement of weathered surficial material (*regolith*) and rock represents a large mass flux on hillslopes in areas of moderate to high relief. Although mountainous areas, particularly in climatic zones of high rainfall, are subject to frequent rock avalanches, landslides and debris flows, slope instability as a hazard is particularly acute in areas underlain by permafrost or ground ice. Gradual melting of permafrost due to global warming is likely to cause major problems of slope instability.

The physics underpinning the stability of material on a slope is dealt with in Chapter 6. The stability criterion is essentially determined by the balance between forces tending to cause motion (gravity acting on the mass of rock, regolith or soil) and those tending to oppose motion (coherence, shear strength). A number of processes may lead to instability. A steepening of the slope increases the downslope component of stress due to the weight of the overlying regolith. Increased pore pressures may reduce the frictional resistance of the body of soil, rock or regolith. Increased pore fluid pressures may result from water soaking in during heavy rains, or pumping by the passage of seismic waves.

Landsliding dominates mass transfer on mountainous hillslopes, especially in regions with high rainfall such as Papua New Guinea, Japan, Taiwan, the eastern Himalayas and New Zealand. An analysis of time series of aerial photographs of the Southern Alps, New Zealand, over a period of 60 years showed that the landslide magnitude–frequency distribution takes the form of a power law (Fig. 3.17). Making certain assumptions about the three-dimensional geometry of the landslides, the landslide efflux of the hillslope system was estimated (Fig. 3.18) and found to be very similar to the fluxes of suspended sediment in rivers draining the Southern Alps, suggesting that landsliding alone explains the denudation of the hillslopes of the western flank of the Southern Alps (see also Section 3.5.2).

Modelling hillslopes

We have seen that hillslopes may be dominated by the relatively slow creep of soil and downslope transport of overland flow, or by the rapid but less frequent failure of the slope by landsliding and associated debris flows. The two situations correspond to where stream incision rates are low and high, respectively. In modelling hillslope evolution over long time-scales it is necessary to account for the discrete nature of the

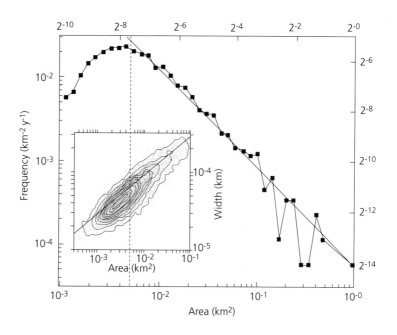

Fig. 3.17 The power law magnitude–frequency distribution of landslides in the Southern Alps (New Zealand), derived from mapping 4984 landslide events east of the Alpine fault which have occurred over the last 60 years. The area to the left of the vertical dashed line is below mapping resolution and should be ignored. The inset graph shows the relationship between the landslide area and the width of the best-fit ellipse, giving information on the plan-view aspect of the landslide scars. After Hovius *et al.* (1997) [24].

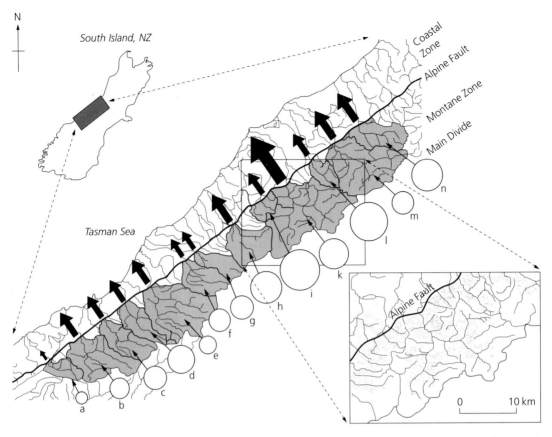

Fig. 3.18 The landslide efflux of mountainous hillslopes, Southern Alps, New Zealand, derived from the same dataset as in Fig. 3.17. Denudation rates of each of 13 catchments draining the western side of the Southern Alps are shown proportional to the areas of circles and vary from 1.8 mm y^{-1} in Moeraki catchment (a) to 18.1 mm y^{-1} in Waitangitaona (j) and Poerua (l) catchments. The sediment discharges are proportional to the size of arrows leaving the catchments, ranging from $1.2 \times 10^5 \, \text{m}^3\text{y}^{-1}$ in Moeraki catchment (a) to $5.1 \times 10^6 \, \text{m}^3\text{y}^{-1}$ in Whataroa catchment (k). After Hovius *et al.* 1997 [24].

individual geomorphic events but to time-average them over the time-scale of interest.

Diffusion A continuous evolution of hillslopes can be modelled as diffusive [25]. Diffusion models are based on the conservation of mass (sediment continuity equation; see Section 5.1) which states that the spatial variation in the sediment transport rate is proportional to the vertical erosion or aggradation rate of the substrate, assuming there to be no changes in the concentration of suspended sediment. Mathematically,

$$\frac{\partial Q}{\partial x} = -\rho_b \frac{\partial y}{\partial t} \qquad (3.11)$$

Q is the discharge of mass per unit width of hillslope,

ρ_b is the bulk density of the mobile regolith, x is the horizontal distance coordinate, y the elevation and t is time. Assume now that the discharge is proportional to the local topographic slope

$$Q = -k\frac{\partial y}{\partial x} \qquad (3.12)$$

where k is a transport coefficient encompassing all of the geomorphic processes acting on the hillslope which contribute to its efficiency. Combining equations (3.11) and (3.12) gives the familiar diffusion equation

$$\frac{\partial y}{\partial t} = \kappa \frac{\partial^2 y}{\partial x^2} \qquad (3.13)$$

where the diffusion coefficient $\kappa = k/\rho_b$. The geometry of the hillslope depends on the diffusion coefficient, but also on the erosion rate of the incising streams which border the hillslopes (Fig. 3.19). If the horizontal length of the hillslope is L, the erosion rate of the bounding streams is $\dot{e}$, and the x coordinate is set to zero at the hillcrest (range divide, interfluve ridge), the steady-state profile of the hillslope, obtained by integrating the diffusion equation twice, becomes

$$y = \frac{\dot{e}}{2\kappa}(L^2 - x^2) \tag{3.14}$$

which shows that the hillslope profile is parabolic. The maximum relief of the hillslope, between the hillcrest and the bounding valley bottom, is the value of y at $x = 0$ minus the value of y at $x = L$:

$$R = \frac{\dot{e}L^2}{2\kappa} \tag{3.15}$$

The slope of the hillslope at the hillcrest is zero, and the maximum slope is of course at the boundary with the stream channel, at $x = L$, where it has the value

$$\left. \frac{\partial y}{\partial x} \right|_{\text{max}} = \frac{\dot{e}L}{\kappa} \tag{3.16}$$

Diffusive problems, such as the thermal diffusion from a mid-ocean ridge, have a characteristic time-scale. In the mid-ocean ridge case (Section 1.5.3) there was a thermal diffusion distance equal to $\sqrt{\kappa t}$. Keeping the same analogy, the time constant for the hillslope system should be given by

$$\tau = L^2/\kappa \tag{3.17}$$

Hillslopes affected by landsliding It is unlikely that hillslopes dominated by discrete failures in the form of landslides and debris flows can be treated as a strictly diffusional problem. This is principally because landsliding depends on a threshold condition being overreached (Chapter 6). This threshold condition is controlled by the local slope and the groundwater level.

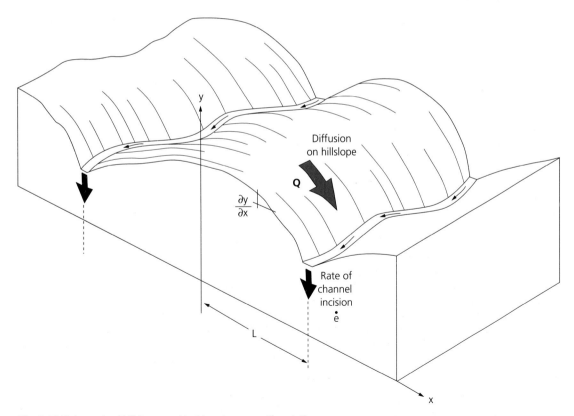

Fig. 3.19 Schematic of hillslopes and incising streams, with notation.

Practical exercise 3.1: Diffusive hillslopes

1 Two channels located 100 m apart incise at a rate of 0.5 mm y⁻¹, producing an interfluve with two symmetrical hillslopes. Calculate the profile of the hillslope system, assuming the denudation to be diffusional with a diffusion coefficient of $50 \times 10^{-3} \, \text{m}^2\text{y}^{-1}$. What is the maximum gradient of the hillslope?

2 What is the maximum relief between the hillslope crest and the channel bed?

3 Calculate the time constant τ.

Solution

1 The hillslope profile can be calculated using equation (3.14). Note that the profile is parabolic (Fig. 3.20). The maximum slope is given by equation (3.16) as 1 : 2, or 26°. This maximum slope is found immediately next to the incising channel.

2 The maximum relief can be found from equation (3.15) as 12.5 m.

3 The value of τ from equation (3.17) is 50 ky. If you are working with a spreadsheet program, now experiment by increasing the width of the hillslopes, the rate of channel incision and the value of the diffusion coefficient. This will give you a good appreciation of the response times expected.

Weathering provides loose material available for diffusion down the hillslope. If this rate of weathering is small compared to the rate at which relief is being created by channel incision, diffusion alone cannot accomplish a steady-state hillslope profile. Instead, the hillslope gradually steepens until the failure criterion is reached, triggering landslides. The rate of channel incision therefore also exerts a strong control on the occurrence of landsliding. In terrains experiencing a great deal of tectonic uplift, such as the Southern Alps of New Zealand, the Finisterre Range of Papua New Guinea or the Santa Cruz Mountains of California, channel incision causes landsliding to dominate the mountainous hillslopes. These hillslopes are characteristically straight (Fig. 3.21). In less uplifting/incising terrains, such as the green and pleasant hills of my native England, hillslopes are dominated by diffusive processes and take on the characteristic parabolic shapes.

Landslides typically involve the entire regolith together with considerable quantities of bedrock. We might expect the efficiency of landslide-dominated hillslopes to be high compared with diffusive hillslopes. If the typical path length for a landslide is denoted by L, the *effective diffusivity* of the landslide-dominated hillslope κ_ℓ becomes

$$\kappa_\ell = \frac{L^2}{\tau} \qquad \text{(3.18)}$$

Slope failure should take place when the slope S exceeds some critical slope S_c for landsliding. In this case at the threshold condition

$$Q = S_c(k_d + k_\ell) \qquad \text{(3.19)}$$

where k_d and k_ℓ are the transport coefficients due to normal diffusion and landsliding respectively, and

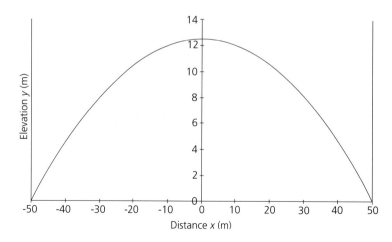

Fig. 3.20 Parabolic diffusive hillslope profiles derived from Practical Exercise 3.1.

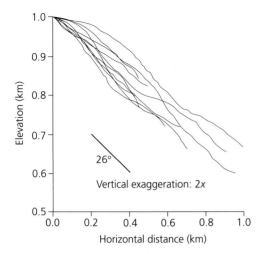

Fig. 3.21 Hillslope profiles from Santa Cruz Mountains. The long straight segments indicate that diffusion is overwhelmed by landsliding. The mean slope is about 26°. After Anderson (1994) [25].

relate to the normal diffusivity and the landslide effective diffusivity.

The maximum possible hillslope relief due to landsliding results from simple trigonometry

$$R = L \tan S_c \qquad (3.20)$$

although actual values measured in the field are likely to be lower because of the diffusional rounding of hillslope crests and the probability of occurrence of landslides at slope angles lower than critical where other factors are favourable.

It is now possible to estimate when landsliding will take over from diffusion as a function of slope. Since the maximum slope in the diffusive case is as given in

equation (3.16), we can assume that landsliding will dominate at slopes higher than the maximum slope for diffusion. That is, if

$$\dot{e} > \frac{\kappa S_c}{L} \qquad (3.21)$$

Stream incision

As noted previously, channel incision drives hillslope processes by maintaining upslope gradients. A key question is therefore what determines rates of channel incision. One approach is to model stream incision as determined by the local stream power, ω, which may be expressed in terms of the discharge of water, Q_w and the slope of the stream bed, S:

$$\omega = \rho g H \bar{u} \sin \alpha = \rho g Q_w S \qquad (3.22)$$

where $\bar{u}$ is the average velocity of the stream and H is its depth. The water discharge must depend on the drainage area above the channel, A, and the average precipitation over that area, $\bar{P}$. Consequently, the rate of channel incision can be expressed as

$$\frac{\partial y}{\partial t} = -c_1 A \bar{P} S \qquad (3.23)$$

where c_1 is a coefficient determined empirically to represent the proportion of stream power expended in incision as opposed to generating heat and transporting loose sediment. In steep reaches of the river, debris flows are likely to play a role in scouring the floors of channels, in which case the slope needs to be raised by a power probably between 1 and 2.

The drainage area upstream of a particular point in the channel will obviously increase as one goes down the river system. This can be approximated by a power law (see p. 126)

Practical exercise 3.2: Landslide-dominated hillslopes

1 Take the same interfluve as in Practical Exercise 3.1, with incising channels located 50 m either side of the hillslope crest. The diffusion coefficient is $50 \times 10^{-3} \, \text{m}^2\text{y}^{-1}$. At what value of the channel incision rate will the hillslope be dominated by landsliding? Assume that the critical slope for failure by landslides is approximately 50°.
2 What is the maximum possible relief at the onset of landsliding as the dominant hillslope process?

Solution

1 The maximum slope for diffusion is given by $S_{max} = \dot{e} L / \kappa$. Setting $S_{max} = \tan 50°$, this gives a critical value of channel incision rate of 1.2 mm y^{-1}. Above this value landsliding dominates. The sensitivity of the critical channel incision rate to variations in the critical slope angle can be explored by varying the value of S_c from 40° to 60°, as shown in Fig. 3.22.
2 The maximum possible relief is given by equation (3.20). With a critical slope of 50°, this is 59.6 m. Note that this is larger than in the case of simple diffusion.

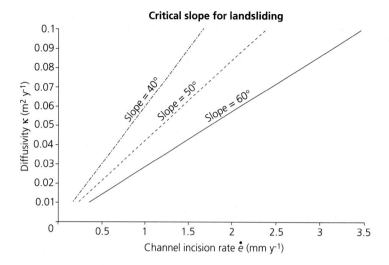

Critical slope for landsliding

Fig. 3.22 Plot of the channel incision rate against the diffusivity to show sensitivity to the critical slope for landsliding. From Practical Exercise 3.2.

$$A = x^{5/3} \qquad (3.24)$$

Now if the precipitation over the drainage basin is assumed to be constant, the rainfall term can be incorporated within the incision efficiency coefficient c, and we can calculate the evolution of the longitudinal profile due to channel incision. Our area–slope product therefore becomes

$$\frac{\partial y}{\partial t} = -c_2 x^{5/3} S \qquad (3.25)$$

The effect of the slope–area product rule can be demonstrated in the evolution of longitudinal profiles shown in Fig. 3.23. The evolution of these river profiles shows that channel incision causes the region of greatest concavity to migrate upstream with time. The maximum downcutting is where the slope–area product is greatest. This is displaced downslope of the drainage divide because of the effect of increasing drainage area, but displaced upslope from the river base level because of the effect of increasing slope.

The profiles generated in Fig. 3.23 bear striking resemblance to observed longitudinal profiles of rivers. The result of the compilation of a very large number of longitudinal profiles of rivers within the erosional sectors of mountain belts reveals certain universal characteristics. Longitudinal profiles after normalization show two distinct segments:

1 A steep upper segment which is universal to rivers draining a large number of mountain belts. This upper segment with universal form is dominated by the contribution of mass flows (landslides, debris flows) to the channel.

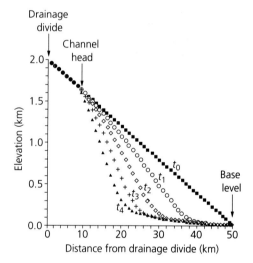

Fig. 3.23 Longitudinal profiles produced by the bedrock incision theory involving the slope–area product rule. The initial profile is a linear slope with gradient 0.004, with a channel head taken arbitrarily to be 10 km from the drainage divide. The river profiles have a base level at 50 km from the drainage divide, causing low slopes in the lower reaches. In this example there is no headward retreat of the channel head. Four subsequent time steps are shown. The region of greatest concavity migrates upstream in response to the upstream migration of the maximum slope–area product. After Anderson (1994) [25].

2 A less steep lower segment which varies in its length considerably between mountain belts. The channel is dominated by fluvial incision.

The observed variation in longitudinal profiles can be explained by considering the balance between the rate of incision and the tectonic uplift rate [26] (Fig. 3.24). It is possible to discern two trends across Fig. 3.24. With increasing precipitation rate, which acts as a surrogate for rate of channel incision, relative to the tectonic uplift rate, rivers extend their lower segments, causing a retreat of the steep upper segment. Sediment delivered to the mountain front should be relatively well sorted and relatively fine-grained where the fluvial transport distance in long lower segments is large. In contrast, where the tectonic uplift rate dominates, steep upper segments comprise most of the longitudinal profile. The sediment delivered to the mountain front should be dominated by mass flows in the form of coarse-grained and poorly sorted material. The second trend is from low uplift and erosional activity to high. This can be recognized in the difference between the profile for the interfluve crest and that for the channel bed.

In a real drainage basin, the hillslope and channel subsystems need to be dynamically linked. It is important to know the response time of the hillslope system relative to the response time of the channel system. We have already seen that the hillslope response time is scaled on L^2/κ. The response time of the channel

must be influenced by the length of the channel and the average velocity of the knickpoints generated when the channel adjusts to a new base level. Such knickpoints migrate headwards, but slow down in their upstream velocity as the stream power decreases due to an upstream decrease in the contributing drainage area.

Assuming the processes in a drainage basin to be controlled by diffusion on the hillslopes and by the stream power rule within the channel, we have the following relation for the response time of the channel [25, 27]:

$$\tau = -\frac{1}{2Wc}\ln\left(\frac{A_0}{A_{\text{total}}}\right) \tag{3.26}$$

where W is the half-width of the drainage basin, A_0 is the drainage area necessary to generate a channel, and which therefore contributes to the channel head, and A_{total} is the total drainage basin area. The channel incision constant c has been estimated as $2 \times 10^{-7}\,\text{m}^{-1}\text{y}^{-1}$ [25]. The calculated response time varies in relation to the size of the drainage basin, which determines the knickpoint velocity. The response time will in most circumstances be several tens of thousands of years. This is significantly longer than the landslide response

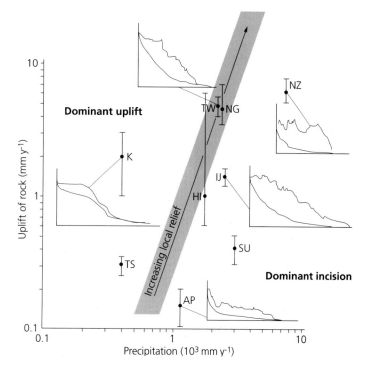

Fig. 3.24 River profile geometry conceptually related to mean annual precipitation and rate of uplift of rock. To the left of the stippled zone rock uplift dominates, to the right stream incision dominates. Along the stippled zone, valley relief increases upwards. TW, Central Range, Taiwan; NG, Finisterre Range, New Guinea; NZ, Southern Alps, New Zealand; K, Kirghizia; TS, northern Tien Shan, China; HI, Central Himalayas; IJ, Irian Jaya; SU, Sumatra; AP, Apennines. After Hovius (1995) [26].

time, but comparable to the hillslope response time. This suggests that the hillslopes characterized by the high effective diffusivities produced by landslides are always in equilibrium with the incising channels. Less steep hillslopes, however, may not be in equilibrium with channel incision and may display morphologies which are transient rather than steady-state.

3.4 Sediment and solute fluxes in drainage basins

A mass balance of the inputs and outputs of a drainage basin involves an import of mass primarily from bedrock weathering, transfer from the atmosphere, and groundwater inflow, and outputs mainly as fluvial export, groundwater outflow, and transfer back to the atmosphere. Participation in the various fluxes depends on the elements concerned. For example, oxygen, hydrogen, carbon, nitrogen and sulphur are exchanged as gases between the Earth's surface and the atmosphere and do not play a major part in mass transfer through the drainage basin. But elements that do not take part in gaseous exchanges move through and out of the drainage basin by fluvial transport as particulate solids and as dissolved load.

The fluvial mass transport through a drainage basin generally dominates the mass balance, most of the mass being obtained by weathering. The total mass of material involved in fluvial transport can be constrained reasonably well by measurements of the dissolved and suspended particulate concentrations of rivers .

Rivers transport particulate sediment in different modes. Coarse particles (sand and gravel) are transported along and very close to the bed as *bedload*. Some particles are occasionally lifted high into the flow and are transported by the fluid turbulence for a certain distance downstream, before settling on the bed and resuming their transport as bedload. This is known as *intermittent suspension*. The finest particles (silt, clay, colloids) are transported in suspension by fluid turbulence. Very fine sediment remains effectively permanently in suspension, when it is known as *washload*. Mass is also transferred through dissolved components. This is known as *solute load*. In general, suspended load constitutes the bulk of the total sediment load, but the relative proportions may vary widely. Different techniques must be used to measure the different types of sediment transport. An understanding of the techniques involved is important in an appreciation of the likely uncertainties in sediment load measurements. Further details on the physical processes involved can be found in Chapter 5.

The denudation of a drainage basin involves a number of processes and fluxes.

• Solid denudation of a drainage basin includes the *liberation* of material by physical hillslope erosion, the *storage* of eroded sediment as colluvium at the base of slopes, as alluvium in river channels and floodplains, and as deposits in lakes and reservoirs, and the remobilization of temporarily stored sediment. There is therefore not a direct link between measured solute and suspended loads in rivers and the rate of physical denudation of the drainage basin.

• Solute denudation is due to chemical leaching throughout the upland and lowland parts of the drainage basin. Solute storage is far less important than with solids, occurring in soils and alluvium as crusts, efflorescences and pans, in groundwater and in lakes.

This section describes the measurements made of dissolved and solid loads and the associated uncertainties, and the factors responsible for interregional variations.

Compilation of the global pattern of sediment loads in rivers gives important information on the total annual sediment delivery to the ocean, and the regional variations in sediment load and thus the factors likely to be responsible for such variations. Sediment load data can be used to calculate the annual loss of sediment per square kilometre of eroding area, the *sediment yield*. Unfortunately, sediment load data are subject to a large number of errors and uncertainties.

• **Data quality variations** caused by differences in measurement techniques, in duration of observation periods, and in sampling procedures. A particular problem is the measurement of sediment loads in large floods when sediment discharge is likely to be at a maximum. Gauging stations, at which hydraulic and sediment data are collected, may also be very widely spaced and subject to strong local influences. The poorest data often originate from less-developed countries, but is in these countries that the majority of high-discharge rivers are found.

• The importance of **small rivers**: the impact of small, mountainous rivers draining directly to the ocean has probably been underestimated.

• The **type of sediment load**: measurements generally account for suspended load. Bedload is difficult to measure and is usually ignored. Yet some vigorous rivers may transport a large amount of relatively coarse-grained sediment as bedload. Similarly, the global pattern of solute load is difficult to interpret in terms of solutional denudation.

• **Sediment may be stored** as colluvium on slopes,

alluvium in valleys and floodplains, and in lakes and desert areas. Ninety per cent of the sediment currently being eroded from the land surface of the conterminous United States is being stored in river systems between the uplands and the sea. The sediment discharges measured at river mouths do not therefore relate directly to rates of soil loss or land surface denudation.

• Present-day sediment loads have been strongly **affected by humans**, both through the construction of reservoirs, dams and irrigation systems, and through vegetational changes and land-use practices [28]. The change to agricultural land use typically accelerates erosion rates by factors of 10–100 (Section 3.6).

3.4.1 Bedload

Bedload is extremely difficult to measure satisfactorily, and our understanding of global patterns of river bedload transport is primitive. The standard technique is the lowering of a bedload trap on to the river bottom. The mass of sediment recovered in the sampling trap (e.g. Helley–Smith bedload sampler) over a sampling interval gives the bedload sediment transport rate. Alternatively, bedload movement can be estimated using tracers, such as radioactive (e.g. labelled with iridium 192), painted or dyed bed material. However, there are a couple of problems. First, all or most of the beadload transport commonly takes place along a narrow band of the total width of the river cross-section. The position of this narrow band varies with variations in hydraulic and sediment characteristics. Closely spaced (ideally 0.5–15 m) sampling stations are therefore required across a river cross-section. This is unlikely to be achieved in large rivers. Second, temporal variations in bedload transport rates may also be very large, bedload appearing to move in 'slugs', causing distinct cyclical trends with time. Consequently, a knowledge of the temporal trends in bedload transport rates is necessary to estimate the time-averaged rate. Sampling times may not be long enough to permit this kind of analysis. Because of the difficulties of mounting an adequate sampling programme, bedload transport rates are commonly estimated from established bedload sediment transport formulae. These formulae rely on a relationship existing between the hydraulic properties of the river, which must be measured at a carefully selected site, and its ability to transport particles of certain size classes as bedload. A number of bedload transport formulae are available. The accuracy of any particular formula appears to depend strongly on local conditions, and there is not one universally applicable formula.

Because of the difficulty of measuring bedload discharges, it is common to assume that it is less than 10% of the suspended load. This would be a very unsafe assumption in coarse-grained, steep, mountainous streams.

3.4.2 Suspended load

Suspended loads are measured at a number of points in the vertical cross-section of the river, as well as at different stations across the river. It is possible to calculate the suspended sediment concentration at any point in the vertical profile by knowing its concentration at a reference depth (Chapter 5). The cheapest and easiest strategy is to take a single sample near the surface. This may be reasonably accurate for silt and clay but is prone to large errors for sand. More elaborate schemes involve the automatic sequential sampling of river water by pumping the sample to a set of storage bottles on the bank. Equipment is also available to submerge, usually by a winch from a bridge or boat, into the river to varying depths, and at various points of the river crossing. Finally, suspended sediment loads may be calculated using permanently installed instruments which measure the light or gamma ray attenuation caused by the passage of suspended grains.

The total suspended sediment load delivered by all the world's rivers to the ocean has been estimated as $13.5 \times 10^9 \, t \, y^{-1}$ [29]. The contribution of bedload may account for a further $1–2 \times 10^9 \, t \, y^{-1}$. The global mean specific sediment yield based on suspended loads of rivers is in the region of $116 \, t \, km^{-2} \, y^{-1}$ [8]. For comparison, the mean annual solute load is $32 \, t \, km^{-2} \, y^{-1}$, of which perhaps half is non-denudational (Section 3.4.3). If the average density of surface rocks is $2700 \, kg \, m^{-3}$, this represents an average lowering of the continental surface of something slightly in excess of $0.05 \, mm \, y^{-1}$. The specific sediment yield varies widely from an average of more than $1000 \, t \, km^{-2} \, y^{-1}$ for mountainous regions to less than $50 \, t \, km^{-2} \, y^{-1}$ for many lowland areas. A key question is what controls these patterns. The global pattern of suspended sediment loads in rivers (Plate 3.1, facing p. 204) shows the following features.

• The highest sediment loads, and the highest specific sediment yields, come from an arc extending from northern China and Japan through south-east Asia to Pakistan and the western Himalayas. Other high-sediment loads are found in the equatorial Andes and Amazon basin, central America, and much of the Pacific north-west of the USA, Canada and Alaska. The highest yields are found in small drainage basins in mountainous terrains and wet climates, such as New Zealand, Papua New Guinea and Taiwan.

• In contrast, despite the very large size of some Asiatic rivers draining to the Arctic Ocean (Lena, Ob, Yenisei), their sediment loads are very small, and the specific sediment yields of their catchments are less than $10 \, t \, km^{-2} y^{-1}$. Together they deliver a small amount of sediment to the Arctic Ocean. Essentially the same is true of rivers draining the Canadian Shield. These regions of the Eurasian and Canadian Arctic are typified by low gradients and low temperatures. Low-sediment loads and yields are also found in the temperate zones of north-west Europe.

• Low sediment loads are also found in much of north and north-east Africa and the Middle East which can be linked to climatic factors. Low precipitation rates are also responsible for the low sediment discharges of Patagonian and north-east Brazilian rivers.

It is clear that there are a number of climatic, tectonic and topographic factors which have some influence on sediment loads and sediment yields. One way of investigating the importance of these factors is to compile a database involving not only sediment load information, but also data on a range of parameters describing the physical and climatic characteristics of a large number of the Earth's major drainage basins. Multivariate analysis of such a global database may provide valuable clues as to the main factors explaining sediment discharge and yield variability (Section 3.5).

3.4.3 Solute load

The task of collecting solute load is simpler than in the case of either bedload or suspended load, because dissolved solids are thoroughly mixed by the turbulence of the water and cross-stream variations are likely to be smaller. Samples are collected either manually by wading or from a bridge, or automatically by transfer of a river water sample to storage vessels on the bank by use of a pump or vacuum. Automatic samplers can be programmed with sampling strategies, such as regular sampling intervals, or intervals dependent on flow velocity or discharge. Continuous records of electrical conductance as a proxy for solute concentration can also be obtained.

Despite the relative ease of collecting solute load data at gauging stations, less is known of the solute load than of the suspended load. More than 80% of the total dissolved load of the world's rivers is made of four components, HCO_3^-, SO_4^{2-}, Ca^{2+} and SiO_2.

Material in solution comes not only from the weathering of soil and bedrock, but also from non-denudational sources such as atmospheric contributions (precipitation, dust), the mineralization of organic matter, plant metabolism and human-made pollution. Consequently, dissolved loads cannot be directly linked to chemical denudation. We know from Section 1.3.5 that rivers may be categorized by the source of the solute load (Fig. 1.19): (i) dominated by atmospheric precipitation, with low salt concentrations ($20–30 \, mg \, l^{-1}$), principally Na^+ and Cl^-; (ii) dominated by rock and soil-derived solutes, with intermediate concentrations, characterized by dominance of calcium bicarbonate; or, (iii) dominated by evaporation and subsequent precipitation, with high salt concentrations ($1000–2000 \, mg \, l^{-1}$), reflecting the precipitation of $CaCO_3$, and relative concentration of Na^+ and Cl^-.

In general, solute concentrations increase with aridity, so that rivers in Kazakhstan, central Asia, have solute concentrations of $6000–7000 \, mg \, l^{-1}$, compared to less than $10 \, mg \, l^{-1}$ for rivers in the humid Amazon basin. These high concentrations are mainly due to the small amount of runoff being highly charged with solutes from accumulations of salt in soil. At a more local scale, solute concentrations also reflect the underlying geology. Most of the world's dissolved load comes from carbonates (45% of total solute load) though they only constitute 16% of the total global outcrop. Evaporites contribute 18% from just 1% of outcrop. The reverse is true of clastic sediments, which, despite occupying 49% of the total global outcrop, contribute just 25% of the total dissolved load, and crystalline rocks, which contribute 12% from 34% of outcrop.

The solute load of a river is the result of the product of the dissolved concentration of the runoff and the volume of that runoff. Low concentrations are commonly accompanied by large volumes of runoff, giving a moderate to high solute load or discharge. High concentrations are often accompanied by low runoff volumes, resulting in low to moderate solute discharges. This is reflected in the global patterns of solute load [30, 31].

The denudational component of the solute load

To what extent can the solute load of a river be viewed as due to chemical/solutional denudation? If the answer to this question were known, it would be possible to use the global solute load dataset to add to the sediment yield information to assess total denudation rates of the continental surface. To answer the question we need to know the contribution to the solute load from atmospheric sources, plant and soil activities, agricultural (fertilizer) and industrial (pollution) sources. For total dissolved solids, a number of studies indicate atmospheric contributions of up to 50%, whereas for individual solutes the proportion may range from almost zero (in the case of silica) to 100% (ammonium, nitrate). Plant metabolism and bacterial activity in soils causing carbonation (Section 3.2.2) are thought to be responsible for the bulk of the dissolved bicarbonate content of water, yet these concentrations are non-denudational. To evaluate accurately the denudational component of the measured solute load is therefore an exceedingly complex task. As a rule of thumb, we can expect

perhaps at least half of the total dissolved solids to have originated from non-denudational sources.

3.4.4 Relation between solute and suspended load

The global average of solute load is roughly one-fifth of the global average of suspended load, but the relative proportions vary strongly from continent to continent and from drainage basin to drainage basin (Figure 3.25). In Europe, for instance, the solute load is greater than the suspended load. In comparing solute and suspended load data, it should immediately be noted that whereas suspended sediment is subject to storage in floodplains, lakes and human-made reservoirs, solute load suffers negligible transmission losses. Furthermore, solute and suspended loads have different concentration–discharge relationships. Whereas suspended sediment concentrations increase rapidly over several orders of magnitude with rising flood discharge, solute concentrations are more stable through time, and generally decrease due to dilution during peak discharges. Consequently, suspended loads are highly susceptible to the infrequent but high-magnitude flood event, whereas solute loads are relatively evenly distributed in time. These two considerations lead one to view the globally averaged

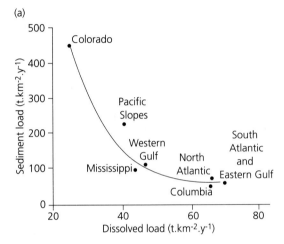

(a)

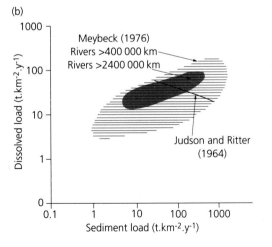

(b)

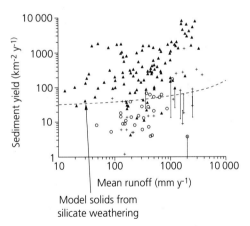

Model solids from silicate weathering

Fig. 3.26 Relation between mean runoff and sediment yield for rivers throughout the world. Open circles, coastal plain (0–100 m headwaters) and lowland (100–500 m); crosses, upland rivers (500–1000 m); solid triangles, mountain (1–3 km) and high mountain (>3 km). Model curve is prediction for solids derived by silicate weathering (after Stallard (1995) [33]), showing that mountainous catchments provide high solid yields for a given runoff, whereas lowland rivers provide low solid yields for the same runoff.

Fig. 3.25 Relationships between the dissolved and suspended sediment components of the total river load in US rivers (a) after Judson & Ritter (1964) [32] and (b) based on world-wide data (after Meybeck (1976) [30]).

ratio of suspended to solute load of roughly 5:1 as a minimum.

The balance between sediment yield and solute yield clearly varies in relation to topography of the drainage basin. Mountainous areas have very high sediment yields compared to solute yields, whereas lowlands have a roughly equal balance (Fig. 3.26).

3.4.5 Sediment rating curves

The transport of particulate sediment and dissolved solids in some rivers may have little to do with erosion of the river bed and much more to do with derivation from overland flow, saturated throughflow and in some instances bank collapse. We should, therefore, not expect there to be any simple relationship between the shear stress of a river and its sediment concentration. Sediment loads in rivers can rarely be shown to be transport-limited, but rather reflect the sediment

availability. Consequently, models relying on universal relationships of sediment flux to slope (diffusive models), or to slope-dependent parameters such as shear stress (see Chapter 5) or stream power, should be treated with caution. Nevertheless, for a given reach of a given river, there are useful relationships between sediment concentration and discharge embodied in so-called *sediment rating curves* (Fig. 3.27). These have the general form

$$Q_s = DC = AD^b \tag{3.27}$$

where Q_s is the sediment transport rate, D is the water discharge (the product of mean velocity and cross-sectional area of stream), C is the sediment concentration, and A and b are coefficients.

During a typical storm the sediment concentration and water discharge are not linearly related. Instead, during the period of rising water discharges sediment

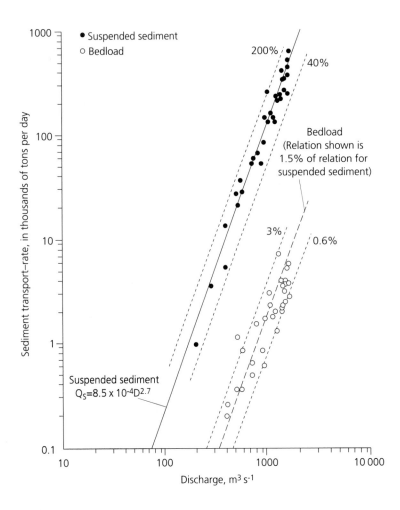

Fig. 3.27 Example of a sediment rating curve for bedload and suspended load transport rate as a function of discharge, Tanana River, close to Fairbanks, Alaska. Data from Burrows *et al.* (1979) [34].

concentrations are very high as overland flow and/or saturated throughflow supply large amounts of weathering products to the river, but declines as rainfall decreases. In addition, water surface slopes are greatest during the rising limb since water levels are at their maximum upstream during the rising limb and downstream for the falling limb. The sediment discharge is therefore in advance of the water discharge, the two commonly peaking at different times. The sediment load is lower on the falling limb for the same water discharge as water slopes decrease and hillslope contributions decline. There is therefore a loop or *hysteresis* in the relationship of sediment concentration to water discharge. A knowledge of this looped relationship is important in the interpretation of sediment load measurements.

There has been considerable debate about the characteristic rating curves for solute transport and the relationship between solute and suspended loads in streams. We have already seen that the dilution effect causes solute concentrations to fall during rising discharge, with a hysteresis between the rising and falling limbs. This characteristic behaviour has been observed in the concentration of particular solutes, for example calcium carbonate (but excluding silica), or in the total dissolved load. The rating curves for solute concentration therefore take the form

$$I = AD^{-b} \qquad (3.28)$$

where I is the total ionic concentration, D is the water discharge and A and b are coefficients. It is important to emphasize, however, that the effect of increasing discharge is invariably to cause a net increase in the solute load, since the decrease in concentration is outweighed by the increase in water discharge (Fig. 3.28).

3.5 Sediment yield and landscape models

Numerous attempts have been made to explain the global pattern of sediment yield in terms of both topographic and climatic factors. Models linking sediment and solute yield to topographic and climatic factors appear to work reasonably well in catchments composed of essentially one bedrock lithology, and in large drainage basins where the effects of a varied hinterland geology cancel out. At intermediate scales, however, it is likely that sediment yields are strongly influenced by lithological contrasts and by land-use practices.

Sediment yield models commonly use the sediment loads of the world's major rivers as an index of

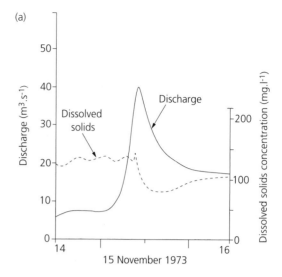

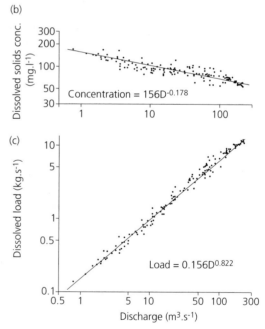

Fig. 3.28 Solute load response to a flood event in a small English river (River Exe at Thorverton). Drop in dissolved solids concentration during peak flood discharges (a) results in the inverse relation between solute concentration and discharge shown in (b), despite the fact that the total dissolved load has a positive relation in the rating curve shown in (c). After Walling (1984) [35].

sediment yield. These sediment load data are biased towards measurements taken in the downstream reaches of drainage basins, and therefore reflect the sum total of denudational and depositional processes taking place in the drainage basin. In particular, the river mouth fluxes are affected by the extent to which sediment is stored in the river floodplain, as we have seen in the case study of the Indus sediment routeing system (Section 3.3.1). The process of storage is likely to be particularly important in large catchments with substantial transportational and depositional components to the sediment routeing system, which is primarily why a number of workers have noted an inverse relationship between the sediment yield (denudation per unit surface area averaged over the entire drainage basin) and drainage basin area (Fig. 3.29).

Our analysis of weathering indicates that temperature and water availability are important factors in controlling the development of regolith, and thereby the provision of the raw material for erosion. In addition, our understanding of hillslope processes indicates that erosivity is determined by a combination of climatic (especially rainfall amount and intensity)

and topographic (e.g. slope) factors. In the following analysis the impact of these factors on river sediment loads is assessed.

Table 3.4 gives the discharges and suspended solid and particulate loads for 97 major rivers from a variety of tectonic, topographic and climatic settings. The sediment and solute loads were measured close to river mouths and the impact of human activity has been assessed and corrected for where possible. No bedload sediment transport data have been included. Although bedload may constitute a significant proportion of the total sediment load (up to 25%), bedload transport rates are generally low in the downstream reaches of the world's major rivers. Small, mountainous drainage basins have not been included. The sediment load data of Table 3.4 are examined in the light of data on the general characteristics of the catchment (surface area, depositional area, main stream length, drainage basin length), relief (mean height, maximum height, slope angle of river bed), climate (total annual precipitation, maximum monthly precipitation, precipitation peakedness, mean annual temperature, annual temperature range) and discharge (mean flow rate, maximum flow rate, discharge peakedness, runoff). Suspended sediment loads have been converted to minimum sediment yield for the drainage basin, neglecting bedload and sediment storage upstream.

3.5.1 The relation between sediment yield and environmental factors
Rates of precipitation

A number of workers have suggested a relationship between mean annual precipitation and sediment yield, strongly mediated by the effects of vegetation cover. Although there are differences between the conclusions of the various workers, they agree on a maximum of sediment yield in semi-arid areas, where vegetation cover is sparse, and a second maximum in regions where mean annual precipitation exceeds 1000 mm. Most hillslope processes are sensitive to the intensity of rainfall rather than the annual total. It has been suggested, therefore, that rainfall peakedness (ratio of average monthly precipitation and maximum monthly precipitation) is one of the main factors governing sediment yield [36]. However, both indices of the rainfall regime only account for a small amount of the variance in the global dataset of major rivers.

Topography

It has been suggested from data from 280 rivers,

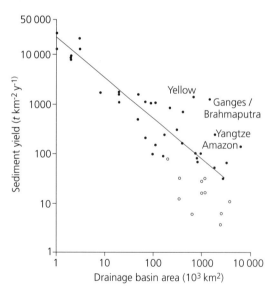

Fig. 3.29 Sediment yield versus drainage basin area for the world's major sediment discharging rivers (over 10 Mt y^{-1}). The open circles are the low yield rivers of Africa and the Eurasian Arctic. The named large rivers all have very high yields compared to their drainage basin area. The slope of the correlation line reflects the area averaging involved in calculating sediment yield, with larger basins having relatively larger proportions of lowland and smaller basins having relatively larger proportions of highland. After Milliman & Meade (1983) [29].

including small mountainous streams, that sediment yield is a log-linear function of basin area and maximum elevation [37]. Similarly, others have suggested a strong control by the ratio of maximum basin elevation to basin length (a kind of basin slope envelope) and runoff [38]. Still others have suggested that mean height is the single most important factor controlling sediment yield [39, 40]. None of these parameters appears to explain the variance in the sediment yield data presented in Table 3.4, suggesting that although topographic characteristics are clearly influential, and may dominate in certain local settings, they do not explain the global patterns of sediment yield revealed from the world's largest catchments.

3.5.2 The importance of tectonic activity

A multivariate analysis of sediment yield data and associated parameters reveals that a combination of environmental and topographic factors explains approximately half of the variance in the global sediment yield data [41]. Although a good deal of the additional variance might be explained by poor data quality, it is more likely that one or more additional factors are involved in determining global patterns of sediment yield.

We have previously learnt that the regolith thickness reflects a balance between regolith loss by denudation and regolith generation by weathering of bedrock. We also know that the permeability of the regolith decreases with depth, causing a downward decrease in reaction rates. Thick regoliths have low weathering rates at the weathering front. Consequently, the

optimal conditions for denudation are rapid removal combined with rapid uplift of fresh rock into the weathering zone. These conditions are particularly well satisfied in tectonically active mountain belts. Active tectonics is important, therefore, not only in generating high topography, but also critically in the continuing supply of bedrock into the weathering zone. This hypothesis can initially be tested by comparing sediment yield from a database including 285 drainage basins of varying area and climatic and topographic characteristics to the tectonic setting of the source area of the drainage basin (Fig. 3.30). It can be seen that sediment yields increase for a given drainage area from cratonic settings characterized by very low rates of tectonic uplift of rocks (sediment yields typically less than $100\,t\,km^{-2}\,y^{-1}$) through to contractional mountain belts experiencing high rates of tectonic thickening (sediment yields typically $100-10\,000\,t\,km^{-2}\,y^{-1}$). Ideally, therefore, a factor expressing the tectonic uplift of rocks should be added to the sediment yield equation for estimating regional patterns of denudation.

On the scale of a continental transect, geomorphic and tectonic data have been combined with both solute and sediment yield data in South American rivers [42]. The highest sediment and dissolved loads were found in mountain belts, evaporites and carbonates providing most of the dissolved loads, and weakly consolidated clastic sediments providing the bulk of the sediment load. Erosion on low-lying shield areas and low-lying alluvial plains is very low because as a regolith/soil develops the rate of chemical weathering drops (Section

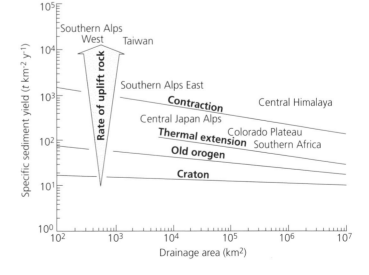

Fig. 3.30 Sediment yield versus drainage area with a least-squares regression of the data according to tectonic setting. The line for thermal extension, meaning 'mantle involvement' or 'active' rifting, is tentative because of the small number of data points. After Hovius (1997) [41].

Table 3.4 Discharges and loads of the world's 97 largest rivers. From the Oxford Global Sediment Flux Database.

River	A (km²)	A_d (km²)	L_r (km)	L_b (km)	S_s (Mt y⁻¹)	S_d (Mt y⁻¹)	E (t km⁻² y⁻¹)	D (mm ka⁻¹)	H (m)	H_{max} (m)	P (mm y⁻¹)	P_{max} (mm month⁻¹)	T (°C)	T_{range} (°C)	Q (m³ s⁻¹)	Q_{max} (m³ s⁻¹)	R (mm y⁻¹)	R_{cf}
Amazon	6 150 000	4	6299	3310	1150	223	187	69	426	6768	1490	260	18	3	200 000	211 838	1026	0.67
Amudar'ya	309 000	10	2620	1380	94	27	304	113	–	7459	222	41	9	28	1450	3 195	148	0.70
Amur	1 855 000	12	4416	2455	52	20	28	10	571	2499	455	119	–2	46	10 300	19 413	175	0.38
Apalachicola	51 800	–	880	521	0.17	1	3	1	–	1458	1223	150	18	19	641	1 177	390	0.32
Brahmaputra	610 000	21	2840	1270	520	61	852	316	2734	7736	2661	664	14	14	19 300	45 700	998	0.38
Brazos	114 000	–	1400	1020	31	3	272	101	–	950	810	119	19	22	222	472	61	0.07
Burdekin	131 000	7	680	520	3	–	23	9	339	1277	640	140	23	11	476	–	115	0.17
Chao Phraya	160 000	22	1200	700	11	3	69	26	–	2300	1246	283	26	7	824	–	162	0.13
Chari	880 000	7	1400	920	4	3	5	2	–	3071	1233	263	27	6	1320	3 177	47	0.04
Colorado (CA)	640 000	–	2333	1300	150	15	234	87	1652	4730	310	47	12	24	634	–	2	–
Colorado (TX)	100 000	–	1450	790	13	5	130	48	–	1440	598	75	18	17	634	–	200	0.33
Columbia	670 000	–	1950	1200	15	35	22	8	1329	3748	568	84	7	24	7930	11 939	373	0.62
Colville	60 900	–	662	520	6	1	98	36	469	2320	–	–	–	–	492	–	255	–
Copper	61 800	–	360	325	70	–	1133	420	–	5952	–	–	–	–	1240	–	633	–
Danube	815 000	16	2860	1250	70	60	86	32	501	3087	960	127	7	21	6660	8 921	258	0.29
Delaware	22 900	–	518	350	0.68	1	30	11	–	1360	900	103	10	27	329	653	453	0.50
Dnepr	504 000	11	2200	1045	2.1	11	4	1	152	325	526	87	7	27	1650	3 255	103	0.20
Dnestr	72 100	20	1350	695	2.5	4	35	13	–	2058	635	97	7	25	379	586	166	0.27
Don	422 000	11	1870	770	6	14	14	5	–	367	527	74	8	27	856	2 797	64	0.12
Ebro	86 800	–	930	470	21	–	242	90	402	3404	632	80	12	16	492	2 475	179	0.32
Elbe	148 000	9	1110	705	0.84	–	6	2	–	1603	694	86	7	19	690	1 053	147	0.22
Fly	64 400	24	744	475	70	–	1087	403	1140	3993	–	–	–	–	4760	–	2331	–
Fraser	220 000	–	1110	736	20	11	91	34	1140	4043	300	50	10	26	3550	7 014	509	–
Ganges	980 000	28	2510	1560	524	75	535	198	890	8848	1573	430	20	15	11 600	34 358	373	0.25
Garonne	86 000	6	650	330	2.2	–	26	10	–	3308	933	114	11	16	600	1 010	220	0.24
Godavari	287 000	1	1500	920	170	–	592	219	413	1300	1122	321	26	13	2920	11 816	321	0.29
Haiho	50 800	–	650	460	81	–	1595	591	–	2870	463	166	12	32	63	–	39	0.09
Indigirka	360 000	2	1726	1120	14	2	39	14	713	3147	131	33	–16	69	1740	5 568	152	–
Indus	960 000	10	3180	1610	250	41	260	96	1855	8611	543	132	17	25	7610	10 128	250	0.49
Irrawaddy	410 000	12	2300	1420	260	92	634	235	758	5881	1878	375	20	9	13 600	–	1046	0.56
Jana	238 000	–	872	815	3	1	13	5	703	3000	163	43	–12	38	920	3 738	122	0.76
Kemijoki	37 800	3	600	320	0.15	–	4	1	–	807	437	73	–2	27	534	1 605	446	0.96
Kizil Irmak	75 800	8	1151	375	23	–	303	112	–	3916	410	65	10	23	192	292	80	0.18
Kolyma	647 000	3	3513	1150	6	–	9	3	564	3147	249	62	–12	54	2250	10 102	110	0.45
Krishna	256 000	0	1290	860	65	–	254	94	–	1892	834	227	26	9	1607	6 253	198	0.24
Kura	188 000	2	1360	650	36	5	191	71	–	4480	668	147	8	26	515	904	86	0.13

River																	
Kuskokwim	116 000	–	1080	700	7.5	–	65	24	–	–	490	90	-4	38	–	–	–
Lena	2 430 000	4	4400	2525	12	88	5	2	602	6194	355	94	-5	47	16 200	74 361	210
Liao He	170 000	–	1350	515	41	–	241	89	496	2029	640	170	7	39	190	–	35
Limpopo	440 000	–	1600	840	33	–	75	28	766	2322	520	107	21	10	160	1425	11
Loire	120 000	5	1110	540	1.5	–	13	5	–	1885	795	84	11	15	–	–	–
Mackenzie	1 448 000	–	4240	2270	125	64	86	32	634	3955	390	58	-1	37	9830	–	214
Magdalena	260 000	–	1530	1050	220	28	846	313	1203	5493	2670	311	19	11	6980	–	847
Mahakam	75 000	20	420	–	12	–	160	59	–	2988	–	–	–	–	2000	–	841
Mahanadi	133 000	1	858	630	60	–	451	167	330	1027	1456	425	25	15	1970	6933	467
Mekong	810 000	22	4500	2950	160	60	198	73	1062	6000	1800	393	24	6	14 900	–	580
Meuse	29 000	–	925	440	0.70	–	24	9	–	692	968	106	9	15	331	596	360
Mississippi	3 344 000	–	5985	2220	400	125	120	44	656	4400	612	99	9	28	18 400	22 730	174
Mobile	57 000	–	1064	580	2.3	4	40	15	–	1360	1349	178	17	20	1590	–	880
Murray	910 000	24	3490	1000	30	9	33	12	266	2239	582	71	17	14	698	855	24
Niger	1 112 700	7	4160	1950	32	10	29	11	429	2918	937	254	28	8	6020	–	171
Nile	2 715 000	9	6670	3600	125	18	46	17	662	5110	832	177	25	7	317	1711	4
Ob	2 500 000	7	5570	2530	16	50	6	2	301	4506	406	68	0	41	12 200	32 421	154
Oder	112 000	35	909	515	0.13	7	1	0.4	–	1603	723	95	7	19	539	776	152
Orange	102 000	0	1860	1285	91	12	89	33	1241	3482	415	75	17	14	2890	–	89
Ord	46 000	1	400	400	22	–	478	177	297	1000	530	150	25	13	165	645	113
Orinoco	945 000	–	2740	1550	150	39	159	59	456	5493	1300	216	25	2	34 900	58 822	1165
Parana	2 600 000	–	4500	2175	112	56	43	16	564	6720	1027	188	19	9	18 000	–	218
Pechora	322 000	5	1810	760	6.1	7	19	7	147	1894	498	71	-2	36	3360	13 810	329
Po	75 000	19	691	480	18	10	240	89	793	4810	1202	162	10	20	1490	1936	626
Red (Song Koi)	120 000	9	1200	860	123	–	1025	380	420	3000	2768	470	15	12	3810	–	1001
Rhein	225 000	2	1360	725	0.72	17	3	1	–	4158	1121	129	7	18	2243	2909	314
Rhone	99 000	6	810	540	60	–	606	224	754	4810	960	113	9	18	1550	1898	494
Rio Colorado (Arg)	65 000	–	1000	1460	6.9	–	106	39	–	6960	342	65	15	17	–	–	–
Rio Grande	670 000	–	2870	1725	30	2	45	17	1279	4295	160	67	15	19	95	240	4
Rio Grande (Arg)	125 000	19	960	650	1	–	8	3	–	4577	774	201	18	7	308	940	78
Santiago	–	–	729	880	13	–	100	37	745	4800	423	76	12	15	951	–	231
Rio Negro (Arg)	130 000	–	1400	625	17	–	96	36	912	2959	940	218	18	6	285	–	50
Rufiji	178 000	–	610	385	3	22	41	15	–	3187	1052	212	15	19	678	1170	293
Sacramento	73 000	–	–	–	–	–	–	–	–	–	–	–	–	–	9510	–	923
Salween	325 000	5	3060	1725	100	1	308	114	–	6070	–	–	–	–	–	–	–
San Joaquin	80 100	–	560	450	1	1	12	4	–	4420	525	105	10	20	123	211	48
Sanaga	135 000	0	860	660	5.9	–	44	16	–	2000	1690	724	24	2	2069	5658	483

River	
Kuskokwim	–
Lena	0.60
Liao He	0.07
Limpopo	0.02
Loire	–
Mackenzie	0.56
Magdalena	0.32
Mahakam	0.32
Mahanadi	0.33
Mekong	0.37
Meuse	0.26
Mississippi	0.65
Mobile	0.05
Murray	0.18
Niger	0.005
Nile	0.38
Ob	0.23
Oder	0.22
Orange	0.21
Ord	0.90
Orinoco	0.21
Parana	0.67
Pechora	0.56
Po	0.37
Red (Song Koi)	0.28
Rhein	0.52
Rhone	–
Rio Colorado (Arg)	0.01
Rio Grande	0.10
Rio Grande (Arg)	0.56
Santiago	0.56
Rio Negro (Arg)	0.05
Rufiji	0.28
Sacramento	–
Salween	0.09
San Joaquin	0.29

Continued on p. 138.

Table 3.4 Continued.

River	A (km^2)	A_d (km^2)	L_r (km)	L_b (km)	S_d (Mt y^{-1})	S_d (Mt y^{-1})	E (t km^{-2} y^{-1})	D (mm ka^{-1})	H (m)	H_{max} (m)	P (mm y^{-1})	P_{max} (mm per month)	T (°C)	T_{range} (°C)	Q (m^3 s^{-1})	Q_{max} (m^3 s^{-1})	R (mm y^{-1})	R_{cf}
Sao Francisco	640 000	2	2800	1510	6	–	9	3	609	1800	1145	232	22	5	3080	5139	152	0.13
Seine	78 600	6	780	370	1.1	12	14	5	–	902	711	77	9	15	685	–	154	0.39
Senegal	441 000	2	1430	900	1.9	1	4	1	–	1000	665	235	29	1	761	3184	54	0.08
Sepik	81 000	30	825	425	80	–	988	366	–	4500	–	–	–	–	2440	–	950	–
Sevemaya Dvina	350 000	–	–	850	4.5	14	13	5	119	200	514	72	–1	32	3360	13 626	303	0.59
Shatt al Arab	1 050 000	19	2760	1475	103	18	98	36	669	4168	498	97	17	26	1460	–	44	0.10
St Lawrence	1 185 000	–	3060	1650	4	59	3	1	265	1917	794	99	6	31	14 300	–	381	0.48
Susitna	50 300	–	454	370	25	–	497	184	1031	6190	–	–	–	–	1270	–	796	–
Susquehanna	72 500	–	733	445	1.8	–	25	9	–	950	1060	120	14	21	1034	2087	450	0.42
Syrdar'ya	219 000	12	2210	1440	12	12	55	20	–	5880	295	51	9	28	581	1011	84	0.30
Tana	91 000	–	720	470	32	–	352	130	555	5200	317	99	27	4	171	372	59	0.19
Terek	43 200	17	623	390	24	3	556	206	–	5642	497	84	9	25	–	–	–	–
Ural	237 000	10	2430	1020	3	3	13	5	–	1000	364	69	2	37	301	1409	40	0.11
Uruguay	240 000	–	–	1085	11	8	46	17	–	2000	1534	182	17	1	5010	7530	658	0.43
Vistula	198 000	33	1014	600	2.5	13	13	5	–	2499	740	114	7	13	1044	1910	166	0.25
Volga	1 350 000	7	3350	1640	26	77	19	7	–	1638	489	68	2	32	8400	22 421	196	0.42
Volta	394 000	5	1600	980	19	3	48	18	–	500	1046	287	28	7	1270	5120	102	0.10
Wester	46 000	12	724	375	0.33	–	7	3	–	1142	848	86	7	17	313	475	215	0.26
Xi Jiang	464 000	5	2129	1150	80	132	172	64	670	2500	1314	230	19	20	9510	15 985	646	0.49
Yangtze	1 940 000	–	5520	2730	480	226	247	91	1688	6800	1173	235	11	20	28 500	47 300	463	0.40
Yellow (Huang He)	980 000	–	4670	2070	120	22	122	45	1885	5500	484	144	11	29	1550	2858	50	0.10
Yenisei	2 580 000	9	5550	2250	13	65	5	2	749	3492	439	84	–3	41	17 800	77 671	218	0.50
Yukon	855 000	–	3000	2140	60	34	70	26	741	6194	236	49	–5	36	6180	18 132	228	0.95
Zaire	3 700 000	–	4370	2020	32.8	36	9	3	740	4507	1586	277	25	3	40 900	57 200	349	0.22
Zambezi	1 400 000	8	2660	2040	48	15	34	13	1033	2606	957	231	22	8	6980	–	157	0.16

Definition of variables

General characteristics drainage area

A drainage area (km^2):   The surface area from which a river collects surface runoff, including the peripheral zones which only contribute runoff in case of exceptional rainfall.

A_d proportion depositional (%):   The surface area of the part of the drainage basin covered with Holocene fluvial sediments, expressed as a percentage of the total drainage area.

L_r river length (km):   The length of the flowline of the main stream of the catchment from its headwaters to the river mouth.

L_b basin length (km):   The maximum length of the drainage basin in the direction of the main stream.

Sediment

S_s annual suspended load (M t y⁻¹): The total amount of material carried in suspension by the river annually, as measured at the mouth of the river.

S_d annual dissolved load (M t y⁻¹): The total amount of material carried in solution by the river annually, as measured at the mouth of the river.

E annual sediment yield (t km⁻² y⁻¹): The amount of material eroded mechanically from a unit surface area annually, averaged over the drainage area, calculated from the suspended load and the drainage area.

D mechanical denudation rate (mm ky⁻¹): The average lowering of the surface per unit time resulting from mechanical denudation, as calculated from the specific sediment yield.

Relief

H mean height (m): The average height of the drainage basin as calculated by Summerfield and Hulton [38] and Pinet and Souriau [40].

H_{max} maximum height (m): The elevation of the highest point in the drainage basin.

H_{pk} relief peakedness: The ratio of the mean height and the maximum height of the drainage basin.

H_r relief ratio* (m km⁻¹): The ratio of the maximum height of the drainage basin and the basin length.

a_{riv} slope river bed (m km⁻¹): The average slope angle of the bed of the main stream of the catchment, calculated as the ratio of the maximum height of the drainage basin and the length of the main stream.

Climate

P total annual precipitation (mm y⁻¹): The total annual precipitation, averaged over a number of meteorological stations distributed more or less equally over the erosional part of the catchment.

P_{max} maximum monthly precipitation (mm per month): The average amount of precipitation for the wettest month of the year.

P_{pk} precipitation peakedness*: The ratio of the average monthly precipitation and the maximum monthly precipitation.

T mean annual temperature (°C): The mean annual daytime temperature, averaged over a number of meteorological stations distributed more or less equally over the erosional part of the catchment.

T_{range} annual temperature range (°C): The difference between the average daytime temperatures for hottest and coldest months.

Discharge

Q mean flow rate (m³ s⁻¹): The long-term average water discharge per unit time, at the mouth of the river.

Q_{max} maximum flow rate (m³ s⁻¹): The average flow rate per unit time, for the month with the greatest discharge.

Q_{pk} discharge peak*: The ratio of the average flow rate to the maximum flow rate.

R specific runoff (mm y⁻1): The total annual height of the water column on a unit surface area which leaves the catchment as surface runoff, as calculated from the mean flow rate and the drainage area.

R_{cf} runoff coefficient: The ratio of the total annual precipitation in the catchment and the total annual water discharge at the river mouth.

* Marked variables have been calculated from other variables in the data set. Some of the calculated variables here not been included in the table.

3.2.2), the weathering rate and erosion rate eventually balancing. Erosion is therefore *transport-limited*, since the rate of physical removal of material is determined only by the erosivity of the transport processes. But in steep terrains, physical transport processes are almost always capable of removing loose material from the regolith—that is, erosion is *weathering-limited*. This is why the tectonic uplift rate is so important in explaining sediment yields. It has been estimated [1] that rock uplift rates of at least $1\,mm\,y^{-1}$ are required to sustain combined chemical and physical erosion. We return to these concepts in relation to a coupled erosion–tectonic model of mountain belts later in this section.

It is a very daunting task to monitor the sediment and solute loads in thousands of river drainage basins around the world. Instead geochemistry can be used to develop a model of stead-state erosion [1] wherein the statistical properties of a regional landscape, such as the magnitude–frequency characteristics of landslides or debris flows, or drainage density, do not significantly change. The predicted steady-state yield of solids does not increase as steeply with increasing runoff as does solute yield (Fig. 3.26). One can conclude from this that the deposition rates in basins, which reflect solid yields, are less sensitive indicators of climate change than *sediment composition*, which reflects varied leaching rates of metals in the source area.

Figure 3.26 clearly distinguishes between mountainous drainage basins with high sediment yield and lowland areas with low sediment yields. This division also corresponds to weathering-limited terrains characterized by rapid erosion and transport-limited terrains characterized by soil development and sediment storage.

Coupled tectonic–erosion model of mountain belts

The debate about the relative contributions of erosion and tectonics to topography goes back a century to Davis and Penck. Clearly the three-dimensional pattern of mechanical uplift attributable to tectonic shortening in collision zones is strongly coupled to the erosional unroofing of the orogen. Rather than erosion acting to degrade an instantaneous tectonic event, the two processes act fully interactively at similar time-scales.

The basic template for the coupled tectonic–erosion model of mountain belts is the doubly vergent wedge shown in Fig. 3.31 [43]. This orogenic wedge has an asymmetrical distribution of rainfall which causes different rates of erosion on the windward and leeward flanks of the wedge. This drives more rapid uplift of the windward flank.

The Southern Alps of New Zealand provide a good natural example. Precipitation west of the main divide is very high (over $12\,m\,y^{-1}$) whereas the eastern flank is in a rain shadow with less than $1\,m\,y^{-1}$. The drainage density on the wet western flank is high, with rivers in steep gorges spaced about 5–$10\,km$ apart and landslide-dominated interfluve hillslopes. The drier eastern flank is characterized by broadly spaced braided rivers. The gross pattern of uplift of rocks (relative to a stationary datum) is of very high rates along the Alpine Fault (over $5\,mm\,y^{-1}$), moderate rates (about $1\,mm\,y^{-1}$) along the main divide, and zero at the east coast. This pattern of tectonic uplift is a response to the relative plate motion between the obliquely convergent Pacific and Australian plates.

Early attempts to model the mass balance in collisional mountain belts relied on the diffusion approach developed for the decay of fault scarps and hillslopes (see Section 3.3). In such a one-dimensional scheme the change in elevation can be related to the local curvature of the topographic surface

$$\frac{\partial h}{\partial t} = \kappa \left(\frac{\partial^2 h}{\partial x^2} \right) + V(x) \tag{3.29}$$

where h is the elevation above sea level, x is the horizontal coordinate, κ is the effective erosional diffusivity and $V(x)$ is the tectonic uplift rate. When the landscape is in steady state, that is when there is no elevation change and the uplift is balanced by the erosion, this equation can be integrated to give the maximum elevation as a function of the uplift velocity

$$h_{max} = \frac{L^2 V}{2\kappa} \tag{3.30}$$

which is identical to the case for hillslopes given in equation (3.17), where L is the ridgecrest to valley bottom distance for the transverse drainage characterizing the Southern Alps (that is, the stream spacing is $2L$). Taking the stream spacing for the western flank of the Southern Alps as $6000\,m$, the maximum uplift rate of $5\,mm\,y^{-1}$, and the maximum elevation as $1500\,m$, the effective erosional diffusivity becomes $15\,m^2\,y^{-1}$. Note that stream spacings must always be small if erosion is able to achieve a steady-state topography with high tectonic uplift rates. The

(a)

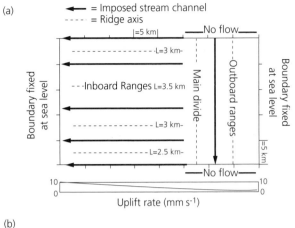

(b)

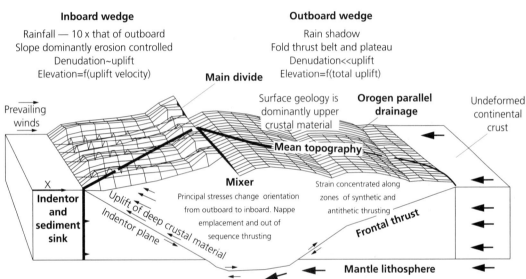

Fig. 3.31 The doubly vergent orogenic wedge. (a) The boundary and initial conditions for the numerical models in Koons (1989) [43]. (b) Plate velocities outside the wedge, tectonic transport directions within the wedge, and surface topography for a doubly vergent wedge with precipitation from the left. The marked asymmetry in denudation promotes a strong asymmetry in uplift within the wedge.

aspect ratio of the range is h_{max}/L, which for the parameter values given above represents a steady-state slope of 26° (aspect ratio of 0.5). This corresponds quite well with the observed topography of the Southern Alps.

The aspect ratio can also be written as $VL/2\kappa$, which, by inspection of equation (3.2), can be seen to have the form of a Peclet number. It is unlikely that erosional processes can keep pace with tectonic uplift rates to produce a steady-state topography when the erosional Peclet number is greater than about 0.6.

In contrast to the narrowly spaced, deep and efficient drainage of the western flank, the rivers of the eastern flank are widely spaced. Combined with low erosional diffusivities expected in the rain shadow of the main divide, this produces high mountains but with low tectonic uplift rates. The elevations of interfluve ridges in the eastern flank are a function of the total uplift rather than being coupled to uplift and erosion rates.

It can reasonably be objected that whereas hillslope erosion can be modelled diffusively (Section 3.3), fluvial transport is a very different and more efficient process. Consequently, the effective erosional diffusivity should become larger at the larger spatial scales where fluvial transport dominates. Apart from the problem of scale, there is the additional problem that erosion is dependent on weathering providing material in a transportable state. Weathering may therefore be the rate-limiting factor in the evolution of some topography. Thus in arid climates where weathering rates are low, we should expect the topography to be weathering-limited rather than diffusive. In the wet western flank of the Southern Alps, slopes are soil-covered because of the rapid rates of weathering. The development of steady-state topographies in mountain belts is therefore unlikely to be achieved unless weathering is sufficiently rapid to ensure that transport processes are not limited.

The linkage of topography and tectonics of a deforming wedge was initiated by considering the wedge to build to a critical taper angle. Beyond this critical slope the wedge deforms by frontal imbrication (forward propagation), extensional collapse or by erosion at its surface to remain subcritical. Building of the critical taper takes place by ductile thickening and thrust-related shortening within the deforming wedge. The analogy is with a pile of sand pushed by a bulldozer. In the critical taper theory the topographic slope can be related directly to the shear stress and normal stress, and the mean topographic slope is a measure of the strengths of the deforming materials comprising the wedge.

Mountain belts are not like a bulldozed pile of sand, however, in one important respect. They are two-sided. One needs to envisage, therefore, that instead

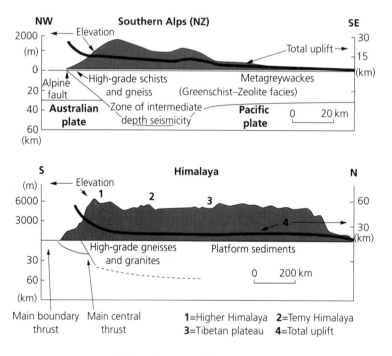

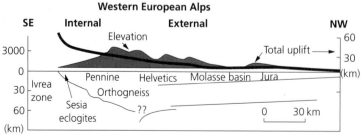

Fig. 3.32 Comparison of the Alps, Himalayas and Southern Alps in terms of their asymmetry and uplift history. All three orogens are asymmetrical in topography, with deeper crustal rocks exposed on the left-hand sides because of higher uplift (exhumation) rates. After Koons (1989), with permission from [43].

of one infinitely high indentor, we have a low indentor over which material may pass. This produces (Fig. 3.31) an orogen with a wedge adjacent to the indentor and an oppositely vergent wedge extending on to the undeformed material. High rates of erosion on the windward flank keep this side of the orogen steep and short, whereas the leeward flank is broad. Since the two flanks are linked by the mechanical behaviour of the rocks in the deforming wedge, the erosion on the steep windward flank controls the topography of the leeward flank.

A number of mountain belts share these large-scale topographic features, such as the Himalayas and European Alps in addition to the Southern Alps of New Zealand. In all cases the outcrops of once deeply buried rocks show that the greatest denudation (exhumation) has taken place asymmetrically (Fig. 3.32), being concentrated adjacent to the indenting plate. On the leeward (outboard) flank the topography shows a similar pattern to the total tectonic uplift, whereas on the windward (inboard) flank the topography is independent of the total uplift. Since erosion rates are low on the leeward flank, widely spaced rivers are commonly focused to run along strike by tectonic structures. This is in contrast to the narrowly spaced steep transverse rivers of the windward flank.

This short study has looked at the coupling between erosion and tectonics in collisional orogens. Similar coupling, however, must take place in other settings. For example, denudation of the flexurally uplifted rift margin of South Africa has also been investigated using a coupled erosion–tectonic model. At a somewhat smaller scale, there are also exciting developments in the linkage of drainage basin development and fault propagation in extensional terrains such as the Basin and Range province of the western USA.

3.6 Human impact on sediment yield

3.6.1 Human impact in the drainage basin

A very wide range of human activities have impact on the drainage basin through engineering projects, land-use management and planning policies in general. Many human activities affect the hydrology of drainage basins, and thus the sediment transfer through them. A number of possibilities have already been sketched out in previous chapters:

• deforestation may cause an increase in surface runoff and soil erosion;
• dam construction may pond sediment upstream and increase erosion downstream;
• major engineering projects may reduce sediment supply to deltas, leading to coastal inundation.

The focus of this section is the interface between man and the drainage basin. There are a number of key issues at this interface, one of which will be concentrated upon, namely, the impact of humans on sediment transfer.

The loss of soil from the land is associated with a set of environmental problems including decreased agricultural productivity, increased runoff and therefore flood hazard, and siltation of reservoirs downstream. Soil erosion may be accelerated by human activities such as deforestation, overgrazing, construction activity, and poor agricultural management. Although soil erosion presents us with a scientific problem, therefore, it is fundamentally also a hazard involving social and economic conditions.

The continuing problem of soil erosion is less to do with a lack of understanding of the physics involved in soil loss and conservation, and more to do with failed agricultural programmes and policies, many of which are applied without due appreciation of local cultural and pastoral practices [44]. The success of soil conservation programmes depends on the taking into account of the social, political and economic dimensions of the proposed changes to land use.

The use of sediment flux data in evaluating the global patterns of denudation was treated in Sections 3.4 and 3.5. On a more local scale, sediment transfer through the drainage basin from eroding hillslopes, through storage systems on slopes (colluvium), floodplains and channels to the output point of the catchment provides significant challenges in environmental management. Sediment transfer affects drainage basin management in a number of ways.

• The transfer and storage of sediment in alluvial valleys affects river planforms and the rates at which river courses migrate or avulse (rapid major change of course rather than gradual migration). These changes affect the navigability of channels, their likelihood of flooding, the loss of property adjacent to channel boundaries, and the stability of bridges, embankments, piers and jetties.

• The high sediment concentrations of rivers may clog up channels, irrigation schemes and especially ponded basins behind dams. Underestimation of the sediment concentration of rivers may lead to a shortening of the lifetime of reservoirs without frequent, extremely costly dredging operations. The building of a series of dams on the Damodar River, India, around 1950 resulted in the rapid silting up of reservoirs because the sediment load of the river had been seriously underestimated. The associated decrease in the water discharge through the river has resulted in excessive sedimentation downstream and the port of Calcutta now needs constant dredging to keep it navigable from the Bay of Bengal. High sediment concentrations also affect the fish and other biota in the river system and the water quality for industrial and domestic usage.

• Contaminants attach themselves to particulate matter. In this way, pesticides, toxic heavy metals and radionuclides may be transported through the drainage network by adsorbtion on sediment. This is particularly important where land-use changes—for example, the development of mining projects or intensive agriculture—take place.

We have previously considered the factors responsible for sediment yields in drainage basins from a broad regional to global perspective. Within individual basins, or within particular climatic–physiographic settings, one or two factors may appear to be more important than others. However, at the scale of the individual catchment, the most important control is probably land use, including urbanization, deforestation, overgrazing, fire, and off-road vehicle use.

The impact of land use on sediment yield is well illustrated by the study of 61 catchments in Kenya shown in Fig. 3.33. The proportion of cultivated land, in addition to runoff, is used as a standard input to sediment yield estimates in parts of the USA. The response of a catchment to increased erosion as a result of changed land use may be to increase the volumes of sediment stored on slopes as colluvium or in channels and floodplains, so that the effects of erosion are not felt at the river mouth. This emphasizes the difficulty

of using sediment load data obtained from gauging stations located near river mouths to estimate denudation rates in distant uplands.

On a local scale, the formulations aimed at explaining the global patterns of sediment yield, such as those described in Section 3.5, are most inappropriate since they gloss over important local details such as infiltration/runoff characteristics. In small catchments the local geology and soil types are extremely important in controlling infiltration capacities and therefore the amount of hillslope erosion. There are a large number of equations designed for local purposes that express sediment yield as a function of rainfall characteristics, topographic characteristics, soil and land-use characteristics.

3.6.2 Deforestation

The human impact on vegetation has important implications for sediment yield and soil conservation, and perhaps none is more important than the deliberate clearance of forest for agriculture or other purposes. Deforestation is not a Third World problem of the second half of the twentieth century, although the current unprecedented awareness of environmental issues has tended to highlight the particular problem of the destruction of tropical rainforests. Deforestation has been carried out for as long as humans have indulged in agriculture. The greatest losses of forest since pre-agricultural times have been, in descending order, temperate forests (32–35%), subtropical woody savannas and deciduous forests (24–25%), tropical evergreen forests (4–6%) [46]. These percentage losses reflect the accessibility of the different types of forest and human population densities. Britain has been almost completely deforested in a number of phases since 3900 BP. The landscape of Europe was transformed in the Middle Ages as growing populations moved into virgin lands. The colonization of North America resulted in a dramatic loss of forest (Fig. 3.34), with temperate forest being reduced to about 6% of its area when the first European settlers arrived just 300 years ago. In the USA, the process has been halted and slightly reversed since the 1930s and 1940s.

The current focus is on the loss of tropical rainforest. Food and Agriculture Organization estimates put the total annual deforestation of tropical rainforest in 1990, for 62 rainforest countries making up about 80% of the total rainforest area of the world, as 16.8×10^6 ha. This is more than the total area of temperate forest remaining in North America. The rate of rainforest deforestation has also increased

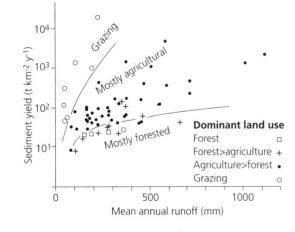

Fig. 3.33 Impact of land use on sediment yield based on 61 catchments in Kenya. Modified from Dunne (1979) [45].

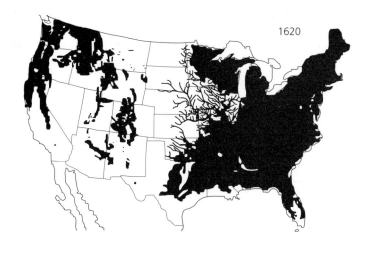

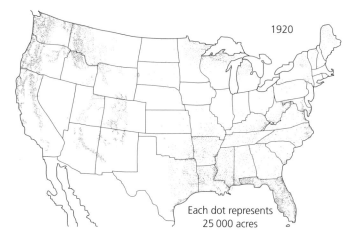

Fig. 3.34 Loss of forest due to human colonization of North America between 1620 and 1920. After Goudie (1993) [47].

Each dot represents 25 000 acres

Table 3.5 Problems associated with deforestation. From Grainger (1992) [48].

1 *Reduced biological diversity*
 (a) Species extinctions
 (b) Reduced capacity to breed improved crop varieties
 (c) Inability to make some plants economic crops
 (d) Threat to production of minor forest products

2 *Changes in local and regional environments*
 (a) More soil degradation
 (b) Changes in water flows from catchments
 (c) Changes in buffering of water flows by wetland forests
 (d) Increased sedimentation of rivers, reservoirs, etc.
 (e) Possible changes in rainfall characteristics

3 *Changes in global environments*
 (a) Reduction in carbon stored in the terrestrial biota
 (b) Increase in carbon dioxide content of atmosphere
 (c) Changes in global temperature and rainfall patterns by greenhouse effects
 (d) Other changes in global climate due to changes in land surface processes

Table 3.6 Impact of change in land use on runoff and erosion in parts of Africa. From Goudie (1993) [47].

Locality	Average annual rainfall (mm)	Slope (%)	Annual runoff (%)			Erosion ($t\,ha^{-1}y^{-1}$)		
			A	B	C	A	B	C
Ouagadougou (Burkina)	850	0.5	2.5	2–32	40–60	0.1	0.6–0.8	10–20
Sefa (Senegal)	1300	1.2	1.0	21.2	39.5	0.2	7.3	21.3
Bouake (Ivory Coast)	1200	4.0	0.3	0.1–26	15–30	0.1	1–26	18–30
Abidjan (Ivory Coast)	2100	7.0	0.1	0.5–20	38	0.03	0.1–90	108–170
Mpwapwa (Tanzania)	c. 570	6.0	0.4	26.0	50.4	0	78	146

A, forest or ungrazed thicket; B, crop; C, barren soil.

considerably over the last couple of decades, and is now thought to represent about 2% of the total rainforest area per year [49, 50]. The problems resulting from tropical deforestation are listed in Table 3.5. Those of particular interest here are soil deterioration, leading to laterization (Section 3.2.3), soil erosion (Section 3.3.2), and accelerated slope instability. We have already seen that extensive deforestation also changes the regional climate, as models of the Amazon basin show (Section 2.5.1).

There is a clear link between deforestation, conversion to agriculture and soil loss. Forest cover protects the ground surface from the physical impact of rain; the humus, roots and rich fauna of forest soils increase infiltration capacities; forest soils are also well aggregated and held together by roots, thereby increasing their shear strength. Removal of forest, therefore, may have serious consequences for soil loss and slope instability. Order of magnitude increases in erosion rates and surface runoff following forest clearance are common (Table 3.6). Deforested areas are also subject to increased mass flows in the form of landslides, debris avalanches, and mudflows. Forest clearance, especially the construction of logging roads, accelerates mass movements. The clearance of forest by burning may lead to especially high rates of erosion for a number of years after clearance because of the damage caused to soil structure by the fire.

Further reading

Topics in weathering and soils are covered in the following texts:

J.E. Andrews, P. Brimblecombe, T.D. Jickells & P.S. Liss (1995) *An Introduction to Environmental Chemistry*. Blackwell Science, Oxford.

R.W. Christopherson (1994) *Geosystems*, 2nd edn, Macmillan College Publ. Co., NJ USA, pp. 547–83.

M.J. Kirkby and R.P.C. Morgan (1980) *Soil Erosion*. Wiley, Chichester.

M.J. Singer & D.N. Munns (1987) *Soils—An Introduction*, 2nd edn. Macmillan.

Sediment routing systems, denudation, sediment loads of rivers, sediment yield, and deforestation are treated by:

M.A. Carson & M.J. Kirkby (1972) *Hillslope Form and Process*. Cambridge University Press, Cambridge.

A. Goudie (1993) *The Human Impact on the Natural Environment*, 4th edn. Blackwell Publishers Oxford.

A. Grainger (1992) *Controlling Tropical Deforestation*. Earthscan, London.

M.J. Selby (1985) *Earth's Changing Surface*. Clarendon Press, Oxford.

I. Statham (1977) *Earth Surface Sediment Transport*. Clarendon Press, Oxford.

M.A. Summerfield (1991) *Global Geomorphology*. Longman, London.

References

1 P.O. Koons (1995) Modeling the topographic evolution of collisional belts. *Annual Reviews of Earth and Planetary Sciences* **23**, 375–408.

2 R.J. Carson (ed.) (1969) *Water, Earth and Man*. Methuen, London, Fig. 4.11.4.

3 R.S. Taylor & S.M. McLennan (1985) *The Continental Crust: its Composition and Evolution*. Blackwell Scientific Publications, Oxford.

4 J.E. Andrews, P. Brimblecombe, T.D. Jickells & P.S. Liss (1995) *An Introduction to Environmental Chemistry*. Blackwell Science, Oxford, p. 53, Fig. 3.3.

5 R.W. Raiswell, P. Brimblecombe, D.C. Dent & P.S. Liss (1980) *Environmental Chemistry*. Edward Arnold, London.

6 R. Gill (1989) *Chemical Fundamentals of Geology*. Unwin Hyman, London.

7 S.S. Goldich (1938) A study of rock weathering. *Journal of Geology* **46**, 17–58.

8 M.A. Summerfield (1991) *Global Geomorphology.* Longman, London.

9 L.G.M. Bass Becking *et al.* (1964) Limits of the natural environment in terms of pH and oxidation-reduction potentials. *Journal of Geology* **68**, 243–84.

10 N.M. Strakhov (1967) *Principles of Lithogenesis*, Vol. 1. Oliver & Boyd, Edinburgh.

11 US Soil Conservation Service (1975) *Soil Taxonomy: A Basic System of Soil Classification for Making and Interpreting Soil Surveys*, Agricultural Handbook 436. Government Printing Office, Washington, DC.

12 E.K. Berner & R.A. Berner (1987) *The Global Water Cycle: Geochemistry and Environment.* Prentice Hall, Englewood Cliffs, NJ.

13 M.A. Velbel (1993) Temperature dependence of silicate weathering in nature: how strong a negative feedback on long-term accumulation of atmospheric CO_2 and global greenhouse warming? *Geology* **21**, 1059–62.

14 P.V. Brady (1991) The effect of silicate weathering on global temperature and atmospheric CO_2. *Journal of Geophysical Research* **96**, 18101–6.

15 M.E. Raymo, W.F. Ruddiman & P.N. Froelich (1988) Influence of late Cenozoic mountain building on ocean geochemical cycles. *Geology* **16**, 649–653.

16 M.E. Raymo & W.F. Ruddiman (1992) Tectonic forcing of late Cenozoic climate. *Nature* **359**, 117–122.

17 J.D. Milliman, G.S. Quraishee & M.A.A. Beg (1994) Sediment discharge from the Indus River to the ocean: past, present and future. In: *Marine Geology and Oceanography of the Arabian Sea and Coastal Pakistan* (eds B.U. Haq & J.D. Milliman). Van Nostrand Reinhold Co., New York, pp. 65–70.

18 B.R. Naini & V. Kolla (1982) Acoustic character and thickness of sediments of the Indus Fan and the continental margin of Western India. *Marine Geology* **47**, 181–95.

19 F. Coumes & V. Kolla (1994) Indus Fan: seismic structure, channel migration and sediment thickness in the Upper Fan. In: *Marine Geology and Oceanography of the Arabian Sea and Coastal Pakistan* (eds B.U. Haq & J.D. Milliman). Van Nostrand Reinhold, New York, pp. 101–10.

20 M.J. Kirkby & R.P.C. Morgan (1980) *Soil Erosion.* Wiley, Chichester.

21 R.E. Horton (1945) Erosional development of streams and their drainage basins; hydrophysical approach to quantitative morphology. *Bulletin of the Geological Society of America* **56**, 275–370.

22 M.J. Kirkby (1969) Infiltration, throughflow and overland flow: and erosion by water on hillslopes. In: (ed. R.J. Charley) *Water, Earth and Man* (ed. R.J. Chorley). Methuen, London, pp. 215–238.

23 T.C. Atkinson (1978) Techniques for measuring subsurface flow on hillslopes In: *Hillslope Hydrology* (ed. M.J. Kirkby). Wiley, Chichester, pp. 73–120.

24 N. Hovius, C.P. Stark & P.A. Allen (1997) Sediment flux from a mountain belt derived by landslide mapping. *Geology.*

25 R.S. Anderson (1994) Evolution of the Santa Cruz Mountains, California, through tectonic growth and geomorphic decay. *Journal of Geophysical Research* **99**, 20 161–79.

26 N. Hovius (1995) Macro-scale process systems of mountain belt erosion and sediment delivery to basins. D. Phil. dissertation, University of Oxford.

27 M.A. Seidl & W.E. Dietrich (1992) The problem of channel erosion into bedrock. *Catena Supplement* **23**, 101–24.

28 M.H. Meade, T.R. Yuzyk & T.J. Day (1990) Movement and storage of sediment in rivers of the United States and Canada. In: *Surface Water Hydrology* (eds M.G. Wolman & H.C. Riggs), The Geology of North America, O-1. Geological Society of America, Boulder, CO, pp. 255–80.

29 J.D. Milliman & R.H. Meade (1983) World-wide delivery of river sediment to the oceans. *Journal of Geology* **91**, 1–21.

30 M. Meybeck (1976) Total mineral dissolved transport by world major rivers. *Hydrological Science Bulletin* **21**, 265–84.

31 M. Meybeck (1987) Global chemical weathering of superficial rocks estimated from river dissolved loads. *American Journal of Science* **287**, 401–28.

32 S. Judson & D.F. Ritter (1964) Rates of regional denudation in the United States. *Journal of Geophysical Research* **69**, 3395–401.

33 R.F. Stallard (1995) Tectonic, environmental and human aspects of weathering and erosion: a global review using a steady-state perspective. *Annual Reviews of Earth and Planetary Sciences* **23**, 11–39.

34 R.L. Burrows, B. Parks & W.W. Emmett (1979) *Sediment Transports in the Tanana River in the Vicinity of Fairbanks, Alaska, 1977–78*, Open File Report 79–1539. US Geological Survey, Washington DC.

35 D.E. Walling (1984) Dissolved loads and their measurement. In: *Erosion and Sediment Yield* (eds R.F. Hadley & D.E. Walling). Geobooks, Norwich, pp. 111–78.

36 F. Fournier (1960) *Climat et érosion: la relation entre erosion du sol par l'eau et les precipitations atmosphériques.* Presses Universitaires de France, Paris, 201 pp.

37 J.D. Milliman & J.P.M. Syvitski (1992) Geomorphic/tectonic control of sediment discharge to the ocean: the importance of small mountainous streams. *Journal of Geology* **100**, 525–44.

38 M.A. Summerfield & N.J. Hulton (1994) Natural controls of fluvial denudation rates in major world drainage basins. *Journal of Geophysical Research* **99**, 13 871–83;

39 F. Ahnert (1984) Local relief and the height limits of mountain ranges. *American Journal of Science* **284**, 1035–55.

40 P. Pinet & M. Souriau (1988) Continental erosion and large scale relief. *Tectonics* **7**, 563–82.

41 N. Hovius (1997) Controls on sediment supply by large rivers. In: *Relative Role of Eustacy, Climate and Tectonics in Continental Rocks* (eds K.W. Shanley & P.J. McCabe). Spec. Publ. Society of Economic Paleontologists and Mineralogists, Tulsa, O.K.

42 R.F. Stallard, L. Koehnken & M.J. Johnsson (1991) Weathering processes and the composition of inorganic material transported through the Orinoco River system, Venezuela and Colombia. *Geoderma* **51**, 133–65.

43 P.O. Koons (1989) The topographic evolution of collisional mountain belts: a numerical look at the Southern Alps, New Zealand. *Annual Review of Earth and Planetary Sciences*, **23**, 375–408.

44 P. Blaikie (1985) *The Political Economy of Soil Erosion in Developing Countries*. Longman, London.

45 T. Dunne (1979) Sediment yield and land use in tropical catchments. *Journal of Hydrology* **42**, 281–300.

46 World Resources Institute (1992) *World Resources 1990–91*. Oxford University Press, New York and Oxford.

47 A. Goudie (1993) *The Human Impact on the Natural Environment*. Blackwell Publishers, Oxford.

48 A. Grainger (1992) *Controlling Tropical Deforestation*. Earthscan, London.

49 N. Myers (1992) Future operational monitoring of tropical forests, an alert strategy. In: *Proceedings of the World Forest Watch Conference. São Jose dos Campos, Brazil* (eds J.P. Mallingreau, R. da Cunha & C. Justice) pp. 317–43.

50 J.P. Lanly, K.D. Singh & K. Janz (1991) FAO's 1990 reassessment of tropical forest cover. *Nature and Resources* **27**, 21–6.

Part two
Acting locally:
fluid and sediment dynamics

4 Some fluid mechanics

Science is built up of facts, as a house is built of stones:
but an accumulation of facts is no more a science than
a heap of stones is a house.

Henri Poincaré (1854–1912) *Science and Hypothesis* [1905]

Chapter summary

Materials at the Earth's surface, such as rock, sediment, ice and water, deform in response to applied force systems. Their flow laws can be posed as a set of constitutive equations. Although a natural material may not obey a flow law perfectly, nevertheless considerable insights can be gained, and predictive power harnessed, if the basic dynamical model can be identified. Most materials of interest to us are: viscous substances, such as clear water, which deform permanently and indefinitely when acted upon by a stress; plastic substances, such as soil and debris, which do not deform at low levels of applied stress, but which flow once a critical yield stress has been exceeded; or a combination of the two flow laws. The relationship of the strain, or strain rate, to the applied stress is of particular importance.

Many problems in mechanics can have light shed on them by focusing on the dimensions of the parameters involved. Dimensional analysis is introduced and applied to a number of particular fluid–sediment problems. Dimensional analysis is a powerful tool because it allows the investigation of a problem involving an unmanageably large number of parameters by breaking it into a smaller number of parameter groupings, each of which is dimensionless. One such common grouping is the Reynolds number.

The simplest description of the deformation of a viscous fluid is that the strain rate, or velocity gradient, is related to the stress by a coefficient known as the viscosity. When a grain falls through a still fluid it experiences a drag force from the surrounding fluid. This drag force, or drag coefficient, is related to the Reynolds number. At low Reynolds numbers, where

viscous forces dominate, the relation between the Reynolds number and the drag coefficient is linear, and the settling velocity of a grain is given by Stokes' law. This law predicts that fine sediment entering the ocean from rivers should take exceedingly long times to reach the sea bed by settling. Higher rates of settling appear to take place in practice because of ingestion of sediment by zooplankton in surface waters and excretion as larger pellets which settle relatively rapidly. At higher Reynolds numbers the relation to drag coefficient strongly deviates from Stokes' law due to the onset of turbulence. The precise fluid movement around and behind an obstacle in a fluid, and the drag force associated with this fluid motion, depend on the Reynolds number and the geometry and orientation of the obstacle. Blunt obstacles experience more drag than highly streamlined obstacles like aeroplane wings because of increased turbulence.

The resultant drag force on an immersed particle or object in a flow is made up of normal pressure forces and tangential shear forces. The Swiss scientist Daniel Bernoulli used the principle of the conservation of energy to deduce that for a constant potential energy an increase in pressure results in a concomitant decrease in shear stress, and vice versa. This has important implications for any object immersed in a flow, such as a moving car, an aeroplane wing, a baseball or cricket ball in flight, or a sand grain on the bed of a river.

Flows can be characterized by their variability in time, known as steadiness, and their variability in space, known as uniformity. For a steady uniform flow down an inclined plane, the basal shear stress balances the downslope component of the fluid weight, giving the basic resistance equation in fluid mechanics which is widely applied in studies of rivers.

The major distinction between laminar and turbulent flows is that in the latter there is considerable transfer of momentum in three dimensions by turbulent eddies. The instantaneous three-dimensional motions can be measured by probes immersed in the fluid, or by non-invasive techniques such as laser-Doppler anemometry, and the intensity of the motion calculated as Reynolds stresses. Flow visualization techniques have been highly successful in identifying the coherent motions associated with turbulence. Turbulent regions of fluid affected by the presence of a boundary (turbulent boundary layers) are known to comprise a sublayer very close to the boundary in which the flow is streaky in the direction of the average fluid motion; an overlying zone of turbulence generation characterized by the ejection of parcels of fluid from the region adjacent to the boundary, known as bursts; and an outer region with larger scales of turbulence. Conservation is achieved by the inrush of fluid from the outer region to the boundary in so-called sweeps. Burst–sweep cycles are important in sediment transport and the generation of bedforms in a cohesionless substrate.

The thin layer of fluid immediately adjacent to the boundary is dominated by viscous rather than turbulent Reynolds stresses. The presence or absence of this viscous sublayer defines hydraulically smooth and hydraulically rough flow, respectively. In physical terms, smooth boundaries are where the viscous sublayer fully encloses any roughness elements on the boundary, whereas in rough flow the roughness elements protrude through the viscous sublayer, thereby disrupting it.

The velocity profile in a turbulent flow is essentially logarithmic. Extrapolating the logarithmic profile down towards the bed reveals that there is a notional height above the bed at which the flow velocity appears to be zero. This roughness length is one indicator of the flow resistance of the boundary.

When fluid moves over a curving or angular boundary, the outer fluid stream may separate from a lower body of fluid along a surface of intense shearing. This process of flow separation is one of the most fundamental attributes of turbulent flow over a boundary, and is an essential aspect of a wide range of engineering and sedimentological problems such as the flow around a pier in a river, or over an aeroplane wing, or over a dune in the desert.

4.1 Introduction: the mechanics of natural substances

In order to understand the processes resulting in the movement of rock, ice, sediment (and their mixtures) at the surface of the Earth, it is necessary to know something of the response of natural materials to applied force systems. In other words, there is a characteristic style of deformation of a natural material when acted upon by a force. The relation between the applied force and the resulting deformation (strain) is embodied in a constitutive equation. There are a range of constitutive laws applicable to natural materials (Fig. 4.1).

• **Linear elastic solids** deform a finite amount which is directly and linearly proportional to the applied stress. The deformation, or strain, is recoverable when the applied stress is removed. The upper part of the lithosphere is thought to behave in this way when loaded by applied forces. However, we will not need to consider linear elastic behaviour a great deal in studying surface processes.

• **Plastic materials** do not deform at low values of applied stress. If the applied stress in increased,

however, the *yield strength* of the material may be exceeded, at which point the material deforms permanently by flow, with no additional increase in the applied stress. Plastic behaviour has application in the failure of slopes and the flow of ice sheets.

• **Viscous substances** ideally have a linear relation between the rate at which they deform and the applied stress, but unlike linear elastic solids, viscous substances deform indefinitely and permanently. Unlike plastic materials, viscous substances deform at any values of stress and do not possess a yield strength. Viscous behaviour is central to the consideration of surface sediment transport under dilute fluids. Viscous fluids may deform by an internal shear between fluid layers with little mixing between layers (laminar flow) or by a highly three-dimensional mixing (turbulent flow).

Some materials show a behaviour that does not conform to the simple constitutive models outlined above. For example, a common behaviour is a combination of elastic and perfectly plastic. Such materials have a yield strength, but once this is exceeded they

(a)

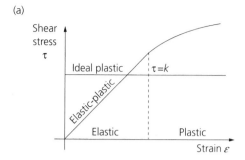

(b)

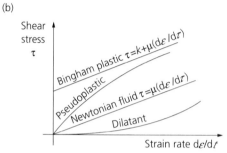

(c)

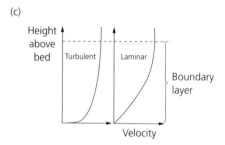

Fig. 4.1 Schematic relationships between (a) shear stress and strain for elastic, plastic and ideal plastic (perfectly plastic) materials; (b) shear stress and strain rate for a number of different fluids; (c) laminar and turbulent Newtonian fluids.

do not fracture like an elastic substance, instead continuing to deform by flow.

We shall see in Section 4.1.2 that in a viscous fluid there is a relationship between the force causing deformation, and the velocity of the fluid. The rate of strain is represented by a velocity gradient. Fluids which exhibit a linear relation between the rate of strain (velocity gradient) and the shearing force per unit area (shear stress) show *Newtonian* behaviour (Fig. 4.1). We shall also see in Section 4.2.1 that the constant relating the applied force and the resulting flow is the *dynamic viscosity*. Some viscous fluids are, however, *non-Newtonian*, obeying flow laws in which the apparent viscosity varies as a function of other

factors. One such factor might be the previous deformational history of the material, which may weaken or strengthen the material to later deformation. Another factor might be the rate at which the material is deforming. For example, a *pseudoplastic* rheology is one where the apparent viscosity falls with increasing rate of shear, which may be applicable to glacier ice or mudflows, whereas in a *dilatant* rheology the apparent viscosity increases with rate of shear, which might apply to the flow of concentrated sand. Many naturally occurring non-Newtonian materials obey flow laws in which the strain rate (rate of shear) is a function only of the stress (though not necessarily a linear function) (Fig. 4.1). A *Bingham plastic* has a linear relation between the strain rate and the stress but with a finite yield strength, and can be used successfully to model muddy debris flows (Chapter 6). Although these more complex constitutive laws find application in a number of Earth surface processes, the emphasis in this chapter is on the transport of particulate sediment by Newtonian fluids. One can think of this as the sediment transport by dilute flowing water.

Before going on to investigate the mechanics of viscous fluids such as water, it is necessary to introduce a technique which is very familiar to engineers and physicists but which tends to remain shrouded in mystery for physical geographers and geologists: *dimensional analysis*. It is very unlikely that you will survive more than a few pages in a text dealing with fluid mechanics at any level without encountering some form of dimensional analysis. For this reason, the fundamentals of the technique are introduced here and applied throughout the rest of the chapter.

4.1.1 Dimensional analysis

There are many physical problems that cannot be solved analytically, that is, by writing the equations and solving them for various initial and boundary conditions. Some problems can be solved numerically using computers by repeating calculations many times (*iteration*). But a further way of studying a physical system is to focus on the dimensions of the parameters involved. We know that the dimensions (in mechanics, principally mass, length and time, denoted henceforth by [M], [L] and [T]) on the opposite sides of an equation must be the same. However, we can go further by grouping parameters which we believe are important in a problem based on their dimensions, even without understanding the physics that links the parameters together. This proves to be useful in solving multi-parametric physical problems, but also in helping to design scaled experiments. By using dimensional analysis [1, 2] we can ensure that experiments carried out at one scale are applicable, in terms of the relative importance of forces, to a smaller or, more commonly, larger scale in nature. This is known as *dynamical similarity*.

In Earth surface processes it is very common for relationships to be derived empirically, that is, from observation. An example can be previewed from Section 4.4 in the form of the resistance equation for river flow. The basic resistance equation for the flow of a river over its bed relates the flow velocity to the average depth and slope of the river by means of a friction coefficient known as C, the *Chézy coefficient* [2]. Yet the Chézy coefficient has dimensions of $[L^{1/2}T^{-1}]$. This means that if the units of flow depth and velocity are changed (for example, from centimetres and seconds to metres and seconds), the value of C will also change. This is clearly unsatisfactory. However, it is possible to divide C by a term so that it no longer has dimensions. The square root of an acceleration, such as that due to gravity, $\sqrt{g}$, would satisfy this requirement, since acceleration has the dimensions $[LT^{-2}]$, and the square root of acceleration has dimensions $[L^{1/2}T^{-1}]$. Consequently, a dimensionless parameter describing flow resistance in a river has the general form C^2/g (or g/C^2).

Dimensional analysis will be used in following sections to study problems that are complex or analytically intractable. The analysis relies on the parameters that are important in a physical system being specified. An example is the viscous flow down an inclined plane [2] (Fig. 4.2), where we wish to know something about the distribution of velocity with depth in the flow. In nature this is likely to be analogous to the flow of lava down the slopes of a volcano, rather than the flow of river water, since

in the latter fluid turbulence rather than viscous deformation dominates flow behaviour (see Section 4.2.3). The same problem is solved analytically in Practical Exercise 4.2 in Section 4.2.2.

The parameters and their dimensions required to specify this problem are shown in Table 4.1. These parameters act in some kind of functional set, and they can be arranged in combinations, or products, which have no dimensions. A commonly used theorem in dimensional analysis states that a problem can be reduced to a relationship among a set of independent dimensionless products. If there are N original variables involving n dimensions, it should be possible to reduce the problem to a set of $N-n$ dimensionless products. In our problem there are five initial parameters and three dimensions, so we should be able to express it in terms of two dimensionless products.

In order to find the two dimensionless products we first choose three variables (known as repeating variables) with independent dimensions, which means that it is not possible to combine two variables in any way that will produce the dimensions of the third variable. In addition, the variable in which we are most interested should not be included. Since we are primarily interested in the flow velocity, these three repeating variables could be $d([L])$, $\rho g S([ML^{-2}T^{-2}])$ and $\mu([ML^{-1}T^{-1}])$. These three variables should be combined with the other variables to make them dimensionless. The three repeating variables above could be combined with u to make a product of the form

$$\mu^a(\rho g S)^b d^c u$$

with dimensions

$$[ML^{-1}T^{-1}]^a[ML^{-2}T^{-2}]^b[L]^c[LT^{-1}]$$

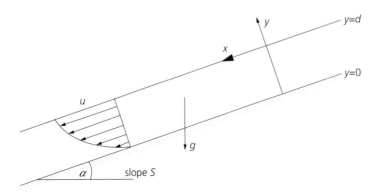

Fig. 4.2 Notation for viscous flow down an inclined plane of slope $S = \tan \alpha$.

Table 4.1 Parameters and dimensions relevant to the study of viscous flow down an inclined plane.

Variables	Dimensions
Velocity, u	$[LT^{-1}]$
Height above basal surface, y	$[L]$
Flow depth, d	$[L]$
Downslope component of weight of unit volume of viscous fluid, $\rho g S = \gamma S$	$[ML^{-2}T^{-2}]$
Absolute viscosity of fluid, μ	$[ML^{-1}T^{-1}]$

By choosing the exponents so that the units are dimensionless, the product becomes

$$\mu u/(\rho g S)d^2$$

By a similar method, it is necessary to combine y with a variable to obtain a dimensionless product. This is simple, since d also has the dimensions of $[L]$. Consequently, the second dimensionless product is (y/d).

The value of this approach is that we can now understand the physical problem of viscous flow down an inclined plane in terms of just two dimensionless variables, instead of having to carry out a very large number of experiments to discover the relationships between the original five variables. Plotting the 'dimensionless velocity' $(\mu u/(\rho g S)d^2)$ against the 'dimensionless height' (y/d) using data from experiments should produce a curve with little scatter if the dimensional analysis has been done correctly and the experiments have been carried out accurately. Excessive scatter would indicate that a parameter that is affecting the results has been left out. This might be the case if, for example, the flow became turbulent rather than viscous, as in the flow of river water.

4.1.2 The mechanics of clear fluids undergoing shear

We have already learnt that viscous fluids deform instantly under the action of a force such as a shear stress, however small, and continue to deform over time without an increase in the shear stress. This means that the constitutive equations of motion for viscous fluids must relate the stress to the rate of strain rather than to the strain itself. This can be demonstrated by considering the shear of a fluid between two parallel plates immersed in a fluid, separated by a distance L. If the upper plate is accelerated to a velocity U by the application of a force F, while the lower plate remains stationary, the fluid in the channel between the plates deforms so that there are no velocity gradients other than that in the plane of the paper in Fig. 4.3. This is known as *Couette flow*. The fluid in contact with the upper plate moves with the same velocity as the upper plate (U), and that in contact with the lower plate moves with the same velocity as the lower plate (zero), a statement of the *no-slip condition*. Consequently the average velocity gradient between the plates is U/L, which must be proportional to the applied force F. We can now generalize the notation to apply to any point in the viscous fluid undergoing shear:

$$\tau = \mu(du/dy) \tag{4.1}$$

where τ is the shear stress, du/dy is the velocity gradient, and the coefficient μ is a constant for a fluid at a given temperature known as the *dynamic viscosity*. The viscosity therefore is a coefficient linking the shear stress to the velocity gradient at a point in a fluid. Fluids exhibiting a flow behaviour dominated by viscosity are *laminar*. The velocity gradient or shear stress in a *turbulent* fluid is not described solely by the dynamic viscosity (Section 4.2.3).

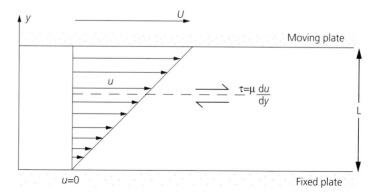

Fig. 4.3 The deformation of a thin layer of fluid between two parallel plates immersed in the fluid. The upper plate moves with a velocity U, whereas the lower plate is held stationary.

4.2 Settling of grains in a fluid

There are a whole range of Earth surface processes where grains fall through a fluid under their own weight. The falling of ash ejected from a volcano, the settling of clay particles through the water of a lake or ocean supplied from a river mouth, the rain of the tiny skeletons of marine micro-organisms from the ocean surface to the deep ocean floor, and the gradual deposition of mud and silt from the slack river waters ponded on floodplains after floods; all of these problems involve the settling of grains in a fluid. In this section the theory is introduced, a dimensional analysis performed, and the results applied to the settling of river-derived fine-grained particles in the eastern Pacific Ocean off the Peruvian coast.

4.2.1 Fluid resistance or drag

The settling of a grain through a still fluid, or the flow of fluid past a stationary grain, provides a relatively simple illustration of the *fluid resistance* exerted by a solid body immersed in the fluid. We might equally well consider it an example of a fluid exerting *drag* on a solid obstacle.

Consider a solitary particle immersed in a fluid as sketched in Fig. 4.4. The fluid approaching the particle (or the fluid into which the particle falls) is decelerated by the particle from its *free stream velocity*, u. If the cross-sectional area of the particle is A, the volume of fluid undergoing deceleration in a unit time interval is uA, and the mass of the same fluid is ρuA. The loss of kinetic energy ($mu^2/2$, where m is mass) caused by the presence of the grain is therefore

$$(\rho uA)u^2/2 = \rho u^3 A/2 \qquad (4.2)$$

Power is the rate of doing work, and is equal to the rate of mechanical energy loss (assuming that energy is not lost by heat). If the velocity of the fluid (u) and the force on the grain (F) do not change in time, we know that the power (P) is given by

$$P = Fu \qquad (4.3)$$

Since this must equal the kinetic energy loss, the *drag force* acting on the grain (F_D) in the direction of the flow can be written

$$F_D = (\rho u^2/2)A \qquad (4.4)$$

However, particles may be variously shaped, and different shapes have different effects on the fluid motion near the particle (Section 4.2.3). From experiments it is known that the drag force is related to $(\rho u^2/2)A$ by a coefficient, known as the *drag*

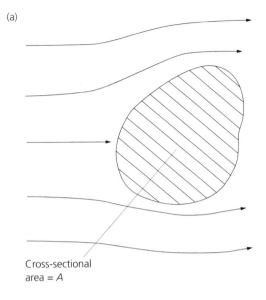

(a)

Cross-sectional
area = A

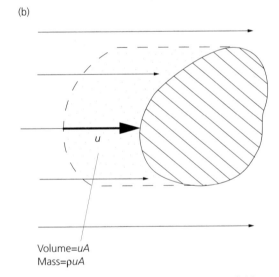

(b)

u

Volume=uA
Mass=ρuA

Fig. 4.4 A solitary particle immersed in a viscous fluid causes deceleration of fluid from its free stream velocity.

coefficient, which varies according to shape variations, so that

$$F_D = C_D(\rho u^2/2)A \qquad (4.5)$$

Highly streamlined bodies have low drag coefficients in the region of 0.1, roughly spherical bodies have values of about 0.5, and extremely blunt obstacles may have drag coefficients as high as 1.0. This is the basis for a great deal of design technology by motor manufacturers. Even relatively young readers will be aware of changes in car body design intended to achieve reductions in drag. We shall see in Figure 4.5 (p. 158) that the drag coefficient is constant over a wide range of flow conditions. However, there is a factor that makes the use of the drag coefficient, as defined above, inexact. This is the fact that the paths of the fluid particles bend around the obstruction. As a result, the viscosity of the fluid must be an important parameter, since viscosity describes the resistance to deformation of a fluid. We shall discover in Section 4.2.2 the conditions in which the viscous deformation of the fluid is important, and the conditions in which it can be neglected.

It is interesting at this point to investigate whether the problem of fluid resistance or drag is susceptible to dimensional analysis (Section 4.1.1). The parameters relevant to the study of fluid resistance are the drag force, the velocity of the fluid, properties of the fluid such as its viscosity (resistance to deformation) and density, and the size of the particle immersed in the flow (Table 4.2).

Following the method outlined in Section 4.1.1, we choose the drag force as the variable we are most interested in, then designate three other variables (repeating variables) to combine in dimensionless products. These three variables should have independent dimensions. This is easily satisfied by choosing density, velocity and particle size, since it is not possible to construct the dimensions of particle size from any combination of density and velocity. To make drag force dimensionless, we can eliminate [M] by dividing drag force by density ([ML^{-3}]), then divide by u^2 to eliminate [T], then finally divide by D^2 to eliminate [L]. The resulting dimensionless product is $F_D/\rho u^2 D^2$.

The second dimensionless product must be a combination of viscosity with the repeating variables. We divide viscosity by density to eliminate [M], divide

Table 4.2 The variables and dimensions for the problem of fluid resistance for a particle immersed in a fluid.

Variables	Dimensions
Drag force, F_D	[MLT^{-2}]
Fluid velocity, u	[LT^{-1}]
Viscosity of fluid, μ	[ML^{-1}T^{-1}]
Density of fluid, ρ	[ML^{-3}]
Size of particle, D	[L]

by velocity to eliminate [T], and finally divide by particle size to eliminate [L], to give the dimensionless product $\mu/\rho u D$.

Since the cross-sectional area of a roughly spherical particle is proportional to D^2, we can write the equation for the drag force (equation (4.5)) in terms of the drag coefficient and D

$$C_D = \frac{F_D}{\rho u^2 D^2} \tag{4.6}$$

which allows us to study the problem of fluid resistance in terms of two dimensionless numbers, C_D and $\mu/\rho u D$. The second dimensionless product when inverted to avoid the use of very small numbers is known as the *Reynolds number, Re*. We shall return to its importance in studying the transition from laminar to turbulent flow in Section 4.2.3.

Experiments show the relationship between drag coefficient and Reynolds number (Fig. 4.5). The important feature to note at this stage is that at low Reynolds numbers there is a sloping relationship between drag coefficient and Reynolds number, but that at Reynolds numbers above 10^2 or 10^3 there is very little variation in C_D.

4.2.2. Stokes' law

In the analysis above we considered the fluid resistance of a flow moving past a stationary particle, or the fall of a particle through a still fluid. An important sedimentological application is in the calculation of the *settling velocity* of a grain of a given size (D) and density (ρ_s) through a fluid of a given density (ρ_f) (Fig. 4.6). Essentially the same considerations would apply to the settling of crystals in a magma chamber. But it is difficult to find the settling velocity from the graph of Re against C_D, since u occurs in both dimensionless products. Furthermore, it is not possible to find the settling velocity directly, because we do not have prior knowledge of the value of the drag force or drag coefficient. It is necessary to attack the same problem from a slightly different angle.

When a sphere falls through a still fluid, the drag force is opposed and balanced by gravity acting on its submerged weight ($\pi D^3 \gamma'/6$), where $(\rho_s - \rho_f)g = \gamma'$ is the submerged weight per unit volume, or submerged specific weight of the particle, and $\pi D^3/6$ is the volume of the spherical particle. The drag force over the surface area of the sphere is therefore given by

$$F_D = \frac{\pi D^3 \gamma'}{6} \tag{4.7}$$

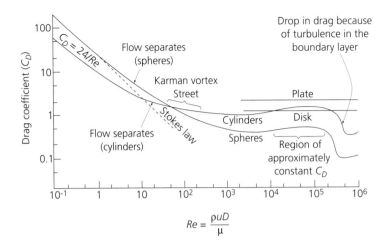

$$Re = \frac{\rho u D}{\mu}$$

Fig. 4.5 Relationship between the drag coefficient C_D and the particle Reynolds number $\rho u D/\mu$ for the steady settling of variously shaped particles. At particle Reynolds numbers less than about 1, the relationship is given by Stokes' law in equation (4.9). The turbulent structure behind cylindrical particles at higher Reynolds numbers is illustrated in Fig. 4.11. After Middleton & Southard (1984) [3].

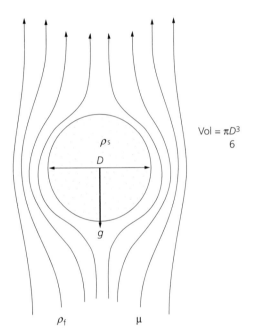

Fig. 4.6 Notation for a solitary spherical particle falling through a still fluid.

If we had neglected fluid density when considering fluid resistance, on the basis that the fluid was not undergoing accelerations, the dimensional analysis would have produced just one dimensionless product. It is simple to deduce that this must be of the form $F_D/\mu u D$. Since only one dimensionless product describes fully the relationships between the four variables relevant to the physical problem, it must be a constant. This constant has the value 3π. Substituting into equation (4.6) for the drag coefficient, we have

$$\frac{F_D}{\mu u D} = 3\pi = \frac{\pi D^3 \gamma'}{6} \tag{4.8}$$

from which can be obtained *Stokes' law*

$$u = \frac{1}{18} \frac{D^2 \gamma'}{\mu} \tag{4.9}$$

The likelihood of fluid density being unimportant applies to flows which are laminar. When flows become turbulent, fluid particles take part in a rapidly varying three-dimensional motion in which local accelerations are very important. For Stokes' law to apply, the Reynolds numbers must be very small—less than 1, where the length term in the Reynolds number refers to the grain diameter.

Practical exercise 4.1: The settling of fine particles in the deep sea

Sediment is brought to the Pacific Ocean by short, swift Peruvian rivers and by wind transport from coastal deserts such as the Atacama. Parallel to the Peruvian coast is the Peru Trench, separating the narrow (<150 km) continental shelf from the deep oceanic basins to the west (Fig. 4.7). The surface waters are dominated by a WNW-directed surface current that has velocities of about 0.5 m s^{-1}.

Calculate the settling velocities for the following grain sizes: clay (<4 µm: take $D = 0.002$ mm); fine silt (8–11 µm: take $D = 0.01$ mm); and coarse silt (32–62 µm: take $D = 0.05$ mm). Assume that the average water temperature is 4°C, with a viscosity of 1.567×10^{-3} kg s^{-1} m^{-1} and a density of *Continued.*

Practical exercise 4.1: *Continued*

$1030 \, \mathrm{kg \, m^{-3}}$. The grain densities can be assumed to be $2650 \, \mathrm{kg \, m^{-3}}$. Gravity is $9.8 \, \mathrm{m \, s^{-2}}$.

If the 6 km-deep Peru Trench is about 150 km offshore, calculate whether these grain sizes would be carried beyond this bathymetric barrier before settling on to the sea bed or not. Assume that the particles settle by Stokes' law.

Solution

The settling velocity (u) of the three grain sizes is as follows, using Stokes' law:

$$u = \begin{cases} 2.25 \times 10^{-3} \, \mathrm{mm \, s^{-1}} & D = 0.002 \, \mathrm{mm} \\ 5.63 \times 10^{-2} \, \mathrm{mm \, s^{-1}} & D = 0.01 \, \mathrm{mm} \\ 1.41 \, \mathrm{mm \, s^{-1}} & D = 0.05 \, \mathrm{mm} \end{cases}$$

The depth (y) to which these grains have fallen by the time they have reached the axis of the Peru Trench is

$$y = \begin{cases} 0.675 \, \mathrm{m} & D = 0.002 \, \mathrm{mm} \\ 16.9 \, \mathrm{m} & D = 0.01 \, \mathrm{mm} \\ 423 \, \mathrm{m} & D = 0.05 \, \mathrm{mm} \end{cases}$$

These results illustrate the exceedingly slow settling rates of fine particulate sediment in the ocean, and the very long transport distances of such material before eventual settling on to the sea floor. Clearly, all of these grain sizes would easily be advected over the trench and into the Peru basin beyond. However, sea bed and water samples indicate that river-derived sediment is restricted to the region east of the trench axis [4]. There must be some process at work which is causing the fine particulate sediment to be deposited more rapidly than Stokes' law would suggest. The main reason is believed to be the ingestion of the sediment by zooplankton inhabiting surface waters, and the subsequent rapid settling of larger aggregates as faecal pellets. The Peruvian coast is a region of upwelling of nutrient-rich water, supporting high biological productivity. Where the oxygen-minimum zone impinges on the sea bed, faecal aggregates have been found in samples. Elsewhere, they are broken down into constituent fine minerals. This supports the idea that ingestion and pelletization are responsible for the river-derived sediment dispersal patterns observed. The detrital sediment on the sea bed beyond the Peru Trench is mostly of wind-blown origin.

4.2.3 Pressure and shear forces on a particle

What does the drag force really represent in terms of the forces on the surface area of an immersed particle settling through a still fluid, or suspended in a moving fluid? The drag force is the sum total (resultant) of all of the tangential (shear) forces per unit area and normal (pressure) forces per unit area over the entire surface area of the particle. Consider a fluid moving past a cylinder at very low Reynolds numbers (Fig. 4.8). We can imagine the parcels of fluid to be moving along tube-like paths such that particles of fluid do not cross between such *streamtubes*. Instead of envisaging tubes of flow, it is more convenient to draw *streamlines* marking the axes of streamtubes. The streamlines show a bunching around the top and bottom of the cylindrical solid, and an expansion along the mid-section. The distributions of pressure and shear stress are sketched in Fig. 4.9. Where the pressure is greatest, as at the stagnation point P, the shear stress is least, and vice versa. This principle has many applications in fluid mechanics, and was developed by the eighteenth-century Swiss scientist Daniel Bernoulli.

Bernoulli's theorem

It is an axiom of physics that energy cannot be lost from a mechanical system, although it may change from one form to another, so in a flowing fluid the total energy, E, is constant. This is embodied in *Bernoulli's equation*, which states that the total energy is made up of the kinetic energy ($\frac{1}{2}\rho_f \, u^2$), potential energy ($\rho_f gh$), pressure energy (p) and frictional heat loss, the latter being taken to be negligibly small for most fluid problems. Here ρ_f is the fluid density, u its velocity, h is elevation, and g gravitational acceleration.

To prove Bernoulli's theorem, consider a tube of fluid defined by fluid streamlines, so that no fluid enters or leaves the streamtube (Fig. 4.10). If fluid moves a small distance ∂x_1 down the tube by the application of a pressure force p_1 on the cross-sectional area of the tube A_1, the work done is the total force on the end of the tube times the distance moved, or $p_1 A_1 \partial x_1$, with units of joules in the SI system. Since $A_1 \partial x_1$ equals the small change in volume ∂V, the work done is $p_1 \partial V$.

Assuming the fluid to be incompressible, the same volume must be discharged from the other end of the tube. The work done at this end of the tube is $p_2 A_2 \partial x_2$, or $p_2 \partial V$. The change in energy between the two ends of the tube is therefore

$$p_1 \partial V - p_2 \partial V = \partial V (p_1 - p_2) \tag{4.10}$$

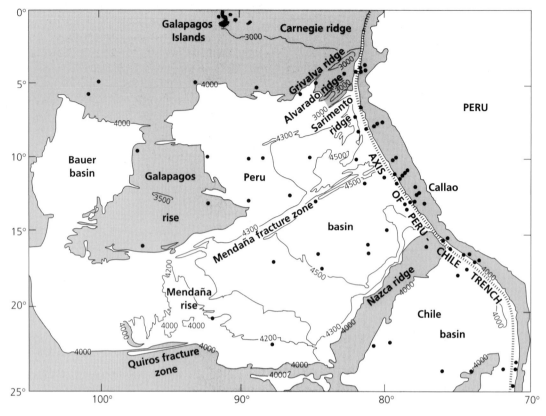

Fig. 4.7 Location map for Practical Exercise 4.1, showing the Peruvian coast, Peru–Chile Trench, deep ocean floor, and locations of sea bed samples collected by Scheidegger & Krissek (1982) [4].

Fig. 4.8 Laminar flow is dominated by viscous effects. This can be visualized by water, streaked with dye, flowing at 1 mm s^{-1} through a narrow gap between two glass plates 1 mm apart (Hele–Shaw flow) past a circular object.

(a) Pressure

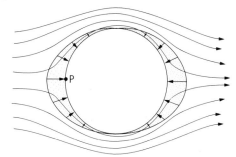

(b) Shear stress

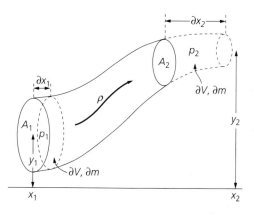

Fig. 4.9 Flow of a real fluid past a cylinder at low Reynolds numbers in the Stokes range, with distribution of pressure (relative to the free stream pressure) and shear stress on the surface of the sphere.

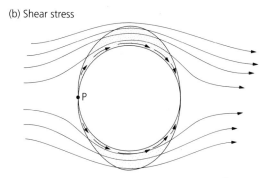

Fig. 4.10 Sketch for the derivation of Bernoulli's theorem for flow through a streamtube.

We can also assume for an incompressible fluid that the volume change ∂V is equivalent to an increment of mass ∂m disappearing from cross-section 1 and

appearing at cross-section 2 with different kinetic and potential energies. The change in the energy can therefore be expressed in terms of the kinetic and potential energies of this small increment of mass. The change in the kinetic energy is

$$\frac{1}{2}\partial m\left\{\left(\frac{\partial x_2}{\partial t}\right)^2 - \left(\frac{\partial x_1}{\partial t}\right)^2\right\} \tag{4.11}$$

where $\partial x_2/\partial t$ and $\partial x_1/\partial t$ are the flow velocities at cross-section 2 and cross-section 1 respectively. The change in potential energy is

$$\partial m\, g(y_2 - y_1) \tag{4.12}$$

where y_2 and y_1 are the heights of the centres of the two ends of the streamtube. The total energy change is therefore the sum of the kinetic and potential energy components in expressions (4.11) and (4.12). Equating this sum with equation (4.10), dividing through by ∂V, and separating the terms for the cross-sections 1 and 2 of the streamtube gives

$$p_1 + \frac{\partial m}{2\partial V}\left(\frac{\partial x_1}{\partial t}\right)^2 + \frac{\partial m}{\partial V}\,gy_1$$
$$= p_2 + \frac{\partial m}{2\partial V}\left(\frac{\partial x_2}{\partial t}\right)^2 + \frac{\partial m}{\partial V}\,gy_2 \tag{4.13}$$

Now since $\partial m/\partial V$ is simply fluid density ρ_f, $\partial x_1/\partial t$ is velocity u_1, and $\partial x_2/\partial t$ is velocity u_2, this simplifies to

$$p_1 + \frac{1}{2}\rho_f u_1^2 + \rho_f gy_1 = p_2 + \frac{1}{2}\rho_f u_2^2 + \rho_f gy_2 \tag{4.14}$$

Generalizing for any streamtube (or, indeed, where the diameter of the streamtube becomes so small that it is in effect a streamline) in a steady flow, we have Bernoulli's familiar equation

$$p + \frac{1}{2}\rho_f u^2 + \rho_f gy = \text{constant} \tag{4.15}$$

The important implication of Bernoulli's equation is that, *for a constant potential energy*, an increase in flow velocity results in a decrease in pressure, and vice versa.

The simple symmetrical pattern in Fig. 4.8 is not found at higher Reynolds numbers. As the Reynolds number increases, the flow becomes complex in the region behind the obstacle (Fig. 4.11). First, paired vortices are shed from the rear of the obstacle, and then, as *Re* increases further, a long turbulent wake is

(a)

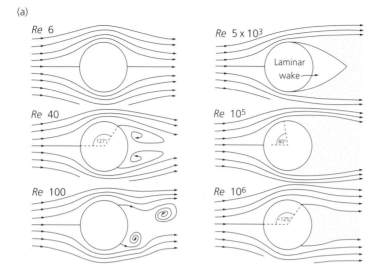

(b)

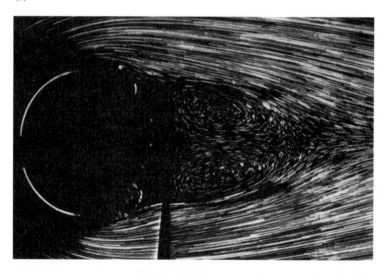

(c)

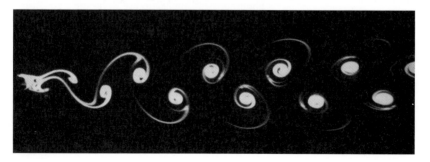

(d)

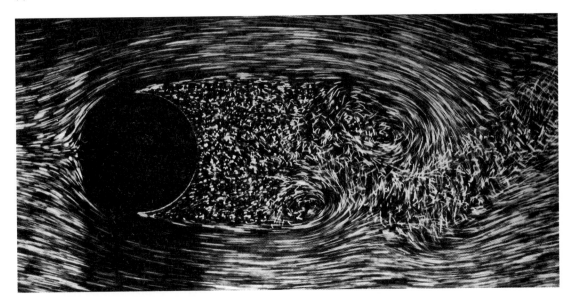

(e)

Fig. 4.11 *Above and opposite.* (a) Flow around a cylinder at progressively higher Reynolds numbers produces characteristic patterns of vortices behind the obstacle dependent on the Reynolds number. After Middleton & Southard (1984) [3]. The sequence of flow visualization results shows (b) paired vortices ($Re = 24$); (c) the development of two parallel rows of staggered vortices known as a von Karman vortex street ($Re = 105$); (d) a fully developed wake of isotropic turbulence ($Re = 2000$); (e) despite the fivefold increase in the Reynolds number ($Re = 10\,000$), the flow pattern is largely unchanged and the drag coefficient remains constant. Flow visualisation results from Van Dyke (1982) [5].

produced behind the obstacle. These effects profoundly alter the distribution of shear and pressure forces on the particle, and thereby the drag force. The onset of turbulence can be detected in the graph of drag coefficient against Reynolds number (Fig. 4.5) as well as in the visualization of the flow behind the obstacle. We shall return to this topic in considering the entrainment of grains from a substrate in Chapter 5 after a little more is known of the structure of turbulent flows.

Natural flows of viscous fluids may be considerably more complex than the situations so far described. Such complexities may arise from the different rheologies of the fluid–sediment mixtures. But complexities may also arise from the variation of the flow both in space and in time. A flow which varies in space is said to be *non-uniform*, and a flow which varies in time is said to be *unsteady*. Most natural flows contain an element of non-uniformity and unsteadiness. However, in the problems encountered in this chapter such effects can be ignored. Non-uniformity and unsteadiness are discussed in relation to submarine highly concentrated flows in Chapter 6. For the moment, we continue in our blissful state of simplicity by considering uniform and steady flows.

4.3 Flow down an inclined plane

We have already seen that a fluid moving past a stationary particle, or a particle falling through a still fluid, causes fluid resistance. The concepts of drag force and drag coefficient were introduced by considering the kinetic energy loss of fluid approaching the immersed particle. An equally important concept relates to the fluid resistance or drag caused by a fluid, such as a river, running over its bed, a topic further developed with a practical exercise (5.1) in Section 5.3.2. In this case, the stream power, that is, the rate at which the fluid does work in running over its bed, is best thought of as the rate at which the potential energy of the water is converted into kinetic turbulent energy [6]. For a river flowing down a slope S (Fig. 4.12a), the rate of loss of gravitational potential energy per unit area of the stream bed is

$$\rho g(\bar{u}S)d \tag{4.16}$$

where $\bar{u}S$ is the drop in elevation of the river in unit time, and $d(1)(1)$ is the volume of water overlying a unit area of stream bed. Since $\rho g Sd$ is the downslope component of the gravity force acting on the volume of water of depth d and unit area, which must be opposed by an equal and opposite drag force exerted on the fluid by the unit area of bed τ_0, the rate of loss of potential energy, or power, is simply given by

$$P = \tau_0 \bar{u} \tag{4.17}$$

This is a result of very wide application. Stream power, or the rate of doing work on the bed, is one of the most important variables in determining the rate of sediment transport, the size of particles deposited on the river bed and the types of bedform (Chapter 5).

Practical exercise 4.2: The velocity profile in a laminar flow

We know that the boundary shear stress in a steady uniform flow down an inclined plane is the force balancing the downslope component of gravity acting on the fluid. This comprises the basic resistance equation

$$\tau_0 = \rho g dS \tag{4.18}$$

But it would also be useful to be able to calculate the shear stress and velocity at some distance from the boundary. To do this, imagine a plane parallel to the boundary at a height y from it (Fig. 4.12b). The shear stress on this plane must be equal to the downslope component of fluid weight from the overlying column of fluid with height $d - y$. This can be written

$$\tau_y = \rho g S(d - y) \tag{4.19}$$

Since $\tau_0 = \rho g dS$, equation (4.19) becomes

$$\tau_y = \tau_0 \left(1 - \frac{y}{d}\right) \tag{4.20}$$

which shows that the shear stress in the flow varies linearly from a maximum at the bed to zero at the surface.

For a laminar flow $\tau = \mu(du/dy)$, so the resistance equation can be written

$$\frac{du}{dy} = \frac{\rho g S(d - y)}{\mu} \tag{4.21}$$

Integrating this equation gives the velocity at any point above the boundary. Since fluid density and viscosity do not vary with height,

$$u = \frac{\rho g S}{\mu}\left(yd - \frac{y^2}{2}\right) + C \tag{4.22}$$

The no-slip condition tells us that we have a boundary condition of $u = 0$ at $y = 0$, so the constant of integration C is equal to 0. The velocity profile
Continued.

Practical exercise 4.2: *Continued*

is therefore parabolic, with velocity increasing towards the free surface. We shall see in Section 4.4.4 that the velocity profile in a turbulent flow is roughly logarithmic rather than parabolic.

1 The Hawaiian islands are built of basaltic shield volcanoes. There has been a history of recent eruptions from Mauna Loa (4484 m) and Mauna Kea (4533 m) (Fig. 4.13). The average slope of the volcanoes can be obtained from the contours on Fig. 4.13. Assuming the lava flows to be laminar, what is the average velocity (measured at 0.4 of the flow depth from the bed) of a lava flow 0.30 m thick running due westwards down the flank of Mauna Loa from the Mokuaweoweo caldera? Assume that the basaltic lava is erupted at 1200°C and has

density 2700 kg m^{-3} and viscosity 100 kg s^{-1} m^{-1} at this surface temperature.

2 What is the basal shear stress of the lava flow?

Solution

1 The contoured map shows that there are roughly constant slopes westward from the summit of Mokuaweoweo caldera. The average slope can be found to be tan $\theta = 0.121$, so $\theta = 6.9°$, and sin $\theta = 0.120$. The velocity at 0.4 of the flow depth, $y = 0.12$ m, can be found from equation (4.22) to be 0.91 m s^{-1}. If the lava flow travelled at this mean velocity from the point of its eruption at the caldera, it would take 12 hours to reach the coast.

2 The basal shear stress can be found from equation (4.18) to be 953 N m^{-2} (953 Pa).

(a)

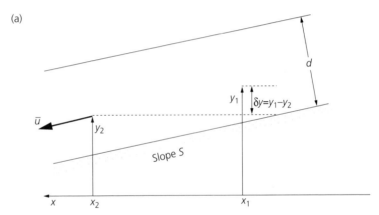

Change in potential energy $= \rho g d\, \delta y$
Rate of loss of potential $= \rho g d(\bar{u}S)$
energy per unit time

(b)

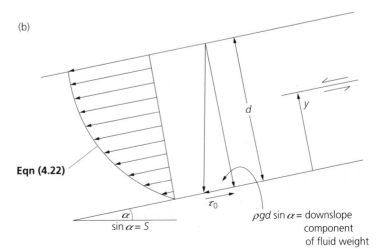

Fig. 4.12 (a) Schematic of the energy balance for a river flowing down a slope. (b) Notation used to derive the velocity profile in Practical Exercise 4.2.

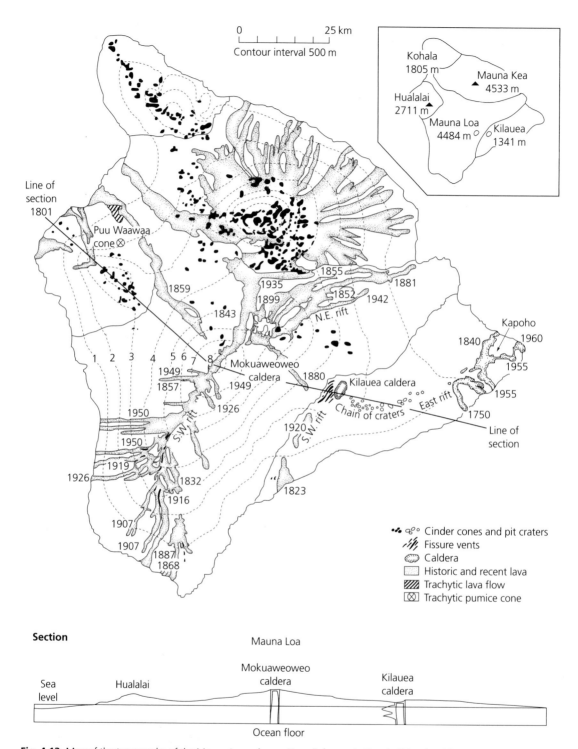

Fig. 4.13 Map of the topography of the Mauna Loa volcano, Hawaii, for use in Practical Exercise 4.2.

4.4 Turbulent flow
4.4.1 The experiments of Reynolds
It has already been remarked upon that in some flows fluid accelerations can be neglected and the deformation depends fundamentally on the viscosity. In other flows, the viscosity, which operates on the molecular scale, is less important and the motion is controlled by the growth and decay of highly three-dimensional structures known as turbulent eddies. This is the fundamental difference between laminar and turbulent flows. The transition between the two flow behaviours was studied by Osborne Reynolds in the 1880s. He conducted some experiments using a tank of still water into which was injected water carrying a dye (Fig. 4.14). The dyed water flowed into a tube and out of the tank through a valve. The temperature of the water allowed the viscosity to be varied. In addition, the pipe diameter and the flow velocity were varied. Reynolds observed that the streak of dyed water always changed from a straight line to a mass of eddies at a particular value ($c. 12\,000$) of a dimensionless number $\rho u D/\mu$, now known as the Reynolds number. (The term 'turbulent' was not used by Reynolds himself, but was introduced at a later date by Lord Kelvin.)

This should not be a great surprise, because we can come to a similar conclusion simply by performing a dimensional analysis of the problem. The parameters involved in this physical problem are as shown in Table 4.3. There are four original variables and three dimensions, so we should be able to construct just one dimensionless product. If we choose viscosity as a dependent variable, the other three are independent repeating variables. The viscosity can be divided by the velocity to eliminate [T], then divided by the fluid density to eliminate [M], and finally divided by pipe diameter to eliminate [L]. This gives the dimensionless product $\mu/\rho u D$. As noted above, in order to avoid very small numbers, the inverse is used as the Reynolds number $\rho u D/\mu$.

Table 4.3 Parameters and dimensions relevant to Reynolds's experiments.

Variable	Dimensions
Mean flow velocity, u	$[LT^{-1}]$
Fluid density, ρ	$[ML^{-3}]$
Fluid viscosity, μ	$[ML^{-1}T^{-1}]$
Pipe diameter, D	$[L]$

Since Reynolds's experiments the utility of the Reynolds number in describing the laminar–turbulent transition has been demonstrated over a much wider range of parameter values. The value of $12\,000$ is not particularly significant. A repeat of Reynolds's experiments in the 1980s found the critical Reynolds number to be less than $12\,000$, perhaps because of the added traffic vibration outside the laboratory in busy downtown Manchester.

4.4.2 The description of turbulence
The major contrast between laminar and turbulent flow is that in laminar flow the particles of fluid move in parallel layers, and the only mixing between different layers takes place slowly at the molecular level by the process of diffusion. Laminar flows are therefore highly predictable as long as we know something about their viscosity. The same is not true of turbulent flows. There is considerable transfer of momentum in three dimensions because of the presence of turbulent eddies. An instantaneous velocity measured at any point in the flow might be very different from the time-averaged velocity of the flow. The length scale of turbulence is much greater than that of molecular diffusion, which means that turbulent fluids are strongly mixed.

The most obvious way of describing turbulence is to measure the instantaneous velocities u, v and w at a point in the fluid in the directions x, y and z (Fig. 4.15). If the time-averaged velocities in these three directions are denoted by $\bar{u}$, $\bar{v}$, $\bar{w}$, then the deviations of the instantaneous velocity components from the time-averaged velocities give an indication of the turbulent fluctuations, denoted by u', v', w'. The mean of the velocity fluctuations over time is zero, so we need a measure of the intensity of these velocity fluctuations such as their root mean square

$$\text{rms } u' = \sqrt{\overline{(u')^2}}$$

It is also very useful to know something about the turbulence in particular planes of the flow. This can be measured using cross-correlations of the velocity components. For example, in the (x, y) plane, the intensity of the velocity fluctuations can be measured by taking the time average of the instantaneous velocity fluctuations in those directions, that is $\overline{u'v'}$.

It is possible to measure the instantaneous velocity fluctuations at a point in the flow using hot wire or hot film anemometers which measure the flow velocity from the change in electrical resistance caused by

(a)

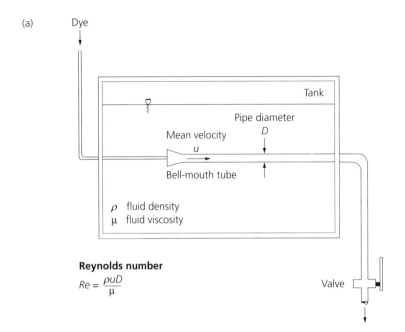

Dye

Tank

Pipe diameter
D

Mean velocity
u

Bell-mouth tube

ρ fluid density
μ fluid viscosity

Reynolds number

$Re = \dfrac{\rho u D}{\mu}$

Valve

(b)

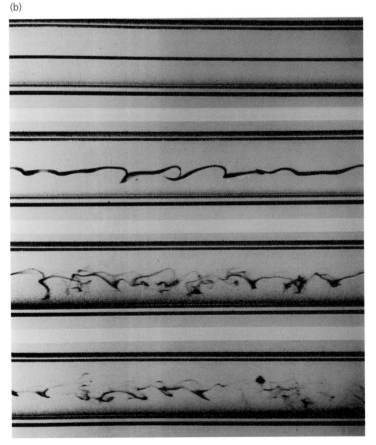

Fig. 4.14 Set-up used by Osborne Reynolds in 1883 to study the onset of turbulence. (a) Water streaked with dye is drawn through a pipe of circular diameter. A repetition of his experiments by N.H. Johanneson and C. Lowe is shown in the photographs in (b) taken at progressively higher Reynolds numbers, from laminar, undisturbed flow (top), transition (second), and fully turbulent flow (bottom two photographs). After Van Dyke (1982) [5].

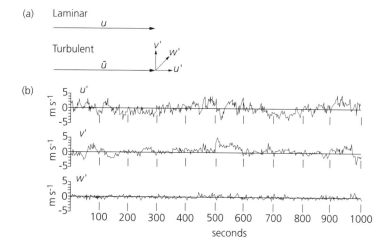

Fig. 4.15 (a) Whereas laminar flow can be described by one vector u, turbulent flow involves a complex three-dimensional motion described by turbulent fluctuations u', v' and w'. Over sufficiently large intervals of time, the fluctuating instantaneous velocities average out to the time-average velocity $\bar{u}$. (b) Measured turbulent velocity fluctuations in a wind at 5.6 m above the ground surface. Data from Panofsky & Dutton (1984) [7].

the cooling effect of the fluid moving across the orientated wire or film. A non-intrusive device is the laser-Doppler anemometer which is based on the reflection of laser light penetrating the flow. Computer statistical packages give a detailed breakdown of the fluid velocities and their variation, from which can be calculated the contribution of different frequencies of motion to the total kinetic energy of the turbulence.

It could be argued, however, that the breakthrough in understanding turbulence came not primarily from its statistical description, still less from its mathematical development, but from its visualization using new techniques such as high-speed photography of flows 'labelled' with dye, reflective particles, or lines of hydrogen bubbles. These flow visualization techniques have revealed a great deal about both the fine-scale and large-scale structures within turbulent flows [8].

Under some flows, the *wall region*, that is, very close to the boundary, appears to be characterized by a streakiness to the flow, the streaks representing alternate lanes of high- and low-velocity fluid moving parallel to the mean flow (Fig. 4.16). These are called *wall layer streaks*. They have a transverse spacing, λ, dependent on flow properties such that the boundary Reynolds number $\rho\bar{\lambda}\,u_*/\mu$ (with the average transverse spacing of the streaks as the length term and the shear velocity ($u_* = \sqrt{\tau_0/\rho}$, see Section 4.4.4) as the velocity term), is constant, at about 100. The streaks are produced by the presence of longitudinal vortices very close to the bed which are constantly exchanging fluid with each other (Fig. 4.16). The lineations on bedding planes of sandstone may be due to this fine-scale structure in the wall region.

The wall layer streaks are subject to intermittent ejection from the wall layer region into the outer part of the flow, a process known as *bursting*. The low-speed lanes of fluid appear to be lifted up, transformed into a horeshoe, then a hairpin vortex, and finally broken down into general turbulence (Fig. 4.17). The end of the burst sequence is marked by the inrush of high-speed fluid towards the bed, known as a *sweep*. The burst–sweep process therefore generates considerable three-dimensional motion and is the main creator of turbulence. The mean burst period appears to be constant when scaled by variables referring to the outer flow

$$\overline{T}_\mathrm{b} = \frac{\bar{t}_\mathrm{b}U_\mathrm{o}}{\delta} \cong 5 \tag{4.23}$$

where $\overline{T}_\mathrm{b}$ is the average burst period scaled by the outer flow variables, $\bar{t}_\mathrm{b}$ is the measured average burst period, U_o is the flow velocity outside the turbulent boundary layer and δ is the boundary layer thickness. For a flow with a free surface such as a river, the flow velocity is the surface velocity and the boundary layer thickness the flow depth. For a river 5 m deep flowing at its surface at 1 m s^{-1}, the burst period should be of the order of 25 s. For a fast (2 m s^{-1}) shallow (0.5 m) flow, the average burst period reduces to little more than 1 s. The inrush of high-velocity sweeps may be particularly important in locally exceeding the threshold of sediment motion on an erodible bed (Chapter 5).

Flow visualization shows that there are also larger coherent flow structures which are similar in scale to the flow itself. These comprise bulges of outward-

(a)

(b)

LSS—low-speed streak
HSS—high-speed streak

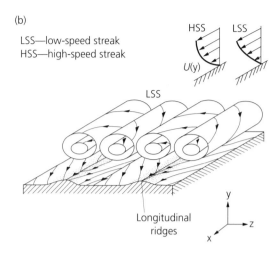

Fig. 4.16 (a) Streakiness in the near-wall region of a turbulent boundary layer revealed by sand grains on a dark boundary. Mean flow shown by arrow; scale bar is 10 mm. After Grass (1971) [9]. (b) Idealized fluid motion associated with the streaks consists of longitudinal vortices which cause locally converging flow where heaps of grains in longitudinal ridges accumulate. This longitudinal fabric produces a primary current lineation on bedding planes. After Allen (1985) [10].

moving turbulent fluid separated by upstream (about 20°) sloping zones of outer fluid reaching almost to the bed (Fig. 4.18). The average time between the passage of bulges over a fixed point on the bed is similar to the burst period. The bed experiences a variation, perhaps a twofold variation, in the shear stress during the passage of the bulges. This should affect bedload transport rates (Chapter 5) and therefore promote the formation of waviness on the deformable bed. Such waviness is likely to be extremely subtle, however, and to travel with the velocity of the large flow structures, which is close to the mean velocity of

the flow. Their passage may be responsible for the millimetre-scale laminae of plane beds (Chapter 5).

The origin of turbulence is invariably linked to the presence of a boundary. The effects of the boundary are felt in the motion of the fluid over a certain distance from the boundary. This is known as the boundary layer, and the structure of turbulence identified from flow visualization techniques commonly refers to that found in a boundary layer.

4.4.3 Structure of turbulent boundary layers
The boundary layer is the layer of fluid whose

Fig. 4.18 (*Opposite.*) Large coherent structures in a turbulent boundary layer on a plane wall based on experimental evidence. (a) The large-scale coherent structures are on the scale of the boundary layer thickness. Their flow pattern is shown as if one is travelling with the flow structure. Smaller vortices caused by the bursting of streaks are seen on the upper surface of the larger coherent structures. (b) Structure of the turbulent boundary layer visualized from a sheet of light through a fog of tiny oil droplets. After Falco (1977) [11].

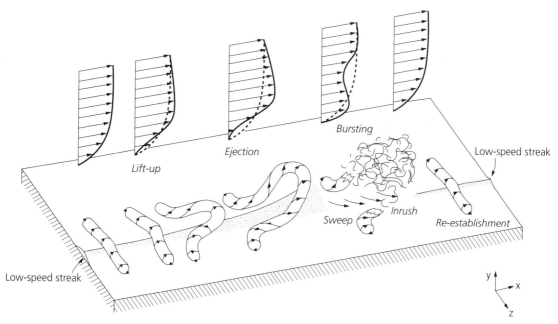

Fig. 4.17 Model of the ejection of low-velocity streaks from the wall region during the process known as bursting. The velocity profiles are plotted with respect to the ground surface, whereas the lift-up, ejection and bursting of the horeshoe, then the hairpin vortex, is shown as if one is travelling with the vortex. After Allen (1985) [10].

(a)

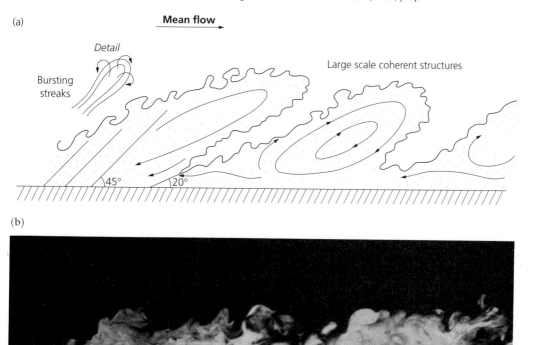

(b)

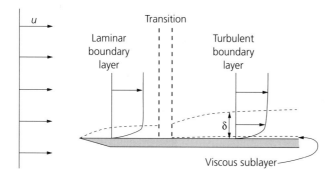

Fig. 4.19 Development of a boundary layer on a flat plate with a zero angle of attack.

motion is affected by the resistance of a boundary. Its development can be considered with reference to a flat plate immersed in a flow (Fig. 4.19). As the flow approaches the leading edge of the plate a boundary layer is developed. Initially, the boundary layer is very thin because the fluid directly in contact with the plate has a zero velocity (the no-slip condition) whereas fluid close to the boundary (but not in contact with it) is moving with its undisturbed free-stream velocity. The shearing is restricted to a very thin zone in which velocity gradients are large. Because the boundary layer is very thin, there is no large-scale three-dimensional transfer of energy and the flow can be treated as laminar. The laminar boundary layer grows downstream by molecular diffusion of momentum until the boundary layer reaches a critical thickness, or more accurately, a critical Reynolds number. It then becomes turbulent and the larger length-scale of the turbulent eddies compared to the molecular viscous effects causes momentum to be transferred more quickly, leading to a rapid expansion of the boundary layer. However, the turbulent boundary layer contains within it a thin *viscous sublayer* next to the boundary in which viscosity dominates.

The total stress in a turbulent boundary layer must be made of two components—a viscous stress and a turbulent stress. Taking the x direction of the flow (parallel to the boundary), the viscous stress is immediately familiar:

$$\mu \frac{\partial \overline{u}}{\partial y}$$

The turbulent stress, known as the *Reynolds stress*, is

$$\rho \overline{u'v'}$$

This turbulent Reynolds stress can be thought of as the mean vertical motion (due to v') of momentum transferred horizontally in the x direction (the inertia $\rho u'$).

Plotting of Reynolds stresses against distance from the boundary illustrates that they reach a maximum at approximately 5% of the boundary layer thickness [9]. Closer to the boundary viscous stresses are increasingly important (Fig. 4.20). The magnitude of the Reynolds stresses varies according to the co-variance being used (most commonly $\rho \overline{u'v'}$) and the nature of the boundary.

Some boundaries contain roughness elements which are entirely enclosed within the viscous sublayer (Fig. 4.21). They are termed *hydraulically smooth*. Other boundaries may contain roughness elements that project through the viscous sublayer so that it only potentially exists in pockets between the roughness elements. Such boundaries are termed *hydraulically rough*. The root mean square of the vertical velocity fluctuations ($\sqrt{\overline{v'^2}}$) increases near the boundary as the roughness increases, whereas that of the horizontal velocity fluctuations $\sqrt{\overline{u'^2}}$ decreases with increased roughness. This gives a good illustration of the fact that rough boundaries promote the ejection upwards of parcels of fluid into the outer part of the boundary layer. Above a height from the boundary of about 20% of the flow depth the effects of smooth and rough boundaries are not picked up strongly in the vertical and horizontal root-mean-square fluctuations, nor in the Reynolds stress (Fig. 4.20).

The Reynolds stress can be expressed in a similar manner to the viscous stress as the product of a viscosity term and a velocity gradient:

$$\tau = (\mu + \eta) \frac{\mathrm{d}\overline{u}}{\mathrm{d}y} \tag{4.24}$$

From this it can be seen that

$$\eta \frac{\mathrm{d}\overline{u}}{\mathrm{d}y} = -\rho \left(\overline{u'v'} \right) \tag{4.25}$$

The term η is known as the *eddy viscosity*. It is much larger than the molecular viscosity μ except in the

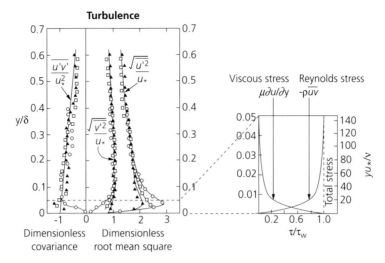

Fig. 4.20 Distribution of stress in a turbulent boundary layer. The values of the root mean square of the velocity fluctuations u' and v' are made dimensionless by dividing by the shear velocity u_*. Their covariance is also made dimensionless by dividing by the square of the shear velocity. Circles refer to a smooth boundary, the squares and triangles are for a progressively rougher boundary. Note that the effects of the different types of boundary are felt in the lower 5% of the boundary layer thickness. The inset for this region (for $Re = 70\,000$) shows that viscous stresses increase at the expense of Reynolds stresses as the bed is approached. τ_w is the stress on the wall (boundary), δ is the boundary layer thickness. The viscous sublayer extends to a dimensionless height of $yu_*/v = 11.5$. After Grass (1971) [9] and Tritton (1977) [12].

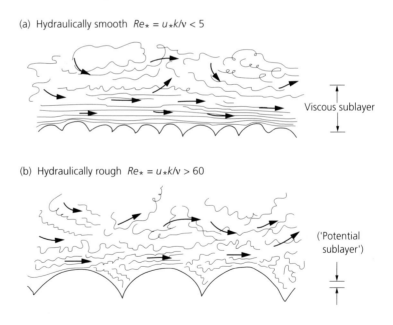

Fig. 4.21 Hydraulically smooth and rough flow in terms of the boundary Reynolds number u_*k/v.

viscous sublayer where the molecular viscosity dominates. Whereas the viscosity μ is constant for a given temperature, the eddy viscosity η varies as a function of flow conditions such as Reynolds number.

Let us return momentarily to the equation for the molecular transport of momentum which applies to the deformation of laminar flows. This is customarily expressed:

$$\tau = \mu \frac{du}{dy} \tag{4.26}$$

Instead, we can express the same equation in a form more explicit for the transport of momentum (inertia):

$$\tau = \frac{\mu}{\rho} \frac{d(\rho \bar{u})}{dy} \tag{4.27}$$

where the term $\rho \bar{u}$ is the fluid momentum per unit volume. Now *Fick's first law of diffusion* states that the mass flux (per unit area) of a substance (Q) is proportional to the concentration gradient by a coefficient known as the diffusivity (κ). Expressed mathematically,

$$Q = -\kappa \frac{\partial C}{\partial y} \tag{4.28}$$

Consequently, τ can be viewed as the flux of the vertical component of the momentum directed along x (down a concentration gradient along y), and the term μ/ρ is the equivalent of the diffusivity. This latter term is known as the *kinematic viscosity v*.

For turbulent momentum transport, if the molecular viscosity can be neglected, the diffusional equation would contain a term for the turbulent momentum flux which is the Reynolds stress $\rho(\overline{u'v'})$ and the equivalent of the diffusivity would be η/ρ, which is called the *kinematic eddy viscosity*. So the eddy viscosity can be seen to determine the rate of transport of momentum in a turbulent fluid. It cannot be determined directly, but can be estimated from the measured result of momentum transport—the velocity profile.

4.4.4 Velocity profiles in turbulent flows

It has previously been stated that within the turbulent boundary layer there is a thin viscous sublayer. Whereas the transfer of momentum is very efficient in the outer part of the boundary layer because the kinematic eddy viscosity is large, it is small in the viscous sublayer because it relies on molecular diffusion. Consequently, the fluid is well mixed in the outer part of the boundary layer, producing a small velocity gradient, but is laminar with a high-velocity gradient close to the bed. Equation (4.26) indicates that the velocity (u) in the viscous sublayer should be a function of the shear stress on the boundary (τ_0), the viscosity (μ) and the distance from the boundary (y). Fluid density is not important since we assume that the flow is laminar and there are no accelerations. Performing a dimensional analysis, we have four initial

parameters and three dimensions. There is therefore just one dimensionless product which must be a constant. This product must be $\mu u / \tau_0 y$. It is found from experiments that this dimensionless constant is roughly unity. If we express the shear stress at the boundary in the dimensions of velocity, termed the *shear velocity* (u_*),

$$u_*^2 = \tau_0 / \rho \tag{4.29}$$

then the velocity, u, at any height in the flow, y, can be written in terms of the shear velocity

$$\frac{\mu u}{\rho u_*^2 y} = 1 \tag{4.30}$$

or, keeping both sides of the equation dimensionless,

$$\frac{u}{u_*} = \frac{u_* y}{v} \tag{4.31}$$

where v is the kinematic viscosity. This is an extremely useful way of expressing the velocity in the viscous sublayer scaled by the shear velocity, because the right-hand side is a form of Reynolds number. The ratio v/u_* has the dimensions of [L] and represents a viscous length-scale. We can compare this viscous length-scale to a length-scale due to the roughness of the boundary k_s in order to assess the relative effect of the boundary roughness on the velocity distribution. The ratio is a Reynolds number applicable to the boundary:

$$Re_* = \frac{\text{roughness length-scale}}{\text{viscous length-scale}} = \frac{u_* k_s}{v} \tag{4.32}$$

At small boundary Reynolds numbers of less than about 5 (Fig. 4.21) the boundary has effectively no influence on the velocity distribution, corresponding to the condition of hydraulically smooth flow. At boundary Reynolds numbers greater than about 60 (Fig. 4.21), the boundary roughness effects far outweigh those due to viscosity, the viscous sublayer effectively ceases to exist, and the flow is hydraulically rough. In between is a transitional field.

The thickness of the viscous sublayer is about 12 times the viscous length-scale. For most natural flows, the viscous sublayer is very thin, generally less than 1 mm. In the core of the flow, relatively far from the boundary and from any free surface, we can assume that the velocity gradient is only dependent on the shear stress at the boundary (or shear velocity), not on viscous or roughness length-scales. So we have three parameters—velocity gradient, shear velocity and height above the boundary. They form one dimensionless product which is equal to a constant, *von Karman's constant* (k), equal to a value of about 0.4 in clear flows:

$$\frac{u_*[\mathrm{LT^{-1}}]}{y[\mathrm{L}]\dfrac{du}{dy}[\mathrm{T^{-1}}]} = k \tag{4.33}$$

This equation can be integrated to find the velocity at any depth y. First, it is rearranged to give

$$\frac{1}{y}\frac{1}{k} = \frac{1}{u_*}\frac{du}{dy} \tag{4.34}$$

When integrated with respect to y, this becomes

$$\frac{1}{k}\ln(y) + C = \frac{u}{u_*} \tag{4.35}$$

Clearly, we wish to remove the constant of integration. The constant requires a boundary condition at which $u = 0$, but this is not satisfied at the bed where $y = 0$. Instead, it is achieved at a height y_0, that is, by letting $C = -(1/k)\ln(y_0)$. Consequently, $-Ck = \ln(y_0)$. Therefore, $y_0 = \exp(-Ck)$, and it follows that

$$\frac{1}{k}\ln\!\left(\frac{y}{y_0}\right) = \frac{u}{u_*} \tag{4.36}$$

This is generally known as the *logarithmic velocity law* or *law of the wall* (Fig. 4.22). When $y = y_0$, then $\ln(y/y_0) \to 0$, so y_0 corresponds to the height above the bed at which the velocity appears to be zero. This is only a notional height, since we know that the velocity distribution in the viscous sublayer is not logarithmic. So although the velocity gradient is not determined by the roughness of the boundary, the value of the velocity itself in the outer part of the flow depends on processes in the viscous sublayer. The *roughness length* y_0 varies according to whether the flow is smooth, transitional or rough.

The velocity profile can also be used to calculate the dynamic eddy viscosity, a parameter that is very difficult to estimate in any other way. The eddy viscosity can then be used to study the turbulent diffusion of suspended sediment in a turbulent boundary layer. This will be of use in estimating suspended sediment concentrations at any depth in a turbulent flow (Section 5.3.4).

We start by recalling that the shear stress in a turbulent viscous fluid is related to the local velocity gradient by the sum of the dynamic and eddy viscosities (equation (4.24)). Far from the viscous sublayer the viscous effects are likely to be minimal compared to the effects of turbulent momentum transfer, so we can neglect the dynamic viscosity. Defining the kinematic eddy viscosity as

$$\varepsilon = \eta/\rho \tag{4.37}$$

we have

$$\varepsilon = \frac{\tau}{\rho(\partial \overline{u}/\partial y)} \tag{4.38}$$

The shear stress at any point in a steady, uniform open-channel flow of depth d (derived in Practical Exercise 4.2) is linear with depth

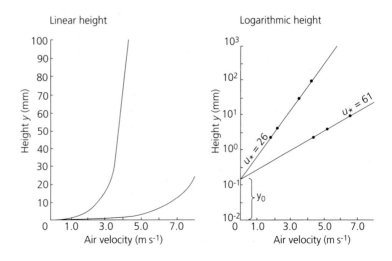

Fig. 4.22 Illustration of the logarithmic velocity law for two air flows of different shear velocity. Plotting of the velocity data against the logarithm of the height allows an extrapolation of the straight lines to a notional height at which the velocity is zero, the roughness length y_0. After Leeder (1982) [13].

$$\tau = \tau_0 \left(\frac{d - y}{d} \right) \tag{4.39}$$

Making use of the fact that $(du/dy) = u_*/yk$ (from the combination of equations (4.26) and (4.29)), the kinematic eddy viscosity can be expressed

$$\varepsilon = u_* \left(\frac{d - y}{d} \right) ky \tag{4.40}$$

showing that it has a parabolic form with a maximum at half of the flow depth and a value of zero at the bed ($y = 0$) and at the surface ($y = d$). Although the weakness of this method is that the dynamic eddy viscosity has been estimated from an assumed velocity law (which strictly applies close to the boundary rather than throughout the flow), it is nevertheless useful in considering the diffusion of suspended sediment (Section 5.3.4).

4.4.5 Flow separation

So far we have considered the flow of a turbulent fluid in a boundary layer. However, the greatest amounts of turbulence are caused by shear associated with the separation of the boundary layer away from the solid boundary. This is known as *flow separation*. It is important to be able to predict when flow separation will take place. Flow separation occurs when a positive (adverse) pressure gradient is set up in the flow (i.e. there is a downstream increase in pressure), causing the boundary layer to be separated from the

---

Practical exercise 4.3: The roughness length derived from a velocity profile

The velocity data in Table P4.1 were obtained from the Columbia River, USA [14]. The Columbia River at the time of measurement had a total depth of 6.4 m. The velocity measurements given are 66 minute averages, that is, although an instantaneous velocity profile may depart significantly from the values given, short time-period variations, such as those due to burst–sweep events and to the advecton of larger eddies, are averaged out over longer time periods. The grain size, D, of the bed material of the Columbia River is gravel with a maximum size of 0.28 m and a mean size of 0.09 m.

1 Plot the velocity profile using a linear scale for velocity, u; and a logarithmic scale for height above bed, y. Draw a best-fit line and extrapolate it towards the bed to calculate the roughness length, i.e. the notional height above the bed where the velocity appears to be zero.

2 The roughness length, y_0, due solely to skin friction (as if the bed were perfectly plane) is often given as

$$y_0 \approx D/30$$

Comparing your result from the Columbia River with the roughness length due to skin friction, what do you conclude about the nature of the Columbia River bed in terms of its resistance?

Solution

1 The best-fit line for the velocity data is straight when linear velocity is plotted against logarithmic height from bed. This suggests that the velocity profile is roughly logarithmic away from the immediate vicinity of the bed. Extrapolation of the line downwards gives a roughness length of 0.023 m.

2 The roughness length due to skin friction ($D_{max} = 0.28$ m, $D_{ave} = 0.09$ m) is smaller than the estimated roughness length from the velocity profile, but by less than an order of magnitude. It must be assumed, therefore, that the bed of the river exerts some, but little, *form drag* due to the presence of larger roughness elements. Such roughness elements might be low gravel bars or clusters of pebbles.

Table P4.1

Height above bed, y (m)	Fluid velocity, u (m s^{-1})
6.4 (surface)	1.19
5.7	1.16
4.8	1.13
4.3	1.11
3.6	1.08
3.0	1.03
2.3	0.98
1.7	0.92
1.1	0.82
0.5	0.65

solid boundary by a region of slow, upstream-moving fluid. This can best be understood by considering the distribution of pressure on a curved wall (Fig. 4.23a).

As fluid approaches the curved wall streamlines converge, causing enhanced shear stresses but reduced pressure because of the Bernoulli effect (Section 4.2.3). On passing the curved wall the streamlines diverge, reducing shear stress and increasing pressure. Outside the thin boundary layer the negative pressure gradient and acceleration on the approach of

the curved wall are balanced by the positive pressure gradient and deceleration as the curved boundary is passed. Within the boundary layer the negative pressure gradient causing acceleration of fluid on the front of the wall easily overcomes the opposing effects of viscous friction. But on the rear of the curved wall, the positive pressure gradient and viscous friction act together to strongly decelerate the flow. If the pressure gradient is sufficiently large, this causes at one point on the boundary the velocity in the boundary layer to be reduced to zero, and downstream from

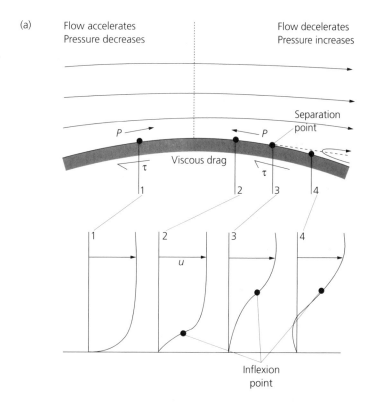

Fig. 4.23 Flow separation over a gently curved wall. (a) Streamlines and velocity profiles at points on the curved wall. After Middleton & Wilcock (1994) [2]. (b) Flow visualization (using air bubbles in water) of flow separation from a curved wall. After Van Dyke (1982) [5].

this point the velocity to be negative, representing a slow reverse flow in the boundary layer. Flow must therefore separate where the positive pressure gradient is sufficiently large for the velocity and shear stress at the boundary to be zero. A great deal of turbulence is generated along the contact between the fast downstream-moving fluid in the outer part of the boundary layer and the slow upstream-moving fluid in the inner part of the boundary layer. The vortices generated along this zone of high shear expand downstream to produce a turbulent wake. Such wakes are familiar to the observer standing at the stern of a boat.

The process of flow separation is crucial to the understanding of the generation of bedforms in a cohesionless substrate under a shear flow (Fig. 4.23b), a topic developed in Chapter 5. In such a case the positive pressure gradient is caused by a sharp change of orientation of the boundary. It is also fundamental to the understanding of the turbulent diffusion of jets, where the positive pressure gradient is caused by a sudden expansion of the flow (in depth and width) (Chapter 7). This occurs, for example, where distributary channels of deltas enter the ocean. Expansions of river channels may also lead to flow separation. In all of these cases, the region beneath the separated flow is a zone of sediment deposition because the reduced or negative fluid velocities are unable to transport the full sediment load.

On a lighter note, some readers may have pondered on why baseballs and cricket balls have stitched seams which protrude above the general level of the ball. The answer, of course, is that the seam causes flow separation. By throwing or bowling the ball in a particular orientation, at a particular speed and spinning at a particular rate, the pitcher or bowler is able to vary the flight path of the ball quite curiously [15]. Understanding the physics of flow separation and drag will not necessarily make you into a better cricketer or baseball player, however.

Further Reading

J.R.L. Allen (1985) *Principles of Physical Sedimentology*. George Allen & Unwin, London.

M.R. Leeder (1982) *Sedimentology Process and Product*. George Allen & Unwin, London, Chapter 4.

G.V. Middleton & J.B. Southard (1984) *Mechanics of Sediment Movement*, 2nd edn, Short Course Notes 3. SEPM, Tulsa, OK.

G.V. Middleton & P.R. Wilcock (1994) *Mechanics in the Earth and Environmental Sciences*. Cambridge University Press, Cambridge.

References

1 [The full consequences of dimensional analysis were not fully realised until the development of the π-theorem by Edgar Buckingham in 1914.] E. Buckingham (1914). On physically similar systems: illustrations of the use of dimensional equations. *Physics Reviews* (IV) **4**, 345. [A useful account is also found in G.V. Middleton & P.R. Wilcock, *Mechanics in the Earth and Environmental Sciences*, Chapter 3, pp. 69–99, CUP, Cambridge, 459pp.]

2 G.V. Middleton & P.R. Wilcock (1994) *Mechanics in the Earth and Environmental Sciences*. Cambridge University Press, Cambridge. [The example of the Chézy equation is given excellent treatment.]

3 Middleton, G.V. & Southard, J.B. (1984) *Mechanics of Sediment Movement*, 2nd edn, Short Course Notes 3. SEPM, Tulsa, OK.

4 K.F. Scheidegger & L.A. Krissek (1982) Dispersal and deposition of eolian and fluvial sediments off Peru and Chile. *Bulletin Geological Society of America* **93**, 150–62.

5 M. Van Dyke (1982) *An Album of Fluid Motion*. Parabolic Press, Stanford, CA.

6 R.A. Bagnold (1960) *Sediment discharge and stream power: a preliminary announcement*. Circular 421. US Geological Survey, Washington, DC.

7 H.A. Panofsky & J.A. Dutton (1984) *Atmospheric Turbulence: Models and Methods for Engineering Applications*. John Wiley and Sons, New York.

8 J.R.L. Allen (1985) Loose boundary hydraulics and fluid mechanics: selected advances since 1961. In: *Sedimentology: Recent Developments and Applied Aspects* (eds P.J. Brenchley & B.P.J. Williams), Blackwell Scientific Publications, Oxford, pp. 7–28.

9 A.J. Grass (1971) Structural features of turbulent flow over smooth and rough boundaries. *Journal of Fluid Mechanics* **50**, 233–55.

10 J.R.L. Allen (1985) *Principles of Physical Sedimentology*. George Allen & Unwin, London.

11 R.E. Falco (1977) Coherent motions in the outer regions of turbulent boundary layers. *Physiology Fluids* **20**, S124–32.

12 D.J. Tritton (1977) *Physical Fluid Dynamics*, 2nd edn. Clarendon Press, Oxford, 519pp.

13 M.R. Leeder (1982) *Sedimentology Process and Product*. George Allen & Unwin, London, Chapter 4.

14 J. Savini & G.L. Bodhaine (1971) *Analysis of current meter data at Columbia River gaging stations, Washington and Oregon*, Geological Survey Water Supply Paper 1869-F. US Geological Survey, Washington, DC.

15 R.K. Adair (1995) The physics of baseball. *Physics Today* **48**(5), 26–31. [A very readable account of the aerodynamics of baseball.]

5: Sediment transport

The Walrus and the Carpenter
Were walking close at hand:
They wept like anything to see
Such quantities of sand:
'If this were only cleared away,'
They said, 'it would be grand!'

Lewis Carroll (1832–98), *Through the Looking-Glass* [1872]

There's sand in the porridge and sand in the bed,
And if this is a pleasure we'd rather be dead.

Noel Coward (1899–1973), *The English Lido* [1928]

Chapter summary

Albert Einstein is reputed to have warned his son against becoming a river engineer, saying that the physics of sediment transport was too complex. One of the reasons for this complexity is the generally turbulent nature of fluid flow above sediment substrates. But of perhaps greater importance is the fact that the boundary is not only movable, but also commonly deforms into a variety of shapes and sizes which feed back and affect the overlying fluid motion, what we might term loose boundary hydraulics. In studying sediment transport we must consider a number of interrelated questions. What causes the threshold of motion of grains resting on a bed under a moving fluid, and can the threshold be predicted? Once the threshold is exceeded, how are sediment particles transported by the flow? When sediment transport is taking place, what are the characteristic configurations taken up by a cohesionless (sandy) bed, and how can these bedforms be explained dynamically?

The dislodgement of a grain on a bed at the threshold of sediment motion can be investigated by considering the balance of moments about a pivot point, and yields the result that the threshold is predicted in terms of a dimensionless stress and a Reynolds number. The Shields diagram and its derivatives embody these results. Following dislodgement, grains may move along the bed as bedload, or be swept into the flow as suspended load. Bedload transport rates are generally related to the stream power, which is proportional to the cube of the velocity, but general bedload sediment transport formulae are notoriously poor in predicting actual transport rates in rivers. The distribution of suspended sediment in a flow can be treated as a problem of diffusion in which the key dimensionless group is the Rouse number. However, the diffusional approach relies on the entire sediment concentration originating from upward diffusion from the bed, whereas in nature the suspended sediment load may result from pumping from upstream.

Fluids in natural streams experience energy losses by friction on the bed and banks—that is, they experience flow resistance. It is necessary to know something about the flow resistance of a river channel in order to investigate the relation between the mean flow velocity and the shear stress. The flow resistance is measured in the form of a dimensionless variable known as the friction factor, but it can also be expressed as the Chézy coefficient or Manning's roughness coefficient. The concept of flow resistance can be used to estimate palaeodischarges of rivers, such as the bedload-dominated meltwater streams issuing from ice sheets. Here, the flow resistance can be approximated from the roughness produced by clasts on the river bed.

When the fluid velocity over a sandy bed is gradually increased in a laboratory flume, a sequence of different bed states develops. In very fine to fine sands, small ripples are succeeded by flat beds, before the surface of the water and the sediment surface both become wavy. In medium sands, the ripples give way to larger dunes, before they are washed out into flat beds and then turn into standing waves. In coarse sands, a lower flat bed is succeeded by dunes and then upper flat beds and standing waves. These bedforms have existence fields related to Reynolds number and boundary roughness. The initiation and evolution of bedforms can be viewed in terms of the growth or decay of instabilities of particular wavelengths along a deformable boundary between two different 'fluids' undergoing shear—the overlying flow and the underlying deforming granular boundary. Whether interfacial instability will develop is given by the Richardson number.

The migration of bedforms gives rise to characteristic stratification patterns where there is net aggradation on the bed through time. Increasing amounts of vertical accretion relative to downstream migration result in steeper angles of climb of bounding surfaces, and consequently increasing portions of bedforms being preserved. Linear crested ripples and dunes produce planar cross-stratification, whereas highly sinuous or irregular crested ripples and dunes deposit trough cross-stratification.

5.1 Introduction

That transport of particulate matter takes place over the Earth's surface is a commonplace observation. Sitting on a beach trying to eat some sandwiches on a windy day should convince one that sand is capable of being transported relatively easily in air. A paddle at the water's edge demonstrates that sand is also capable of being transported in water by the motion of waves, and by the powerful but shallow swash and backwash currents. We are familiar with the astounding and beautiful ornamentation of the desert floor caused by sediment transport, the regular patterning of an estuarine shoal, and we have, most of us, walked on the dried-up course of a river and seen the corrugations in its bed. The range of structures, in scale and geometry, on the river bed or desert floor is at first bewildering. Geologists can recognize in sedimentary rocks the very same structures. It is clear that there is an interactive link between the form of the substrate and the overlying fluid, perhaps best termed *loose boundary hydraulics* [1]. The complexity of the link is one of the reasons why sediment transport is an exceedingly difficult problem, and certainly not one which attracts scientists looking for neat and tidy solutions.

In Part One of this book, sediment transport was placed in the context of the denudation of the continents and the global working of the hydrological cycle. At this point we turn from the holistic to the reductionist in order to gain further insights into the physics of sediment transport. Over the last two to three decades there has been a preoccupation among sedimentologists with depositional environments. This preoccupation has largely been driven by the need to correctly interpret past environments and palaeogeographies based on the preserved sedimentary record. In some notable cases, this has resulted in a certain discontinuity, perhaps one might say dysfunction, between those investigating the present-day processes of sediment transport, and those interpreting the sedimentary record. However, sediment transport is of importance to a range of disciplines beyond sedimentary geology. An understanding of the physics of sediment transport, and of the present-day fluxes of sediment in natural systems, is of crucial importance in fields such as river engineering, hazard assessment, and floodplain and coastal management.

In this chapter many of the fluid mechanical principles introduced in Chapter 4 are developed in the specific area of sediment transport. The two chapters should ideally be read and used as a pair.

There is a wide range of problems in the physical sciences where use is made of the idea of conservation. We have encountered this principle in previous chapters. In Chapter 4, for instance, we made use of the conservation of energy in formulating Bernoulli's theorem. The fundamental principle underpinning a great deal of work in sediment transport and deposition also relies on conservation—the conservation of sediment mass. The principle is expressed as the sediment conservation equation, or, perhaps more commonly, the *sediment continuity equation*. Its derivation is included below to highlight its fundamental usefulness.

5.1.1 The sediment continuity equation

The principle of conservation of mass which underlies the continuity equation is very usefully applied to the transport of sediment over a substrate that is capable of erosion and aggradation.

Let us consider an area of stream bed with unit width normal to the flow and length dx parallel to the flow, with height y above some reference datum (Fig. 5.1). The area of this piece of stream bed is

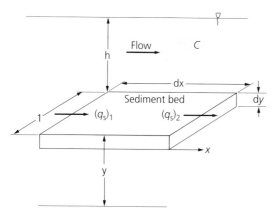

Fig. 5.1 Derivation of the sediment continuity equation.

therefore dx. A fluid flow with suspended sediment volume concentration C has a sediment transport rate (per unit width of stream bed) at the upstream end of the reference area $(q_s)_1$ and at the downstream end of the reference area $(q_s)_2$. We can denote the difference in the transport rate at the upstream and downstream ends of the reference area as dq_s. Any change in the bed elevation due to either erosion or deposition must be reflected in a downstream change in the sediment transport rate as long as the sediment concentration, C, does not change. The amount of sediment deposited over the reference area associated with a downstream decrease in the sediment transport rate is simply the product of the difference between the upstream and downstream sediment transport rates (dq_s) and the time interval dt. This sediment produces a layer of thickness dy with a porosity λ over an area dx. Expressed mathematically,

$$\mathrm{d}y\,\mathrm{d}x = -\frac{1}{1-\lambda}\,\mathrm{d}q_s\mathrm{d}t \tag{5.1}$$

A change in bed elevation may also result from a temporal change in the volume concentration of suspended sediment throughout the flow. For example, if the suspended sediment concentration throughout the flow is decreasing, the bed must be accreting irrespective of downstream variations in the sediment transport rate. The bed elevation change due to this effect in a flow of depth h is

$$\mathrm{d}y\,\mathrm{d}x = -\frac{1}{1-\lambda}\Big(h\mathrm{d}C\Big)\mathrm{d}x \tag{5.2}$$

The net change in bed elevation over a time interval dt is therefore

$$\frac{\mathrm{d}y}{\mathrm{d}t} = -\frac{1}{1-\lambda}\left(\frac{\mathrm{d}q_s}{\mathrm{d}x} + h\frac{\mathrm{d}C}{\mathrm{d}t}\right) \tag{5.3}$$

In other words, the change in bed elevation over time (erosion or deposition) is related to the downstream change of the sediment transport rate and to the change of the suspended sediment concentration in time. If the suspended sediment concentration does not vary in time, that is, d$C/\mathrm{d}t = 0$, or if the suspended sediment concentration is zero, then the sediment continuity equation simplifies to

$$\frac{\mathrm{d}y}{\mathrm{d}t} = -\frac{1}{1-\lambda}\frac{\mathrm{d}q_s}{\mathrm{d}x} \tag{5.4}$$

This equation can be used to study the dynamics of bedforms (Section 5.4). It has also been used to model the transport of sediment in fluvial systems by combining the continuity equation with the equation for uniform flow down an inclined plane derived in the previous chapter. Such a combination yields a result linking the sediment flux to the local topographic slope, the sediment diffusion equation (see also Section 3.3.2).

5.2 The threshold of sediment movement under unidirectional flows

In considering the flux of sediment under a fluid flow it has been suggested that the rate of doing work on the bed, or stream power, is a parameter of fundamental importance. Nevertheless, the sediment transport rate and the stream power cannot be exactly in phase. This is simply because as stream power increases from zero there is at first no sediment transport. Only once a threshold is reached is sediment transported downstream by the fluid. This point is termed the threshold of sediment motion. If the shear stress on the bed at the threshold of sediment motion is τ_c (where the subscript refers to critical), then the *effective stress* driving sediment transport is $\tau - \tau_c$. The effective stress will be discussed further in considering sediment transport formulae.

5.2.1 Forces on a particle resting on a bed

It is now a familiar concept that streamlines converge as they are deflected over an obstacle and diverge beyond the obstacle. Consider a cylindrical object on a horizontal bed (Fig. 5.2a). In an ideal, indeed fictitious, fluid that has no viscosity (an *inviscid fluid*) and where the rate of change of momentum is solely due to pressure forces, the pattern of streamlines would be perfectly symmetrical and there would be

(a)

Resultant lift

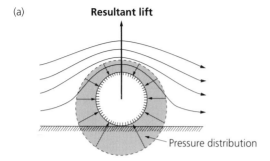

Pressure distribution

(b) Point of flow   Lift   **Resultant pressure
     separation   component   force**

Pressure
distribution

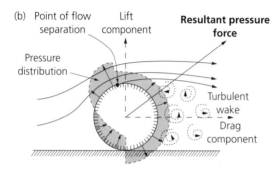

Turbulent
wake
Drag
component

Fig. 5.2 Flow pattern and pressure distribution over a
cylinder lying on a flat bed (a) for an inviscid flow, and
(b) for a viscous fluid at moderately high Reynolds numbers.
After Middleton & Southard (1984) [2].

no drag force on the cylinder. Since the cylinder
is resting on a bed, however, there is an unequal
pressure distribution between the top and bottom of
the cylinder, resulting in a net upward force. This is
known as the *lift force*. The more realistic case for the
flow of a viscous fluid at moderate to high Reynolds
numbers past the same cylinder involves flow separation (Fig. 5.2b). Flow separation behind the cylinder
causes a substantial drag force as well as a lift force.
The resultant pressure force will therefore act with
both downstream and upward components.

Opposing the fluid pressure forces on a particle are
those tending to keep the particle in place on the bed.
The former have been introduced above; for a solitary
grain, the latter are due to the submerged weight of
the grain. The grain will only be lifted into the flow if
the fluid forces exceed the gravitational forces. On a
natural bed individual particles rest among variously
shaped, sized and packed neighbours. Entrainment
occurs therefore by the rotation of the grain about a
pivot point. Movement begins when there is a balance
of moments about this pivot (Fig. 5.3). The direction

of easiest movement of a particle makes an angle α to
the mean bed surface, passing through the pivot. An
'average' value of α for submerged grains in water is
equivalent to their angle of repose, about 35°. This
angle depends on the packing arrangement of the bed,
particularly the extent to which a grain is exposed to
the flow. A grain perched high on a bed is much more
likely to be entrained than one embedded tightly
among other particles. Consequently, large perched
grains may be more easily entrained than smaller, less
exposed grains. The packing arrangement of particles
on a bed, as well as their size, consequently has a major
role to play in the threshold of sediment movement.

We can therefore write, for the initiation of movement, a balance of moments about the pivot.

$$F_G(\sin \alpha)a_1 = F_F(\cos \alpha)a_2 \tag{5.5}$$

where F_G is the gravitational force, F_F is the resultant
fluid pressure force on the particle being made up of
lift and drag components, and a_1 and a_2 are their
respective moment arms. The problem is that it is very
difficult to formulate the gravitational force except in
the case of perfect shapes such as spheres, the pivot
angle α may vary considerably with natural beds, and
the fluid force is affected by a number of factors such
as the degree of enclosure within the viscous sublayer,
the variations of fluid forces due to turbulence, the
area of particle subjected to fluid forces and the drag
coefficient. (Although lift and drag coefficients can be
calculated under controlled experimental conditions,
the prediction of the threshold under natural conditions is far from straightforward.)

Let us introduce coefficients to take into account
most of the effects causing variability. For example,
the gravitational force can be written in terms of a
coefficient which accounts for shape and sorting
effects:

$$F_G = c_1 D^3(\rho_s - \rho_f)g = c_1 D^3 \gamma' \tag{5.6}$$

Likewise, considering the drag force only (parallel to
the bed), a coefficient can be introduced to account
for the packing of the grains and their shape effects
influencing the drag coefficient. If we assume that the
drag force is proportional to the product of the shear
stress on the bed and the area of the grain exposed to
this shear stress, the drag force becomes

$$F_D = c_2 D^2 \tau_0 \tag{5.7}$$

Balancing moments at the threshold of sediment
motion,

$$a_1 c_1 D^3 \gamma' \sin \alpha = a_2 c_2 D^2 \tau_0 \cos \alpha \tag{5.8}$$

Forces acting on a grain in a bed

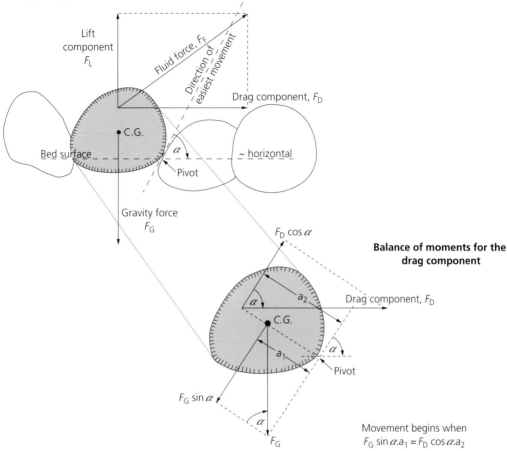

Fig. 5.3 Forces acting on a grain (shaded) resting in a bed of similar grains, with a balance of moments for the drag component at the onset of grain movement. After Middleton & Southard (1984) [2].

Grouping the coefficients, we have

$$\left[\frac{a_1 c_1}{a_2 c_2} \tan \alpha\right] = \frac{\tau_0}{D \gamma'} \qquad (5.9)$$

The coefficients on the left-hand side are conventionally lumped together and denoted by β. Its value reflects a large number of factors, including the characteristics of the grains comprising the boundary and the local flow above it. It should therefore be expected to vary according to the boundary Reynolds number.

It may be objected that lift forces have been ignored. The lift force originates from the sheltering effect on the lower part of the grain. Once the grain is entrained, therefore, we should expect the lift force to vanish quickly as the grain becomes supported by the fluid. Drag forces alone then transport the grain downstream. While in the bed, lift forces are therefore important. Inclusion of the lift component, however, merely results in a change in the value of β. Furthermore, the lift force effect should also be roughly proportional to the boundary Reynolds number. However, the definition of the threshold condition inevitably relies on direct experimentation because of the large number of unknown factors contributing to β.

5.2.2 Dimensional analysis of the threshold problem

The line of argument above, based on a balance of

Table 5.1 Variables and dimensions in the problem of the threshold of sediment motion.

Parameter	Dimensions
Critical shear stress on the bed, τ_c	$[ML^{-1}T^{-2}]$
Fluid density, ρ_f	$[ML^{-3}]$
Fluid viscosity, μ	$[ML^{-1}T^{-1}]$
Grain diameter, D	$[L]$
Submerged specific weight of grain, γ'	$[ML^{-2}T^{-2}]$

moments, is supported by a dimensional analysis of the problem. The variables important in the threshold of motion of uniformly sized spherical spheres on a flat bed are given in Table 5.1. There are five initial variables and three dimensions, so there should be two dimensionless products. It is by no means clear what the repeating variables should be, but guidance is provided by the fact that a Reynolds number of some form and a dimensionless product involving the balance between gravitational and fluid forces would be convenient.

Let us choose (somewhat arbitrarily) D, ρ_f, τ_0 as the repeating variables. We can then combine the submerged specific weight with the repeating variables as follows:

$$(\gamma')^a(\tau_0)^b(D)^c(\rho_f)^d$$

with the dimensions given in Table 5.1. For the product to be dimensionless, the sums of the exponents for [M], [L] and [T] must be zero. So we have the three simultaneous equations for mass, length and time:

[M] $a + b + d = 0$
[L] $-2a - b + c - 3d = 0$
[T] $-2a - 2b = 0$

from which it can be immediately established that $d = 0$. Arbitrarily setting the exponent $c = -1$ so that D appears in the denominator, it can be found that $a = -1$ and $b = 1$. The dimensionless product is therefore $\tau_0 / \gamma' D$.

The viscosity can now also be combined with the repeating variables to obtain the second dimensionless product. It should have the form

$$(\mu)^a(\tau_0)^b(D)^c(\rho_f)^d$$

Taking the dimensions listed in Table 5.1, we have the three simultaneous equations referring to mass, length and time:

[M] $a + b + d = 0$
[L] $-a - b + c - 3d = 0$
[T] $-a - 2b = 0$

Setting a arbitrarily at -1, it can be easily found that $b = 1/2$, $c = 1$, and $d = 1/2$. The dimensionless product therefore becomes

$$\frac{D\sqrt{\rho_f}\sqrt{\tau_0}}{\mu} = \frac{\rho_f u_* D}{\mu} = \frac{u_* D}{v} \qquad (5.10)$$

which is recognizable as the boundary Reynolds number (Chapter 4). The dimensional analysis therefore confirms that the threshold of sediment movement can be expressed in terms of two dimensionless products corresponding to a ratio of fluid and gravitational forces (β) and a boundary Reynolds number. It should be possible to carry out experiments to discover the precise form of this relationship.

5.2.3 The Shields diagram

The German engineer Shields collected data on the threshold of sediment movement for a range of different-density materials (granite, coal, amber, barite) combined with the results of previous experiments using quartz sand (Table 5.2) . The data show considerable scatter and occupy a relatively narrow range of boundary Reynolds number (Fig. 5.4a). It is important to understand how Shields obtained these results in comparing them with later compilations.

Shields experimented with different flow depths, velocities and slopes in order to vary the shear stress on the bed. He measured the sediment transport rate in the flume and extrapolated this back to zero to obtain the critical shear stress at the threshold of sediment motion. However, it is not certain whether Shields was measuring the critical shear stress for a flat bed or a rippled bed. Nor were turbulent velocity fluctuations, rather than time-averaged conditions, taken into account.

Subsequent workers have considered the probabilistic (stochastic) basis for the threshold of sediment motion by assessing the variability in both instantaneous shear stresses on the bed under a turbulent boundary layer and the susceptibility of the bed to dislodgement, reflecting its grain size characteristics and packing. In particular, the burst–sweep process (Section 4.4.3) must have an effect on the entrainment condition. Others have focused on the definition of the threshold by direct microscopic observation of the bed.

(a)

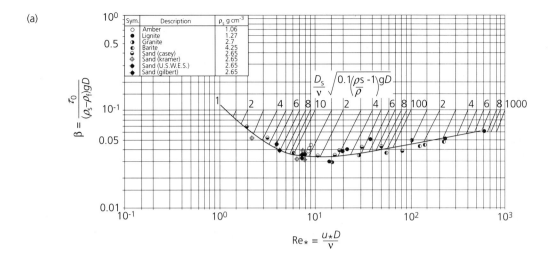

(b)

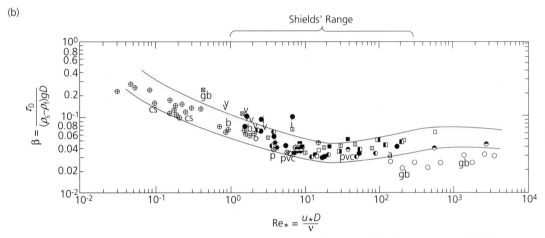

Fig. 5.4 (a) The Shields diagram, originally proposed by Shields (1936) [3], and modified by Vanoni (1964) [4], showing a dimensionless stress against the grain or boundary Reynolds number. In order to calculate the shear stress required to initiate movement of a sediment of given density ρ_s and size D immersed in a fluid of density ρ_f, it is first necessary to calculate the value of $(D/\nu)\sqrt{0.1\ (\rho_s/\rho_f - 1)gD}$, then to find the intersection with the Shields curve, and finally to read the value of β from the ordinate. (b) The Shields curve extended to smaller and larger boundary Reynolds numbers (grain sizes). After Miller *et al.* (1977) [5]. Key to symbols in Table 5.2.

The Shields diagram has been extended at both lower and higher boundary Reynolds numbers (Fig. 5.4b). At low boundary Reynolds numbers (less than 10), corresponding to fine grain sizes, there is a relationship of increasing β for decreasing boundary Reynolds number, reflecting the fact that fine, well-packed sediments with cohesive bonds, entirely enclosed within the viscous sublayer, are more difficult to entrain than fine sands. In coarser sands at higher boundary Reynolds Numbers (over 100) the original

Shields diagram shows a rather flat relation between β and boundary Reynolds number.

Finally, the threshold may be affected by the sorting of the bed material, and in particular the state of *armouring* of the bed by a layer of coarse grains. An 'exposed' large grain may roll or hop along a bed of finer grains, into which it is not able to fall for protection, a process known as *overpassing*. Conversely, the removal of finer grains may leave a protective layer of coarser particles at the surface which armour

Table 5.2 Granular materials used in studies of threshold of motion (all flows in water except as marked).

Source	Symbol	Material	Density (g cm^{-3})
Casey (1925) [in Tison (1954) [6]	◑	Sand	2.65 (assumed)
Shields (1936) [7]	▬	Amber	1.06
	■	Lignite	1.27
	▯	Granite	2.70
	◧	Barite	4.25
(Casey)	⊻	Sand	2.65
(Kramer)	⊡	Sand	2.65
(USWES)	✦	Sand	2.65
(Gilbert)	□	Sand	2.65
Vanoni (1964) [4]	⊠	Quartz sand	2.65
	⊠gb	Glass beads	2.49
Neill (1967) [8]	○	Gravel	2.53
	○a	Cellulose acetate	1.31
	○gb	Glass balls	2.50
Grass (1970) [9]	V	Fine Mersey sand (quartz)	2.65
White (1970) [10]	⊕	Natural sediment	2.52
	⊕b	Lead-glass ballotini	2.60
	⊕cs	Crushed silica	2.55
	⊕p	Polystyrene	1.05
	⊕pvc	Welvic PVC	1.54
mentor oil ⎰	⊗	Natural sediment	2.52
	⊗b	Lead-glass ballotini	2.60
	⊗p	Polystyrene	1.05
⎱	⊗pvc	Welvic PVC	1.54
Everts (1973) [11]	●	Quartz sand	2.65
	●i	Ilmenite	4.70
Paintal (1971) [12]	◓	Quartz sand	2.65

the bed against further erosion. The principle of overpassing and armouring may explain a considerable amount of scatter in the Shields diagram. The concept of bed armouring is discussed again in Section 5.3.2 in relation to flow resistance and palaeohydrology of bedload-dominated streams. Its application to deflation by the wind can be found in Chapter 10.

As it stands, the Shields diagram is difficult to use, because shear stress (or shear velocity) and grain size occur on both axes. This problem has been alleviated in two ways. One is the inclusion of an additional scale which combines fluid and grain characteristics in a dimensionless form (Fig. 5.4a). Another is the combination of the grain Reynolds number and Shields's β in such a way as to eliminate the shear velocity. The *Yalin parameter* is

$$\Phi = \frac{Re_*^{2}}{\beta} = \frac{(\rho_s - \rho_f)gD^3}{\rho_f v^2} \tag{5.11}$$

where $v = \mu/\rho$ is the kinematic viscosity. A plot of β against Φ has the same general from as the Shields curve (Fig. 5.5).

5.3 Modes of sediment transport

Once the threshold of sediment movement is exceeded, loose particles on a substrate begin to move in various modes (Fig. 5.6). Some grains move by a slow sliding or rolling motion in contact with the bed. Other particles are lifted a few grain diameters into the fluid and are transported downstream in ballistic trajectories before impact on the bed. These particles therefore move in short hops or jumps, a process known as *saltation*. Other grains are entrained higher in the flow and are advected downstream by fluid turbulence as *suspended load*. Some of the suspended load falls to the bed to reside there until it is entrained once again, which is therefore termed *intermittently suspended load*, whereas other particles virtually never

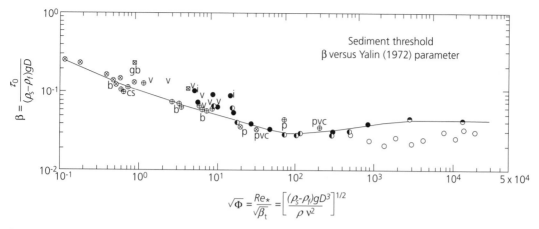

Fig. 5.5 The Yalin parameter against Shields's β, modified from Yalin (1977) [13]. Key to symbols in Table 5.2.

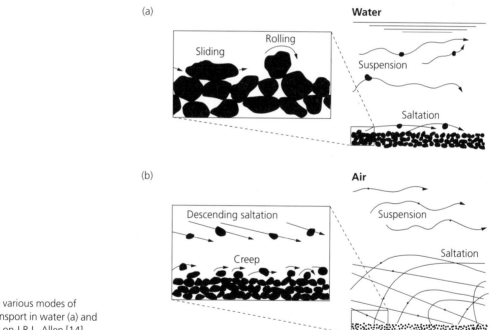

Fig. 5.6 The various modes of sediment transport in water (a) and air (b). Based on J.R.L. Allen [14].

touch the bed, being permanently suspended in the flow. This is known as *washload*.

Whether a grain moves as bedload or suspended load depends on its size and density and on the hydraulic properties of the flow. A sand grain may move in suspension in a swift mountainous river, by bedload in a river flowing through foothills, and may be static in a sluggish lowland stream. Particles may alternate between suspended and bedload transport during periods of high and low discharge of a river. On a still shorter length- and time-scale, a particle may be transported in suspension while caught up in powerful turbulent eddies, but move by bedload, or reside temporarily in the bed material, in sheltered zones such as the base of lee slopes of bedforms under the separated flow. The distinction between bedload and suspended load transport is therefore not an easy one.

A useful concept is that of the *transport stage*, defined as the shear velocity (u_*) driving downstream sediment transport divided by the settling velocity (w) causing deposition on the bed. It can be plotted against the relative proportion of grains moving in different transport modes (Fig. 5.7a). Experiments show that grain velocities are low compared to the mean flow velocity when the transport stage is also low. However, when the shear velocity exceeds the settling velocity, the grains move with approximately the velocity of the mean flow (Fig. 5.7b). This point, where $u_* = w$, approximates to the saltation–suspension threshold.

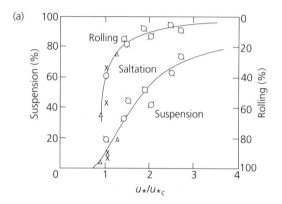

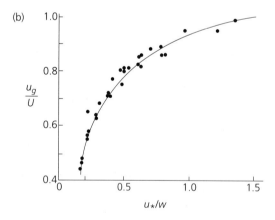

Fig. 5.7 (a) Modes of movement of a solitary grain moving over a fixed bed as a function of the dimensionless shear velocity u_*/u_{*_c} using the experimental data of Francis (1973) [15]. The different symbols are for grains of nominal diameter 8.4 mm but of different density in a flow 48 mm deep. (b) Dimensionless grain velocity u_g/U versus the dimensionless shear velocity u_*/w for grains between 0.5 mm and 1.0 mm but of different densities, using the experimental results of Francis (1973) [15]. U is the velocity of the fluid.

In the following sections the essentials of bedload and suspended load transport are described more or less in isolation from the considerable changes taking place on the mobile or 'loose' boundary itself. These changes, manifested in the occurrence of bedforms of various geometries and sizes, are discussed in Section 5.4. It is important, however, to appreciate that the mechanics of bedload and suspended load transport is linked to the characteristics of the boundary itself.

5.3.1 Bedload

The proportion of bedload relative to the total load varies considerably between different drainage basins (see also Chapter 3). It may also vary strongly within a single drainage basin according to the nature of the flood hydrograph. It is clearly important to be able to estimate bedload sediment transport rates from an engineering point of view (bridge and pier erosion; reservoir filling), and also from a geomorphological point of view, since bedload-dominated rivers have different planforms and rates of migration than those dominated by suspended load.

That the bedload sediment transport rate could be related to the shear stress exerted by the river was proposed in the second half of the nineteenth century by P. Du Boys (1879). Subsequently, a number of empirical equations have been developed based primarily on the collection of data from laboratory flume experiments. These equations generally incorporate characteristics of the size and roughness of the bed material and the shearing force of the fluid expressed in terms of either shear stress or velocity.

Other attempts at predicting the bedload sediment transport rate involve a certain amount of theory. One approach is to treat grain transport as a probabilistic phenomenon, whereby the probability of entrainment is calculated for a number of different size classes. Another approach is to consider sediment transport as the result of the stream doing work on its bed [16]. This is briefly considered below.

The immersed weight of sediment being transported by a fluid over a unit area of bed is a force per unit area, or stress, measured in Newtons per square metre. Since the transport rate of sediment is the product of the immersed weight (a stress) and the transporting velocity, the transport rate has the dimensions of a rate of doing work, or power. However, the sediment load acts downwards on the bed and not parallel with the bed, so it cannot be thought of as directly equivalent to stream power.

Nevertheless, we can think of a sediment-transporting flow as doing work at a particular rate, and its rate of doing work (or power) as being related through conversion factors to the rates of transport of the various transport modes (principally bedload and suspended load).

The only way in which such an approach can be used to predict sediment transport rates is by deriving parameter and coefficient values from empirical data. These empirical data allow the relation between an index of stream power and observed transport rate to be defined.

Stream power has the dimensions $[ML^2T^{-3}]$ over a unit area of stream bed $[L^{-2}]$, so is proportional to the cube of velocity. At low flow velocities just above the threshold, the *excess velocity* is likely to be more appropriate. This can be incorporated in a number of ways. A usage which takes full account of the threshold condition is

$$Q \propto (\acute{U}^2 - U_c^{\,2})(\acute{U} - U_c) \qquad \textbf{(5.12)}$$

where Q is the sediment transport rate, $\acute{U}$ is the velocity averaged over the flow depth and U_c is the critical (threshold) velocity for grain transport. It is more common, however, to express the transport rate in terms of the force (per unit area) driving sediment motion, the shear stress. In equation (5.12) the first bracketed term is then equivalent to the excess stress.

Despite over a century of study of the problem, bedload transport rate predictions are still notoriously difficult [15]. Bedload transport equations developed from one particular river generally perform very badly in other rivers [17]. This is because the transport of bed material of mixed sizes and shapes in rivers of highly variable depth, geometry and discharge behaviour is an extremely complex problem. A great deal of variation in actual bedload sediment transport rates may result from processes such as the following.
- **Armouring and consolidation in gravel-bed rivers:** we have encountered this effect previously in relation to the threshold of sediment movement. Winnowing of finer particles leaves a protective, coarser pavement or armoured layer. Partial breaching of the armour exposes the sub-armour gravel to rapid erosion. Healing of the armour layer during waning flow causes bedload sediment transport rates to fall rapidly. Freshly deposited sediments tend to be 'overloose' and vulnerable to renewed bedload transport. With time, grains interlock and fine matrix infiltrates into void spaces, making the bed more resistant to entrainment.

- **Resistance of bedforms (microforms) in sand- and gravel-bed rivers:** flow resistance affects the efficiency of a stream and therefore its bedload transport rate. Changes in the configuration of the sand (flat, rippled or duned bed) or gravel bed (pebble clusters, transverse ribs) therefore affect sediment transport rate.
- **Difficulties in assessing the threshold of sediment transport:** the Shields curve is appropriate for planar beds, rather than natural beds with microform roughness. In addition, clast shape affects the pivot angles of particles on the bed. The smallest pivot angles, and therefore the greatest ease of erosion, occur in spherical populations. Greater pivot angles occur in bed populations that deviate progressively strongly from an ideal spherical shape. Since bedload equations commonly rely on the excess stress or velocity (above the critical stress or velocity at the threshold), variations in the entrainment condition affect predictions of the bedload sediment transport rate.
- **Unsteadiness in flood and bedload discharges:** bedload transport commonly takes place in distinctive pulses not directly related to the flood hydrograph. This unsteadiness may be due to the passage, formation or break-up of microforms (bedforms) and macroforms (bars) along the bed.

5.3.2 Flow resistance and palaeohydrology in bedload rivers

We have seen in the practical exercise on steady uniform flow down an inclined plane (Practical Exercise 4.2) that the driving force for the movement of water from high altitudes to low altitudes is the downslope component of gravity acting on the mass of the water. In that practical exercise we balanced the downslope force against an upslope-acting frictional force at the bed. In this section we look at the frictional energy losses during fluid flow, that is, *flow resistance*. Once we know something about the frictional losses of energy to the boundaries of an open-channel flow, the channel bed and banks, it is then possible to advance our understanding of fluvial hydrology and palaeohydrology. A palaeohydraulics problem is provided in Practical Exercise 5.1.

We have seen the application of Bernoulli's theorem to the generation of lift and drag forces on a particle in a flow in Section 4.3.1, and seen the results of this in terms of the threshold of sediment motion in Section 5.2. We can now build on the idea of conservation of energy in considering the frictional energy losses in rivers.

In the streamtube example used in the derivation of Bernoulli's theorem, there were no frictional losses of energy at the boundaries of the streamtube. However, in natural streams energy is expended by friction on the channel bed and banks. The parameters that are likely to be important in describing the physics of this situation are already familiar—flow depth (h) and fluid density (ρ_f) are important in determining the basal shear stress (τ_0), and flow velocity (u) is also affected by the fluid viscosity (μ). Where the bed of the stream is rough, energy losses should in some way be related to the length scale of the roughness, which we can call the *roughness height* (k_s). There are therefore two length-scales in problems of flow resistance, the flow depth (or pipe diameter) and the roughness height. These can be combined into a dimensionless variable h/k_s, the *relative roughness*. We are now quite used to expressing fluid flows in terms of a Reynolds number $Re = \rho_f uh/\mu$. If we could convert the shear stress into a third dimensionless variable, we could then express it in terms of the other two dimensionless groups, the relative roughness and the Reynolds number. Shear stress is measured in pascals (kg m^{-1} 5^{-2}), so to make τ dimensionless we could divide by ρu^2. In the somewhat stricter language of dimensional analysis, we have six variables involving three dimensions—$h[L], \rho_f[M\ L^{-3}], \tau_0[M\ L^{-1}T^{-2}], u[L\ T^{-1}], \mu[M\ L^{-1}T^{-1}], k_s[L]$—so the problem can be reduced to a set of three dimensionless products. The third dimensionless product is usually given as

$$\frac{8\tau_0}{\rho_f u^2} = f \qquad (5.13)$$

where f is called the *friction factor*. Flow resistance, therefore, as measured by the friction factor, is a function of the Reynolds number and the relative roughness.

How are friction factors estimated for natural streams? The roughness of a natural stream is clearly highly variable, and presents us with a far more complex problem than, for instance, the roughness of the inside of a pipe carrying fluids such as water or sewage. River beds are commonly covered with dunes and bars and occupied by channels and pools. As a result, friction factors are calculated for each individual river where the mean flow velocity has been measured. This is possible because we can combine the equation for steady uniform flow down an inclined plane (4.18) with the equation for the friction factor (5.13) to obtain

$$\rho_f gh \sin \alpha = \frac{1}{8} \rho_f u^2 f \qquad (5.14)$$

which, solving for u, gives

$$u = \frac{\sqrt{8gh \sin \alpha}}{f} \qquad (5.15)$$

This is usually written in the form

$$u = C(h \sin \alpha)^{0.5} \qquad (5.16)$$

where $C = (8g/f)^{0.5}$ is the Chézy coefficient introduced at the beginning of Chapter 4. It is a widely used index of flow resistance in rivers and, of course, is proportional to the friction factor f.

In an open-channel flow the discharge of water is the product of the cross-sectional area of the channel and the mean flow velocity. However, only the bed and banks exert friction on the river. If we call the bed and banks the *wetted perimeter*, we can define the *hydraulic radius*, R_H, as the cross-sectional area divided by the wetted perimeter (Fig. 5.8). In rivers that are very broad compared to their depth, the flow depth is approximately equal to the hydraulic radius. In narrow but deep streams, the hydraulic radius is a better length-scale for the calculation of flow resistance. It is used in an alternative to the Chézy equation,

$$u = \frac{R_H^{2/3} \sin \alpha^{1/2}}{n} \qquad (5.17)$$

where n is *Manning's roughness coefficient*. Typical values for natural channels are given in Table 5.3. It can be seen that $n = (1/C)R_H^{4/3}$.

It is possible to make estimates of the former discharges of rivers by combining the relationship for shear stress for steady uniform flow down an inclined plane with a resistance equation. A promising area to look at is the gravelly outwash of glaciers [18,19] (see also Chapter 11). Since different generations of channels are preserved in terrace sequences in areas undergoing deglaciation, it might be possible to test the idea that there have been substantial changes in the discharges of these meltwater river systems during deglaciation.

There are a large number of assumptions and sources of error. The main ones that we should consider are as follows.

• **Use of slope:** former steep meltwater channels may have been subjected to rapid velocity fluctuations, large-scale turbulence and local effects of funnelling

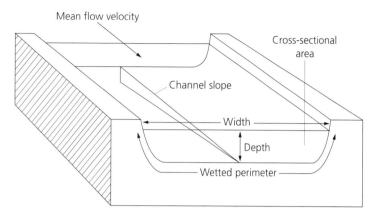

Fig. 5.8 The wetted perimeter and hydraulic radius in a river.

Discharge = Cross-sectional area × mean flow velocity
Hydraulic radius = Cross-sectional area/wetted perimeter

Table 5.3 Values of Manning's roughness coefficient for various types of natural channel [20].

Channel type	Normal value	Range
Small channels (width <30 m)		
Low-gradient streams		
Unvegetated straight channels at bankfull stage	0.030	0.025–0.033
Unvegetated winding channels with some pools and		
shallows	0.040	0.033–0.045
Winding vegetated channels with stones on bed	0.050	0.045–0.060
Sluggish vegetated channels with deep pools	0.070	0.050–0.080
Heavily vegetated channels with deep pools	0.100	0.075–0.150
Mountain streams (with steep unvegetated banks)		
Few boulders on channel bed	0.040	0.030–0.050
Abundant cobbles and large boulders on channel bed	0.050	0.040–0.070
Large channels (width >30 m)		
Regular channel lacking boulders or vegetation	—	0.025–0.060
Irregular channel	—	0.035–0.100

and expansion, so the assumption of steady uniform flow down an inclined plane may be a cause of errors in the use of slope to estimate the basal shear stress. Differential isostatic uplift (Chapter 1) may also have tilted the terraces, making estimates of former slopes difficult.
• **Particle size as a means of estimating flow resistance**: the largest clasts are assumed to represent the maximum discharge conditions. Measurements of the largest clasts include errors—those due to the choice of the sampling sites; the choice of how many clasts are measured over how large an area in order properly to represent the clast population; and the choice of measurement to represent the amount of

projection of the clast above the mean bed level and therefore a good estimate of the roughness. It is assumed that all clast sizes are available in the catchment, so that the size transported is a reflection of the strength of the river. It is possible that the geology of the source region influences clast size significantly, irrespective of river discharge. Finally, we also need to assume that the observed deposits were not emplaced by debris flows, bank slumping, ice-rafting, or some mechanism other than a Newtonian fluid flow.
• **Particle size as a means of estimating the critical bed shear stress for entrainment**: this is usually carried out using the Shields function, which expresses

the critical shear stress (τ_c) as a linear function of the clast size (D):

$$\tau_c = \beta \gamma' D \qquad (5.18)$$

where β is the Shields function, $\gamma' = (\rho_s - \rho_f)g$ is the submerged specific weight of the particle, and D is the representative grain size. The value of β should vary according to whether the gravel bed is 'underloose' (i.e. closely packed or imbricated; $\beta \approx 0.1$), 'normal' ($\beta \approx 0.056$) or 'overloose' (i.e. perched clasts, 'quick' bed; $\beta \approx 0.02$).

Further ideas on the transport of grains in concentrated dispersions can be found in Chapter 6. The process of saltation is dealth with in more detail in Chapter 10 in the context of sediment transport by wind.

5.3.3 Suspended load

In Chapter 3 we considered the measurement of suspended sediment concentrations in rivers, and considered the significance of these results for the global pattern of weathering and erosion. Here we

---

Practical exercise 5.1: Palaeohydrology of meltwater streams

1 A series of eight terrace forms with good evidence of palaeochannel courses bound the valley of the Watson River about 12 km from the present margin of the Greenland ice sheet [21]. This area of Søndre Strømfjord is thought to have become deglaciated between about 7000 and 6500 BP. Maximum deglaciation was attained at about 6000 BP, followed by a small readvance. During the latter part of the Holocene there have only been small ice margin oscillations. Calculate the palaeodischarges of rivers in seven terraces. The input data are given in Table P5.1.

Discharges can be calculated as follows:
(a) Calculate the critical bed shear stress using the Shields function for a 'normal' boundary.
(b) Calculate the critical flow depth from the equation for steady uniform flow down an inclined plane.
(c) Calculate the flow resistance, assumed to be solely due to the gravelly bed, rather than to the presence of large roughness elements such as bars, in the form of Manning's n. An empirical method based on multiple regression of steep, coarse-grained streams is

$$n = 0.32 S^{0.38} h^{-0.16} \qquad (5.19)$$

where S is the slope.
(d) Calculate the resistance using the friction factor. This can be done in a number of ways. An empirical method derived from measurements in steep, braided channels with large roughness elements gives

$$\frac{1}{\sqrt{f}} = \left\{ 1 - 0.1 \left(\frac{k_s}{R_H} \right) \right\} 2 \log_{10} \left(\frac{12 R_H}{k} \right) \qquad (5.20)$$

Table P5.1

Terrace (date)	Slope	Clast size (m)	Flow width (m)
1 (7500–6500 BP)	0.0117	0.177	205
2 (7500–6500 BP)	0.0259	0.303	355
3 (7500–6500 BP)	0.0204	0.306	380
4 (7500–6500 BP)	0.0032	0.240	23.2
5 (6500–5500 BP)	0.0204	0.313	28.3
6 (6500–5500 BP)	0.0140	0.148	36.2
7 (700–300 BP)	0.0131	0.092	552.7

where the roughness height is best represented by $k_s \approx 4.5 D$.
(e) Calculate the critical mean flow velocity U_c based on Manning's n and on the friction factor f.
(f) Calculate the maximum total discharge Q from

$$Q = w h_c U_c \qquad (5.21)$$

where w is the width of the channel.
2 A recently recorded flood resulting from a glacier dam burst (*jökulhlaup*) [22] in 1984 had a discharge of about $1.2 \times 10^3 \, \text{m}^3 \text{s}^{-1}$, and the normal discharge of the Watson River is about $0.1 \times 10^3 \, \text{m}^3 \text{s}^{-1}$. Interpret your palaeodischarge data in terms of the changes taking place during deglaciation and in the light of both the present-day discharge of the Watson River and the catastrophic *jökulhlaup* discharge.
3 The glacial advance and retreat over the last 7500 years is given in Fig. 5.9. How do the palaeodischarges relate to the position of the ice front?

Continued.

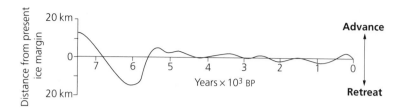

Fig. 5.9 Glacial advance and retreat at Søndre Strømfjord over the last 7500 years, for use in Practical Exercise 5.1.

Practical exercise 5.1: *Continued*

Solution
The calculations for this exercise are shown in Table P5.2 and require little comment. Note, however, that we could have used a number of ways of estimating the resistance coefficients n and f. Each method may be the more appropriate given local conditions (slope, channel size, clast size). Our method of calculating the friction factor is strictly applicable for slopes $S \geqslant 0.02$. This is not always satisfied in our dataset, but the errors in using the equation for friction factor outside its recommended limits are relatively small.

By plotting the range of palaeodischarge for the seven terraces against time, it can be seen that the largest discharges are associated with the phase of maximum retreat of the ice cap in west Greenland. Since then, discharges appear to have been much smaller, and the estimates for terraces 5 and 6 are similar to the present-day discharge of the Watson River. The jökulhlaup in 1984 had a discharge that compares with the larger palaeodischarge estimates for earlier in the Holocene. This suggests that although discharges have certainly fallen since the period of maximum deglaciation about 6500 years ago, the continued proximity of the ice sheet since that time has acted to maintain peak meltwater discharges at a high level. This also demonstrates that the rare event, operating on a short time-scale, may mask longer-term discharge trends in the catchment.

Table P5.2

Terrace	1	2	3	4	5	6	7
Slope	0.0117	0.0259	0.0204	0.0032	0.0204	0.0140	0.0131
Clast size (m)	0.177	0.303	0.306	0.240	0.313	0.148	0.092
Flow width (m)	205.0	355.3	380.0	23.2	28.3	36.2	552.7
Hydraulic radius (m)	1.38	1.07	1.38	4.34	1.29	0.93	0.65
Critical shear stress ($N\,m^{-2}$)	160.4	274.7	277.4	217.5	283.7	134.2	83.4
Critical flow depth (m)	1.40	1.08	1.39	6.93	1.42	0.98	0.65
Manning's n	0.056	0.079	0.069	0.026	0.069	0.063	0.066
Friction factor, f	0.162	0.346	0.264	0.093	0.291	0.194	0.175
Mean flow velocity (based on n) ($m\,s^{-1}$)	2.39	2.13	2.57	5.79	2.45	1.79	1.30
Mean flow velocity (based on f) ($m\,s^{-1}$)	2.82	2.52	2.90	4.33	2.80	2.36	1.95
Discharge (based on n) ($m^3\,s^{-1}$)	686	817	1358	931	99	64	467
Discharge (based on f) ($m^3\,s^{-1}$)	809	967	1532	696	113	84	701

are more concerned with the physics underpinning sediment suspension in turbulent flows such as rivers, and particularly with ways in which we might model the sediment concentration in a flow deriving its sediment from its own bed.

Grains are suspended when their settling velocity is exceeded by the root mean square of the vertical instantaneous velocity of the fluid. The grains suspended in a fluid transmit a force to the fluid equal to their weight, so high suspended sediment concentrations cause the pressure in a fluid to increase. The increase in pressure will be dependent on the suspended sediment concentration and the submerged specific weight of the particles $(\rho_s - \rho_f)g = \gamma'$. It is commonly termed the *buoyant weight* of the grains.

The distribution of suspended sediment in a flow can be treated as a problem of diffusion. The diffusion takes place from a region of high concentration at the bed, to a region of low concentration near the surface. We know that the mass flux of a diffusant (Q), such as suspended sediment, is linearly proportional to the gradient of its concentration C. This is expressed in Fick's' first law of diffusion

$$Q = -\kappa \frac{\partial C}{\partial y} \tag{5.22}$$

where κ is a coefficient known as the diffusivity (see Section 4.4.3). Assuming conservation of mass, the flux of diffusant into and out of an infinitesimal control volume must be equal to the rate of change of mass in that volume. In one dimension this requirement for continuity is

$$\frac{\partial C}{\partial t} = -\frac{\partial Q}{\partial y} \tag{5.23}$$

Substituting for Q from Fick's first law,

$$\frac{\partial C}{\partial t} = \frac{\partial}{\partial y}\left(\kappa \frac{\partial C}{\partial y}\right) \tag{5.24}$$

which is known as *Fick's second law*. If the diffusivity does not vary with y, that is, it does not vary with the concentration of the diffusant, κ can be taken outside of the partial differential to give

$$\frac{\partial C}{\partial t} = \kappa\left(\frac{\partial^2 C}{\partial y^2}\right) \tag{5.25}$$

If there is no net change in the concentration in time (steady conditions), then $\partial C/\partial t = 0$ and

$$\frac{\partial^2 C}{\partial y^2} = 0 \tag{5.26}$$

This is the one-dimensional *Laplace equation*. We can now work out the concentration profile by integrating the Laplace equation for certain boundary conditions. Integrating once

$$\frac{dC}{dy} = K_1 \tag{5.27}$$

and then again

$$C = K_1 y + K_2 \tag{5.28}$$

where K_1 and K_2 are constants of integration. These constants can be determined from the boundary conditions

$$\begin{aligned} C &= C_0 &&\text{at} \quad y = 0 \\ C &= 0 &&\text{at} \quad y = h \end{aligned}$$

where $y = 0$ at the bed and $y = h$ far from the bed where the concentration is very low. From the first boundary condition $C_0 = K_2$, and from the second

$$0 = K_1 h + C_0$$

$$\therefore K_1 = -\frac{C_0}{h}$$

Consequently, the solution for the concentration as a function of y is

$$C = C_0\left(1 - \frac{y}{h}\right) \tag{5.29}$$

Inspection of equation (5.29) shows that we have produced a concentration profile which is linear from a maximum at the bed to zero at a height equal to the flow depth. Observed sediment concentrations in rivers (see Fig. 5.12) clearly are not linear. This suggests that a key process has been ignored from the foregoing analysis. This process is the settling of grains. We examine the balance between the upward diffusion and gravitational settling of grains in the next section (5.3.4).

5.3.4 A diffusion model for suspended sediment concentrations

The kinematic eddy viscosity was derived from an assumed logarithmic velocity profile in Section 4.4.4 and shown to vary parabolically in an open-channel flow as a function of height above the bed. The problem needing to be addressed here is that of the diffusion of suspended sediment in a turbulent flow. Let us first make an assumption that the diffusion of momentum in a shear flow, given by the kinematic eddy viscosity, is closely related to the diffusion of a tracer such as suspended sediment, given by the *turbulent diffusivity*

$$\kappa_t = \beta \varepsilon \tag{5.30}$$

where β is approximately equal to 1. Consequently, from equation (4.40),

$$\kappa_t = \beta u_* \left(\frac{h-y}{h} \right) ky \tag{5.31}$$

where as a reminder u_* is the shear velocity, d is the flow depth, y is the height above the bed, and k is von Karman's constant. For steady conditions, that is, where the sediment concentration in the flow is not changing in time, any loss of sediment by downward settling must be balanced by the upward supply of sediment by diffusion from the region of high concentrations near the bed (Fig. 5.10). The flux of downward settling is the product of the sediment concentration and the settling velocity, whereas the flux of upward diffusion is given by Fick's first law. Balancing the two gives

$$wC = -\kappa_t \frac{dC}{dy} \tag{5.32}$$

$$\frac{dC}{C} = -\frac{w\,dy}{\kappa_t} \tag{5.33}$$

and, substituting equation (5.31) into equation (5.33),

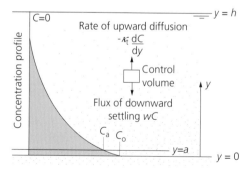

Fig. 5.10 The concentration of suspended sediment is a result of the upward diffusion from the bed balanced by the downward gravitational settling.

$$\frac{dC}{C} = \frac{w h\, dy}{\beta k u_* (h-y) y} \tag{5.34}$$

We can introduce a boundary condition that the concentration at a reference height a close to the bed is C_a and then integrate the left-hand side of the equation from C_a to C, and the right-hand side of the equation from a to y. In other words,

$$\int_{C_a}^{C} \frac{dC}{C} = \left(\frac{w}{\beta k u_*} \right) \int_{a}^{y} \frac{dy}{\frac{y}{h}(h-y)} \tag{5.35}$$

which gives

$$\frac{C}{C_a} = \left[\frac{h-y}{y} \times \frac{a}{h-a} \right]^{w/\beta k u_*} \tag{5.36}$$

Using an equation of this sort [23], the concentration of suspended sediment can be found at any depth in the flow in relation to the concentration at a reference depth, usually taken as very close to the bed, representing the thickness of the moving bedload layer, and approximately equal to twice the mean grain size [24]. The relative concentration of suspended sediment where $y = 0.05h$ is given in Fig. 5.11. To obtain the actual fractional suspended sediment concentration as a function of depth, one needs to know the concentration at the reference height C_a. This could be calculated assuming that the suspended grains have a concentration in the moving bedload layer predicted by a bedload sediment transport equation of the general form given in equation (5.12). This computation is carried out, for instance, in the FORTRAN subroutine included in [25].

The grouping $w/\beta k u_*$ is known as the *Rouse number*. Since $\beta \approx 1$ and $k \approx 0.4$, a Rouse number of 2.5 corresponds to the criterion that $w = u_*$. This is a rough criterion for suspension. If the settling velocity exceeds the shear velocity, the Rouse number should exceed 2.5 and bedload transport rather than suspension should take place. It can be seen from Fig. 5.11 that the region where the Rouse number exceeds 2.5 is very close to the bed, supporting the view that $w = u_*$ is a reasonable criterion for suspension. But it should be borne in mind that the method outlined above is expected to break down very close to the bed because the balance between upward turbulent diffusion and downward settling is likely to be violated by the effects of fluid lift and drag forces

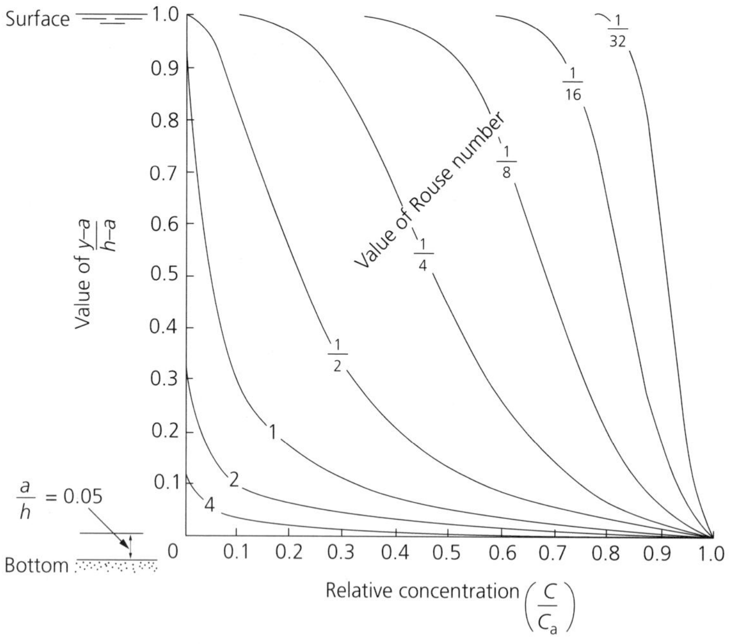

Fig. 5.11 The distribution of suspended sediment assuming upward diffusion from the bed, expressed as a relative concentration C/C_a versus the relative height above the bed, where $a = 0.05h$. After American Society of Civil Engineers (1963) [26].

and grain interactions where sediment concentrations are very high. Below the reference height a, a bedload sediment transport formula should be used.

High suspended sediment concentrations are favoured by increases in the bed roughness, since this increases the shear velocity. Such an increase might come about through the development of dunes on the bed of the river. An increase in the viscosity of the flow would also hinder settling rates and therefore increase suspended sediment concentrations. Increased viscosity would occur in cold-water streams and in river water with large amounts of colloidal washload.

The suspended load and its transport rate can now be found from a knowledge of the distribution of the sediment concentration. The suspended load transport rate, Q_s, is simply the product of the mass of suspended load over a unit area of bed and the depth-averaged velocity of the flow taken at the same location as the suspended sediment measurement. In terms of dry mass of material, m_s,

$$Q_s = m_s \acute{U} \tag{5.37}$$

where

$$m_s = \rho_s \int^b C(y) \mathrm{d}y$$

The transport rate is usually expressed in immersed weight of sediment. The term ρ_s is then replaced by

Practical exercise 5.2: Suspended sediment concentrations in the Mississippi River

The Mississippi River at St Louis, Missouri, USA, is approximately 8.5 m deep. The velocity profile measured on 24 April 1956 is given in Table P5.3 (see Fig. 5.12a). The suspended sediment concentrations for a range of grain sizes were also measured at a number of different heights above the bed [20]. We can therefore test whether the suspended sediment loads in this river are approximated by the diffusion model of Rouse.

A reference depth of about 5% of the total flow depth gives $a = 0.5$ m. The sediment concentrations at this reference depth for three grain size classes are given in Table P5.4. The average settling velocity for grains in these size classes is also provided. For fine grain sizes the settling velocity is given by Stokes' law (Section 4.2.2). For the larger grain sizes, the values are taken from results from experiments in settling tubes [28,29].

1 Plot the velocity data on log-linear graph paper with linear velocity against logarithmic depth. Extrapolate the best-fit straight line towards the bed. It intercepts the depth axis (where $\bar{u} = 0$) at the roughness length y_0. The logarithmic velocity law

Continued.

Practical exercise 5.2: *Continued*

Table P5.3

Distance above bed, y (m)	Velocity, $\bar{u}$ (m s–1)
7.7	1.25
5.2	1.20
3.2	1.05
1.8	1.00
0.9	0.90
0.6	0.80

Table P5.4

Grain size (mm)	0.002–0.016	0.125–0.25	>0.25
Concentration at			
y = a (ppm)	145	270	100
Settling			
velocity (mm s⁻¹)	0.6	30	70

(equation (4.36)) can be adapted to calculate the shear velocity u_*.

$$u_* = k(u_1 - u_2)/2.3(\log y_1 - \log y_2)$$

2 The sediment concentration at any depth relative to the concentration at the reference height $y = a = 0.5$ m is given by equation (5.36). Calculate the Rouse number for the three grain size classes, assuming the von Karman coefficient to be $k = 0.4$ and $\beta = 1$. What are the calculated sediment concentrations at heights 2.5 m and 5.0 m above the bed for the three grain size classes? How do your results compare with the measured suspended sediment concentrations given in Fig. 5.12b?

Solution

The velocity profile allows the shear velocity to be estimated as 0.245 m s⁻¹.

The Rouse number simplifies to $w/0.4u_*$. The ratio C/C_a can then be simply found for the three grain size classes chosen, and the predicted sediment concentrations compared with those observed (Table P5.5). Note that, as expected, the coarser grains are located very close to the bed, whereas the fine particles are diffused upwards by turbulent momentum, reaching high concentrations far from the bed. We should beware, however, of the effects of high sediment concentrations on the velocity profile. The value of von Karman's coefficient decreases from 0.4 for highly concentrated flows.

This exercise indicates that the suspended sediment concentration can be calculated for different grain size classes using the turbulent diffusion approach. However, it should be stressed that the exercise has been simplified by taking the sediment concen-tration at a reference depth as a measured value. If this were not available, it would be necessary to estimate it at a depth corresponding to the top of the moving bedload layer, as explained in the text.

Table P5.5

Grain size (mm)	0.002–0.016	0.125–0.25	>0.25
Rouse number	0.0061	0.31	0.71
$\dfrac{C}{C_a}$, y = 2.5 m	099	0.56	0.26
$\dfrac{C}{C_a}$, y = 5 m	0.98	0.30	0.06

$(\rho_s - \rho_f)g$. Since the depth-mean velocity and the suspended load vary with position along a cross-section of the river, the suspended load transport rate must be integrated across the entire width of the flow to obtain the total suspended sediment discharge.

Although the diffusion approach has a certain physical elegance, there are reasons why it should be questioned as a predictor of actual suspended sediment concentrations. Perhaps the most serious is the probability that the suspended load is not entirely derived by turbulent diffusion from the bed. Much suspended material originates from outside of the channel and is contributed by overland flow from hillslope erosion. The suspended sediment concentration therefore is supply-limited rather than hydraulically limited. Consequently, as in the case of bedload prediction, much effort has gone into the collection of empirical data with which to calibrate suspended sediment transport rate formulae. The most common expression of the relation between water discharge (Q) and suspended sediment concentration (C) is the sediment rating curve (Section 3.4.5). The coefficient exponent in the sediment rating/curve

(a)

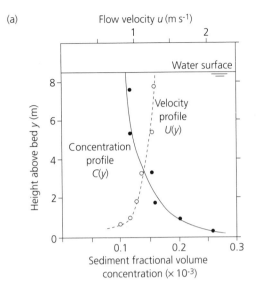

(b)

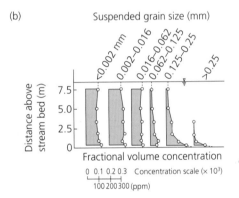

Fig. 5.12 (a) Meaured velocity and suspended sediment concentration profiles in the Mississippi River at St Louis, Missouri, on 24 April 1956. (b) Sediment concentration profiles for a range of grain size intervals. After Colby (1963) [27].

$$C = aQ^b \qquad (5.38)$$

vary from river to river, and particularly between rivers of markedly different hydrological behaviour—for example, between flashy desert streams, on the one hand, and perennial, equable, temperate zone streams, on the other. We have previously considered the hysteresis between flood and sediment discharges in drainage basins in Chapter 3.

5.4 Bedforms in a cohesionless substrate

So far we have considered the threshold of sediment

motion and the different modes in which sediment is transported, but we have not considered the interaction of the bed with the overlying flow, other than as a provider of sediment for transport or in causing flow resistance. Although it is a commonplace observation that the sandy desert floor and the dry beds of rivers are moulded into geometrical, periodically recurring irregularities, it is by no means clear why such irregularities should occur. These geometrical forms are called bedforms and they are responsible for a particular structure in the sediments formed by their migration. The preservation of the constructional faces of migrating bedforms results in a form of bedding with inclined internal surfaces, known as *cross-stratification*. It is possible to view such cross-stratification in ancient sediments and to attempt to reconstruct the flow conditions responsible for its generation.

Many bedforms are orientated with their faces and crestlines transverse to the main sediment-transporting flow. Under water, they occur at two scales: ripples (small) and dunes (large). Although the flow of fluid and transport of grains are similar under both ripples and dunes, they are not entirely similar dynamically, and it is common for ripples to be superimposed on the faces of dunes. Ripples and dunes formed under unidirectional flows have shallow-dipping upstream faces known as the *stoss*, and steeply sloping downstream faces known as the *lee*. When the ripple or dune is migrating under a fluid flow, individual grains of sand can be seen rolling or saltating up the stoss side, and tumbling down the lee side or being transported in the separated flow. This pattern is shown schematically in Fig. 5.13. Other flow-transverse bedforms have low rounded surface profiles. Some bedforms are aligned with their faces, crestlines or axes parallel to the main sediment-transporting flow. The best examples are the linear, longitudinal dunes of the desert (Chapter 10).

The bedforms associated with oscillating or tidally reversing flows are discussed in Chapter 8. The bedforms found under aerodynamic flows are discussed in Chapter 10. The brief discussion here is restricted to the case of relatively clear unidirectional flows and the focus will be on an understanding of their kinematics. The reader is referred to other books listed in the Further Reading section at the end of this chapter for a broader descriptive and dynamical coverage of bedforms.

5.4.1 Froude number

In open-channel flows such as rivers, it is not possible

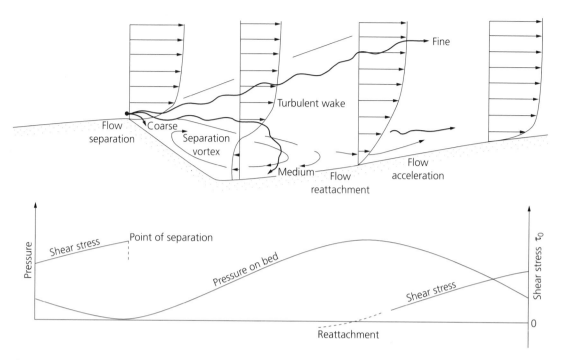

Fig. 5.13 Schematic pattern of flow and sediment dispersal over a ripple or dune. Based on Middleton and Southard (1984) [2].

to ignore the deformation of the free surface of the fluid in studying sediment-hydraulic problems. It is well known that the surface of rivers is rarely entirely flat, especially when flowing swiftly. In problems where a free surface, or an internal surface between fluids of different density, plays a role, it is necessary to include the effects of gravity in tending to flatten any deformation of that surface away from perfect planarity.

The parameters involved in this problem include the length-scale between the undisturbed level of the fluid and the surface of interest, which in this case is the flow depth (d), the average velocity of the flow, and the acceleration due to gravity, since this counteracts any disturbance of the surface. The three relevant parameters can be combined into one dimensionless product known as the Froude number.

$$Fr = \frac{\acute{U}}{\sqrt{gd}} \tag{5.39}$$

When a free surface flow approaches an irregularity in its bed, or an artificial shallowing such as a weir, the free surface may be deformed. In order to conserve discharge, two possibilities exist: the flow may shallow and speed up; or deepen and slow down. Similarly, when water from a tap strikes a flat plate it spreads radially as a sheet (Fig. 5.14). The water spreads initially as a thin sheet, but then suddenly deepens at a certain distance from the centre. By dropping some dye into the water at a number of different locations, it would be easy to show that the water has a higher velocity where it is shallow, and a lower velocity where it is deeper. For the same discharge of water, therefore, there are two flow behaviours, corresponding to supercritical and subcritical. Between the two, where the water deepens suddenly, is a *hydraulic jump*. Hydraulic jumps can be observed in swift rivers and commonly in the backwash of beaches.

Whether one flow behaviour or the other takes place depends on the Froude number, the transition from subcritical to supercritical flow taking place at a Froude number of 1. This has direct relevance to the existence of bedforms under an open-channel flow. Consider a flow which has shallow water waves on its surface (Fig. 5.15) (see Chapter 8 for more on wave theory). The phase velocity (celerity) of a shallow water wave in a water depth h is given by

$$C = \sqrt{gh} \tag{5.40}$$

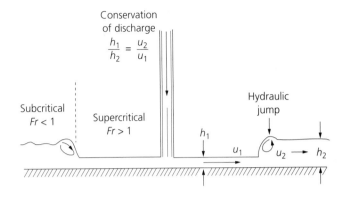

Fig. 5.14 Hydraulic jump observed when water spreads as a thin sheet from a tap.

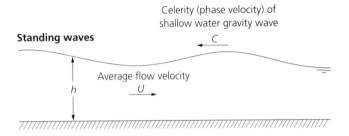

Fig. 5.15 Standing waves develop when the shallow water wave celerity equals the mean flow velocity. This corresponds to a Froude number of 1.

Consequently, it follows that when the average flow velocity, U is equal to the phase velocity, C, the surface waves are stationary with respect to the bed, that is, they appear to be 'standing' (Fig. 5.15). This condition occurs when $C = \sqrt{gh} = \hat{U}$, which is when the Froude number is equal to 1. Flows with a Froude number greater than 1 are termed supercritical; those with a Froude number less than 1 are subcritical.

5.4.2 Dimensional analysis of bedforms under a shear flow

We have already seen the value of a dimensional analysis in studying problems involving large numbers of variables. If a successful dimensional analysis of the problem of the creation of a succession of distinctive bed configurations could be carried out, it might help our physical understanding as well as guide experimental programmes in flumes.

It is assumed that the flow does not vary in time or in space (i.e. it is steady and uniform) and that the bedforms are in equilibrium with the flow [22]. It is further assumed that the bed material can be described by one variable, the mean grain size, rather than including information on sorting, packing and grain shape. The relevant variables must include those pertaining to the bed material (its mean grain size, D,

and its density, ρ_s); those referring to the fluid, such as density (ρ_f) and viscosity (μ); the acceleration due to gravity because of its effect on particle weight and on free surface waves; and those variables representing characteristics of the flow (flow depth, h, and either flow velocity or boundary shear stress).

It is debatable whether flow velocity or boundary shear stress should be used as an indication of flow strength. They are, of course, related through the frictional resistance of the flow, but when bedforms are present in a mobile bed the frictional characteristics may change markedly with changes in the bed state. It is found that when the mean flow velocity is increased over a sandy bed, the shear stress first increases rapidly as ripples and dunes exert increasing drag (Fig. 5.16). The shear stress then decreases abruptly as the dunes are washed out and replaced by a flat bed, before increasing again with the formation of standing waves and antidunes (see also Section 5.4.4). As a result, there may be more than one bed state for a particular value of the boundary shear stress, and mean flow velocity is therefore preferred as an indication of flow strength in dimensional analysis.

The variables involved are therefore as given in Table 5.4. Since there are seven variables and three dimensions, there should be four dimensionless

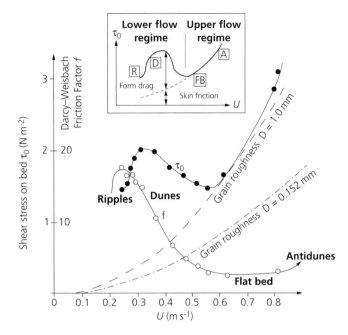

Fig. 5.16 Variation of the bed shear stress, τ_0, and the Darcy–Weisbach friction factor, f with mean flow velocity, U. The resistance above the line for grain roughness is due to 'form drag' caused by larger roughness elements on the bed. Experimental data from Vanoni & Brooks (1957) [30] for flow over a fine sand bed. After Raudkivi (1990) [1].

Table 5.4 Variables and dimensions in the problem of bedforms under an open-channel flow.

Variable	Dimensions
Mean grain size, D	[L]
Sediment density, ρ_s	[ML^{-3}]
Fluid density, ρ_f	[ML^{-3}]
Fluid viscosity, μ	[ML^{-1}T^{-1}]
Acceleration due to gravity, g	[LT^{-2}]
Mean flow depth, h	[L]
Mean flow velocity, U	[LT^{-1}]

products. The variables of greatest interest are flow velocity, flow depth and grain size. The other four are therefore treated as repeating variables, but since the fluid and sediment densities have identical dimensions, we immediately combine them to produce a dimensionless density ratio ρ_s/ρ_f. We can combine mean flow velocity, flow depth and grain size with the repeating variables (but excluding the sediment density) to form three additional dimensionless products,

$$U\left(\frac{\rho_f}{\mu g}\right)^{1/3}, h\left(\frac{\rho_f^2 g}{\mu^2}\right)^{1/3}, D\left(\frac{\rho_f^2 g}{\mu^2}\right)^{1/3}$$

representing a dimensionless velocity, a dimensionless flow depth and a dimensionless grain size. Note that because of the choice of these three variables as those of principal interest, the dimensionless products are different to the familiar Reynolds number and Froude number. If the water temperature is assumed constant between different experiments (not always justifiable, but the viscosity effect can be standardized to say, 10°C), the grain density used is restricted to quartz sand and the fluid used is always plain water, then we can not only neglect the density ratio, but also ignore the bracketed terms above as constants. This is the justification for plotting mean flow velocity against mean grain size for a certain flow depth, or flow velocity against flow depth for a given sediment size. Both plots are common.

5.4.3 The existence fields of bedforms

The investigation of the various types of bedform found under a range of substrate and flow conditions has been carried out mostly through a large number of flume experiments [31]. Although it might be argued that the results of experiments should properly be shown on a graph with dimensionless axes such as Reynolds number and a dimensionless stress such as Shields's β [32], it has become common for the reasons given in Section 5.4.2 to show the existence fields of bedforms in a diagram of mean flow velocity

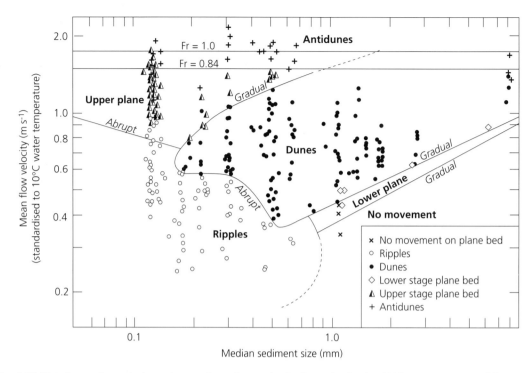

Fig. 5.17 Plot of mean flow velocity against median sediment size (both standardized to 10°C water temperature) for a mean flow depth of 0.25–0.40 m to show the existence fields of bedforms. After Southard (1991) [31].

against grain size (Fig. 5.17). The important features of this diagram are as follows.

• There is a large field of ripples in small grain sizes and low flow velocities. This field attenuates towards coarser grain sizes and is replaced at mean grain sizes greater than about 0.6 mm by a field of plane beds. The disappearance of ripples in coarse sands has been linked to the lack of a viscous sublayer over hydraulically rough boundaries [33].

• Dunes occur in a large field at moderate flow velocities and medium to coarse grain sizes. In grain sizes finer than about 0.2 mm, dunes are replaced by the upper part of the ripple field or by a field of plane beds. These plane beds are thought to be due to the occurrence of very high sediment concentrations near the bed which dampens turbulence and prevents the propagation of bedforms [34].

• At all flow velocities ripples or dunes are replaced by plane beds, and then by bed waves in phase with surface waves known as *standing waves* and *antidunes*.

5.4.4 The flow regime concept

It has already been stated that in flows with loose boundaries there is not a simple relationship between flow velocity and boundary shear stress because of changes in the frictional characteristics of the boundary. A plot of boundary shear stress against mean flow velocity reveals two segments offset by an abrupt decrease in boundary shear stress. For a boundary with fixed roughness (such as sand grains glued to a flat plate or to the inside of a pipe), we should expect a monotonic relationship between τ_0 and U. The pattern observed for a loose boundary shows that the boundary shear stress is always above the fixed roughness curve (Fig. 5.16). This is attributed principally to the presence of bedforms, and specifically the enhanced drag caused by the pressure variations due to flow separation over bedform crests. It is possible, therefore, to speak of two regimes of flow. In the *lower flow regime* there is a relatively steady increase in boundary shear stress as flow velocity is increased corresponding to the occurrence of ripples, then dunes, on the loose boundary. There is another trend of steadily increasing boundary shear stress at higher velocities, not far from the curve for fixed plane roughness, corresponding to plane beds, then standing waves, then antidunes. This is known as the *upper flow regime*. Between the two is the abrupt fall caused by the loss of form resistance as dunes are washed out into plane beds and flow separation ceases. It is important to note that the transition from the lower and upper flow regimes does not occur at a Froude number of 1, which marks the start of supercritical flow at the onset of standing waves.

5.4.5 Flow over ripples and dunes

Inspection of the transport of particulate sediment, or of dyed fluid, in a laboratory flume shows that sediment transported up the stoss side avalanches down the lee side or is advected in suspension beyond the brink point of the ripple or dune (Fig. 5.13). There is therefore a process of erosion of the stoss side and deposition on the lee side which causes the migration of the ripple or dune form. This can be understood from the sediment continuity equation (Fig. 5.18).

As the flow runs up the stoss side the streamlines converge, causing an increase in the shear stress on the boundary and therefore an increase in the sediment transport rate. There is therefore a positive gradient in the sediment transport rate on the stoss side. This should, from the continuity equation, result in erosion of the stoss side. Flow separates at the brink point, leaving a region of sluggishly recirculating fluid in the lee of the ripple or dune. The shear stress is low and the pressure relatively high beneath the separation zone, so the sediment transport rate is low. There is a negative gradient in the sediment transport rate which should result in deposition on the lee side. The combination of erosion of the stoss and deposition in the lee results in the migration of the ripple form.

5.4.6 Stability theory

What controls the periodicity of bedforms? One way of viewing the problem is to consider the bedforms as waves developing on a deformable granular boundary as an instability between two different 'fluids' undergoing shear. A stable state would be a plane bed, whereas an instability would be a ripple or dune. The ripple wavelength would then be the wavelength of perturbation showing the fastest rate of growth [28]. The instabilities developing at the interface between two fluids undergoing shear are known as *Kelvin–Helmholtz* instabilities.

Figure 5.19 illustrates the velocities of two fluids of the same density and viscosity, but moving with different velocity. The velocities of the two fluid streams relative to each other are shown by the streamlines. This pattern of relative motion may result from the fluids moving in opposite directions, or if one fluid

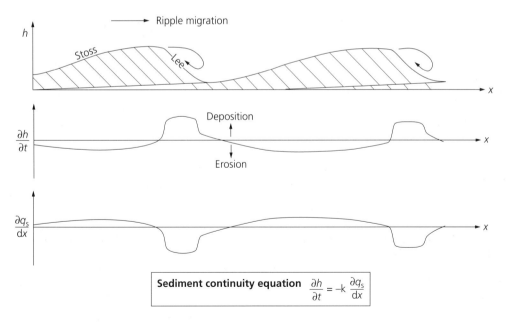

Fig. 5.18 Schematic illustration of the change of bed elevation with time $\partial h/\partial t$ and downstream variation in the bedload sediment transport rate $\partial q_s/\partial x$ over a ripple bedform. After Middleton & Southard (1977) [19].

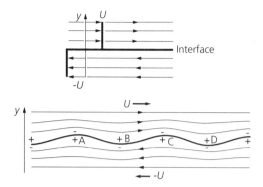

Fig. 5.19 The relative velocities of two fluid streams and pressure distribution along the interface.

is moving faster than the other. If a wavelike perturbation is introduced to the interface, a streamtube close to the interface will vary in its width with the same wavelike form. The Bernoulli equation suggests that where the streamtube widens the velocity will decrease and the pressure will increase. Since the streamtube widens under the crests and over the troughs (A to D, Fig. 5.19), the pressure increases at these localities. Conversely, where the streamtube narrows, as under troughs and over crests, the pressure will decrease. Consequently, the wavelike perturbations should continue to grow unless there is some additional force that opposes this pressure distribution. In real fluids the force opposing growth is viscosity. The balance between the pressure forces tending to promote growth and the viscous forces retarding growth is a Reynolds number. This should not come as a great surprise, since the situation of the damping of instabilities is precisely the condition of laminar flow, at low Reynolds numbers, and the situation of growth of instabilities is the case of turbulent flow at high Reynolds numbers.

The damping or growth of wavelike perturbations depends not only on the Reynolds number (where the length term is the initial amplitude of the perturbation, and the velocity term is the velocity difference between the two fluid streams) but also on the frequency of the perturbation. The growth of instabilities between two opposing streams typically results in the formation of rolls of fluid at the interface (Fig. 5.20).

(a)

(b)

Fig. 5.20 Kelvin–Helmholtz instabilities develop along an interface between water above coloured brine caused by the tilting of the long rectangular tube (a), and between superimposed streams of water travelling at different speeds illuminated by a vertical sheet of laser light (b). Examples from Van Dyke (1982) [36].

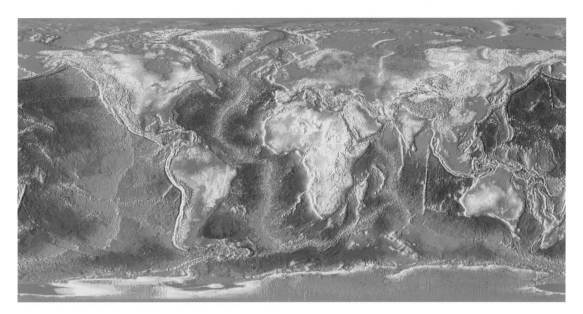

Plate 1.1 World topography and bathymetry using the ETOPO5 database. The hue and saturation of colours are controlled by the depth or elevation. Intensity is controlled by shaded relief calculation. In the oceans the darkest blue colour shows depths to about 10.8 km below sea level, while on land the brighter reds go to heights of 5 km, and some very high elevations, such as the Himalayas, go to about 8.7 km and are shown in light grey. Shallow water depths are also shown by light grey. This image was made available by the National Geophysical Data Center, Boulder, Colorado.

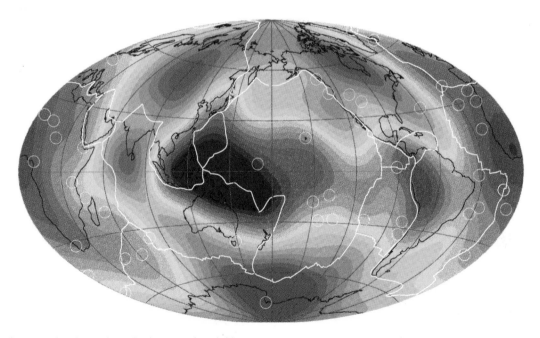

Plate 1.2 The observed geoid at long wavelength (degrees 2 to 10) with hotspot locations superimposed. Contours are at an interval of 10 m and the blue colour indicates up (positive). Image provided by Professor John Woodhouse, Oxford University, Oxford.

[*facing page 204*]

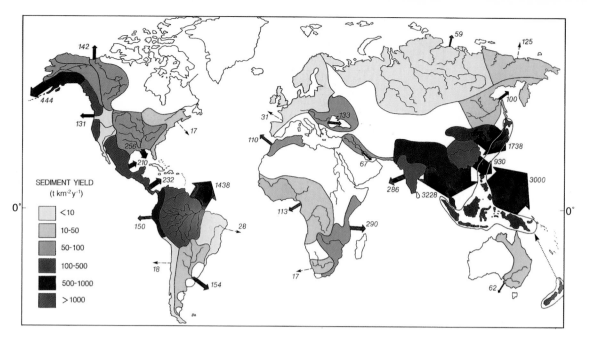

Plate 3.1 Global pattern of sediment discharge to the ocean from the suspended sediment loads of the world's major rivers. Widths of arrows are proportional to the absolute amount of sediment delivered to the ocean (in tons × 10⁶ y⁻¹). After Milliman and Meade (1983) World-wide delivery of river sediment to the oceans, *Journal of Geology* **91**, 1–21.

Plate 6.1 The turbulent cloud of volcanic ash which erupted from Mount Pinatubo in the Philippines in 1991 moves by the action of gravity on its excess density compared to the surrounding air. (Obtained, with permission, from Garcia/Saba/Katz picture agency, London.)

Plate 7.1 View of the delta formed by the River Mississippi as seen from Shuttle Challenger on mission 61A of 30 October to 6 November 1985. This photograph shows clearly the pattern of sediment deposition of this type of river delta. Between the kinks in the river towards the top left is the Town of Buras. From the tip of the Southwest Point (left) to North Point (right) is a distance of around 50 km. Photograph courtesy of the Science Photo Library, London; originally from the NASA/Science Photo Library, USA.

Plate 7.2 Infrared SPOT satellite image of part of the tide-dominated Ganges delta, India, just south of Calcutta (not visible). North is towards the top. Vegetation appears red, water blue (but paler where rich in sediment), exposed sediment white, mangrove swamp bright red. The main river here is called the Hugli. Note the huge amount of sediment being washed into the Bay of Bengal. The Ganges Delta is a vast flood plain with hundreds of small streams and rivers. Its fertile, low-lying land is intensively farmed and subject to annual floods by the monsoon. Image produced by one of the French SPOT satellites, CNES, January 1987 and obtained, with thanks, from the Science Photo Library, London.

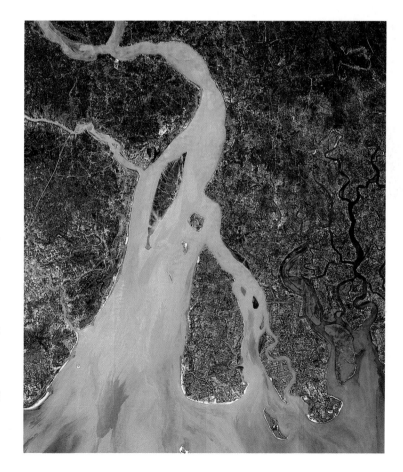

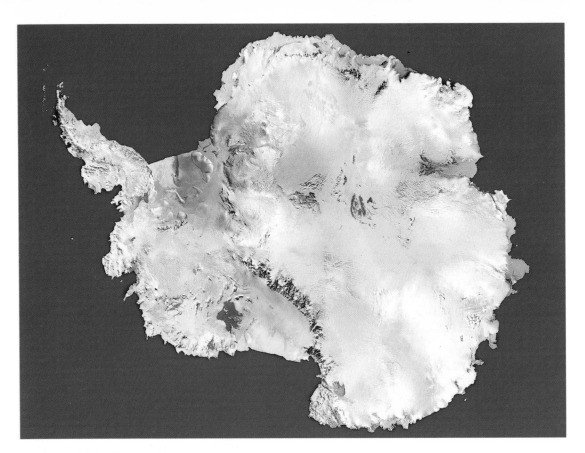

Plate 11.1 This Satellite mosaic image of Antarctica was compiled from 1 km resolution AVHRR data obtained by the NOAA weather satellites. (The image was supplied courtesy of the National Remote Sensing Centre, Farnborough, UK.)

Where the underlying fluid is denser than the overlying fluid, gravity acting on the density gradient is a powerful mechanism for damping instabilities. This might apply to the flow of a turbid current beneath stationary clear water (see Chapter 6). The stability of the interface should depend on the velocity gradient, which promotes instability, and the density gradient, which stabilizes the interface. These opposing tendencies are found in a dimensionless product known as the *Richardson number*:

$$Ri = \frac{(g/\rho)(d\rho/dy)}{(dU/dy)^2} \qquad \textbf{(5.41)}$$

If the Richardson number is small (less than 0.25) interfacial instability is likely to occur.

It has been suggested that ripples and dunes form as instabilities between the fast-moving, low-viscosity overlying fluid and the high-viscosity, granular, cohesionless boundary. A stability analysis suggests that there should be two separate modes of instability, one dependent on bed roughness and one dependent on flow depth. These two modes correspond to ripples and dunes. In such a model, ripples and dunes are bed waves analogous to Kelvin–Helmholtz instabilities.

5.4.7 Defect propagation

It might reasonably be objected that the granular boundary has an infinite viscosity compared to the overlying fluid and also that ripples can be observed to grow in size from initial small defects in the bed before reaching an equilibrium size. This suggests that ripples and dunes are not initiated as instabilities along mixing layers but result from downstream propagation from bed defects. The presence of the defect causes local pressure gradients, leading to flow separation. This causes a local pattern of erosion and deposition which amplifies in time to an equilibrium wavelength in the form of a ripple.

The development of wavy, periodic bedforms from bed defects can be understood by once again referring to the sediment continuity equation. Imagine a wavy bed of low amplitude and an overlying flow with a downstream variation in sediment transport rate. Now introduce a spatial lag between the bed elevation and the sediment transport rate (Fig. 5.21). The sediment continuity equation predicts that there will be an increase in bed elevation where the downstream gradient in sediment transport rate is negative, and vice versa. For a lag distance of between three-quarters and one wavelength the bed wave should amplify and be translated downstream, whereas with a lag

distance of between zero and one-quarter of the wavelength, the bed wave should be damped and translate downstream. This exaplains the generation and degeneration of ripples and dunes, respectively. For lag distances between one-quarter and three-quarters of a wavelength, the bed wave should either amplify or dampen but translate upstream, corresponding to the case of antidunes. The importance of bed defects is therefore that they cause a local shift in the transport rate maximum, resulting in a lag with the bed elevation.

5.4.8 Stratification caused by the migration of bedforms

The type of stratification produced by the migration of bedforms depends primarily on two factors: the three-dimensional geometry of the bedforms; and the ratio between downstream translation and vertical accretion. Bedforms which are linear crested and with a roughly uniform crestal height migrate downstream with a uniform depth of erosion. The downstream migration of linear crested bedforms therefore creates a subplanar, horizontal basal surface in the plane normal to flow and a subplanar but dipping surface in the section parallel to flow (Fig. 5.22). The angle of dip depends on the second factor above. Between successive basal surfaces lie the inclined planes representing successive lee-side slopes (Fig. 5.22).

Bedforms with sinuous crests produce three-dimensional scour pits and spurs as they migrate downstream. In the section normal to flow the basal surfaces are scooped or trough-like due to the downstream passage of scour pits (Fig. 5.22). This distinguishes the stratification due to the migration of sinuous crested bedforms, called *trough cross-stratification*, from that produced by linear crested bedforms, *planar cross-stratification*.

The relative importance of downstream translation to vertical build-up of the bedform controls its *angle of climb*. Where the bedform builds vertically at a very small rate compared to its downstream translation in the same period of time, only the lee sides of the ripples or dunes are preserved and the angle of climb is less than the slope of the stoss side. At higher rates of vertical build-up the stoss-side may be preserved as well as the lee. At very high rates of vertical build-up the entire bedform may be draped by laminae of equal thickness. Stratification is said to be subcritically climbing if no stoss-side laminae are preserved and supercritically climbing where stoss-side laminae as well as lee-side laminae are preserved (Fig. 5.23). Practical Exercise 5.3 develops this idea.

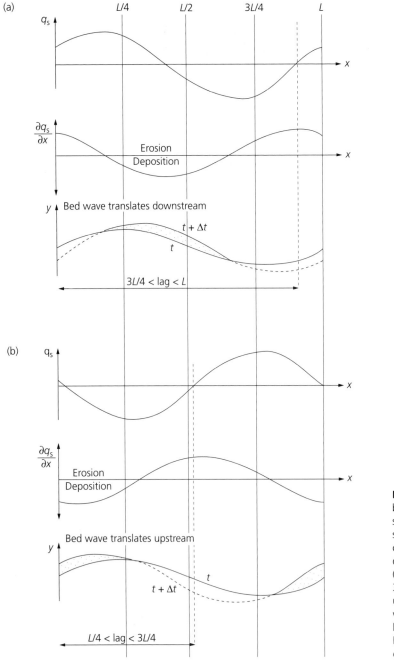

Fig. 5.21 The effect of a spatial lag between the bed elevation and the sediment transport rate. In (a) the spatial lag is between $3L/4$ and L, causing the bed wave to translate downstream like a ripple or dune. In (b) the spatial lag is between $L/4$ and $3L/4$, causing the bed to translate upstream like an antidune. Bed waves will amplify if the lag lies between $L/2$ and L, and dampen if it lies between 0 and $L/2$. Fuller discussion in Allen (1985) [37].

The fact that bedforms climb downstream explains the occurrence of cross-strata, representing old ripple and dune depositional surfaces, between bounding surfaces (cross-set boundaries) which truncate the upper parts of cross-strata. The geometry produced at different angles of climb is a case of simple trigonometry. Let us consider the steady migration of uniform, two-dimensional ripples with simple triangular profiles

(a) Planar cross-stratification

Flow direction

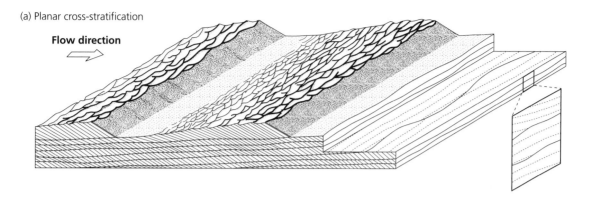

(b) Trough cross-stratification

Flow direction

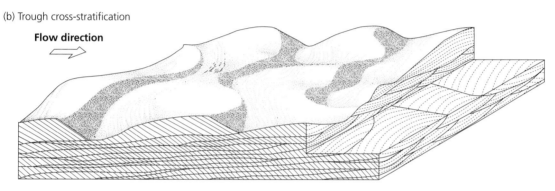

Fig. 5.22 Stratification produced by the migration of linear and sinuous crested dunes or ripples. After Harms *et al.* (1982) [38].

(Fig. 5.24). The ripples have constant stoss slopes ξ, height H and wavelength L. Any changes in the position of the bed must be due to net transfer of bedload with rate q_x and net transfer from the suspended load to the bed with rate per unit area of bed q_y, where the subscripts remind us that the two transport rates operate essentially in the horizontal and vertical directions, respectively. The velocity (or celerity) of the bed wave can be denoted in the horizontal and vertical directions as

$$V_x = \frac{2q_x}{H\rho_b} \qquad (5.42)$$

and

$$V_y = \frac{q_y}{\rho_b} \qquad (5.43)$$

where ρ_b is the bulk density of the sediment. The resultant of these two velocities is the actual velocity of the bedform relative to the original horizontal datum, which climbs at an angle given by

$$\tan \zeta = \frac{V_y}{V_x} = \frac{q_y H}{2q_x} \qquad (5.44)$$

from which it can be seen that the angle of climb is directly proportional to the rate of transfer of sediment from suspension to the bed and the bedform height, and inversely proportional to the bedload transport rate.

The case where stoss-side laminae just begin to be preserved represents the critical condition where $\xi = \zeta$. The thickness of cross-strata between successive bounding surfaces is $L \tan \zeta$, reaching a maximum of $L \tan \xi$ at the critical condition. The relation between

Practical exercise 5.3: The angle of climb of bedforms

A fluid flow is transporting sediment as bedload with a horizontal rate of $0.2\,\mathrm{kg\,m^{-1}\,s^{-1}}$ as an array of ripples with constant stoss slope angle of $10°$, height $0.05\,\mathrm{m}$ and wavelength $0.40\,\mathrm{m}$.

1 At what rate of vertical fall out of suspended sediment will the bedforms begin to climb at the critical angle?

2 What would be the effect of doubling the vertical rate of fall out per square metre of bed?

Solution

1 From equation (5.44) the angle of climb can be found as a function of the vertical fall out rate. Putting the angle of climb equal to the stoss-side angle for the critical angle of climb, we find that $q_y = 1.41\,\mathrm{kg\,m^{-2}\,s^{-1}}$.

2 Doubling the vertical fall out rate causes the ripples to climb supercritically at an angle of $19°$.

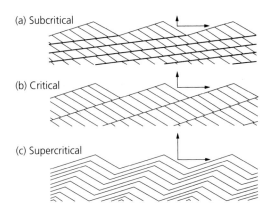

Fig. 5.23 Illustration of the preservation of lee- and stoss-side laminae under different conditions of bedload transport and suspended sediment fall out rate.

the stoss slope and the angle of climb therefore determines the style of preserved stratification. This analysis is, of course, simplified, because bedforms are not all uniform in size. If non-uniform, the migration of these bedforms under conditions of net deposition should give rise to sets of variable thickness compared to bedform height. The geometries of cross-stratification resulting from the migration of bedforms of varying geometry, and of the migration of two or more sets of bedforms in varying directions, lends itself to computer simulation [39]. The range of

patterns in [32] are purely geo-metrical constructions, but give a strong visual impression of the complexities to be expected under a range of flow and bed conditions.

Further reading

J.R.L. Allen (1984) *Sedimentary Structure: Their Character and Physical Basis*, Developments in Sedimentology 30. Elsevier, Amsterdam.

J.R.L. Allen (1985) *Principles of Physical Sedimentology*. George Allen & Unwin, London.

J.D. Collinson & D.B. Thompson (1982) *Sedimentary Structures*. George Allen & Unwin, London.

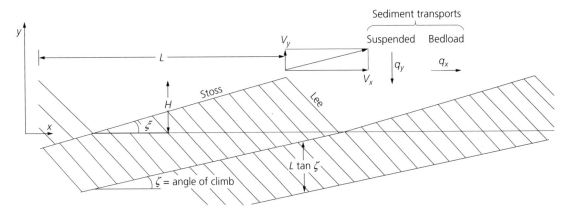

Fig. 5.24 Sketch to illustrate the parameters controlling the angle of climb of ripple or dune cross-sets.

G.V. Middleton & J.B. Southard (1977) *Mechanics of Sediment Movement*, Short Course Notes 3. SEPM, Tulsa, OK.

A.J. Raudkivi (1990) *Loose Boundary Hydraulics*, 3rd edn. Pergamon Press, Oxford.

D.M. Rubin (1987) *Cross-Bedding, Bedforms and Paleocurrents*, SEPM Concepts in Sedimentology and Paleontology 1. Tulsa, OK.

J.B. Southard (1991) Experimental determination of bedform stability. *Annual Review of Earth and Planetary Sciences* **19**, 423–55.

M.S. Yalin (1977) *Mechanics of Sediment Transport*, 2nd edn. Pergamon Press, Oxford.

References

1 A.J. Raudkivi (1990) *Loose Boundary Hydraulics*, 3rd edn. Pergamon Press, Oxford

2 G.V. Middleton & J.B. Southard (1984) *Mechanics of Sediment Movement*, 2nd edn, Short Course Notes 3. SEPM, Tulsa, OK.

3 A. Shields (1936) Anwendung der Ähnlichkeitsmechanik und der Turbulenzforschung auf die Geschiebebewegung. *Mitteilungen der Preussischen Versuchsanstalt für Wasserbau und Schiffbau* **26**, 26pp. Translated by W.P. Ott & J.C. van Uchelen, United States Department of Agriculture, Soil Conservation Service, California Institute of Technology.

4 V.A. Vanoni (1964) *Measurements of critical shear stress for entraining fine sediments in a boundary layer*, Report KH-R-7. W.M. Keck Lab., Hydraulics Water Resources, California Institute of Technology, Pasadena, California.

5 M.C. Miller, I.N. McCave & P.D. Komar (1977) Threshold of sediment motion under unidirectional currents. *Sedimentology* **24**, 507–27.

6 L.J. Tison (1953) Studies of the critical shear stress for the entrainment of bed materials. In: *Proceedings of the International Association for Hydraulic Research*, Minneapolis (transl. I.N. McCave), pp. 21–35.

7 A. Shields (1936) Application of similarity principles and turbulence research to bed-load movement. In: *Mitteilungen der Preussischen Versuchsanstalt für Wasserbau und Schiffbau*, Berlin. (trans. W.P. Ott & J.C. Uchelen. California Institute of Technology, W.M. Keck Laboratories of Hydraulics and Water Resources, Report No. 167.

8 C.R. Neill (1967) Mean velocity criterion for scour of coarse, uniform bed material. *International Association of Hydraulic Research, 12th Congress Proceedings*, Fort Collins, Colorado, **3**, 46–54.

9 A.J. Grass (1970) Initial instability of fine bed sand. *Proceedings of the American Society of Civil Engineers* **96**, 619–32.

10 S.J. White (1970) Plane bed thresholds of fine grained sediments. *Nature*, **228**, 152–3.

11 C.H. Everts (1973) Particle overpassing on flat granular boundaries. *Proceedings of the American Society of Civil Engineers* **99**, 425–8.

12 A.S. Paintal (1971) A stochastic model for bed load transport. *Journal of Hydraulic Research* **9**, 527–3.

13 M.S. Yalin (1977) *Mechanics of Sediment Transport*, 2nd edn. Pergamon Press, Oxford.

14 J.R.L. Allen (1994) Fundamental properties of fluids and their relation to sediment transport processes. In: *Sediment Transport and Depositional Processes* (ed. K. Pye). Blackwell Scientific Publications, Oxford, pp. 25–60.

15 J.R.D. Francis (1973) Experiments on the motion of solitary grains along the bed of a water stream. *Proceedings of the Royal Society of London* **332A**, 443–71.

16 R.A. Bagnold (1966) *An approach to the sediment transport problem from general physics.* Professional Paper 422–I. US Geological Survey, Washington, DC.

17 G. Parker, P.C. Klingeman & D.G. McLean (1982) Bedload and size distribution in paved gravel-bed streams. *Proceedings of the American Association of Civil Engineers, Journal of the Hydraulics Division* **108**, 544–71.

18 J.K. Maizels (1983) Proglacial channel systems: change and thresholds for change over long, intermediate and short time scales. In: *Modern and Ancient Fluvial Systems* (eds J.D. Collinson & J. Lewin), International Association of Sedimentologists Special Publication. Blackwell Scientific Publications, Oxford. **6**, 251–66.

19 M. Church (1978) Palaeohydrological reconstructions from a Holocene valley fill. In: *Fluvial Sedimentology* (ed. A.D. Miall), Memoir 5. Canadian Society of Petroleum Geologists, Ottawa, pp. 743–72.

20 V.T. Chow (ed.) (1964) *Handbook of Applied Hydrology*. McGraw-Hill, New York.

21 J. Maizels (1986) Modeling of paleohydrologic change during deglaciation. *Géographie Physique et Quaternaire* **40**, 263–77.

22 D.E. Sugden, C.M. Clapperton & P.G. Knight (1985) A jökulhlaup near Søndre Strømfjord, west Greenland, and some effects on the ice sheet margin. *Journal of Glaciology* **31**, 366–8.

23 H. Rouse (1937) Modern conceptions of the mechanics of turbulence. *Transactions of the American Society of Civil Engineers* **102**, 436–505.

24 H.A. Einstein (1950) The bedload function for sediment transportation in open channel flows. *US Department of Agriculture, Technical Bulletin* **1026**, 1–78.

25 R. Slingerland, J.W. Harbaugh & K.P. Furlong (1994) *Simulating Clastic Sedimentary Basins*. Prentice Hall, Englewood Cliffs, NJ.

26 American Society of Civil Engineers, Task Committee on Preparation of Sedimentation Manual (1963) Suspension of sediment. *Proceedings of the American Society of Civil Engineers* **89**(HY5), 45–76.

27 B.R. Colby (1963) *Fluvial sediments—a summary of source, transportation, deposition, and measurement of sediment discharge*, Bulletin **1181–A**. US Geological Survey, Washington, DC.

28 R.J. Gibbs, M.D. Matthews & D.A. Link (1971) The relationship between sphere size and settling velocity. *Journal of Sedimentary Petrology* **41**, 7–18.

29 W.E. Dietrich (1982) Settling velocity of natural particles. *Water Resources Research* **18**, 1615–26.

30 V.A. Vanoni & N.H. Brooks (1957) *Laboratory Studies of the Roughness and Suspended Load of Alluvial Streams*, Report E-68. Sedimentation Laboratory, California Institute of Technology, Pasadena.

31 J.B. Southard (1991) Experimental determination of bedform stability. *Annual Review of Earth and Planetary Sciences* **19**, 423–55.

32 M. Leeder (1983) On the interactions between turbulent flow, sediment transport and bedform mechanics in channelized flows. In: *Modern and Ancient Fluvial Systems* (eds J.D. Collinson & J. Lewin), International Association of Sedimentologists Special Publication 6. Blackwell Scientific Publications, Oxford, pp. 5–18.

33 M.R. Leeder (1980) On the stability of lower stage plane beds and the absence of current ripples in coarse sands. *Journal of the Geological Society of London* **137**, 423–9.

34 J.R.L. Allen & M.R. Leeder (1980) Criteria for instability of upper stage plane beds. *Sedimentology* **27**, 209–17.

35 K.J. Richards (1980) The formation of ripples and dunes on an erodible bed. *Journal of Fluid Mechanics* **99**, 597–618.

36 M. Van Dyke (1982) *An Album of Fluid Motion*. Parabolic Press, Stanford, CA.

37 J.R.L. Allen (1985) *Principles of Physical Sedimentology*. George Allen and Unwin, London, 272pp.

38 J.C. Harms, J.B. Southard & R.G. Walker (1982) *Structures and Sequences in Clastic Rocks*, Short Course Notes 9. SEPM, Tulsa, OK.

39 D.M. Rubin (1987) *Cross-Bedding, Bedforms and Paleocurrents*, Concepts in Sedimentology and Paleontology 1. SEPM, Tulsa, OK.

6 Hyperconcentrated and mass flows

The sounding cataract
Haunted me like a passion: the tall rock,
The mountain, and the deep and gloomy wood,
Their colours and their forms, were then to me
An appetite.

William Wordsworth (1770–1850), *Lines composed a few miles above Tintern Abbey* [1798].

Chapter summary

Mass flows are common on both subaerial hillslopes and submarine slopes where they take the various forms of soil creep, debris flows, landslides and avalanches. Hyperconcentrated fluid-rich flows also typify some rivers in flood, the slope-hugging pyroclastic surges of volcanoes, and turbid underflows in lakes and seas. Hyperconcentrated flows of grains in air typify the movement of sand flows down the slip-faces of wind-blown dunes. In all of these instances, a primary mechanism for flow is the action of gravitational forces on a static slope or on a body of excess density arising from its high concentration relative to the ambient fluid.

Mass flows and hyperconcentrated flows range widely in speed, integrity and geometry. Soil creep is the slow downslope movement of a thin sheet caused by repeated expansion and contraction. In some climates the cycle of expansion and contraction is caused by freezing and thawing, and is called solifluction.

A range of mass movements such as avalanches, debris flows and landslides involve the overreaching of a threshold which causes slope failure. Slope failure can be studied in terms of the tangential (shear) and normal (pressure) forces on a potential plane of failure, the relation between the two being known as the friction coefficient. The situation is made more complex by the fact that different materials have different stress–strain behaviours, many with fields of non-elastic behaviour. A commonly used formulation for the failure of a body with a strength (apparent cohesion) acted upon by shear and pressure forces is the Navier–Coulomb criterion of failure. This can be used to predict the landsliding of a homogeneous material on a slope. Fluid pressures appear to be important in triggering landslides by causing a reduction of the normal stress on a potential plane of failure.

Debris flows are common on steep terrains on land and are increasingly recognized as primary mechanisms of denudation of even gentle submarine slopes. They can be modelled as a combination of an ideal plastic and a Newtonian fluid (the Bingham plastic model), or as highly concentrated dispersions (the non-Newtonian fluid model), in which there is a nonlinear relation between the shear stress and the strain rate. These flow models can be tested with observations on natural debris flows. The onset of turbulence in a debris flow is predicted by a dimensionless grouping known as the Hampton number. Turbulence is unlikely except in debris flows on very steep slopes.

The subaqueous turbulent fluid–sediment mixtures driven by gravity are known as turbidity currents. Similar, but not identical, dynamics apply to sea breezes and to the hot clouds of ash flowing down the flanks of volcanoes. A densimetric Froude number is useful in explaining a number of properties of turbidity currents, such as the amount of mixing at the upper interface of the flow. Entrainment, which causes dilution of the hyperconcentrated flow with ambient water, and deposition of sediment, which lowers the density contrast with ambient water, both result in flow dissipation. There is therefore a characteristic runout distance for turbidity currents that can be predicted from theoretical fluid mechanics and tested by observations of the sea bed. Some fine-grained turbidity currents may be capable of autosuspension, whereby the gravitational energy of the turbidity current due to its suspended sediment concentration is just enough to overcome the effects of friction, causing the current to flow downslope for very large distances.

Sea bed bathymetry has a number of important effects on turbidity current mechanics and pathways. These effects include those due to the presence of a restricted basin, which causes currents to be funnelled along narrow pathways and to be deflected by lateral basin slopes. Turbidity currents may also be affected by intrabasinal highs, such as sea bed fault scarps.

The paradigm in turbidite sedimentology is that the deposits of turbidity currents result from an unsteady (waning) flow, giving rise to the Bouma and Lowe turbidite models for low- and high-density currents, respectively. Deposition from quasi-steady but non-uniform depletive turbidity currents is also likely, and may, in particular, result in the deposition of massive sands.

Natural density currents can be superbly visualized from the pyroclastic flows and surges emanating from volcanoes. Pyroclastic flows, which are thought to be non-turbulent and to be dominated by fluidization due to the loss of gas, are similar to some debris flows. Pyroclastic surges, on the other hand, are characterized by turbulence and are more analogous to submarine turbidity currents. The recent eruption (1980) of Mount St Helens, USA, showed that pyroclastic surges are also strongly affected by underlying topography.

6.1 Introduction

Chapter 4 focused on the mechanics of Newtonian fluids, and in Chapter 5 these mechanics were applied to the transport of sediment by running water. Sediment can also be transferred over the Earth's surface by movement at very high concentrations. In such cases the sediment particles, or blocks of rock, are transported not by the drag on individual particles from an enclosing fluid, but as a sediment mass or sediment–fluid mixture under the influence of gravity. We can think of a continuum from dilute water floods acting as Newtonian fluids at one extreme, through hyperconcentrated flows, which because of their high sediment concentrations may be better approximated by a non-Newtonian fluid, to mass flows at the other extreme where the solid particles and any interstitial fluid move together as a single viscoplastic body [1]. The distinction between hyperconcentrated and mass flows is therefore not entirely straightforward.

Some parts of the Earth's surface, especially those associated with relatively high slopes, are particularly prone to mass movements and the generation of hyperconcentrated flows. The main denudational process of many of the high parts of mountain belts is transfer of soil and rock by ice movement, landslides, avalanches, debris flows and soil creep. The submarine continental slope, flanks of oceanic islands and regions of oversteepening caused by sediment supply from river mouths are also subject to slope failure, generating slides, slumps, debris flows and hyperconcentrated underflows known as turbidites. Mass flows and hyperconcentrated flows are important, therefore, from the standpoint of the volumetrics of sediment flux through the Earth surface system, but also in terms of their potential as natural hazards. Mass movement deposits have a distinctive sedimentological signature that helps their identification in the rock record.

6.1.1 Variability of mass flows and hyperconcentrated flows

A great deal is known of gravity-driven flows on land. They are extremely variable, according to the coherency of the debris–fluid mixture, its speed of transport and its geometry, notably as expressed by the ratio of the deposit thickness to the length, sometimes referred to as the *Skempton ratio* [2]. Perhaps the most widely used classification scheme is that of Varnes [3], who recognized the primary mechanisms for mass movement as creep, flow, slide, heave, fall and subsidence (Fig. 6.1 and Table 6.1).

Although the variability of mass movements in nature is very wide, the gravity-driven movements affecting both subaerial and submarine slopes can be simplified into three categories, though they are capable of transformation from one category to another (Fig. 6.2):
• rock avalanches and falls, slides and slumps generated by gravitational instability only;
• plastic flows which move as laminar fluids;
• turbulent hyperconcentrated flows.

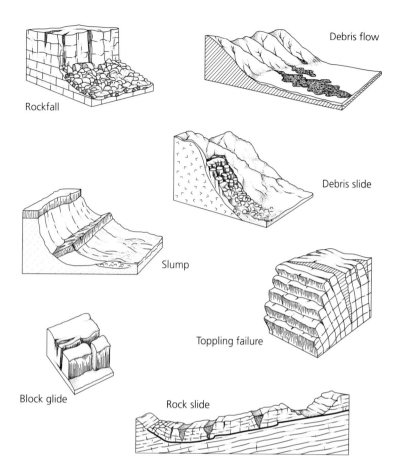

Fig. 6.1 The primary mechanisms for mass movements based on Varnes (1978) [3].

Falls, avalanches, slides and slumps are commonplace on land and also affect large areas of the sea bed. *Rockfalls* are particularly common in steep, but generally sparsely vegetated, mountainous regions affected by periglacial conditions. Many are triggered by earthquakes. *Rock avalanches* are extremely rapid, and involve thorough mixing of particles during transport. They derive some of their energy from their entrapped air, in a manner similar to the hot, hyperconcentrated clouds of pyroclastic material emitted from volcanoes. They are extremely rapid, with velocities in excess of 300 m s^{-1}. Rock avalanches generated from rotational failures are common on volcanic seamounts, such as at Le Réunion and Tenerife in the Atlantic, and in the Hawaiian archipelago.

Subaerial and submarine landslides, however, maintain their integrity as they move downslope over a basal failure surface, so that the amount of internal strain is relatively small. Their transport distances are generally limited. They can be classified according to their failure surface. Slumps are relatively long compared to their thickness (Skempton ratio greater than 0.33), whereas slides are relatively short compared to their thickness (Skempton ratio less than 0.15). Rotational slides over concave basal detachments rarely move far downslope, and are thick compared to their downslope extent. Slumps and slides may be successive, commonly overlapping or progressively cutting back into the slope (retrogressive). On land, landslides cleared 130 km^2 of vegetated slopes in New Guinea in 1935, and it is estimated that 12% of slopes, based on that experience, landslides every century.

Landslides occur in many different environments, but mountainous, vegetated, wet terrains such as Papua New Guinea, Taiwan and New Zealand are particularly prone. Landslides may account for nearly all of the hillslope erosion in tectonically and climatically active mountains. Slides and slumps are also very common beneath the sea. The pro-delta

Table 6.1 Classification and characteristics of the major types of mass movement.

Primary mechanism	Mass movement type	Materials in motion	Moisture content	Type of strain and nature of movement	Rate of movement
Creep	Rock creep and continuous creep	Especially readily deformable rocks, such as shales and clays and soil	Low	Slow plastic deformation of rock, or soil producing a variety of forms including cambering, valley bulging and outcrop bedding curvature	Very slow to extremely slow
Flow	Solifluction	Soil	High	Widespread flow of saturated soil over low- to moderate-angle slopes	Very slow to extremely slow
Flow	Gelifluction	Soil	High	Widespread flow of seasonally saturated soil over permanently frozen subsoil	Very slow to extremely slow
Flow	Mudflow	>80% clay-sized	Extremely high	Confined elongated flow	Slow
Flow	Slow earthflow	>80% sand-sized	Low	Confined elongated flow	Slow
Flow	Debris flow	Mixture of fine and coarse debris (20–80% of particles coarser than sand-sized)	High	Flow usually focused into pre-existing drainage lines	Very rapid
Flow	Debris (rock) avalanche (sturzstrom)	Rock debris, in some cases with ice and snow	Low	Catastrophic low-friction movement of up to several kilometres, usually precipitated by a major rockfall and capable of overriding significant topographic features	Extremely rapid
Flow	Snow avalanche	Snow and ice, in some cases with rock debris	Low	Catastrophic low-friction movement precipitated by fall or slide	Extremely rapid
Slide — Translational	Rock slide	Unfractured rock mass	Low	Shallow slide approximately parallel to ground surface of coherent rock mass along single fracture	Very slow to extremely rapid
Slide — Translational	Rock block slide	Fractured rock	Low	Slide approximately parallel to ground surface of fractured rock	Moderate
Slide — Translational	Debris/earth slide	Rock debris or soil	Low to moderate	Shallow slide of deformed masses of soil	Very slow to rapid
Slide — Translational	Debris/earth block slide	Rock debris or soil	Low to moderate	Shallow slide of deformed masses of soil	Slow
Slide — Rotational	Rock slump	Rock	Low	Rotational movement along concave failure plane	Extremely slow to moderate
Slide — Rotational	Debris/earth slump	Rock debris or soil	Moderate	Rotational movement along concave failure plane	Slow
Heave	Soil creep	Soil	Low	Widespread incremental downslope movement of soil or rock particles	Extremely slow
Heave	Talus creep	Rock debris	Low		Extremely slow
Fall	Rockfall	Detached rock joint blocks	Low	Fall of individual blocks from vertical faces	Extremely rapid
Fall	Debris/earth fall (topple)	Detached cohesive units of soil	Low	Toppling of cohesive units of soil from near-vertical faces such as riverbanks	Very rapid
Subsidence	Cavity collapse	Rock or soil	Low	Collapse of rock or soil into underground cavities such as limestone caves or lava tubes	Very rapid
Subsidence	Settlement	Soil	Low	Lowering of surface due to ground compaction usually resulting from withdrawal of ground water	Slow

Lateral component predominant — Vertical component predominant

Source: Based largely on Varnes (1978) [3].

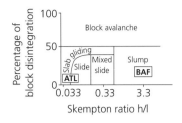

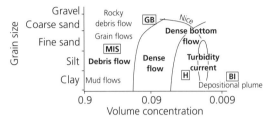

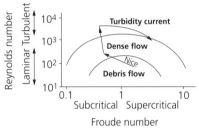

ATL = Atlantic continental margin   GB = Grand Banks
BAF = Baffin Bay, Canada   H = Huanghe, China
MIS = Mississippi Delta   BI = Bute Inlet

Fig. 6.2 A dynamic classification of submarine mass movements. (a) Slides and slumps can be differentiated by their Skempton ratio. As the amount of disintegration becomes higher, slumps and slides become block avalanches. Different types of plastic flow can be differentiated from turbidity currents based (b) on their concentration and grain size, and (c) on their Reynolds and Froude numbers. Actual examples of submarine mass movements are plotted, together with the Nice event, France, which transformed from a debris flow into a turbidity current. After Mulder & Cochonat (1996) [4].

region of the Mississippi is affected by a wide range of submarine landslides [5] which pose a significant hazard for offshore development such as hydrocarbon exploration and production platforms and pipelines (Fig. 6.3).

Plastic flows are typically underconsolidated masses of sediment in which the internal structure is significantly disorganized by the movement but coherency is maintained. A range of plastic flows is known, from the slow creep down slopes, through the more rapid debris flows driven primarily by gravity, to fluidized and liquefied flows whose motion is dependent on the presence of interstitial fluid.

'Creep' refers to the slow, elastic deformation of sediment under a constant load (force system) not involving failure. It is well known in soils on land, but is also recognized widely from the sea bed, commonly on pro-delta slopes.

Debris flows, containing a poorly sorted mixture of rock clasts and silt–mud matrix, generally run down pre-existing channels. Their tongue- or lobe-shaped deposits are thin compared to their downslope length. On land they commonly transform from landslides, and similar transformations appear to take place under the sea. The emplacement of the rock debris on the sea bed, as happens after collapse of volcanic seamounts in the Canaries, for example, may generate large debris flows which travel for hundreds of kilo-metres downslope [6] (Fig. 6.4). Such catastrophic failures of volcanic islands are thought to also be res-ponsible for tsunamis [7]. Debris flows also represent more immediate hazards. The debris flows at Mount Huascarán, Peru, in 1962 and 1970 caused major loss of life. The 1962 event started as an avalanche of ice, then mixed with mud and water to form a mudflow with a volume of $10^7 \, m^3$. The 1970 event was a debris flow with a volume of $5–10 \times 10^7 \, m^3$ which travelled 160 km downstream to the Pacific Ocean. Debris flows may impound lakes by blocking drainage. A massive debris flow $(2.5 \times 10^9 \, m^3)$ generated by an earthquake in the Pamirs (Tajikistan) in 1949 blocked the Murgab River, forming a lake 284 m deep and 53 km long. The debris flows originating from pyroclastic flows entraining water, from eruptions beneath crater lakes, or especially by the downslope resedimentation of freshly deposited tephra after heavy rains, are known as *lahars*.

Highly sediment charged floods (hyperconcentrated flows) are especially common in small mountainous catchments. The particular rheological properties of hyperconcentrated flows on land are thought to initiate at sediment concentrations of 40% by weight (20% by volume). They are less turbulent than stream floods because of the high effective viscosities. At very high sediment concentrations the flow may become non-Newtonian.

Under water, hyperconcentrated currents in which fluid turbulence suspends the sediment load are known as turbidity currents, but they vary greatly in their sediment concentrations. Low-density examples probably originate by resuspension of fine-grained sea bed sediment, by evolution from small-scale slope

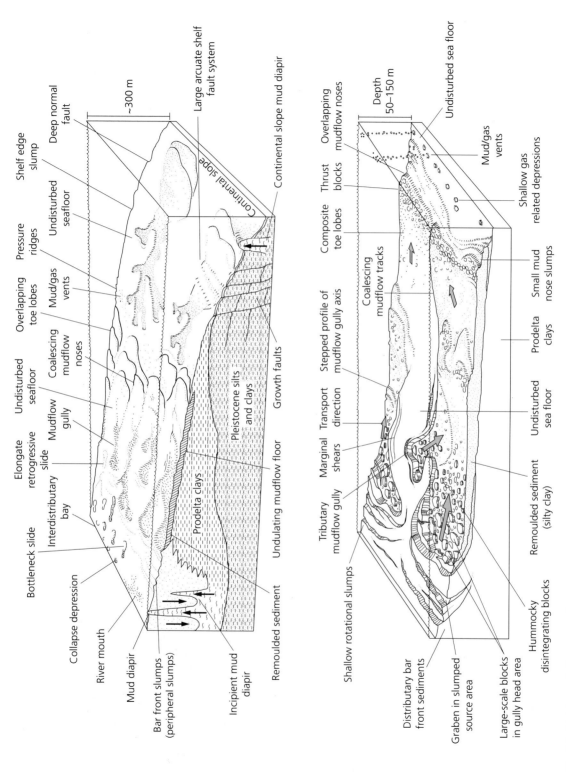

Fig. 6.3 (a) Schematic illustration of the range of sea bed features found between the river mouth and interdistribuary bays and the shelf edge, off the Mississippi delta. (b) Detail showing schematically delta-front mudflow gully systems composed of upslope arcuate slumps to downslope composite lobes. The length of the gully systems has been much shortened for illustrative purposes. After Roberts *et al.* (1980) [5].

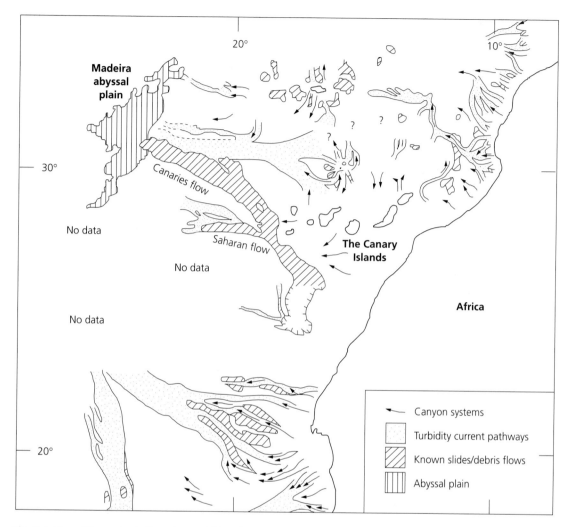

Fig. 6.4 Map of the sea bed off north-west Africa, showing the transport paths of turbidity currents, and giant debris flows. The Canaries debris flow originated in the loading of the sea bed by a rock avalanche caused by the catastrophic collapse of a nearby volcanic seamount. The Saharan debris flow, however, was due to failure of the continental margin. After Weaver *et al.* (1992) [8].

failures, or by transformation from sediment plumes (see Chapter 7). Many low-density turbidity currents have been observed in steep fjords, where they are 'ignited' by slope failure of the fjord walls. The first bottom-hugging (hyperpycnal) (Chapter 7) were described from Lake Mead, USA. The Huanghe River, China, is one of a number of major rivers which produce hyperpycnal outflows more or less continuously. Many more rivers generate low-density turbidity currents as hyperpycnal outflows during flood discharges. Ignitive high-density turbidity currents typically

transform from slides and slumps. The turbidites generated from the Grand Banks slumps of 1929, triggered by an earth-quake, are an example. A more recent example in 1979 took place offshore near Nice airport because of the overloading of coastal embankment materials during construction of the airport (Fig. 6.2). Submarine turbidites are extensive, but thin.

Acidic volcanoes emit large quantities of pyroclastic material into the atmosphere and hot, hyperconcentrated flows down their flanks. Parts of the volcanic cone may be blown apart by the highly explosive activity, as in the eruption of Mount St Helens in 1980. The highly concentrated hot dispersions of particulate material may flow for long distances from the flanks of the volcano, commonly as a result of the gravitational collapse of a cloud of tephra. They are extremely destructive and extremely rapid, the larger flows attaining speeds of 200 m s^{-1}. Hyperconcentrated flows accompanied the Mount St Helen's eruption.

The fundamental difference between soil creep and many of the other types of mass movement is that the latter involve slope failure. In order to study the generation of slides, slumps, debris flows and avalanches we need to consider the yield stress of the static rock or debris. In addition, in order to understand their flow behaviour it is necessary to investigate their dynamic properties. These dynamic properties are generally complex. Through loss of strength by disintegration or dilution one flow type may change into another. For example, translational slides are thought to disintegrate into debris flows, which may themselves transform by dilution with sea or lake water into turbidity currents. On the other hand, hyperconcentrated stream floods may lose water by infiltration and evaporation and gradually turn into debris flows.

Clearly, any classification of mass movements and hyperconcentrated flows should ideally allow for the transformation from one type to another. The general proposal for submarine movements (Fig. 6.2) recognizes this possibility. Slides and slumps can be classified by their thickness–length ratio and the amount of sediment disintegration. Plastic flows and turbidity currents can be classified according to sedimentological parameters such as the grain size of the material carried and the volume concentration of sediment, and according to dynamical parameters, such as Reynolds and Froude numbers. The interpreted transformation behaviour of the event at Nice in 1979 is superimposed on Fig. 6.2c [4].

6.2 Soil creep

Soil creep takes place by a number of possible processes [9].
• The continuous creep caused by the breakage of clay mineral bonds.
• More importantly, expansion and contraction of the soil caused by heating and cooling, wetting and drying (particularly of expandable clays), and freezing and thawing. The expansion, or *heave*, takes place along a resultant of the heaving force, which is normal to the ground surface, and the vertically acting soil

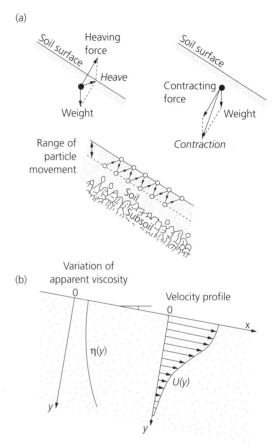

Fig. 6.5 (a) The heave and contraction of particles in a creeping soil results in a zigzag motion downslope. (b) The use of a downward increasing effective viscosity most satisfactorily simulates the observed velocity profiles in creeping soils (after Allen (1985) [10]).

particle weight (Fig. 6.5a). The contraction takes place along the resultant of the vertical weight and the contracting force normal to the slope. The result of the expansion and contraction is a zigzag downslope movement.
• Biological activity—such as that of earthworms and ants—reworks the soil and bring loose particles to the surface which are then prone to downslope movement.
• Throughflow of water is capable of transporting fine soil particles through void spaces in the soil.

The relative importance of these different processes varies according to climate and altitude. The creep in humid, warm climates is likely to be dominated by wetting and drying and biological activity. In mountains and in cold temperate and arctic climates freeze–thaw processes dominate, giving rise to a rapid but variable creep known as solifluction.

The measurement of the displacement of markers in the soil shows that the fastest soil creep takes place at the surface and decreases downwards, but the form of the velocity profile appears to be variable. One way of viewing the mechanics of soil creep is to lump all of the possible creep processes into one parameter, the *apparent viscosity*. If this parameter does not vary with depth, the soil creep should approximate to a laminar flow with a parabolic velocity profile. This is identical to the case of the flow of lava down a slope (Practical Exercise 4.2). There will be a certain depth below the surface where the creep velocity reduces to zero, depending on the value of the apparent viscosity and the surface velocity. Typical zones of soil creep are decimetres to a metre in thickness and surface velocities vary between about 2 and 10 mm y^{-1}. It is more likely, however, that the apparent viscosity is itself a function of depth below the surface (Fig. 6.5b). Moreover, it should increase with depth since below the capillary fringe (Chapter 3) soil moisture contents do not fluctuate to allow wetting and drying, and biological activity is much reduced. This has the effect of inflecting the velocity profile and causing finite soil creep rates at greater depths than in the laminar, parabolic case. The measured velocity profiles in soils undergoing creep support the idea that the apparent viscosity increases downwards.

6.3 The initiation of slope failure

In considering the initiation of a range of mass movement processes including avalanches, debris flows and slides, it is necessary to investigate the state of stress leading to failure. This branch of Earth surface processes therefore shares a great deal of theory with structural geology.

6.3.1 Friction

The sets of forces acting on a pre-existing surface in a rock are of two types. One set acts along the surface, with a frictional force opposing the force tending to cause slip, the shear force. The other force acts normally to the surface, and is due to the load of the overlying rock. This force per unit area is a normal stress or pressure. The relation between the friction force and the normal force is a constant known as the *friction coefficient*. The friction coefficient can also be expressed as the tangent of the *angle of sliding friction*. This relation is commonly known as *Coulomb's law*. It can be illustrated simply by considering the sliding

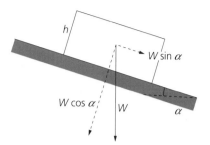

Fig. 6.6 A block of weight W sliding down an inclined plane at an angle α.

of a block on an inclined plane (Fig. 6.6). At the point of sliding the downslope component of gravity on the plane must equal the frictional force, and the component of gravity perpendicular to the plane constitutes the normal force. From Coulomb's law, the ratio is the friction coefficient f, given by

$$f = \frac{\rho g h \sin \alpha}{\rho g h \cos \alpha} = \tan \alpha \qquad (6.1)$$

For a static block, therefore, the friction coefficient can be expressed as the tangent of an angle, where in this simple illustration it is equal to the slope. In a mass of dry sand grains, the *internal friction angle* is a function of the size and packing of the grains. This is also roughly the angle to which a pile of grains can be built up, known as its *angle of repose*.

The sliding of a block over a basal shear plane may be facilitated by changing either the component of gravity acting along the plane or the effective normal stress acting perpendicularly to it. An increase in the downslope gravitational component may result from the ground motion accompanying earthquake activity. A decrease in the normal stress may result from lubrication of the shear plane with fluids, such as water. In both cases, the friction coefficient decreases, leading to slip.

The concept of a friction coefficient can also be applied to a rock mass subject to a stress field rather than a block on an inclined plane. The relation between the normal stress and the friction is then known as *Byerlee's law*.

6.3.2 Strength of natural materials

The strength of a material is measured by the stress required to cause failure and permanent deformation. The stress–strain behaviour of a material depends on a number of factors: the nature of the solid material

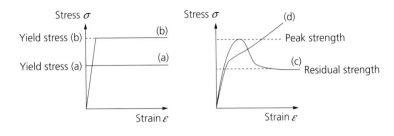

Fig. 6.7 Stress–strain relationships for a variety of materials: (a) ideal plastic; (b) ideal elastic-plastic; (c) material showing work- or strain-softening (such as thick custard powder); (d) material showing work-hardening (such as porous sandstones undergoing cataclasis). After Middleton & Wilcock (1994) [11].

itself; the effect of fluid pressures in pores, cracks and joints; the temperature; the rate of strain; and the *confining pressure*. Together, these factors make up the constitutive law of the material (this topic was introduced in Section 4.1). But it is not always easy to say when failure has actually taken place (Fig. 6.7). The definition of strength is unambiguous in the case of a perfectly elastic substance with a clear yield strength. It is also straightforward in the case of an ideal plastic which deforms instantly and continuously when the yield strength is exceeded for as long as the stress is applied. The same is true of materials which show a combination of ideal elastic and ideal plastic behaviour. However, many real substances exhibit a behaviour in which an elastic deformation merges gradually into a non-elastic deformation. In the non-elastic field the material may continue to deform at a lower stress than its peak stress, a situation called *strain-softening* or *work-softening*; alternatively, it may become stronger with continued strain, called *strain-hardening* or *work-hardening*. In these real materials the meaning of the term 'strength' and its numerical value are far less clear. This is particularly true with the deformation of soil. Complex stress–strain behaviour results from the operation of a number of different deformation mechanisms. For example, a general expansion in volume, or *dilatation*, involves shearing throughout a material rather than along a shear plane. On the other hand, a soil may suffer *consolidation* into a more compact material. Dilatation and consolidation may take place at different strains during the deformation of soil, leading to a highly nonlinear stress–strain behaviour.

Each material responds differently to an applied stress, depending on whether the stress is compressive or tensile. The strengths of materials under tension are in general smaller than under compression.

It is common to distinguish between two main regimes of flow:

1 a material that fails by fracturing is said to be *brittle*;

2 a material that fails by a continuous flow is termed *ductile*.

The theory of stress and strain in two and three dimensions is dealt with in a number of texts listed in the Further Reading section at the end of this chapter. For our purposes it is sufficient to accept that the stress at a point in a material can be expressed in terms of components along three axes. The mean stress in a material is the sum of these principal stresses divided by 3. So when all three principal stresses are equal, we have a homogeneous or isotropic state of stress and the mean stress is equal to any of the principal stresses. This is the situation to be expected in a body of water, when the state of stress is termed *hydrostatic*; when the stress results from the weight of overlying rocks it is termed *lithostatic*. The difference between the actual stress and the mean stress is known as the *deviatoric stress*. If the orientation and magnitude of the principal stresses are known, then it is possible to calculate the magnitudes of stresses acting on any plane at an angle to the principal stresses. This is commonly done graphically for two-dimensional stress fields using the *Mohr diagram*, or its computerized equivalent. In two dimensions, there are two planes along which the shear stress is at a maximum; these planes are orientated at 45° to the directions of the principal stresses.

6.3.3 The Navier–Coulomb criterion of failure

In practice, rocks do not fail on surfaces exactly at 45° to the principal stresses. One way of expressing the angle of the planes along which failure takes place is the *Navier–Coulomb criterion*. This is commonly written

$$\tau_c = C + \sigma \tan \phi \qquad (6.2)$$

where τ_c is the critical shear stress at failure, C is a strength which exists even at zero normal stress (pressure), called the apparent *cohesion*, σ is the normal stress and ϕ is the apparent angle of internal friction. Assuming a linear relation between the shear stress and normal stress on the potential plane of failure, the angle between normals to this plane and the principal stress direction is given by

$$\alpha_c = \frac{90 + \phi}{2} \qquad (6.3)$$

The apparent angle of internal friction varies for rock, soil and sediment types between about $35°$ and $60°$. We shall see below that it is strongly affected by fluid pressures. Some sediments, such as clean dry sand, have virtually no cohesive strength. Small admixtures of silt and clay, however, give sediments cohesive strength. Cohesive and frictional strengths can be looked up in tables for commonly occurring materials.

This theory can now be applied to the failure of a sediment mass under its own weight, leading to sliding on a slope.

6.3.4 Sliding on a slope

This section considers the sliding of a homogeneous material such as poorly bedded sediment on a slope caused by the state of stress below the ground surface (Fig. 6.8). It is further assumed that the slope continues uniformly at an angle α for a long distance compared to the area being studied. The shear stress increases uniformly with depth due to the weight of the overlying homogeneous sediment. At a certain depth below the surface the stress will equal the strength on a plane parallel to the surface. This is the *potential plane of failure*

The forces on a volume of unit width across the slope and unit length down the slope are its weight W and the forces N and T due to the normal and shear stresses σ, τ acting on its lower surface. The force acting normally to the failure plane is given by

$$N = W \cos \alpha \qquad (6.4)$$

Now the weight of the parallelogram-shaped volume of material (Fig. 6.8) is

$$W = y\rho g \cos \alpha \qquad (6.5)$$

Hence the normal force on the sloping plane responsible for the normal stress σ is

$$N = [\rho g y \cos \alpha] \cos \alpha = \rho g y \cos^2 \alpha \qquad (6.6)$$

In a similar fashion the force acting parallel to the potential plane of failure responsible for the shear stress τ is

$$T = W \sin \alpha = \rho g y \cos \alpha \sin \alpha \qquad (6.7)$$

The condition of failure is given by the Navier–Coulomb criterion. At the point of failure the strength is equal to the critical shear stress so that

$$\rho g y \cos \alpha \sin \alpha = C + [\rho g y \cos^2 \alpha] \tan \phi \qquad (6.8)$$

or, dividing throughout by $\rho g y \cos^2 \alpha$, we obtain an expression in terms of the tangent of the critical slope for failure

$$\tan \alpha = \left(\frac{C}{\rho g y \cos^2 \alpha} \right) + \tan \phi \qquad (6.9)$$

which is the fundamental relation for the initiation of a landslide. It can be used to calculate the critical slope for failure if the cohesive and frictional strengths of the accumulated debris, soil or rock are known. It also helps us to understand how a slope might become unstable if there were a change in any of the parameters involved. For example, the slope may increase by sedimentation in its upper regions as on a delta front, or by undercutting by river erosion at its base. The frictional or cohesive strengths may decrease by weathering of the materials close to and at the surface

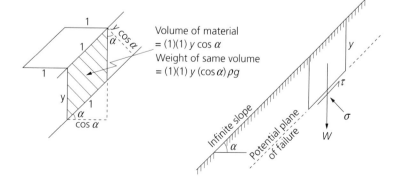

Volume of material
= (1)(1) $y \cos \alpha$
Weight of same volume
= (1)(1) $y (\cos \alpha) \rho g$

Fig. 6.8 Notation for the sliding of a homogeneous material on an infinite slope along a plane of failure.

or by different land use such as deforestation. The weight may also be increased by loading from materials derived from upslope. If the material has no cohesion, as in a pile of dry sand, inspection of equation (6.8) shows that the slope angle is exactly the angle of internal friction.

Failure may make use of a pre-existing weakness in the material on a slope. This might be a fault or major joint in a rock, or a prominent bedding plane in sediments. This can lead to catastrophic rockslides, as at Frank, Alberta, in 1903.

6.3.5 Rotational failures

In Section 6.3.4 a slope was considered which was very long. The situation is more complex on a slope which is short. The presence of the short slope produces a characteristic stress field in the underlying rock, soil or sediment. This stress field can be visualized by drawing lines whose tangents are always in the direction of a principal stress. These are called *stress trajectories*. Under a flat surface the stress trajectories for the maximum principal stress σ_1 should be more or less vertical, and the trajectories for the minimum and intermediate principal stresses should be at right angles to this in the horizontal plane (Fig. 6.9).

We know that the plane of failure should be at an angle to the principal stress given by $\alpha = (90 + \phi)/2$. Assuming that the frictional strength of the material does not vary, a failure surface can be constructed that is always at the angle α to the stress trajectory. For a short slope, as depicted in Fig. 6.9, the failure plane will be at an angle α to the σ_2 trajectory. Consequently it is curved, and the slide is rotational, with the failure surface cutting the ground surface close to the base of the slope.

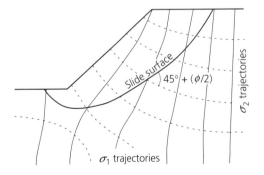

Fig. 6.9 Stress trajectories and failure plane for a short slope. The slide surface is drawn so as to make an angle of $45° + \phi/2$ with the σ_2 trajectories (dashed lines). After Middleton & Wilcock (1994), p.140, [11].

6.3.6 The importance of fluid pressures

Although landslides may be initiated by the mechanisms outlined above, most of them appear to result from increases in pore pressure causing a reduction of the fraction of the normal stress acting on the potential plane of failure.

The stress that is responsible for causing deformation in a porous material is the difference between the total stress and that arising from pore fluid pressure. This is termed the *effective stress*. In a thickness of sediment such as freshly deposited clay, the weight of the overlying material causes an expulsion of pore fluids and a deformation of the framework grains. This is what is understood by *consolidation*. If the pore fluid is expelled slowly as sedimentation continues at the surface, the fluid pressure in the pores will remain hydrostatic. The effective stress is then equal to the submerged weight of the solids. But if the sediment is deposited quickly there is a sudden increase in the total stress. If water cannot escape from the pores easily consolidation does not take place and the additional stress is transferred to the pore fluid, producing an *excess pore pressure*, (i.e. in excess of hydrostatic). A gradual loss of fluid reduces the excess pore pressure and the sediment gradually consolidates. However, consolidation reduces permeability, which may prolong the period during which the sediment is *underconsolidated*.

A different scenario is provided by the erosion of some of the sediment pile, or the removal by other means of an overburden such as the melting of ice. In this case the sediments are *overconsolidated* and have higher frictional strengths than normally consolidated equivalents. The compact muddy tills deposited under ice sheets are typically overconsolidated (Chapter 11). Overconsolidation may also result from the withdrawal of water (or oil) from porous sediments around wells causing a decrease in pore pressure.

If it is the effective stress rather than the total normal stress that determines the deformation of a porous material and its frictional strength, the Navier–Coulomb failure equation should be modified accordingly:

$$\tau_c = C + \sigma_e \tan \phi \qquad (6.10)$$

where σ_e is the effective stress. Consequently, the increase in pore fluid pressure caused by heavy rainfall or rapid melting of snow may so reduce the effective stress compared to the total stress that failure may occur even on very low slopes.

This concludes our short investigation of the failure of static rock, debris, soil or sediment. It is now necessary to study the mechanics of moving natural substances.

6.4 The mechanics of debris flows

It has already been pointed out that the flow of natural materials such as ice, soil, mud and debris does not conform to that of a Newtonian fluid. The mechanics of ice is discussed in Chapter 11. In the present chapter the mechanics of sediment–water mixtures is focused on since these mass flows are major contributors to the denudation of mountains and the submarine erosion of muddy delta fronts and the continental slope. The flow of natural substances such as sediment–water mixtures invariably involves complex rheologies. It is often necessary to combine the ideal behaviours from two different constitutive models to get close to the actual rheology of the natural substance. Some of these constitutive models have previously been outlined in Section 6.3.2. Two models have been used to explain the movement of debris flows: a combination of an ideal plastic and a Newtonian fluid, known as a *Bingham plastic* model (developed by Johnson [12]); and a model based on the flow of highly concentrated dispersions of cohesionless grains [13], which is a non-Newtonian fluid model (summarized by Campbell [14]; see also [15, 16]).

6.4.1 Bingham plastic model

A Bingham plastic does not deform until a critical shear stress is reached, after which it deforms with a linear relation between the shear stress and the strain rate, that is, as a Newtonian fluid (Fig. 6.10). If k is the critical shear stress, the constitutive law is

$$\tau = k + \mu \frac{d\varepsilon}{dt} \tag{6.11}$$

where ε is the strain and μ is the viscosity. For a material of depth h with a Bingham plastic rheology moving with velocity u down a slope α, we can integrate the equation of motion to find the velocity at any depth y, and therefore the velocity profile. The shear stress on any plane parallel to the basal surface at a height y is equivalent to the downslope component of the weight of the overlying material:

$$\tau = \rho g (h - y) \sin \alpha \tag{6.12}$$

Since the strain rate is a velocity gradient, combining equations (6.11) and (6.12), the equation of motion we wish to integrate is

$$\frac{du}{dy} = \frac{1}{\mu} \left[\rho g \sin \alpha (h - y) - k \right] \tag{6.13}$$

Since the viscosity does not vary with height above the bed, and $\rho g \sin \alpha$ is also a constant, integration gives

$$u = \frac{\rho g \sin \alpha}{\mu} \left(hy - \frac{y^2}{2} \right) - \frac{ky}{\mu} + C \tag{6.14}$$

The no-slip condition provides a boundary condition that $u = 0$ at $y = 0$, so the constant of integration, C, is zero. When the depth in the flow is such that $h - y = k / \rho g \sin \alpha$, the first term on the right-hand side is equal in magnitude to the second term on the right-hand side. Above this elevation there is no velocity gradient because the shear stress is less than the yield strength, so the material moves as a *rigid plug* with no internal deformation (Fig. 6.10, see p. 225). The maximum velocity, which occurs at the base of the rigid plug, is given by [11]

$$u_{max} = \frac{\gamma_x \left(h - \dfrac{k}{\gamma_x} \right)^2}{2\mu} \tag{6.15}$$

where $\gamma_x = \rho g \sin \alpha$.

  The Bingham plastic model explains many of the observed features of debris flows. Debris flows on alluvial fans commonly contain large boulders projecting through the top of the flow, suggesting that they are rafted along the top of the flow despite being denser than the matrix in which they lie. These large clasts are supported partly by the matrix strength. The Bingham plastic model also fits the observation that when the shear stress driving flow falls below the matrix strength, the entire debris flow 'freezes' into a poorly sorted deposit. Since the shear stress increases downwards, the debris flow should freeze from the rigid plug downwards. However, some observations suggest that other flow laws may be appropriate.

6.4.2 Non-Newtonian fluid model

The flow law for a laminar uniform Newtonian flow down an inclined plane is given in Section 4.3. This is modified for a non-Newtonian fluid which has a nonlinear relation between the strain rate and the shear stress, to become

$$a\tau^n = \frac{du}{dy} \tag{6.16}$$

For a Newtonian fluid $a = 1/\mu$ and $n = 1$.

One example of a non-Newtonian flow is the shearing of a concentrated dispersion of cohesionless grains such as sand. Bagnold [13] showed that at high rates of shearing and volume concentrations of grains higher than 9% there is a stress directed away from the bed transmitted directly by collisions between grains. This stress, P, which Bagnold termed *dispersive pressure*, is related to the strain rate (or velocity gradient), grain size and density and their volume concentration. This can be deduced from a dimensional analysis of the problem.

If debris flows move as non-Newtonian fluids by the action of dispersive pressure, they should 'freeze' when the upward-directed dispersive pressure is insufficient to keep clasts of a given size and density from sinking. If the dispersive pressure falls below a critical value the flow should settle on to the bed, so the debris flow should 'freeze' from the bottom up, in contrast to the case for a Bingham plastic.

6.4.3 Turbulence in debris flows

It is well known that the criterion for the initiation of turbulence in a Newtonian fluid is the Reynolds number. However, it is doubtful whether this is directly applicable to debris flows since they have different constitutive laws. Debris flows may have a criterion for turbulence that is relevant to Bingham plastic or non-

---

Practical exercise 6.1: Dimensional analysis of the problem of highly concentrated dispersions

This exercise is similar to Review Problem 1 on pp. 91–2 of Middleton & Wilcock (1994) [11]. It is particularly valuable to treat it as a problem of dimensional analysis because the physics is rather complex.

In a concentrated dispersion of grains there is a force per unit area directed away from the bed which supports grains within the flow. This dispersive pressure, P, is produced by collisions between grains. The dispersive pressure should therefore be affected by the mass of the grains, their size and density. The other parameter of relevance is obviously the rate of shearing du/dy. The parameters and dimensions are therefore as in Table P6.1.

There are four variables and three dimensions, so we should be able to form just one dimensionless product (Section 4.1). The dimensionless product must have the form

$$P^a D^b \rho_s{}^c \left(\frac{du}{dy}\right)^d = 0 \qquad \text{(6.17)}$$

Following the method introduced in Section 4.1, and applied in Sections 4.2.1, 4.4.1, 5.2.2, 5.3.2 and 5.4.2 in a variety of different contexts,

$a + c = 0$
$-a + b - 3c = 0$
$-2a - d = 0$

These equations can be solved to give

$a = 1$

Table P6.1

Parameter	Dimensions
Dispersive pressure, P	$[ML^{-1}T^{-2}]$
Grain size, D	$[L]$
Grain density, ρ_s	$[ML^{-3}]$
Rate of shearing du/dy	$[T^{-1}]$

$b = -2$
$c = -1$
$d = -2$

and so the dimensionless product can be written

$$\Pi = \frac{P}{\rho_s D^2 (du/dy)^2} \qquad \text{(6.18)}$$

Consequently, the dispersive pressure is

$$P = K\rho_s D^2 \left(\frac{du}{dy}\right)^2 \qquad \text{(6.19)}$$

where K is a constant. This relationship has been confirmed experimentally by Bagnold [13]. K is shown to be dependent on the grain concentration.

This constitutive law shows that the dispersion of grains shears even if the stress (dispersive pressure) is very small. There is no yield strength. Furthermore, the rate of strain du/dy is proportional to the square root of the stress. Returning to the basic constitutive equation, therefore, the exponent n is equal to $1/2$ for the flow of highly concentrated cohesionless grains.

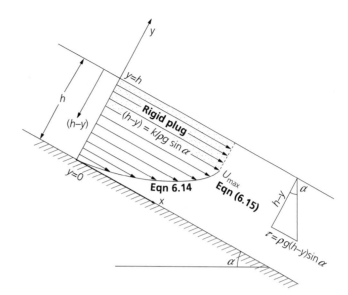

Fig. 6.10 Velocity profile and rigid plug for a Bingham plastic model of debris flows.

Newtonian flow models. To develop this argument, it should be recollected that the Reynolds number in Newtonian fluids is the ratio of inertial forces tending to cause shear and the viscous forces tending to resist deformation. However, the resisting forces in a Bingham plastic involve not only viscosity but also the material strength. Inspection of the constitutive law for a Bingham plastic (equation (6.11)) shows that the strength has the units of stress.

For a laminar Newtonian fluid the viscous stress is given by the now familiar relation

$$\tau = \mu \frac{du}{dy} \qquad (6.20)$$

or for a laminar flow of depth h and average velocity U,

$$\tau = \frac{\mu U}{h} \qquad (6.21)$$

The ratio of the strength to the viscous stress is therefore

$$\frac{kh}{\mu U} = B \qquad (6.22)$$

which is called the *Bingham number*. At very low Bingham numbers (less than 0.1) the viscosity dominates and the transition to turbulence should be controlled by a Reynolds number. At high Bingham numbers the strength dominates. The transition from laminar to turbulent flow then takes place at progressively higher Reynolds numbers as the Bingham number increases (Fig. 6.11). The relation between

the Reynolds number and the Bingham number at the transition from laminar to turbulent flow is

$$\frac{Re}{B} = \frac{\left(\rho Uh/\mu\right)}{\left(kh/\mu U\right)} = \frac{\rho U^2}{k} \approx 1000 \qquad (6.23)$$

This ratio, which is termed the Hampton number, is proportional to the inertial force over the strength. The shear strengths of natural subaerial debris flow materials are typically in the region of 10^3–10^4 Pa [12] and viscosities of between 10^2 and 10^3 Pa s have been measured [17]. Bulk densities typically range between 2000 and 2400 kg m^{-3}. The following exercise demonstrates how the Hampton and Bingham numbers can be combined with the velocity equation for a Bingham plastic to find out whether debris flows are likely to be turbulent, what slopes are required for a debris flow to become turbulent, and the slopes required for the debris flow to come to a halt.

Practical exercise 6.2: Flow behaviour of debris flows

A debris flow of thickness h moves down a slope α with a maximum velocity U. The bulk density ρ is 2200 kg m^{-3}, the shear strength k is 300 Pa, and the viscosity when moving 1000 Pa s. Assume that the debris flow deforms as a Bingham plastic.
1 At what speed should the debris flow become turbulent?

Continued on p. 226.

Practical exercise 6.2: *Continued*

2 Explore the combination of flow depths and slopes that would be required to cause the onset of turbulence. Natural slopes vary between about 0.1 at basin edges and on fault scarps, and less than 0.001 on the basin plains of the ocean. Given these limits, what flow depths are required for turbulence to develop? How likely is it that debris flows become turbulent in nature?

3 If the debris flow thickness is 5 m and the slope is 0.01, what is the thickness of the rigid plug? What is the maximum velocity, measured at the height of the base of the rigid plug, when the debris flow is moving over the 0.01 slope?

4 Assuming that the debris flow will 'freeze' when the shear stress is less than the shear strength of the debris, at what slope will the debris flow finally come to a halt, assuming its flow thickness does not change?

Solution

1 The transition from laminar to turbulent flow is given by a critical Hampton number

$$H = \frac{\rho U^2}{k} \approx 1000$$

from which the critical velocity of $11.7\,\mathrm{m\,s^{-1}}$ can be obtained.

2 The equation for the maximum velocity of a Bingham plastic is found at a depth of $k/\rho g \sin \alpha$ below the surface of the flow. Using equation (6.23), the debris flow thickness required for the onset of turbulent flow is 3.4 m and 46.8 m for slopes of 0.1 and 0.001, respectively. This rules out the occurrence of turbulence on all but the steepest slopes.

3 We know that the shear stress in a Bingham plastic falls below the shear strength at a height of $y = h - (k/\rho g \sin \alpha)$ above the base. The rigid plug has a thickness of $k/\rho g \sin \alpha$, which is 1.39 m. The maximum velocity is measured at a height of 3.61 m above the base. From equation (6.15), it is $1.41\,\mathrm{m\,s^{-1}}$.

4 The debris flow will come to a halt when the shear stress at the base is less than the shear strength of the material. This will occur when $k/\rho g \sin \alpha = h$, that is, when the rigid plug thickness is the entire flow thickness. Solving for the slope for a 5 m thick flow, we have $\sin \alpha = 0.0028$, corresponding to a slope of 0.16°. The debris flow may have travelled a considerable distance before the slope has reduced to such a value.

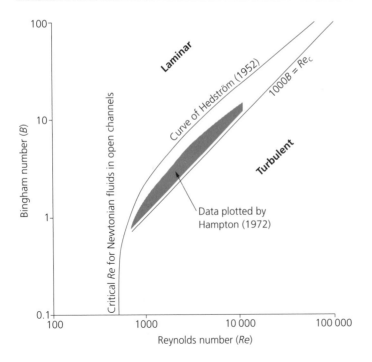

Fig. 6.11 Bingham number versus the critical Reynolds number for the onset of turbulence in a Bingham plastic. Based on Hampton (1972) [18], taken from Middleton & Southard (1977) [19].

6.5 Turbidity currents

6.5.1 Density currents in nature

In a wide range of flows in nature the movement is driven by a density contrast between the current and the ambient fluid (Plate 6.1, facing p. 204). The density difference may result from differences in temperature, salinity, or, for hyperconcentrated flows, by sediment concentration. Such flows are commonly called *gravity currents* or *density currents*. When the density difference can be specifically attributed to the concentration of sediment particles and the hyper-concentrated flow is subaqueous, the density currents are called *turbidity currents*, turbid meaning 'muddy'. There is therefore a distinction between homogeneous density-driven currents and particle-driven currents.

Perhaps the most familiar density currents to observers on the Earth's surface are the dust-laden atmospheric flows at cold fronts in dry areas, and the violent, hot clouds of gas, ash and rock fragments flowing down the flanks of volcanoes. Less dramatic examples are the sea breezes that blow inland as a result of the temperature differences of air heated from the sea and land surfaces. The action of density currents in lakes and the ocean cannot be so easily directly visualized. However, there are a number of lines of evidence to suggest that density currents are extremely important in sediment transport under water. First, the destructive path of individual density currents has been monitored by the successive breaking of sea bottom cables, as in the Grand Banks earthquake of 1929. Second, plumes of sediment-laden water can be observed at river entry points to reservoirs, lakes and seas. Third, underwater instrumentation may record the passage of bottom-hugging currents, though such instrumentation tends to be swept away in all but the weakest density currents. Finally, and perhaps most importantly, the sedimentary deposits of the sea floor, and the consolidated sedimentary rocks of the geological record, demonstrate that sandy and silty sediments have been transported long distances into the deep sea from their source.

Analogue currents can be very easily created in the laboratory by releasing a volume of fluid from a lock gate into a reservoir of fluid with different density. The density current characteristically develops a somewhat bulbous and slightly overhanging frontal region known as the head, and a long, thinner body. A number of gravity current types may be seen in laboratory experiments and inferred from 'natural laboratories' such as lakes (Fig. 6.12).

• If the density of the density current is greater than that of the ambient fluid, it flows along the bottom of the reservoir as an *underflow*. This situation is sometimes termed *hyperpycnal flow*. Sediment-charged, cold glacial meltwater entering lakes represents a good example of underflows.

• If the density of the density current is less than the density of the ambient fluid, it moves along the surface of the reservoir as an *overflow*, the case of *hypopycnal flow*. The plumes of light river water extending into and over more dense, saline seawater are examples (Chapter 7).

• If the reservoir fluid is stratified, with a density interface at a particular depth, and if the density of the density current is intermediate between that of the stratified reservoir fluids, it runs along the density interface as an *interflow*. Lakes are commonly stratified either seasonally or permanently. Sediment-laden currents from rivers may flow along the density interface (*pycnocline*), slowly sprinkling the lake floor with fine sediment.

• Turbidity currents, which derive their excess density from their high sediment concentrations, may lose some of their sediment by deposition during their lifetime as underflows. This reduces their ability to continue flowing along the bottom, resulting in a deceleration. If the fluid transporting the sediment particles is less dense than the ambient fluid, the loss of sediment concentration may result in a lift-off from the bottom at a certain distance from the source

Fig. 6.12 Different types of submarine density current illustrated from a lake basin with a density stratification, based essentially on Lake Brienz, Switzerland.

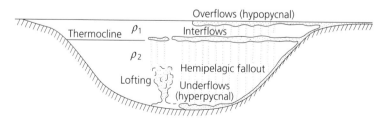

(or lock gate). This process is termed *lofting*. It has been particularly noted in the case of hot, sediment-charged volcanic outbursts which, after dropping a high proportion of their coarsest load, loft into the atmosphere because of their high temperatures.

There are clearly a number of important parameters in studying the dispersal of particulate sediment by turbidity currents. These include the runout distance of the turbidity current, the sediment discharge, and their frequency of occurrence. The sediments deposited by turbidity currents, their grain size and bedforms, give an indication of the mechanics of the flow at the time of deposition. An understanding of the mechanics of density currents in general is important not only because they are agents of environmental change and represent natural hazards, but also because sediments interpreted as turbidity current deposits (turbidites) host considerable natural resources such as oil and gas.

6.5.2 The mechanics of turbidity currents

One of the complicating factors in the consideration of the mechanics of turbidity currents is that they commonly flow down a slope. There are therefore two fundamental forces tending to promote movement. One is a pressure force solely due to the density difference between the density current and the adjacent fluid. The other is a body force due to gravity acting on the slope. We start our analysis of underflows by referring back, therefore, to the case of steady uniform flow down an inclined plane, as in the case of a river. Then the force generating the flow is the downslope component of the immersed weight of the density current. For an element of unit width and length, this driving force is

$$h\Delta\rho g \sin\alpha \qquad (6.24)$$

where h is the thickness of the density current, $\Delta\rho$ is the density difference between the density current and the ambient fluid, and α is the slope. This driving force increases, therefore, if the underflow becomes denser or thicker, or the slope steepens. The driving force is opposed by a frictional force. In the case of a river, this friction originates solely from the bed. But in the case of an underflow, the total friction acts at both interfaces—the bed and the upper boundary of the density current (Fig. 6.13). The resisting force can therefore be written for a unit area of flow

$$\tau_0 + \tau_i \qquad (6.25)$$

where the subscripts refer to the bed and the upper interface of the flow.

We can assume that the shear stresses at the two inferfaces are proportional to the kinetic energy of the moving density current. Consequently,

$$\tau_0 = k_0 \cdot \frac{1}{2}\rho U^2$$

$$\tau_i = k_i \cdot \frac{1}{2}\rho U^2 \qquad (6.26)$$

where ρ is the density of the ambient fluid and k_0 and k_i are coefficients expressing the proportionality between shear stress and kinetic energy at the two interfaces. These coefficients are 'friction', 'drag' or 'resistance' coefficients. For a turbulent channelized flow, such as a river, it is conventional to use a coefficient in the form of $f/4$, where f is known as the *Darcy–Weisbach friction factor*. It is nothing more than a drag coefficient with wide use in river hydraulics (Chapter 5). Typical values of the Darcy–Weisbach friction factor can be obtained from graphs and tables for natural flows. The total resisting force per unit area now becomes

$$\frac{\rho U^2}{8}\left(f_0 + f_i\right) \qquad (6.27)$$

and, equating the driving and resisting forces (expressions (6.24) and (6.27)), we have

$$U = \left[\frac{8g\Delta\rho h \sin\alpha}{\left(f_0 + f_i\right)\rho}\right]^{1/2} \qquad (6.28)$$

which, by slight rearrangement, becomes

$$\frac{U}{\left(g'h\right)^{1/2}} = \frac{8\sin\alpha}{f_0 + f_i} \qquad (6.29)$$

where $g' = g\Delta\rho/\rho$ is a density-modified ('reduced') gravity. Expressed in this way, it can be seen that the left-hand side has the form of a modified Froude number; this is commonly termed the *densimetric Froude number*, Fr'.

Improved imaging of the sea floor over the last decade has proved the existence of channels of varying dimensions and planform geometries. These channels are the sites of major throughput and deposition of sediment *en route* to the deep ocean floor. They must be carved by turbidity currents since they occur in water depths far below those scoured by tidal currents or river currents at times of low sea level. Furthermore, on the upper part of the Amazon cone, the margins of

(a)

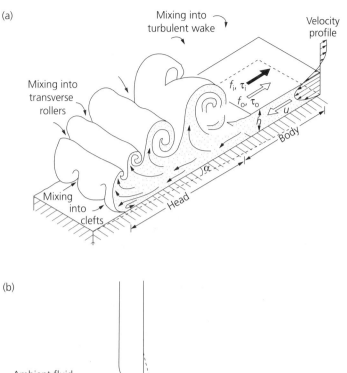

(b)

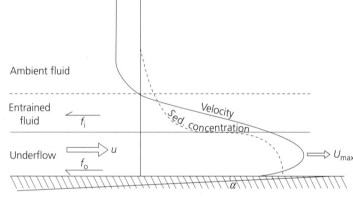

Fig. 6.13 (a) Schematic illustration of the head and body of a turbidity current showing the frictional resistance experienced both at the bed and at the upper interface of the current. (b) Velocity and sediment concentration profiles.

channels up to 200 m deep are elevated relative to the sea floor in the form of *levees* because of the deposition of sediment from overtopping turbidity currents. This suggests that turbidity currents may be at least as thick as the channels are deep, and that they must be powerful enough to be erosive at their bases to scour the channels. Although the friction coefficients are largely unknown, the equation for the velocity of a turbidity current given above indicates that such thick turbidity currents would be extremely vigorous even with small density differences and low submarine slopes.

The friction coefficient at the bed, as in rivers, is affected by the presence of roughness elements on the bed, such as bedforms, and by the concentration of particles near the bed. If the density current is acting as a turbulent fluid, we might expect the basal friction coefficient to behave similarly to that in rivers—that is, to vary only as a function of roughness over a range of Reynolds numbers. The friction coefficient at the upper interface is affected by the entrainment of ambient fluid and the mixing of turbid water with ambient. This in turn depends on the density contrast across the interface and the speed and thickness of the density current. These parameters are included in the densimetric Froude number, Fr', so we should expect f_i to vary with Fr'. Experimental results suggest that the friction at the upper interface remains low for $Fr' < 1$, but that at supercritical densimetric Froude numbers it increases markedly as large amounts of mixing take place at the upper interface. This has the effect of diluting and expanding the flow.

Since the density current experiences resistance at both its top and bottom boundaries, the velocity profile must show a maximum at some level within the flow. The height at which velocity reaches a maximum must depend on the relative magnitude of f_0 and f_i. If they are equal, the maximum velocity is at half of the turbidity current thickness. The velocity and sediment concentration in a steady uniform turbidity current are shown in Fig. 6.13.

Practical exercise 6.3: Use of densimetric Froude number

1 It is thought that the sum of the friction coefficients for the lower and upper interfaces of a turbidity current is in the region of 0.02. At what sea bed slope would you expect the turbidity current to become supercritical?

2 The average slope of the upper part of the Amazon cone is 0.02. Channels on the surface of the cone are typically 100 m deep. Would you expect a density current that has an excess density compared to the density of ambient seawater of 0.05 and has the same flow depth as the maximum depth of the channel to be supercritical or subcritical? Assume that the sum of the friction coefficients for the bottom and top interfaces of the turbidity current is 0.02.

Solution

1 Setting the densimetric Froude number equal to 1, we have

$$1 = \frac{8 \sin \alpha}{f_0 + f_i}$$

from which it can be easily found that the slope is 0.0025, corresponding to 0.14°. Turbidity currents may become supercritical therefore even on low slopes as long as the friction on their interfaces is not excessive.

2 The average velocity of a turbidity current is given by equation (6.28). In the Amazon cone example, this gives an average flow speed of 20 m s^{-1}. This represents a densimetric Froude number of 2.8, suggesting that the flow would be supercritical. Turbidity currents as thick as 100 m are very likely to be supercritical.

Flow dissipation

It can be seen in experimental density currents that the upper surface of the head of the flow is thrown into a number of transverse vortices or billows. These vortices are carried back and pass into a turbulent wake behind the head. They appear to be identical to the Kelvin–Helmholtz instabilities between two fluids undergoing shear (Chapters 4 and 5). The fact that dense fluid is continually being supplied to the vortices and escaping from the head suggests that the average velocity of fluid within the density current must be greater than the velocity of the head, which is therefore continually being replenished with dense material. This has an important implication. A fixed-volume source, such as an instantaneous slope failure, should eventually result in the head becoming under-supplied with dense fluid, causing it to decelerate and eventually dissipate. A constant-flux source such as a river in flood, should be able to maintain replenishment to the head and thereby increase its runout distance. If replenishment exceeds the loss of fluid in the tranverse vortices, the head should grow in size, as has been observed in experiments as the slope is increased.

The transport distance, or *runout distance* of a turbidity current is obviously of major interest in attempting to explain the distribution of particulate sediment in the deep sea. In investigating runout distances and the depositional patterns under turbidity currents, it is important to know something about the causes of dissipation of a turbidity current. These can be summarized as follows (Fig. 6.14).

• **Feeding of the head:** we have already seen that the presence of transverse vortices demonstrates that the average velocity of the density current exceeds the speed of the head. Experiments suggest that for low slopes the ratio of the velocity of the body to the velocity of the head is about 1.16 [20]. If the turbidity current has a finite length, by the time the last sediment is released from the source the fluid initially at the rear will have caught up the head in a distance approximately equal to the initial length. Feeding the head is therefore a potent way of dissipating a turbidity current where constant-volume sources on low slopes are involved.

• **Loss of excess density by entrainment:** the driving force for transport may be reduced by the entrainment of ambient fluid leading to dilution. This may take place at the head by the drawing in of ambient fluid along clefts between bulbous lobes, or along the interface between the body of the flow and the overlying turbulent wake. Dissipation by entrainment is thought to be of minor importance.

• **Loss of excess density by sediment deposition:** sediment deposition should take place if the turbidity

(a) Feeding of head

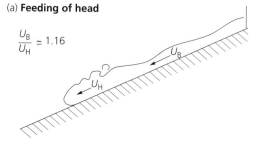

$$\frac{U_B}{U_H} \simeq 1.16$$

(b) Entrainment

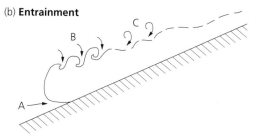

(c) Deposition

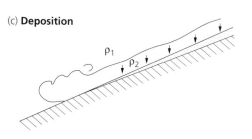

Fig. 6.14 Flow dissipation by: (a) feeding of the head caused by the speed of the body exceeding that of the head; (b) entrainment in clefts at the overhanging head (A), in transverse vortices (B), and in the turbulent wake (C); (c) by sediment deposition reducing the density of the current compared to ambient water.

current has a sediment load which is greater than the theoretically transportable load. The actual mass of sediment in a turbidity current is simply the product of the flow height, sediment density and sediment concentration. However, the theoretically transportable load relies on the use of bedload and suspended load transport formulae. For all but the most dilute turbidity currents, the flow is likely to be heavily overloaded with sediment [10], so dissipation by sediment loss should be an important process.

Autosuspension

The excess density caused by the high concentrations of particles in a turbidity current causes a lateral flow. The flow results in fluid turbulence which in turn keeps the sediment particles in suspension. The high suspended sediment concentration then drives further lateral flow. Ideally, there is a condition whereby the gravitational energy of the turbidity current is just enough to generate turbulence to keep the sediment load in suspension and to overcome friction at the lower and upper boundaries of the flow, so that there is neither erosion nor deposition of sediment as the current moves downslope (Fig. 6.15). This is the condition which Bagnold [21] called *autosuspension*. If autosuspension operates in turbidity currents, it might explain their very long transport distances of over 1000 km in some instances. A critical factor must be the grain size distribution of the sediment load. There should be a certain grain size kept in suspension by turbulence generated by flow down a certain slope. Grains less than this size are easily kept in suspension and add energy to the flow by increasing the excess density. Grains larger than this size take energy from the flow by requiring large amounts of turbulence to keep them suspended.

In rivers, by far the most energy is expended in overcoming the frictional resistance of the boundaries

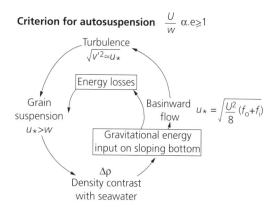

Fig. 6.15 Conceptual model of autosuspension (discussed in full in Pantin (1979) [22]), representing a feedback involving the suspension of sediment which drives fluid motion, causing turbulence, which in turn promotes sediment suspension. The criterion for suspension includes an efficiency factor e describing the way in which the flow uses power to suspend particles relative to overcoming friction on its bed. It can be seen that for a given turbidity current velocity, high efficiencies, high slopes and fine grain sizes favour autosuspension. After Pickering et al. (1989) [23].

of the flow, thereby generating turbulence, and only a small part of the energy in turbulence is used in keeping sediment suspended. Changes to the frictional characteristics of the boundaries, such as the development of bedforms, are likely to have effects on the turbulence that far outweigh the effects of small changes in suspended sediment concentrations. However, in turbidity currents, a much greater proportion of the turbulence must be used to keep the high sediment concentrations in suspension. It is possible therefore, but by no means certain, that autosuspension may take place in turbidity currents with fine-grained loads.

6.5.3 Deposition from turbidity currents
Theoretical fluid mechanics
Theoretical models based on the conservation of mass and momentum [24] and the assumption of moderate to high Reynolds numbers (so that viscous forces can be ignored) allow calculations of the size of the current as a function of time and the thickness of deposited sediment as a function of distance from source, or runout distance. The runout distance for a density current of initial 'volume' per unit width q released instantaneously on a zero slope is

$$x_r = 3 \left(\frac{g'q^3}{\overline{w}^2} \right)^{0.2}$$ (6.30)

where g' is the initial 'reduced' gravity and $\overline{w}$ is the average settling velocity of the suspended particles. The average deposition under the turbidity current is related to the initial sediment concentration and 'volume' per unit width of the density current at source and its final runout length:

$$\overline{\eta} = \frac{\rho C_0 q}{x_\infty}$$ (6.31)

where C_0 is the initial fractional sediment concentration by volume, and x_∞ is the runout distance at $t = \infty$. The actual deposit thickness should show a maximum relatively close to the source and a long tail reducing to zero at the maximum runout distance (Fig. 6.16). These models present sedimentologists with a challenge to map runout distances of individual turbidites.

The time taken to reach the maximum runout distance and therefore the average velocity of the turbidity current can also be calculated. The results can best be illustrated by considering some well-preserved turbidity current deposits (Table 6.2). For example, the Tertiary Marnoso arenacea of the northern Apennines, Italy, includes thick bedded turbidites such as the Contessa bed. The Black Shell turbidite is a deposit up to 4 m thick on the Hatteras abyssal plain in the western Atlantic. As a generalization, the fluid-mechanical model indicates that, across a range of initial sediment volumes, the parent density currents entered their depositional basins at speeds of about 5–10 m s^{-1} and with particle concentrations of 5–10% by volume. The single depositional events resulting from these turbidity currents last from several hours to 1–2 days. The turbidity current responsible for the Black Shell turbidite was probably 100 m in thickness, reducing to 30 m at 400 km from the source. These calculations, which of course serve as rough approximations because of the assumptions involved, give a good idea of the likely speed, thicknesses, duration and runout distances of naturally occurring turbidity currents responsible for the deposition of sandy turbidites up to 4 m in thickness.

Effects of sea bed bathymetry
So far we have considered the flow of a density current down a uniform slope. The sea bed, however, is likely to be irregular at a variety of scales. Two scales are briefly considered here.

Tectonically active basins such as pull-aparts in strike–slip regimes, young rifts where the lithosphere is being stretched, and piggy-back basins perched on thrust wedges, may all be sufficiently narrow to reflect back a turbidity current derived from a source flank from the opposite flank. Axially moving turbidity currents may also be affected by lateral sea bed slopes. That reflection should be important is expected from the fact that the typical runout distances of turbidity currents are measured in hundreds of kilometres, whereas many tectonically active basins have widths of tens of kilometres. Laboratory experiments suggest that as the turbidity current runs up the opposing slope it decelerates and forms an overthickened cloud of sediment which collapses by sending *solitary waves* down the opposing slope. Such reflected waves may be responsible for reworking of the surface of the previously deposited sediment or for dislocations in the overall upward fining in grain size in the turbidity current deposit. The process involved is effectively a transfer of kinetic energy from the moving turbidity current to the potential energy of the overthickened bulge and then to the kinetic energy of the reflected waves.

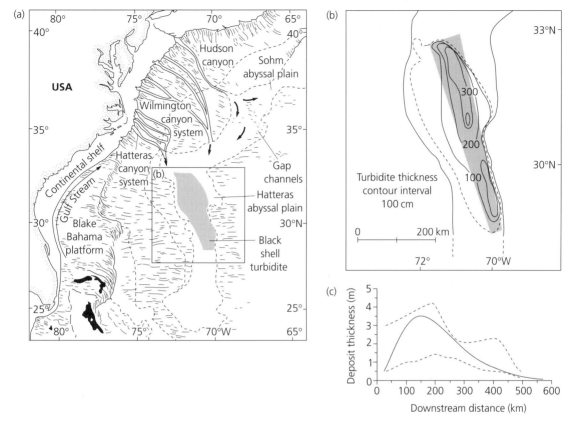

Fig. 6.16 The Black Shell turbidite of the Hatteras abyssal plain. (a) Location map, showing main routes of sediment supply to the Hatteras abyssal plain. The shaded area shows the distribution of the Black Shell turbidite. After Pickering *et al.* (1989) [23]. (b) Isopachs of the Black Shell turbidite. The shaded swathe is the region for which deposit thickness is plotted in (c). (c) Range of present thickness of Black Shell turbidite compared to model prediction of Dade & Huppert (1994) [24], showing a runout distance of 500–600 km and maximum deposit thickness of 3–4 m.

Table 6.2 Application of fluid-dynamical theoretical model to natural turbidity current deposits [24]

Parameter	Marnoso arenacea	Black Shell turbidite
Initial 'volume', q	2 km²	10 km²
Initial concentration, C_0	15%	5%
Runout distance, x_r	330 km	700 km
Runout time, t	10 h	24 h
Mean flow speed, U	5–9 m s⁻¹	6 m s⁻¹

Turbidity currents, and the sediment distribution resulting from their passage, are affected by intrabasinal sea bed highs acting as obstacles. Such obstacles might be individual fault scarps, for example. Experiments in laboratory tanks suggest that significant ponding of the turbidity current can occur even if the height of the obstacle is only 10% of the flow thickness. Ponded currents deposit anomalous thicknesses of sediment. If turbiditic sands are the hosts of natural resources such oil and gas, the positioning of sea bed irregularities in relation to turbidity current pathways is a valuable tool in exploration.

Importance of unsteadiness and non-uniformity

So far we have for the most part considered the density current as a steady uniform flow down an inclined plane. However, our brief exploration of turbidity currents already suggests that we should doubt that this is always the case. We have considered the difference between a fixed-volume source and a continuous feed, we have speculated on the likely effects of sea bed bathymetry, and we have examined the ways in which flow dissipation might take place. All of these factors and others suggest that turbidity currents may not be steady and uniform. As a reminder, flow unsteadiness refers to changes taking place through time at a fixed location, so a manifestation of unsteadiness is waxing and waning flow (Fig. 6.17). Non-uniformity refers to spatial velocity variations at a given time. Non-uniform flows which decelerate downstream are termed *depletive*, those which accelerate *accumulative*. If a fixed volume of sediment-concentrated fluid is released from a reservoir, or is 'instantaneously' dumped into a water body, it is most unlikely that the density current will

maintain a constant velocity through time at a point on the bed—that is, it will most likely be unsteady. On the other hand, a hyperpycnal flow from a river into a lake or sea may be persistent, so that the resulting density current is quasi-steady. A density current that spreads from a confined channel on to a flat plain, or one which loses its integrity by progressive, downslope mixing with ambient fluid (see Section 6.6), will be non-uniform and depletive, though it may be quasi-steady.

The non-uniformity and unsteadiness of density currents affects their tendency to deposit grains transported in turbulent suspension. A given grain carried in suspension will deposit if the flow decelerates in a reference frame moving with the particle [25, 26] (Fig. 6.17, 6.18). The acceleration experienced by this grain, termed the *substantive acceleration*, is made up of the local, *temporal acceleration* which will be zero for steady currents, and the downflow, *convective acceleration*, which will be zero for uniform currents:

$$\frac{\mathrm{d}u}{\mathrm{d}t} = \frac{\partial u}{\partial t} + u\frac{\partial u}{\partial x} \qquad \textbf{(6.32)}$$

The fate of a sediment grain in the flow cannot be determined, therefore, unless the unsteadiness and non-uniformity of the flow are known. The density current resulting from a single surge is likely to deposit sediment rapidly due to the depletive waning flow. However, although many density currents may deposit sediment by waning, this is not a prerequisite. A flow which was steady could deposit sediment by being depletive. The various possibilities are sketched out in Fig. 6.18.

The possibility that density currents may be quasi-steady for much of their lifetime, rather than being waning, has a number of important implications. Most importantly, it implies that the thickness of the sedimentary deposit should bear no relation to the thickness of the density current. A thin but long-lived quasi-steady and depletive current could deposit a thick bed; a thick, waning but accumulative current may deposit nothing at all at a particular location.

The paradigm in turbidite sedimentology is that density currents always conform to the surge-like, highly unsteady, waning type, involving a 'collapse' of the flow on to the bed [27]. The typical vertical sequences found in turbidites have traditionally been interpreted in terms of a waning, unsteady flow of either low density or high density, giving rise to the

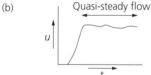
(a)

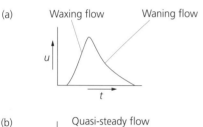
(b)

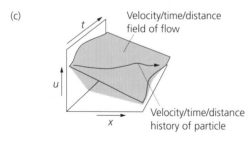
(c)

Fig. 6.17 (a) Plot of velocity versus time at a fixed geographical location for a highly unsteady flow, typical of a single surge-like turbidity current. (b) Quasi-steady current sustained over a longer period of time. (c) History of a particle travelling with the current (i.e. in the Lagrangian reference frame). The shaded surface shows the velocity of a steady, depletive current. The arrow illustrates the motion of a particle within the current. Although the flow is steady, the particle decelerates because of the non-uniformity of the flow. After Kneller & Branney (1995) [25].

(a)

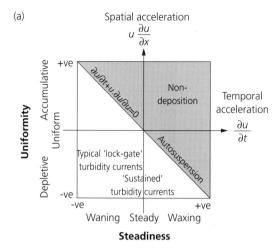

(b)

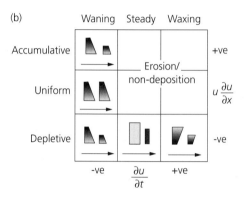

Fig. 6.18 (a) Matrix of flow types based on the spatial and temporal accelerations. (b) The vertical and lateral trends in the associated deposits. Typical surge-like turbidites, such as those produced experimentally by the release of hyperconcentrated fluids from a lock, are waning depletive types and give rise to upward fining and downstream thinning and fining. Quasi-steady 'sustained' turbidity currents may give rise to turbidites with little upward variation in grain size, but with downstream thinning and fining. After Kneller (1995) [26].

Bouma model [28] (Fig. 6.19a) or the *Lowe model* [29] (Fig. 6.19b), respectively. How common quasi-steady flows are in nature is currently uncertain. Facies types such as thick massive sands and massive *ignimbrites* have been attributed to deposition under quasi-steady, depletive flows [30], where the hyperconcentrated zone close to the bed is dominated by non-turbulent grain interactions and hindered

settling. The lowermost A division of the Bouma model probably has a similar origin.

6.6 Pyroclastic density currents

Although geologists have focused strongly on turbidity currents because of the large volume of turbidites preserved in the rock record, the density currents originating from active volcanoes unquestionably have greater impact on human society. Volcanic eruptions have not only profound farfield influences, such as the climatic cooling resulting from high atmospheric dust loadings referred to in Chapter 2, but also dramatic, and often devastating, near-field impacts on human activities. Indeed, the demise of human civilizations has been attributed to eruptions of volcanoes in the past. For example, the eruption of Santorini in about 1500 BC has been linked to the collapse of the Minoan civilization. In more recent times, the eruptions of Mount St Helens, USA, in 1980 and Mount Pinatubo, Philippines, in 1991 attracted great media interest. It is appropriate, therefore, to conclude this chapter by briefly considering the special characteristics of the pyroclastic flows emanating from active volcanoes. These flows are driven essentially by the excess density of a very highly concentrated particulate load. Their special properties include the fact that the interstitial fluid is much hotter and more buoyant than the surrounding air (or water in the case of submarine eruptions). Another characteristic feature is the 'activity' of the particulate matter, which during transport may aggregate to form accretionary particles, become welded together at high temperature, or change shape and density through cooling. These processes are very different to the transport of 'passive' sand, silt and clay by submarine density currents.

Pyroclastic flows form one part of a spectrum of processes that may take place during an eruptive phase of a volcano. Such processes may include eruptions of lava, and the fallout of tephra ejected into the atmosphere during explosive activity, called pyroclastic fall. In the following discussion only the flows driven by gravity acting on the hot, gas-particle suspensions are considered.

There is some variation in the terminology of pyroclastic flows. A distinction is commonly made between two end-member types of pyroclastic density current: those involving vesiculated low-density pumice; and those involving unvesiculated, dense lava clasts, commonly called *nuées ardentes*. The

Low-density   **High-density**

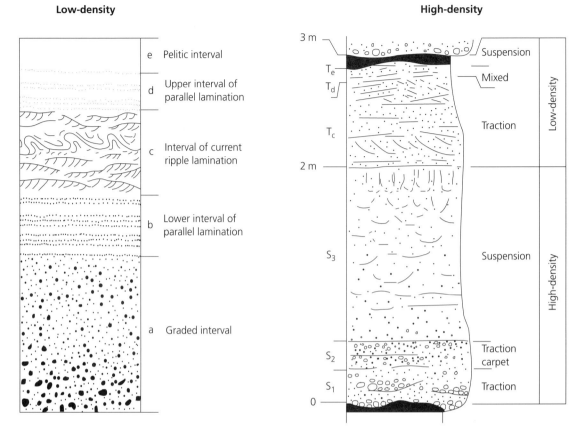

Fig. 6.19 Models for low-density and high-density turbidites based on Bouma (1962) [28] and Lowe (1982) [29].

deposits of the former are known as *ignimbrites*, whereas the deposits of the latter are called *block-and-ash deposits* (Fig. 6.20). Pyroclastic flows of lower density are known as pyroclastic surges:

Many pyroclastic flows result from the collapse under gravity of an eruption column of a volcano [31, 32], the potential energy attained by the upward gas thrust from the volcano being transferred into kinetic energy as the flow moves rapidly over the ground surface. Since the eruption column may be kilometres high, the resulting pyroclastic flows can be exceptionally vigorous. Two independent parameters appear to control the mass eruption rate and thereby the density and stability of an eruption column (Fig. 6.21): the volatile content of the magma, high contents promoting high eruption velocities; and the vent radius, which controls the upward flux of gas and magma. At very high mass eruption rates, the convecting eruption column may become so concentrated in particulates that it loses its

buoyancy, causing gravitational collapse. Pyroclastic fall deposits may therefore be overlain by pyroclastic flow deposits as the volcano evolves from a condition of column stability to one of column instability.

Pumiceous pyroclastic flows and ignimbrite deposits

Pumice flows are thought to be very high-concentration sediment gravity flows which are non-turbulent. Their particles are partly supported by the upward escape of gas, a process known as *fluidization*, and partly from grain–grain contacts giving rise to a dispersive pressure and from the finite strength of the high concentration gas-particle mixture. Pumice flows therefore appear to bear more in common with partially fluidized, partially dilatant debris flows than with turbidity currents. Since particles are not free to move easily relative to each other, pumiceous (ignimbrite) deposits are poorly sorted. The largest eruption forming ignimbrites (sometimes called *ash-flow tuffs*) in

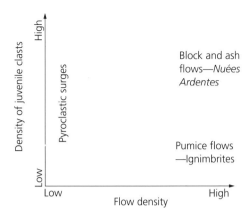

Fig. 6.20 Principal types of pyroclastic flow according to the density of the constituent clasts (vesiculated pumice versus non-vesiculated lava, for example), and the density (or concentration) of the flow. We can distinguish between low-density pyroclastic surges, and high-density pumice flows and block-and-ash flows (*nuées ardentes*). After Wright *et al.* (1980) [33] in Francis (1993) [34].

modern times took place at Mount Katmai in Alaska in 1912, leaving a 22 km long ignimbrite deposit extending down the valley of the Ukak River.

Pyroclastic flows are strongly affected by topography, commonly being channelled along valleys, in contrast to the blanketing effect of pyroclastic fall deposits. Individual ignimbrite flow units may show a distinct vertical zonation (Fig. 6.22) in the concentration of pumice and lithic clasts. The lowermost unit is a thin laminated unit rich in lithics and crystals (layer 1) thought to be due to *ground surge*. The overlying fine-grained basal layer (2a) of the ignimbrite proper is thought to result from intense shearing near the ground surface. The main bulk of the ignimbrite (layer 2b) consists of a poorly sorted mixture of pumice clasts and ash with a variable content of lithics and crystals. It is thought to be segregated by density by means of fluidization. A cap of fine ash (layer 3) represents the final fallout of ash convected out of the pumiceous flow. Such co-ignimbrite ashes were

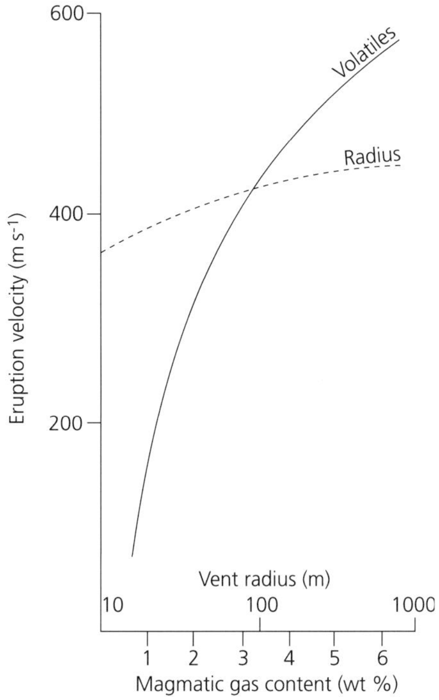

Fig. 6.21 Effect on the eruption column velocity of magmatic gas content (solid line) and vent radius (dashed). After Wilson *et al.* (1980) [35].

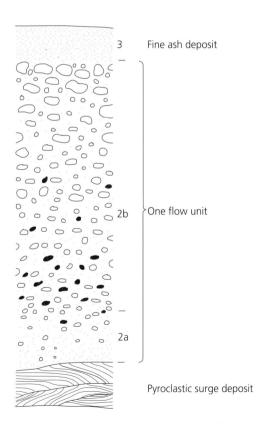

Fig. 6.22 Model for a 'standard' ignimbrite. After Sparks *et al.* (1973) [36].

particularly important in the enormous eruption of Tambora in Indonesia in 1815.

Nuées ardentes *and block-and ash deposits*

The 'glowing clouds' (or, perhaps better, 'glowing avalanches') have been termed *nuées ardentes* since their desctruction of St Pierre in Martinique on 8 May 1902. Three different mechanisms have been proposed, including: the gravitational collapse of lava flows and domes as on Mount Merapi, Indonesia, in 1942–3 (these are essentially hot avalanches); triggering by explosive events on growing lava domes, as on Mount Pelée in Martinique in 1902; and eruption column collapse, as in the case of the 1902 eruption of Soufrière, St Vincent.

Block-and-ash deposits consist of blocks of juvenile magma in an ash matrix. Like pumiceous varieties of pyroclastic flow, they are also typically confined along pre-existing valleys. Reverse grading is present due to the upward concentration of the largest lithic clasts by dispersive pressure, but because no pumice is present, there is no segregation by density in response to fluidization, as seen in ignimbrites.

Pyroclastic surges

Pyroclastic surges are low-concentration sediment gravity currents in which the particles are supported primarily by turbulence. Movement of the surge is also affected by the momentum of expanding gas and the momentum of particles acting as projectiles exploded from the volcano's vent or flanks or collapsing from a tall eruption column. Particle concentrations may become very high close to the substrate, leading to a density stratification. The lower, high-density layer may transform into a denser pyroclastic flow. The deposits of pyroclastic surges are better sorted than those of pyroclastic flows, and typically are well stratified due to the high transport rates of bedload over a plane or wavy surface.

A number of different pyroclastic surges are recognizable.

• **Base surges** are associated with hydrovolcanic explosions and are due to the collapse of eruption columns. Base surge deposits result from the violent interaction of water and magma, and consist of poorly sorted, highly fragmented material. They typically thin rapidly away from the vent. In proximal positions, they characteristically contain large-scale cross-stratification, demonstrating the existence of large flow-transverse bedforms under the base surge.

• **Ground surges** are found immediately beneath ignimbrites (layer 1). They are enriched in lithics, especially crystals, suggesting density segregation with the fine pumice dust elutriated away. They may form at the head of a pyroclastic flow from the forward jetting of ash. The resulting ground surge deposit would then be overrun by the advancing pyroclastic flow depositing an ignimbrite.

• **Ash cloud surges** are similar to ground surge deposits in many ways, but instead are found within or laterally equivalent to ignimbrites, where they appear to represent local dilutions of denser pyroclastic flows, or be associated with pyroclastic fall deposits (as in the eruption of Vesuvius AD 79 which destroyed Pompeii), where they probably reflect variations in the stability of an eruption column.

We have previously noted that pyroclastic density currents are strongly affected by topography. Because of their greater mobility, pyroclastic surges commonly override valley sides and may be traced into valley bottoms where denser pyroclastic flow deposits are ponded. The effect of topography was demonstrated well by the 18 May 1980 surge at Mount St Helens [37]. A major surge took place when a blast removed a huge part of the side of the volcano. The blast surge travelled both to the west over a series of ridges and valleys parallel to the flow direction, and to the east where the topographic grain is at right angles to the flow direction. The extra roughness caused by the terrain along the flow path to the east caused large amounts of friction, which slowed down the surge relative to that moving to the west and caused large variations in patterns of deposit thickness between the thinly mantled ridge tops and ponded valley bottoms. Pyroclastic surges are known to have overtopped mountain barriers of over 1 km relief, spreading deposits many tens of kilometres from their source.

Further reading

J.R.L. Allen (1985) *Principles of Physical Sedimentology.* George Allen & Unwin, London, pp. 234–5.

J.E. Costa & G.F. Wieczorek (1987) *Debris Flows/Avalanches: Process, Recognition and Mitigation,* Reviews in Engineering Geology VII, Geological Society of America, Boulder, CO.

A.M. Johnson (1970) *Physical Processes in Geology.* Freeman, Cooper & Co., San Francisco, Chapters 12–14.

G.V. Middleton & P.R. Wilcock (1994) *Mechanics in the Earth and Environmental Sciences.* Cambridge University Press, Cambridge.

K.T. Pickering, R.N. Hiscott & F.J. Hein (1989) *Deep Marine Environments: Clastic Sedimentation and Tectonics.* Unwin Hyman, London.

References

1 J.E. Costa (1988) Rheologic, geomorphic, and sedimentologic differentiation of water floods, hyperconcentrated flows, and debris flows. In: *Flood Geomorphology* (eds V.R. Baker, R.C. Kochel & P.C. Patton). Wiley, New York, pp. 113–22.

2 A.W. Skempton & J.N. Hutchinson (1969) Stability of natural slopes and embankment foundations, state-of-the-Art report. In: *Proceedings of the 7th International Conference on Soil Mechanics and Foundation Engineering, Mexico City*, Vol. 2 pp. 291–335.

3 D.J. Varnes (1978) Slope movements and types and processes. In: *Landslides: Analysis and Control* (eds R.L. Schuster & J. Krizek), Transportation Research Board Special Report 176. National Academy of Sciences, Washington DC, pp. 11–33.

4 T. Mulder & P. Cochonat (1996) Classification of offshore mass movements. *Journal of Sedimentary Research* **66**, 43–57.

5 H.H. Roberts, J.N. Suhayda & J.M. Coleman (1980) Sediment deformation and transport on low-angle slopes: Mississippi River delta. In: *Thresholds in Geomorphology* (eds D.R. Coates & J.D. Vitek). George Allen & Unwin, London. pp. 131–67.

6 A.B. Watts & D.G. Masson (1995) A giant landslide on the north flank of Tenerife, Canary Islands. *Journal of Geophysical Research* **100**, 24 487–98.

7 J.G. Moore, D.A. Clague, R.T. Holcomb, P.W. Lipman, W.R. Normark & M.E. Torresan (1989) Prodigious submarine landslides on the Hawiian Ridge. *Journal of Geophysical Research* **94**, 17 465–84.

8 P.P.E. Weaver, R.G. Rothwell, J. Ebbing, D. Gunn & P.M. Hunter (1992) Correlation, frequency of emplacement and source directions of megaturbidites on the Madeira Abyssal Plain. *Marine Geology* **109**, 1–20.

9 M.A. Carson & M.J. Kirkby (1972) *Hillslope Form and Process*. Cambridge University Press, Cambridge.

10 J.R.L. Allen (1984) Sedimentary Structures: their Character and Physical Basis. Elsevier, Oxford, 663pp.

11 G.V. Middleton & P.R. Wilcock (1994) *Mechanics in the Earth and Environmental Sciences*. Cambridge University Press, Cambridge.

12 A.M. Johnson (1970) *Physical Processes in Geology*. Freeman, Cooper & Co., San Francisco, Chapters 12–14.

13 R.A. Bagnold (1956) The flow of cohesionless grains in fluids. *Philosophical Transactions of the Royal Society, London* **249A**, 235–97.

14 C.S. Campbell (1990) Rapid granular flows. *Annual Review of Fluid Mechanics* **22**, 57–92.

15 T. Takahashi (1981) Debris flow. *Annual Review of Fluid Mechanics* **13**, 57–77.

16 H.M. Jaeger, S.R. Nagel & R.P. Behringer (1996) The physics of granular materials. *Physics Today* **49**, 32–8.

17 R.P. Sharp & L.H. Nobles (1953) Mudflow of 1941 at Wrightwood, Southern California. *Bulletin of the Geological Society of America* **64**, 547–60.

18 M.A. Hampton (1972) The role of subaqueous debris flow in generating turbidity currents. *Journal of Sedimentary Petrology* **42**, 775–93.

19 G.V. Middleton & J.B. Southard (1977) *Mechanics of Sediment Movement* Short Course Notes 3. SEPM Tulsa, OK.

20 E. Simpson & R.E. Britter (1979) The dynamics of the head of a gravity current advancing over a horizontal surface. *Journal of Fluid Mechanics* **94**, 477–95.

21 R.A. Bagnold (1962) Autosuspension of transported sediment; turbidity currents. *Proceedings of the Royal Society, London* **265A**, 315–19.

22 H.M. Pantin (1979) Interaction between velocity and effective density in turbidity flow: phase plane analysis, with criteria for autosuspension. *Marine Geology* **31**, 59–99.

23 K.T. Pickering, R.N. Hiscott & F.J. Hein (1989) *Deep Marine Environments: Clastic Sedimentation and Tectonics*. Unwin Hyman, London.

24 W.B. Dade & H.E. Huppert (1994) Predicting the geometry channelised deep sea turbidites. *Geology* **22**, 645–8.

25 B.C. Kneller & M.J. Branney (1995) Sustained high-density turbidity currents and the deposition of thick massive sands. *Sedimentology* **42**, 607–16

26 B.C. Kneller (1995) Beyond the turbidite paradigm: physical models for deposition of turbidites and their implications for reservoir prediction. In: *Reservoir Characterisation of Deep Marine Clastic Systems* (eds D.J. Prosser & A. Hartley), Special Publication. Geological Society of London, London.

27 G.V. Middleton & M.A. Hampton (1976) Subaqueous sediment transport and deposition by sediment gravity flows. In: *Marine Sediment Transport and Environmental Management* (eds D.J. Stanley & D.J.P. Swift). Wiley, New York, pp. 197–218.

28 A.H. Bouma (1962) *Sedimentology of Some Flysch Deposits: A Graphic Approach to Facies Interpretation*. Elsevier, Amsterdam.

29 D.R. Lowe (1982) Sediment gravity flows: II. Depositional models with special reference to the deposits of high-density turbidity currents. *Journal of Sedimentary Petrology* **52**, 279–97.

30 Branney, M.J. & Kokelaar, B.P. (1992) A reappraisal of ignimbrite emplacement: progressive aggradation and changes from particulate to non-particulate flow during emplacement of high-grade ignimbrite. *Bulletin of Volcanology* **54**, 504–20.

31 [Vertical column collapse was first identified in the 1929 eruption of Konagatake, Japan. The process was established quantitatively by Sparks and Wilson.]

R.S.J. Sparks & L. Wilson (1976) A model for the formation of ignimbrite by gravitational column collapse. *Journal of the Geological Society of London* **132**, 441–51.

32 R.S.J. Sparks, L. Wilson & G. Hulme (1978) Theoretical modelling of the generation, movement and emplacement of pyroclastic flows by column collapse. *Journal of Geophysical Research* **83**, 1727–39.

33 J.V. Wright, A.L. Smith & S. Self (1980) A working terminology of pyroclastic deposits. *Journal of Volcanology and Geothermal Research* **8**, 315–36.

34 P. Francis (1993) *Volcanoes: A Planetary Perspective.* Oxford University Press, Oxford.

35 L. Wilson, R.S.J. Sparks & G.P.L. Walker (1980) Explosive volcanic eruptions IV. The control of magma properties and conduit geometry on eruption column behaviour. *Geophysical Journal of the Royal Astronomical Society* **63**, 117–48.

36 R.S.J. Sparks, S. Self & G.P.L. Walker (1973) Products of ignimbrite eruptions. *Geology* **1**, 115–18.

37 R.V. Fisher (1990) Transport and deposition of a pyroclastic surge across an area of high relief: the 18 May 1980 eruption of Mt. St. Helens, Washington. *Bulletin of the Geological Society of America* **102**, 1038–54.

7 Jets, plumes and mixing at the coast

The fountains mingle with the river,

and the rivers with the ocean;

The winds of heaven mix forever

With a sweet emotion;

Nothing in the world is single;

All things, by a law divine,

In one spirit meet and mingle.

Why not I with thine?

Percy Bysshe Shelley (1792–1822), *Love's Philosophy* [1819]

Chapter summary

The mixing or mutual interference of fluids of different characteristics is important in a number of different settings and is also at the heart of many pollution problems. The mixing of river water with ocean water at the coast is of particular ecological, geochemical and sedimentological interest. Two dynamical situations involving, first, bouyancy, related to the density difference of the two fluids, and second, inertia from river outflows, are investigated. Mixing between the two fluid reservoirs inevitably leads to dissipation.

A narrow flow issuing into a large reservoir, such as the eruption of a column of hot ash into the atmosphere at a volcano, or the flow of a muddy stream into a deep lake, spreads conically as a jet. The velocity of fluid in a turbulent jet depends on its initial velocity, on the distance from the orifice and on the entrainment of fluid across its boundaries. The velocity structure of the jet controls, for example, the pattern of sediment deposition under it. The Rhein River jet in Lake Constance can be successfully modelled in this way and matched with sediment distributions on the floor of the lake. However, the dynamics at river mouths where they enter the ocean is considerably more complex because of the effects of bouyancy and those due to the bathymetry and energy of the receiving basin.

Where a turbulent jet is not free to expand in all directions because of a shallow sea floor, it is modified into a plane rather than axisymmetric jet. Plane jets are affected by friction on the sea bed; they decelerate more quickly and spread at much wider angles than axisymmetric jets, causing sediment to be restricted to regions relatively close to the outflow.

The most common situation is for the water of the river outflow to move as a light surface layer over the denser, salty seawater (hypopycnal flow). A salt wedge then intrudes up the river mouth and a fresh water layer flows outward as a plume. A densimetric Froude number is useful in explaining the dynamics. The efficiency of mixing of the two fluids allows estuaries to be classified as of salt wedge, mixed and partially mixed types. The low salinity plumes issuing from estuary mouths may travel for astonishing distances as relatively coherent water masses. Low salinity can be measured in surface waters 1000 km from the mouth of the Amazon River. Plumes may be strongly affected by Coriolis effects and by wind-driven currents, causing them to hug the coast for tens of kilometres instead of moving directly offshore.

The Mississippi outflow is presented as a case study of a mixing zone between bouyant fresh river water and a saline receiving basin. Wave and tidal energy in the Gulf of Mexico are both low, so the observed activity at the river mouth can be safely viewed in terms of inertial jet diffusion and buoyancy-driven flow. The outflow at South Pass has a strong salt wedge whose activity depends on the discharge from the river and flood–ebb tidal variations. A significant effect at South Pass is the formation of internal waves between the upper fresh layer and the lower, dense, salt wedge. This phenomenon strongly affects the outflow dynamics and the pattern of sediment deposition.

A large number of processes in the receiving basin may affect outflow dynamics, including tidal currents, surface wind-generated waves, wind-driven and other coastal currents, thermal stratification and, locally, the presence of ice. Strong tidal currents are particularly effective in mixing river and seawater thoroughly and obliterating any density stratification. High wave activity on steep-fronted, high-energy coasts may also enhance mixing and cause sediment deposition to be concentrated nearshore.

The water circulation and sediment deposition in estuaries have important implications for pollutant transport. Both are strongly affected by neap–spring tidal cycles, the flow pathways of flood and ebb currents and the mixing characteristics of the fresh and salt water. High sediment concentrations in estuaries with moderate to high tidal ranges are typically found just landward of the head of the salt intrusion in a so-called turbidity maximum. Its position and intensity are controlled by a number of fluid-mechanical factors, including the effects of mixing on flocculation. Flocculation promotes rapid settling of muddy aggregates during slack water.

7.1 Introduction to mixing phenomena

In many natural environments fluids of different characteristics meet. The extent to which they mix or mutually interfere is of considerable interest. Problems of mixing are encountered in a wide variety of natural environmental settings, such as the meeting of two rivers with different sediment loads at a confluence, the flow into a lake of a meltwater stream, the mixing of fresh and salty water in an estuary, the eruption into the atmosphere of a column of ash from a volcano. Mixing is also a problem in areas of environmental management such as pollution control. How fast, and over what area will a pollutant be carried when entering a water body such as a lake or sea? The main focus of this chapter, however, is the mixing phenomena taking place at the coast, or at the edges of lakes, particularly with respect to the transport and deposition of sediment. Two dynamical situations are studied: one is the bouyancy-driven flow resulting from density differences between fresh and salty water. This gives rise to a *low-salinity plume* moving over a lower salty layer. The second is the inertia-driven flow or river water into a reservoir such as a lake or sea, called a *jet*. Both of these dynamical situations have profound sedimentological implications.

The problem of mixing between two fluids has been met in the consideration of the flow of turbidity currents in the previous chapter. It was seen that under some circumstances instabilities develop at the interface between two fluids as a result of the interfacial shear. These Kelvin–Helmholtz instabilities may grow into vortices which result in a drawing in of ambient fluid, thereby mixing the two fluid masses. Alternatively, the instabilities at the interface may be dampened by the density contrast. The tendency of the gravitationally stable density gradient to stabilize the interface is described by the Richardson number (Chapter 5). Whether two fluids retain their integrity or are completely mixed is of great importance for a number of interrelated reasons:

- mixing leads to a loss of the density differences which drive fluid motion;
- mixing causes flow expansion by the entrainment of one fluid mass into the other, leading to deceleration;
- mixing may cause geochemical changes, leading to processes such as precipitation, solution or flocculation.

Mixing phenomena are complex, and detailed treatment of the physics involved is beyond the scope of this introductory text. Those who are interested may wish to consult more specialized texts in the fields of fluid mechanics and physical oceanography. The approach in this chapter is to introduce some simple physical models so that the important parameters are recognized. These oversimplified physical models are then brought to life with examples from the real world. In so doing, some of the complexities as well as the generalities become clear.

7.2 Model of a turbulent axisymmetric jet

When a narrow flow of fluid enters a large reservoir it expands into a conical shape and mixes with the ambient water. The stream of fluid issuing from a narrow orifice is known as a jet (Fig. 7.1). Jets are of considerable importance to the distribution of sediment from river mouths into lakes, lagoons, seas and oceans, but perhaps the best way to visualize turbulent jets is to study their development as vertical discharges of hot gas and tephra from the narrow vents of volcanoes into the infinitely large atmosphere. Such a flow is termed *axisymmetric*, since the jet is symmetric about an axis, which in this case is vertical. The modelling of an axisymmetric turbulent jet allows us to ignore the complicating effect of a

(a)

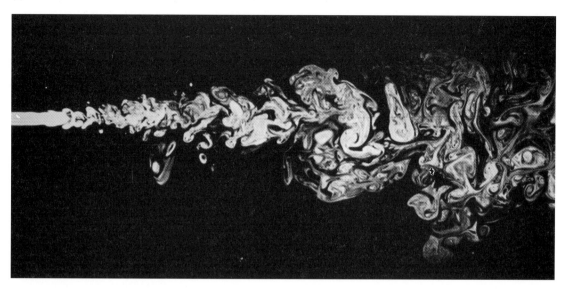

(b)

Fig. 7.1 Visualization of an axisymmetric jet. In (a) the turbulent structure and spreading of a jet of fluid directed downwards into a reservoir of water is shown by laser-induced fluorescence. The Reynolds number is approximately 2300. The development of instability and turbulence in an axisymmetric jet is demonstrated by (b). A laminar stream of air marked with smoke leaving a circular tube at a Reynolds number of 10 000 develops vortices at its edge, then abruptly becomes turbulent. After Van Dyke (1982) [1].

'solid' boundary such as the sea bed. This proves to be a very useful starting point for the more complex situation found in nature at river mouths.

Consider a jet with a circular cross-section of radius r_0 at its origin and a uniform velocity at its origin of U_0 measured along the x axis (Fig. 7.2). For a certain

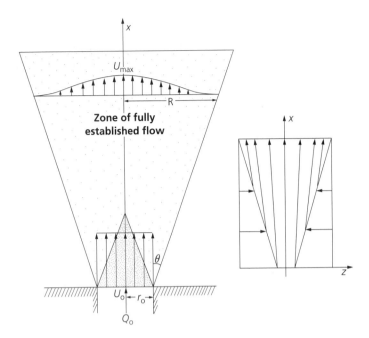

Fig. 7.2 A jet issuing from a circular orifice of radius r_0 with a velocity U_0 spreads with an angle θ. The inset shows time-averaged fluid motion where entrainment of ambient fluid is allowed.

distance from the origin, which is called the *development region* (approximately five times the radius at the origin), the velocity of the fluid comprising the jet remains at the value U_0, but with increasing distance from the origin the fluid moves with a decreasing velocity, but with a maximum along the centreline of the jet. The time-averaged velocity, U, therefore varies both longitudinally with distance from the origin (x) and radially across the jet.

We first assume that there is no flow through the boundary of the jet, in the form of either fluid escaping from the jet or ambient fluid being drawn in. Then we can state that the discharge of the jet must be constant along cross-sections at different distances from the origin along x. The discharge at the origin is the product of the cross-sectional area of the orifice and the velocity at the origin:

$$Q_0 = \pi r_0^2 U_0 \tag{7.1}$$

The discharge at any distance x is the product of the mean velocity, $\acute{U}$, and the cross-sectional area along that cross-section. The radius, R, at x can be found by assuming a constant spreading angle θ, so that

$$R = x \tan \theta = \varepsilon x \tag{7.2}$$

Consequently, equating the two discharge relations,

$$\hat{U} = \frac{r_0^2 U_0}{R^2} = \left(r_0^2 U_0\right)\left(\frac{1}{(\varepsilon x)^2}\right) \tag{7.3}$$

which shows that the mean velocity of fluid in the jet varies inversely with the square of x. That is, there is a rapid deceleration of the jet.

But this model of a jet with impervious boundaries is clearly an oversimplification. Jets become diluted with ambient fluid, so there must be an inflow through the boundary of the jet. That this inflow takes place can be easily visualized from the vortices which commonly develop along the zone of mixing (Fig. 7.1). What effect does the drawing in of surrounding fluid have on the dynamics of the jet?

Rather than conserve discharge, we conserve the flux of momentum of the jet along cross-sections situated at any distance x from the origin. The momentum is the product of the mass of fluid moving through a cross-section and its mean velocity. The momentum flux is the transport rate of the momentum. At the origin the momentum flux is therefore

$$M_0 = (\pi r_0^2 \rho U_0) U_0 \tag{7.4}$$

where ρ is the fluid density. The momentum flux at any distance x is given by

$$M = (\pi R^2 \rho \acute{U}) \acute{U} \tag{7.5}$$

Equating the two fluxes of momentum gives, after some rearrangement,

$$\hat{U} = \left(r_0 U_0\right)\left(\frac{1}{\varepsilon x}\right) \tag{7.6}$$

showing that the mean velocity now varies simply inversely with x. The flux of ambient fluid into the jet causes it to decelerate less rapidly than in the impermeable model above. The difference between the mean velocities as calculated using equation (7.3) and equation (7.6) must reflect the effect of the drawing in of ambient fluid. It is therefore an indication of the relative discharges of the entrained fluid and the orifice fluid. Since the difference must get larger with increases in x, it can be seen that the discharge of entrained ambient fluid must become progressively more important. The relative discharge from the orifice therefore becomes increasingly small with distance from the origin, which explains the progressive dilution of turbulent jets.

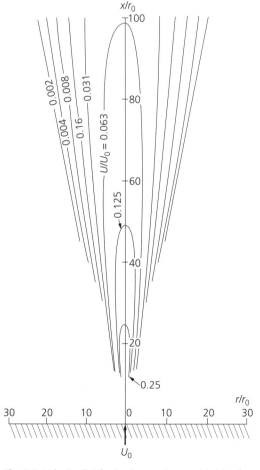

Fig. 7.3 Velocity distribution in an axisymmetric jet, where the distance is scaled by the orifice radius and the velocity is relative to the outlet velocity. After Allen (1985) [2].

If the cross-sectional variation of velocity is, for example, a cosine function or has the form of a normal (Gaussian) probability function, and if the mean velocity can be calculated from equation (7.6) as a function of distance from the origin, the distribution of velocity can be drawn throughout the jet by scaling the distance by r_0 and the velocity by U_0. The pattern for an axisymmetric jet is shown in Fig. 7.3. This clearly has major implications for the ability of the jet to transport its sediment load.

The dynamics of jets will be returned to in considering particulate sediment transport and deposition at river mouths in Section 7.3.1.

7.3 River outflows

The discharge of rivers into the ocean has major effects on coastal waters. These effects can be simplified as those originating from (i) the reduction of salinity as fresh river water is discharged into and mixed to varying degree with ambient seawater; (ii) the introduction of particulate and dissolved matter, such as sediment, pollutants, organics and nutrients; high sediment discharges reduce the optical transparency of coastal waters and cause sediment shoals to develop, thereby affecting navigation. The dynamics of river outflows is therefore crucial to an understanding of the physics, chemistry and productivity of coastal waters. The same dynamics control the fate of the vast quantities of sediment delivered to the ocean from continental denudation (Chapter 3). Conse-quently, outflow dynamics determine the sedimentary record preserved in ancient deltaic deposits.

7.3.1 Dynamics at river mouths

The dynamics of the outflow of water and its sediment load at river mouths are controlled by a large number of factors, chief of which are the characteristics of the outflowing river water, the coastal bathymetry and the energy of the receiving marine basin. Outflow dynamics vary not only between different river mouths, but also at the same river mouth during periods of low and high discharge. The particular dynamics involved at the river mouth controls the way in which the effluent is dissipated and therefore the patterns of sediment deposition on the estuary, fjord or sea bed. Outflow dynamics therefore fundamentally control delta morphologies and their internal sedimentary architecture.

Although effluents rarely can be characterized by one simple process, it is a convenient starting point to think of two primary end-members which may dominate in any river mouth setting where the river

(a) **Inertial jet**

Plan view

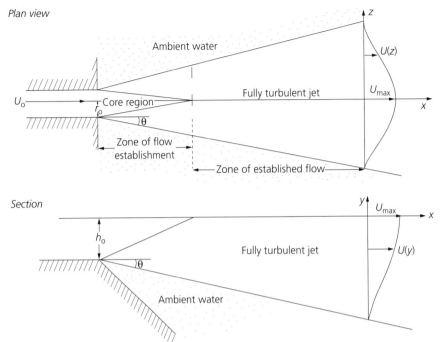

Section

(b) **Buoyant effluent**

Plan view

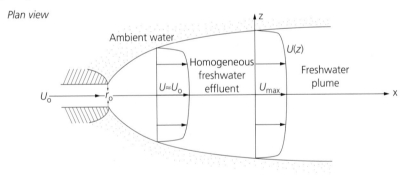

Section

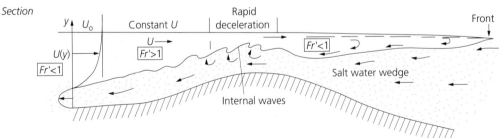

Fig. 7.4 Two end-member situations for outflow dynamics: (a) inertial turbulent jet diffusion; (b) density-driven buoyant plume. Densimetric Froude numbers are added to (b). After Wright (1977) [3].

outflow is more important than the wave and tidal energy of the receiving basin (Fig. 7.4): (i) behaviour as a turbulent jet, driven by *outflow inertia*; and (ii) flow dominated by the density difference (*buoyancy*) between the inflow and the ambient water.

The theory of an idealized axisymmetric jet was introduced in Section 7.2. This is now adapted to the case of a river outflow. The case of a buoyancy-driven flow is similar to the examples of density currents given in Chapter 6. Buoyancy effects are then discussed and are found to be important in the case study of the Mississippi delta that follows in Section 7.3.2.

General considerations

Where there is no density difference between the effluent and the ambient water, the structure of the jet should be influenced by the inertial forces of the outflowing water compared to the retarding viscous forces. That is, jet structure must be controlled at least in part by a Reynolds number. In this case the Reynolds number should have the form

$$Re_o = \frac{\rho U_o \left[h_o (b_o / 2)^{0.5} \right]}{\mu} \qquad (7.7)$$

where the subscripts refer to the outlet, U is the mean outlet velocity, h is the outlet depth, b its width, and the square root is required to give the term in square brackets (half of the rectangular cross-sectional area) the dimensions of length. Submerged axisymmetric jets are fully turbulent at Reynolds numbers greater than about 3000 [3]. Most rivers should have fully turbulent jets, as can be appreciated by inserting some typical parameter values in equation (7.7). The dynamic viscosity of water at 18°C is 1.06×10^{-3} Pa s, and the density of fresh water is 1000 kg m^{-3}. The kinematic viscosity $v = \mu/\rho$ is therefore in the region of 1×10^{-6} m^2 s^{-1}. Since μ and ρ do not vary greatly, except in the case of very high suspended sediment loads, it can be seen that river outflows would need to be very small and sluggish for turbulence not to take place.

However, the dispersion of a river outflow is also affected by its buoyancy with respect to ambient seawater. If the outflowing water moves as a light hypopycnal flow over a denser, intruding *salt wedge*, the dynamics differs considerably from the case of an inertial turbulent jet. Whenever there is a problem in which gravity acts on an interface, such as the free surface of a river, or the upper interface of a density current, we should expect a Froude number to be applicable. In this case the densimetric Froude

number expresses the tendency for inertial forces to overcome the buoyancy forces set up by the density stratification. It has the familiar form

$$Fr' = \frac{U}{\sqrt{g' h_1}} \qquad (7.8)$$

where $g' = (\Delta \rho / \rho_1) g$ is a 'reduced' or density-modified gravity, h_1 is the depth of the density interface, and the subscripts 1 and 2 refer to the plume and underlying ambient water, respectively [4]. If inertial forces are dominant Fr' should be high, and the opposite is true if buoyancy dominates over inertial turbulent jet diffusion. Inspection of equation (7.8) shows that when there is no density stratification, $\Delta \rho \to 0$ and $Fr' \to \infty$. Laboratory experiments suggest that when the densimetric Froude number is less than about 16, buoyancy forces become increasingly important.

As in so many previous examples, therefore, there are two dimensionless products which are valuable in describing the dynamical situation at river mouths: a Reynolds number with a length term referring to the outlet dimensions; and a Froude number modified by the density contrast between the plume and ambient fluids.

We can now proceed to consider the particular cases of, first, turbulent jet diffusion and, second, low-salinity, buoyancy-driven plumes.

Turbulent jets at river mouths

How does the situation at river mouths differ from the simple axisymmetric model presented in Section 7.2? The essential question is whether the jet is free to expand in all directions, or whether the bottom boundary forces the jet to expand only laterally [5]. The latter are termed *plane jets*. We might expect the generally shallow water depths seawards of most river mouths to favour the development of plane jets. Axisymmetric jets may, however, occur where streams debouch into steep-sided lakes or reservoirs.

We already know that an axisymmetric jet spreads at a constant angle and that where entrainment takes place across a permeable jet boundary, the mean velocity also varies linearly with distance from the orifice or outlet. The rate of expansion of the jet is crucial to understanding the lateral distribution of sediment, ignoring for the moment the effects of sediment transport by waves and other marine processes. The rate of expansion for a fully turbulent plane jet ε is a constant equal to 0.22 [6], corresponding to an angle between the centreline and the jet boundary of 12°. Jets of this type, therefore, expand at low rates, and their maximum velocities (along the centreline) also decrease slowly with distance from the outlet (Fig. 7.5a). Consequently, inertial turbulent jets should transport their suspended sediment load relatively far into the ambient water but over a laterally restricted area.

A number of laboratory studies suggest that the transverse velocity profile either side of the centre-line

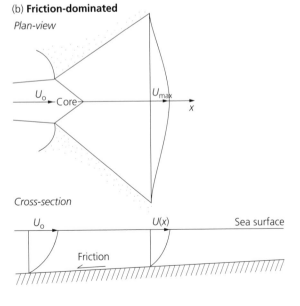

(a) **Inertia-dominated**

Plan-view

Cross-section

(b) **Friction-dominated**

Plan-view

Cross-section

Fig. 7.5 Contrast between (a) an inertia-dominated (near-axisymmetric) jet and (b) a plane jet with sea bed friction. The friction-dominated case is characterized by more rapid deceleration, and a wider spreading angle. After Wright (1977) [3].

of a turbulent jet is approximated by a normal distribution function. Consequently, in the zone of established flow, the longitudinal velocity (directed along x) at a point (x, z) relative to the orifice can be found [7] from

$$U_{x,z} = U_o \alpha \left(\frac{r_o}{x}\right)^{1/2} \exp\left(\frac{-z^2}{2\sigma^2}\right) \qquad (7.9)$$

where r_o is the half-width of the river mouth, σ is the standard deviation of the Gaussian velocity distribution, which experiments show to be a simple function of x ($\sigma = 0.109x$), and α, from inspection of equation (7.6), must be inversely related to the spreading angle, ε, and for fully turbulent jets is equal to 2.275. The maximum velocity is along the centreline of the jet ($z = 0$), so the maximum velocity relative to the outlet velocity is

$$\frac{U_{max}}{U_o} = \alpha\left(\frac{r_o}{x}\right)^{1/2} \qquad (7.10)$$

The Rhein delta The best examples of turbulent jets are likely to be found where rivers enter hydrologically open lakes. This is for two main reasons. First, the density of the river water is likely to be similar to the density of the fresh water of the lake. Second, there is unlikely to be major wave, tide and current activity in the lake to modify the outflow dynamics. An especially interesting case is provided by the outflow of the Rhein River into the south-eastern part of Lake Constance close to the border of Switzerland, Austria and Germany [8] (Fig. 7.6). The entry point of the Rhein River into the lake was artificially changed in 1900, so the rate of sediment deposition and delta progradation is very well constrained. The channel width is 200 m, its depth 4 m, and the average velocity at the mouth 0.28 m s^{-1}. The Reynolds number of the jet suggests that it should be fully turbulent.

The delta front has advanced steadily into the lake from its new outlet since 1900, prograding a distance of nearly 2 km in about 60 years [6] into a water depth of approximately 60 m. The relatively steep nearshore slopes ensure that friction on the lake bed is of minor importance. There is very little riverine sediment beyond about 3 km from the present position of the outlet, suggesting that, if we neglect resedimentation downslope on the lake floor, within this distance the flow velocities in the jet have diminished to levels unable to keep fine sediment in suspension. This distance is 15 times the channel width at the outlet, and water depths are 15 times the average depth at the outlet.

Incorporation of friction on the sea bed The dynamics of plane jets must be affected by friction on their lower boundaries. In plane jets, where there is no density stratification, a balance must be achieved between the inertial forces of the jet and the frictional forces on the bed. If friction on the other (frontal and lateral) boundaries of the jet is neglected, and lateral entrainment is also ignored, the mean velocity (averaged through the water column to the sea bed) of the

---

Practical exercise 7.1: Velocity field of a turbulent jet

1 Using the Rhein River effluent in Lake Constance, calculate the longitudinal velocity at a point 3 km from the outlet along the centreline of the jet. The velocity in a fully established turbulent jet is given in equation (7.9). What is the velocity at a lateral distance of 1 km from the centreline at $x = 3$ km?

2 Along the centreline of the jet, at what distance from the outlet has the maximum velocity decreased to 1/e (0.368) of the outlet velocity?

Solution

1 At 3 km from the outlet, the maximum longitudinal velocity from equation (7.9) with $U_o = 0.28$ m s^{-1} and $r_o = 100$ m is 0.12 m s^{-1}. Velocities should decline transversely from the maximum longitudinal velocity with a normal distribution function. At a radius of 1 km from the axis at a distance of 3 km from the outlet the velocity from equation (7.9) is only 1.2 mm s^{-1}, which is below the settling velocity of all but the very finest sediment. It is expected, therefore, that riverine sediment should be dispersed within an arc up to 2 km wide at a distance of 3 km from the outlet. This agrees reasonably well with the distribution of sediment on the lake floor (Fig. 7.6). Throughout this analysis, however, it should be remembered that it has been assumed that the lake water is entirely inert, that friction can be ignored, and that there are no resedimentation processes taking place on the delta front. Even so, there is sufficient agreement to suggest that the Rhein outflow dynamics is essentially described by the diffusion of a turbulent jet.

2 From equation (7.10) the longitudinal velocity along the centreline of the jet has decreased to 1/e of its outlet velocity in a distance of 3.8 km. We might think of this as the characteristic length-scale of the jet.

(a)

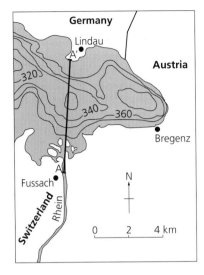

(b)

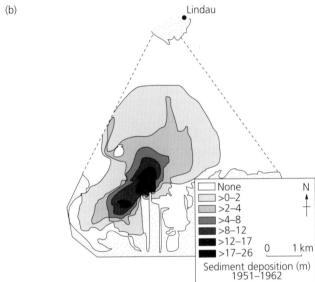

(c)

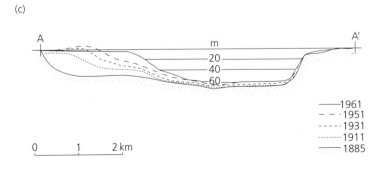

Fig. 7.6 (a) Location map of Lake Constance (Bodensee), showing the entry point of the Rhein. Contours are lake bottom elevation above sea level. A–A' is section shown in (c). (b) Thickness of deltaic sediments supplied by the Rhein River between 1951 and 1961. (c) Cross-section showing progradation from the south of the delta over the period 1885–1961. After Müller (1966) [8].

jet seawards of the outlet will vary according to the distance from the outlet, the frictional resistance of the sea bed and the water depth seawards of the outlet. Scaling the mean velocity at some distance from the outlet by the velocity at the outlet gives [4]

$$\frac{\hat{U}}{U_0} = e^{-K(x/\bar{h})} \qquad (7.11)$$

where K is a friction coefficient related to the Chézy coefficient, C, or alternatively the friction factor, f,

$$K = \frac{g}{C^2} = \frac{f}{8} \qquad (7.12)$$

and $\bar{h}$ is the average water depth out to the horizontal distance x. Similarly, assuming conservation of discharge with no lateral entrainment of ambient fluid, the expansion of the jet can be calculated by scaling the radius at any distance x by the radius at the outlet

$$\frac{r_x}{r_0} = \frac{h_0}{h_x} e^{K(x/\bar{h})} \qquad (7.13)$$

where r_0 and h_0 are the half-width and depth of the outlet, and r_x and h_x are the half-width and water depth at x, showing that the expansion depends on the water depth seawards of the outlet and the frictional characteristics of the sea bed. A typical value of the friction coefficient for the sea bed is 2×10^{-3}.

The fluid motion in the friction-dominated jet is responsible for the transport in suspension of sediment until the point is reached where the settling velocity of the grains exceeds the shearing velocity of the jet fluid. In addition, the shear stress on the sea bed transports sediment seawards as bedload. These two processes result in a seaward fining of deposited sediment under the frictional jet.

The effect of sea bed friction is therefore to cause a marked lateral expansion and increase in deceleration of the plane jet (Fig. 7.5b). Sediment should be distributed laterally over a wide distance, but should be restricted to a zone close to the outlet. Friction-dominated jets are likely to occur where rivers with significant bedload sediment transport rates debouch into areas of low submarine slopes.

Buoyancy-driven outflows

The mixing of seawater and fresh water generally starts in estuaries (Section 7.4). The presence of a salinity and therefore density gradient along the estuary results in a pressure difference within the water. This in turn drives a surface seaward flow of low-salinity water and a deeper landward flow of saline water. Tidal currents flowing through the estuary (Chapter 8) cause a mixing of the different salinity layers. The interaction of the salinity-driven flow and the tidal currents results in the particular types of estuarine circulation discussed in more detail in Section 7.4:
• **salt wedge estuaries** are characterized by the dominance of the density-driven flow, resulting in a surface seaward flow of fresh water over an intruding wedge of salty water;
• **mixed estuaries** are characterized by complete vertical mixing by tidal currents;
• **partially mixed estuaries** have a slight salinity increase with depth and a net seaward flow in the upper layer and net landward flow in the lower layer.

In sufficiently wide estuaries, the seaward and landward flows may be separated by the topographic effects of deep channels and shallow shoals, and by the Coriolis force causing a deflection to the right (in the northern hemisphere, looking towards the sea) of the outflowing surface water and a deflection to the left-hand side of the estuary (still looking towards the sea) of the upstream-moving salty water.

Beyond a salt wedge or stratified estuary, the low-salinity water may flow out as a *low-salinity plume*, separated from the ambient seawater by a pronounced density gradient known as a *front* (Fig. 7.7). Fronts develop in general between water masses of different properties, but are particularly common between seawater and outflowing plumes. Fronts can be recognized by colour differences due to the content of phytoplankton or suspended particles, and by concentrations of foam or floating detritus. Mixed estuaries, however, cannot release low-salinity plumes since the density gradient has already been broken down within the estuary. Estuarine water is then mixed with seawater by turbulent diffusion. Consequently, estuarine circulation strongly affects coastal water and sediment dispersal.

Low-salinity plumes In the mouth of the Mississippi River (see also Section 7.3.2) a salt wedge extends more than 300 km upstream during low-discharge conditions. During high discharges the salt wedge is pushed completely out of the estuary and the river water flows into the Gulf of Mexico as a sediment-laden plume (Plate 7.1, facing p. 204). The plume affects seawater out to the edge of the continental shelf, situated some 50 km from shore. Low-salinity water can be identified as much as 1000 km from the mouth of the Amazon River.

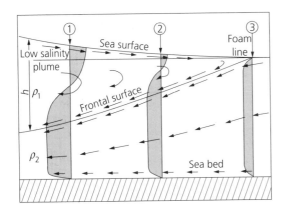

Fig. 7.7 Streamlines and velocity profiles for a low-salinity plume with associated front (schematic). After Bowman & Iverson (1978) [9].

The Columbia River enters the Pacific Ocean from its partially mixed estuary situated in the north-west USA (Fig. 7.8). The upper layer flows as a plume into the ocean and affects salinities over a very extensive area. However, there are marked seasonal variations in the distribution of salinity because the plume is very strongly affected by wind-driven currents (see also Chapter 9). The same is essentially true in the case of the Hudson River entering the Atlantic Ocean. Here, the plume hugs the coast during periods of strong winds when the plume behaviour is almost entirely controlled by prevailing winds.

Clearly, if we wish to know something of the distribution patterns of river-derived sediment or pollutants, it is necessary to study plume dynamics. In reality, this can be a very complex enterprise, since it is necessary to incorporate the effects of jet diffusion as well as to account for the wave, tide and wind-driven motion of coastal waters [7,10,11]. Although Coriolis effects and background energy of the receiving basin are important modifying factors, plume dynamics are controlled fundamentally by density contrasts caused by salinity differences. The low-salinity river water has a free surface which because of the buoyancy of the light water is elevated relative to dense seawater. A sharp front may separate the elevated low-salinity water from the ambient seawater (Fig. 7.7). The main driving force for the lateral spreading of the plume is therefore a horizontal pressure gradient, rather than inertia. In this way plumes are density currents in the sense of Chapter 6, and are quite distinct from inertia-driven jets.

As the plume spreads over seawater it becomes thinner and its elevated surface falls. If we assume that

the plume is driven by the pressure gradient, which is consequent on the lateral density contrasts, and that lateral mixing is small, the spreading of the plume should depend on two parameters: the relative density contrast with the underlying water

$$\zeta = \frac{\rho_2 - \rho_1}{\rho_1} = \frac{\Delta\rho}{\rho_1} \qquad (7.14)$$

where ρ_1 and ρ_2 are the densities of the plume and underlying water, respectively; and the densimetric Froude number

$$Fr' = \frac{U}{\sqrt{g'h}} \qquad (7.15)$$

where $g' = \zeta g$ is the reduced (density-modified) gravity, U is the mean velocity of the upper layer and h is the depth to the interface with the underlying water. The elevation of the plume water above the surrounding water is determined by the relative density difference and the plume thickness (ζh).

This theory has been applied to South Pass, the main distributary channel and associated plume at the Mississippi delta [12]. At the mouth of South Pass during low to normal river discharges, the breadth of the plume was 240 m, its thickness 1–2 m, velocity 0.8–1.2 m s^{-1} and relative density contrast 0.014–0.020. This gives a densimetric Froude number of between 0.8 and 1.2, that is, at about its critical value. As the plume spreads, the plume thickness decreases, causing it to become supercritical, leading to the formation of *internal waves* on the interface and to much increased mixing with seawater. This dilutes the plume and slows it down. At a distance from the mouth of eight times its initial width, the plume had increased its width by a factor of between 6 and 8 and decreased its thickness by a factor of about 4. The Mississippi River outflow at South Pass is therefore a good place to study plume dynamics.

7.3.2 Mississippi outflow case study

The Mississippi River, which drains an area of over 3×10^6 km^2 in North America, is one of the world's largest rivers, discharging between 8400 and 28 000 m^3 s^{-1} of water into the Gulf of Mexico during low and flood stages, respectively. In addition, it delivers approximately 5×10^{11} kg of sediment annually to the sea, mostly in the form of suspended load. It is an excellent example of river outflow dynamics because the tidal energy of the Gulf of Mexico is low, and wave energy reaching the delta front is attenuated by the shallow offshore slopes. Consequently, we can rule

(a)

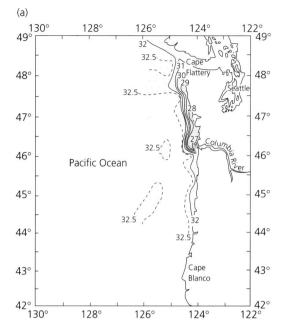

(b)

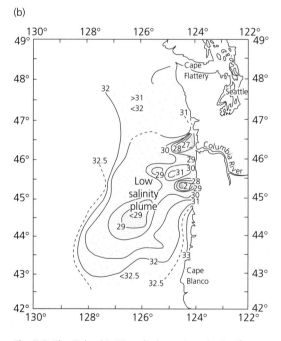

Fig. 7.8 The Columbia River discharge into the Pacific Ocean. Surface salinities vary strongly between winter, when the low-salinity plume is forced northwards tight to the coast (a), and summer, when an extensive plume spreads into the Pacific Ocean (b). After Duxbury (1965) [13].

out processes other than inertial jet diffusion and buoyancy-driven flow as being significant in this case.

The density (salinity) structure and flow velocities at the mouth of South Pass during a period of low river discharge are shown in Fig. 7.9. The zone of rapid density gradient, or *pycnocline*, is clearly visible, separating outflowing fresh water in an upper layer from a landward-moving salt wedge. The density difference is just over 2%. The densimetric Froude number at the channel mouth is about 1, and decreases upstream as the depth of the fresh upper layer increases. The upper layer thickness and density contrast appear to adjust themselves in order to maintain a densimetric Froude number of about 1 at the mouth. This is because at $Fr' > 1$ there is interfacial mixing, which erodes the top of the salt wedge and lowers the density contrast with the upper layer. This self-regulating process always seems to result in densimetric Froude numbers of about 1 when a salt wedge is present in the lower reaches of the South Pass channel. The position of the salt wedge is also affected by tidal (flood–ebb) flow variations, intruding during flood tides, but being displaced out of the channel mouth during ebb tides. Salt wedge intrusion is also hampered by high flood water discharges.

The activity of the salt wedge has a major effect on sediment transport. Suspended sediment is concentrated near the pycnocline, probably as a result of increased turbulence due to interfacial mixing, upward saltwater entrainment and flocculation of colloidal material as it comes into contact with saline water. The intrusion of the salt wedge also appears to be associated with increased discharges through crevasse channels along the main distributary route (Plate 7.1, facing p. 204), producing small crevasse deltas.

At low to normal discharges, the outflow from South Pass is in general agreement with predictions from buoyancy theory. This is expected from the well-developed stratification suggested from the low densimetric Froude number of about 1. Both the seaward expansion and seaward thinning of the plume match theory reasonably well. However, a significant effect found at South Pass, and thought to occur in other river outflows, is that seawards of the river mouth there is increased mixing by the breaking of internal waves where densimetric Froude numbers are highest (Fig. 7.10). This causes a marked seaward deceleration and thickening of the plume, reducing its densimetric Froude number and the attainment of subcritical conditions. This in turn reduces the mixing across the pycnocline and consequently causes a sharpening of the density interface. The plume then moves out into

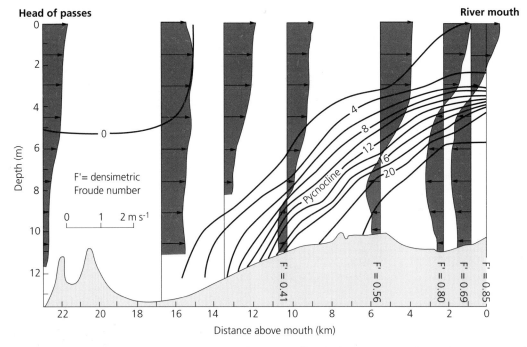

Fig. 7.9 Salinity structure (measured as density in excess of 1000 kg m⁻³) and velocities between the South Pass river mouth and Head of Passes at a time of low to normal discharge. Densimetric Froude numbers are superimposed. At low stage the pycnocline is well developed between the overriding fresh water plume and the underlying salt wedge. After Wright & Coleman (1974) [4].

the Gulf of Mexico, spreading and thinning. Increased mixing may result from higher than average wind, ware and tidal current energies.

The deceleration of the plume causes the sediment load to be deposited. Measurements of suspended sediment concentrations show a major seaward decrease in the region of increased mixing caused by the breaking of internal waves. This effect is particularly well seen in the suspended sediment concentrations at the pycnocline. The coarser suspended sediment is primarily deposited in the zone of initial flow expansion leading up to the zone of internal wave mixing. This is the position of a distributary mouth bar with its crest elevated 5 m above the base of the outlet (Figs 7.9 and 7.10). The finer suspended sediment is carried in the plume over the bar and falls to the sea bed beyond the bar as the plume velocity decelerates.

Densities of plume water are remarkably uniform in the direction transverse to the effluent axis. Although internally rather homogeneous, there is an abrupt velocity gradient at the margin of the plume. This is further support for the idea that the outflow is dominated by a low-salinity buoyant plume rather

than by turbulent jet diffusion during times of low and normal discharges.

During high river discharges the buoyancy effects are dominated by those of inertia and friction. High discharges flush the salt wedge out of the distributary mouth (Fig. 7.11), the spreading and deceleration of the outflow being dominated by the inertia of the jet and friction on the sea bed. Average (depth-mean) velocities during high discharge are up to 2 m s⁻¹. Since the mouth of the distributary at South pass is about 240 m wide and 1 m deep, the outlet Reynolds number from equation (7.7) is greater than 2×10^8, demonstrating the existence of a fully turbulent outflow. Surface water velocities measured close to the bar crest are weak compared to the velocity at the outlet, showing that the jet decelerates strongly as a result of turbulent mixing between the outflow and ambient water. Velocities also decrease transversely across the outflow, as predicted by the turbulent jet model. A major difference between high and low discharges is the shear on the bed in the region between the mouth and the bar. During high discharges there is a seaward bottom current in this region since

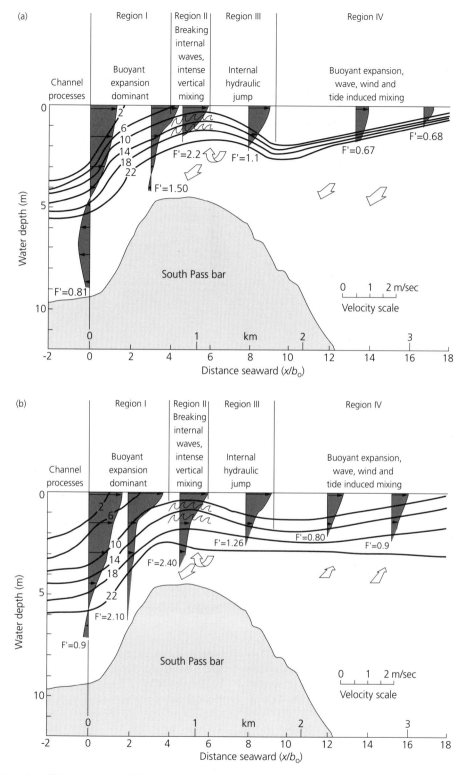

Fig. 7.10 Density (salinity) structure and flow pattern immediately seawards of the distributary mouth of South Pass during low discharge: (a) flooding tide; (b) ebbing tide. Hydrodynamic regions of Wright & Coleman (1974) [4] are shown. Note the intense vertical mixing due to internal waves and supercritical densimetric Froude numbers in region II. This causes deposition of sediment to form the South Pass bar.

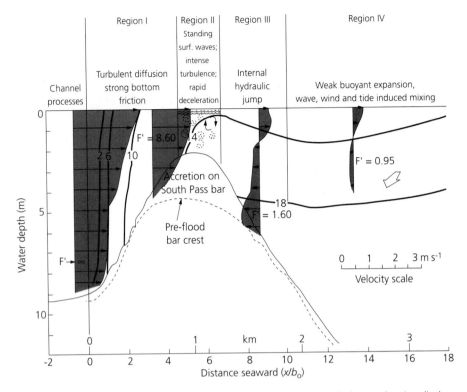

Fig. 7.11 Density (salinity) structure and flow pattern at South Pass during high river discharge, showing displacement of salt wedge out of the river mouth, and accretion of sediment on South Pass bar. After Wright & Coleman (1974) [4].

the salt wedge has been pushed seawards. Consequently, bedload is transported out of the river mouth and on to the landward facing flank of the bar during high discharges. Seawards of the bar, the density stratification is weaker than in the case of low to normal discharge, but the outflow appears to behave as a buoyant plume in this region.

The South Pass outlet of the Mississippi therefore varies in is outflow dynamics between low–normal and high discharge periods (Fig. 7.12).

7.3.3 Modifying marine processes
The preceding discussion applies to the situations at river mouths where the outflow is dominated by either inertia or buoyancy and other effects are negligible. However, a large number of marine processes may affect river outflow dynamics, for example:
• tidal currents;
• surface gravity waves;
• internal waves;
• coastal and nearshore currents;
• wind-driven currents;
• thermal stratification;
• sea ice.
Some of these processes are more important than others. In the following paragraphs the impact of tides and waves on river outflows is considered briefly. Further details can be found in Chapter 8. Wind-driven currents are outlined in Chapter 9.

Tidal currents
Tidal currents have a major role in the mixing of water in estuaries and thereby control plume activity. However, they may also have a strong effect on the amount and pattern of sediment transport in the river mouth region. An index of the strength of tidal currents is the tidal range of the sea into which the river enters. A number of large rivers enter macrotidal (range in excess of 4 m) seas, such as the Ganges–Brahmaputra in the Bay of Bengal (Plate 7.2, facing p. 204), the Klang in Malaysia, and the Ord in Australia. Tidal currents are typically very powerful, commonly far exceeding the river discharge. They mix the sea and river water thoroughly, obliterating any possible

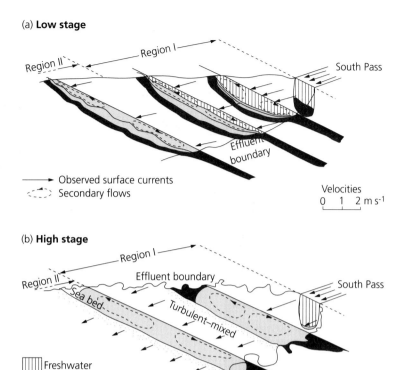

(a) **Low stage**

Region II

Region I

South Pass

Effluent boundary

⟶ Observed surface currents
⟨⌒⟩ Secondary flows

Velocities
0 1 2 m s⁻¹

(b) **High stage**

Region I

Region II

Effluent boundary

South Pass

Sea bed

Turbulent–mixed

Freshwater

Seawater

Mixed freshwater and sea water

Ambient freshwater

Velocities
0 1 2 3 m s⁻¹

Fig. 7.12 Comparison of effluent dynamics at (a) low and (b) high stages. During low stage the effluent behaves as a homogeneous buoyant outflow with a well-developed low-salinity plume and intruding salt wedge. At high (flood) stage, the effluent has characteristics of a frictional turbulent jet with enhanced mixing of seawater and riverine fresh water. Modified from Wright & Coleman (1974) [4].

density stratification.

Sediment transport is strongly bidirectional longitudinally up and down tidal channels deeply penetrating the coast. Sediment is typically deposited in linear sand ridges separated by deep tidal channels.

Surface gravity waves

River outflows entering protected seas, lagoons and lakes suffer little modification by surface waves because of their small fetch, or because shallow offshore slopes attenuate wave energy by friction, leading to low wave energies at the coastline. Some outflows, however, enter large seas with steep offshore slopes and are strongly affected by waves. The Senegal delta of western Africa is an example. Incoming waves interact with outflowing plumes, enhancing mixing between the plume and ambient fluid. This in turn causes deceleration. High wave activity may be so efficient at mixing that low-salinity plumes are destroyed within a few hundred metres of the river mouth. Rapid deceleration leads to deposition of the sediment load

on the sea bed. Shoaling waves may redistribute this sediment shorewards (Chapter 8), causing the river mouth to be blocked, and, where there is a longshore component to the net sediment transport, forcing the main channel to be displaced along the coast.

7.3.4 A note on delta classification

The importance of the energy of the receiving basin is emphasized in a very common classification of deltas which uses a triangular diagram with, at its apices, three end-member types [14] labelled river-dominated, tide-dominated and wave-dominated. Modern deltas are plotted within the triangular space (Fig. 7.13). Although there is some qualitative rationale behind such thinking, the diagram is of questionable value. This is essentially because it is not possible to plot a modern delta rigorously on the diagram, its position being a question of subjective judgement. We have already seen that, in dynamical terms, outflows are driven by buoyancy and/or inertia. Their dynamics are controlled by densiometric

Froude number and Reynolds number. Marine processes such as tidal currents and waves affect delta morphology through their effects on plume and jet dynamics and on sediment redistribution.

A more satisfactory approach is to devise a framework whereby modern, and perhaps ancient, deltas might be compared quantitatively. Deltas have consequently been evaluated in terms of river discharge relative to wave power, expressed as *discharge effectiveness index*, the ratio of the discharge per unit width of river mouth to the wave power per unit width of wave crest at the coast [15]. This facilitates comparison between different modern deltas where discharge and wave climate data are known [16] (Fig. 7.14). If river discharge, wave power and tidal range can be estimated from preserved sedimentary deposits,

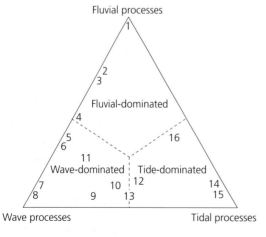

1	Mississippi
2	Po
3	Danube
4	Ebro
5	Nile
6	Rhône
7	São Francisco
8	Senegal
9	Burdekin
10	Niger
11	Orinoco
12	Mekong
13	Copper
14	Ganges-Brahmaputra
15	Gulf of Papua
16	Mahakam

Fig. 7.13 Ternary diagram of present-day deltas, after Galloway (1975) [14].

it is possible to make a comparison of an ancient delta with modern analogues (Fig. 7.15) [17].

7.4 Estuaries

Estuaries are a key element in the routeing of sediment (and pollutants) from terrestrial drainage basins to the sea. Estuaries are not passive funnels through which sediment is passed *en route* for the ocean. They are dynamic entities in which complex physical and chemical changes take place. Their mixing behaviour has a profound influence on the seaward extrusion of low-salinity plumes on to the continental shelf. The focus of this section is the fluid and sediment transport within and through estuaries.

One of the first observations one might make of estuaries is that they are commonly muddy. The Gironde estuary of south-west France and the Severn estuary of south-west Britain are but two examples. Suspended sediment concentrations are generally very high and organic contents similarly high. This has important implications for pollutant transport, since pollutants are absorbed on clay and organic particle surfaces.

7.4.1 Estuary types

The primary control on sediment transport and deposition in estuaries is the water circulation pattern, which is in turn determined by the interaction between river-derived fresh water and salty seawater. Over time-scales longer than individual tides, the net sediment transport reflects the residual flow of water. Such residual flows are strongly affected by the mixing behaviour of the estuary, and this is encapsulated in the classification of estuaries into highly stratified, partially mixed and well-mixed types (Fig. 7.16) (Section 7.3.1). Highly stratified, or salt wedge type estuaries are common in areas of low tidal range (microtidal). They are characterized by strong vertical density–salinity gradients. Although the density interface between the overlying fresh water and the underlying salt wedge moves up and down the estuary with changes in river discharge, it is a relatively stable feature of the estuarine circulation. However, with an increase in tidal range (mesotidal, 2–4 m range), the pycnocline becomes more mobile, advancing and retreating up and down the estuary with a tidal frequency. Turbulent mixing between the two water layers results, leading to a longitudinal density–salinity gradient down the estuary. Because of the higher amounts of mixing, a greater influx of seawater is required to balance the volume lost from the lower layer into the upper layer. This increases the discharges

Practical exercise 7.2: River deltas in relation to river discharge and wave power

1 The Mississippi, Danube, Ebro, Niger, Nile, São Francisco and Senegal deltas have the maximum and minimum monthly discharges and maximum and minimum monthly wave powers at the 30 ft isobath (*c.* 10 m) and shoreline given in Table P7.1. If the discharge effectiveness index is defined as the ratio of the discharge per unit width of river mouth to the shoreline wave power, calculate this parameter for the seven deltas. In addition, calculate the wave attenuation ratio as the ratio of the 30 ft wave power to the shoreline wave power.

2 The shapes of seven deltas are illustrated in Fig. 7.14. How does the information on discharge effectiveness index and wave attenuation ratio help to explain these coastal configurations?

Solution

1 The discharge effectiveness index and wave power

attenuation ratio for the seven deltas is given in Table P7.2.

2 It is clear that the Mississippi, Danube and Ebro deltas all have highly cuspate to arcuate forms which protrude far into the receiving basin. This is correlated with a high discharge effectiveness index, reflecting the strength of river output against wave power. Such deltas also have high values for the attenuation of wave power. In contrast, the São Francisco and Senegal deltas are subtle features of the coastline and are constantly reworked by wave action. This correlates with very low values of the discharge effectiveness index and low values of wave attenuation.

A full analysis would analyse the variation of wave power and discharge throughout the year (see Wright & Coleman [15], especially their Table 3; note that the mean annual values for discharge effectiveness and wave attenuation ratio given in Table 4 of [15] are different than those calculated above). The results suggest that delta

Table P7.1

Delta	Mean monthly offshore (30 ft isobath) wave power		Mean monthly shoreline wave power		Mean monthly discharge (per foot of outlet width) × 10^3 (ft^{3}s^{-1})	
	Low	High	Low	High	Low	High
Mississippi	18.4 (June)	267.1 (Mar.)	0.002 (June)	0.057 (Oct.)	279.8 (Oct.)	1086.9 (Apr.)
Danube	10.8 (July)	75.4 (Apr.)	0.013 (July)	0.052 (Apr.)	149.6 (Oct.)	304.2 (May)
Ebro	4.7 (July)	262.6 (Oct.)	0.007 (July)	0.215 (Oct.)	5.3 (Aug.)	31.3 (Mar.)
Niger	17.4 (Jan.)	444.9 (July)	0.43 (Jan.)	3.23 (July)	63.7 (Apr.)	932.6 (Oct.)
Nile	57.7 (Aug.)	149.82 (Mar.)	4.51 (June)	13.35 (Feb.)	0.8 (Apr.)	205.8 (Sept.)
São Francisco	200.6 (Apr.)	618.1 (Nov.)	11.7 (Apr.)	35.2 (Nov.)	46.5 (Oct.)	197.9 (Mar.)
Senegal	75.5 (Jan)	447.4 (July)	28.0 (Jan.)	186.3 (July)	0.3 (Mar.)	122.1 (Sept.)

Table P7.2

Delta	Discharge effectiveness index	Wave attenuation ratio
Mississippi	19 068–139 900	4 686–9 200
Danube	5 850–11 508	830–1 450
Ebro	146–757	671–1 221
Niger	148–289	40.5–138
Nile	0.18–15.4	11.2–12.8
São Francisco	4.0–5.6	17.1–17.6
Senegal	0.01–0.66	2.4–2.7

morphology depends strongly on the ability of the rivers to build shallow platforms seawards of their mouths, which attenuate deep water wave energy, rather than on the energy of the receiving basin *per se*. Such an ability will depend in part on the tectonic subsidence and compaction of the delta front region and the rate of supply of sediment.

It should be noted that the tidal energy of the receiving basin has not been accounted for in this exercise, nor has the bouyancy of the outflow. Consequently, the actual dynamics at river mouths has been much oversimplified.

(i)

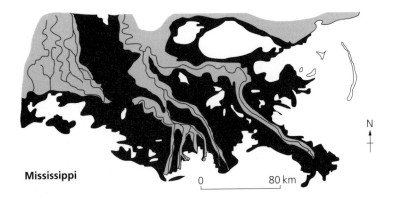

Mississippi

0   80 km

N

(ii)

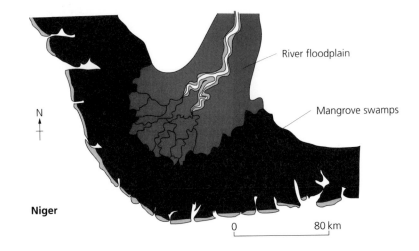

River floodplain

Mangrove swamps

Niger

0   80 km

N

(iii)

Nile

Cultivated floodplain

0   80 km

N

■ Sand-rich deposits of
  river channels, beach ridges
  and aeolian dunes

■ Fine-grained deposits of
  floodplains, salt marshes
  and mangrove swamps

Fig. 7.14 *See p. 262 for caption.*

(iv)

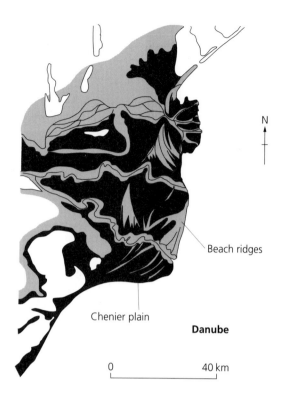

Beach ridges

Chenier plain

Danube

0   40 km

(v)

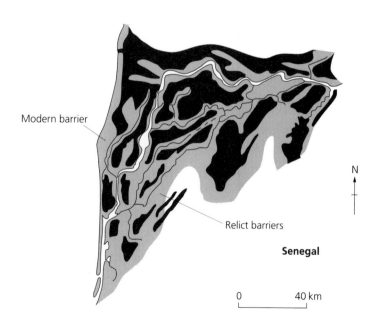

Modern barrier

Relict barriers

Senegal

0   40 km

Continued on p. 262.

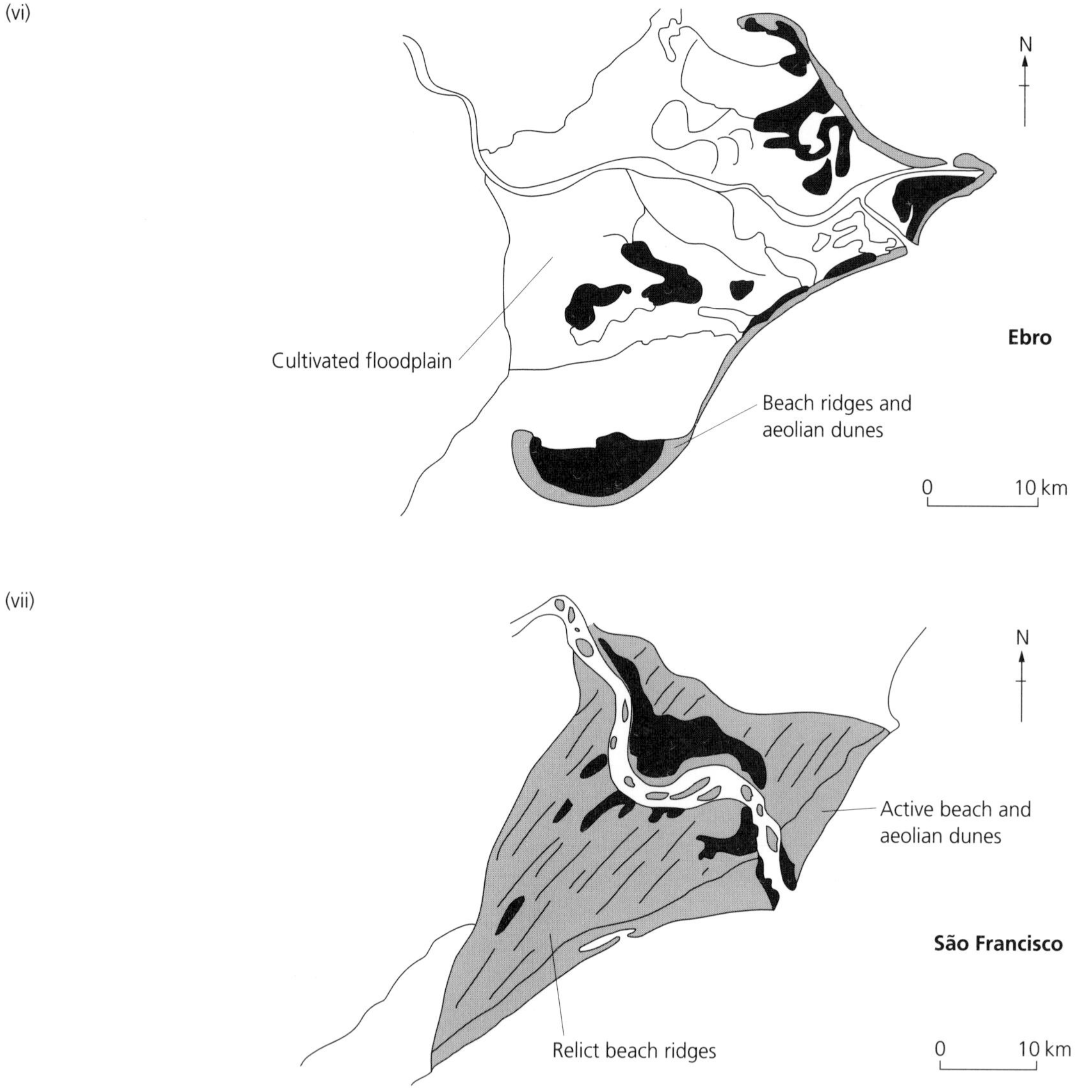

(vi)

Cultivated floodplain

Ebro

Beach ridges and
aeolian dunes

0   10 km

(vii)

Active beach and
aeolian dunes

São Francisco

Relict beach ridges

0   10 km

Fig. 7.14 *See pp. 260 and 261 for parts (i)–(v)*. Parts (i) to (vii) show the gross morphology of the Mississippi, Niger, Nile, Danube, Senegal, Ebro, and São Francisco deltas, respectively. These illustrate the variation in form as a function of discharge effectiveness index. Redrawn from Wright & Coleman (1974) [4].

of water through the estuary mouth. As tidal range varies through a neap–spring cycle (Chapter 8), the degree of mixing and fluid discharges also varies. At even higher tidal ranges (macrotidal, over 4 m), the vertical salinity stratification is broken down completely by vigorous tidal currents flowing up and down the estuary. The residual water circulation is then dominated by the flow paths of the flood and ebb tide, which in large estuaries commonly become

spatially separated because of Coriolis effects. The size and shape of the estuarine basin can also affect the magnitudes and durations of the flood and ebb pulses.

7.4.2 The turbidity maximum and its controls

A common feature of estuaries, particularly in meso-tidal and macrotidal regimes, is for a zone of maximum sediment concentrations to be found in a region

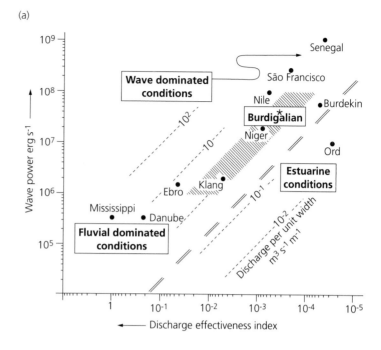

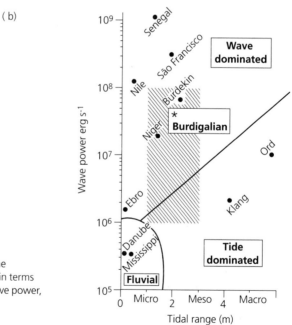

Fig. 7.15 Reconstruction of the early Miocene (Burdigalian) Alpine coastline in Switzerland, in terms of (a) discharge effectiveness index versus wave power, and (b) tidal range versus wave power. After Homewood & Allen (1981) [17].

salinities of 1–5 parts per thousand just landwards of the head of the salt intrusion (Fig. 7.17). These sediment concentrations may reach levels as high as 10 000 ppm in some macrotidal estuaries. This zone is known as the *turbidity maximum*. Since the concentrations in the turbidity maximum appear to remain at the same levels for prolonged periods of time, there must be a flux of suspended sediment through the

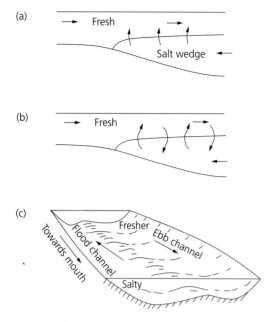

Fig. 7.16 Classification of estuaries according to their mixing behaviour: (a) salt wedge or highly stratified; (b) partially mixed; (c) well mixed. Arrows indicate the tidally averaged, residual water circulation. After Dyer (1994) [18].

zone, with replenishment from river sources and net loss to the sea.

The position and intensity of the turbidity maximum are affected by two processes. First, it is pushed downstream by high river discharges: although there is a greater mass of suspended sediment supplied to the turbidity maximum by the river in flood, the suspended sediment occupies a wider cross-sectional area of estuary and there are likely to be greater losses to the ocean. Consequently, the balance of this trade-off determines wether suspended sediment concentrations will increase or decrease during periods of higher river discharge. The turbidity maximum varies its position by 40 km according to seasonal discharge fluctuations in the Gironde estuary (Figure 7.18).

Second, it oscillates with flood and ebb tides, being located near the head of the estuary at high tide and further down-estuary at low tide. The high-tide maximum is relatively dilute, but as the tide ebbs the concentrations commonly increase, and some deposition may take place by settling out from the water column. The position and intensity of the turbidity maximum will also be affected by the variations in tidal heights and velocities through a neap–spring

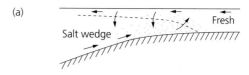

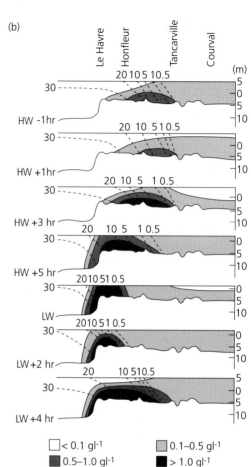

Fig. 7.17 (a) Schematic diagram to show the formation of the turbidity maximum in a partially mixed estuary. (b) The turbidity maximum in the Seine estuary, France, at intervals through a tidal cycle (at springs). The river discharge is 780 $m^3 s^{-1}$. HW, high water; LW, low water; salinities in parts per thousand. After Avoine (1981) [19].

cycle (Chapter 8). The neap–spring cycle causes a variation in the relative magnitude of the river to the tidal volumes. Consequently, an estuary may be fully mixed at spring tides with a highly concentrated turbidity maximum located up-estuary, but partially mixed at neaps when weaker currents, capable of

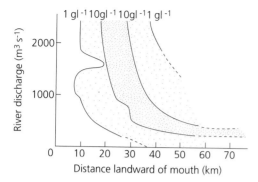

Fig. 7.18 The variation of the position of the turbidity maximum in the Gironde estuary, France, in relation to river discharge. The turbidity maximum is pushed relatively seawards by as much as 60 km by high river discharges. After Allen (1973) [20].

suspending less sediment, result in lower concentrations in a turbidity maximum located down-estuary.

There are a number of possible controls on the intensity and position of the turbidity maximum. These controls determine the ability of estuarine currents to erode, suspend and deposit sediment, and to affect sediment settling behaviour through processes such as *flocculation*.

Density-driven fluid circulation

The convergence of sediment-rich river water and the landward-moving salt intrusion are thought to produce a zone of slow water movement in salinities of 1–5 parts per thousand near the head of the salt intrusion. Mixing in this region between the two water masses maintains high suspended sediment concentrations through the water column. A suspended particle in the surface layer will be transported downstream, be mixed into the lower layer, and transported upstream to the head of the salt intrusion, and returned to the surface layer by turbulent mixing. There is consequently a general circulation with its landward limit close to the head of the salt intrusion.

Pumping by tidal currents

The fact that the turbidity maximum occurs landwards of the salt intrusion suggests that there are processes controlling its position other than the general density-driven circulation. Superimposed on this general circulation is a pumping of sediment through the estuary by the tidal flows. Such pumping is exceedingly complex because of the many non-

linearities (lags) controlling the sediment transport under time-varying flow, and the resultant sediment transport is commonly a small net resultant from large opposing sediment fluxes. Asymmetrical pumping landwards is a feature of many estuaries caused by the flood tide being shorter in duration but more vigorous than the ebb. This would displace the turbidity maximum landwards from the head of the salt intrusion.

Flocculation

There are a number of processes affecting the settling, deposition and erosion of muddy sediment in the estuary. One of the most important is the tendency for clustering of fine particles into aggregates which then settle at a faster rate. This process of flocculation results from the electrical charges on particles [21] but is mediated strongly by organic binding. Flocculation is promoted when clay mineral particles are transported from fresh river water into the salty waters of the coast or estuary. The size of the aggregates, or flocs, is controlled primarily by the suspended sediment concentration and inversely by the shear stress of the fluid. At high shear stresses, flocs are broken up. Flocculation is probably the explanation for the positive relationship between sediment concentration and settling velocity. The flocculation of fine particles promotes the rapid settling out of mud during slack water, which may produce a fluid mud layer above the bed.

Further reading

J.R.L. Allen (1985) *Principles of Physical Sedimentology*. George Allen & Unwin, London, Chapter 11.

S. Pond & G.L. Pickard (1983) *Introductory Dynamical Oceanography*, 2nd edn. Pergamon Press, Oxford.

R. Slingerland, J.W. Harbaugh & K.P. Furlong (1994) *Simulating Clastic Sedimentary Basins*. Prentice Hall, Englewood Cliffs, NJ, Chapter 5.

L.D. Wright (1977) Sediment transport and deposition at river mouths: a synthesis. *Bulletin of the Geological Society of America* **88**, 857–68.

References

1 M. Van Dyke (1982) *An Album of Fluid Motion*. Parabolic Press, Stanford, CA.

2 J.R.L. Allen (1985) *Principles of Physical Sedimentology*. George Allen & Unwin, London, Chapter 11.

3 L.D. Wright (1977) Sediment transport and deposition at river mouths: a synthesis. *Bulletin of the Geological Society of America* **88**, 857–68.

4 L.D. Wright & J.M. Coleman (1974) Mississippi River mouth processes: effluent dynamics and morphologic

development. *Journal of Geology* **82**, 751–78. [Wright & Coleman [4] appear to be incorrect in their definition of the reduced gravity, using ρ_2, the density of the water underlying the plume, instead of ρ_1, the density of the plume, in the denominator of the reduced gravity. This makes very little numerical difference in the case of low-salinity plumes, but is at variance with the normal way of formulating densimetric Froude number.]

5 C.C. Bates (1953) Rational theory of delta formation. *Bulletin of the American Association of Petroleum Geologists* **37**, 2119–61.

6 K.D. Stolzenbach & D.R.F. Harleman (1971) *An analytical and experimental investigation of surface discharges of heated water*, Department of Civil Engineering Report 135. Massachusetts Institute of Technology, Cambridge, MA.

7 R. Slingerland, J.W. Harbaugh & K.P. Furlong (1994) *Simulating Clastic Sedimentary Basins*. Prentice Hall, Englewood Cliffs, NJ, Chapter 5. [There is now a range of computer programs available to model sedimentation at river mouths.]

8 G. Müller (1966) The new Rhine delta in Lake Constance. In: *Deltas in their Geologic Framework* (ed. M.L. Shirley). Houston Geological Society, Houston, TX, pp. 107–24.

9 M.J. Bowman & R.L. Iverson (1978) Estuarine and plume fronts. In: *Oceanic Fronts in Coastal Processes* (eds M.J. Bowman & W.E. Esaias). Springer-Verlag, Berlin, pp. 87–104.

10 J.P.M. Syvitski & S. Daughney (1992) Delta 2: delta progradation and basin filling. *Computers and Geosciences* **18**, 839–97. [There is now a range of computer programs available to model sedimentation at river mouths.]

11 J.P.M. Syvitski & J.M. Alcott (1993) Grain 2: predictions of particle size seaward of river mouths. *Computers and Geosciences* **19**, 399–466. [There is now a range of computer programs available to model sedimentation at river mouths.]

12 L.D. Wright & J.M. Coleman (1971) Effluent expansion and interfacial mixing in the presence of a salt wedge, Mississippi River delta. *Journal of Geophysical Research* **76**, 8649–61.

13 A.C. Duxbury (1965) The union of the Columbia River and the Pacific Ocean—general features. In: *Ocean Science and Ocean Engineering*. Marine Technology Society, Washington DC, pp. 914–22.

14 W.E. Galloway (1975) Process framework for describing the morphologic and stratigraphic evolution of deltaic depositional systems. In: *Deltas, Models for Exploration* (ed. M.L. Broussard), Houston Geological Society, Houston, TX, pp. 97–8.

15 L.D. Wright & J.M. Coleman (1973) Variations in morphology of major river deltas as functions of ocean wave and river discharge regimes. *Bulletin of the American Association of Petroleum Geologists* **57**, 370–98.

16 J.T. Wells & J.M. Coleman (1984) Deltaic morphology and sedimentology, with special reference to the Indus River delta. In: *Marine Geology and Oceanography of the Arabian Sea and Coastal Pakistan* (eds B.U. Haq and J.D. Milliman). Van Nostrand Reinhold, New York, pp. 85–110.

17 P. Homewood & P.A. Allen (1981) Wave-, tide- and current-controlled sandbodies of Miocene Molasse, western Switzerland. *Bulletin of the Association of Petroleum Geologists* **65**, 2534–45.

18 K.R. Dyer (1994) Estuarine sediment transport and deposition. In: *Sediment Transport and Depositional Processes* (ed. K. Pye). Blackwell Scientific Publications, Oxford, pp. 193–218.

19 J. Avoine (1981) *L'Estuarine de la Seine: sédiments et dynamique sédimentaire*. Docteur de Specialité thesis, Université de Caen, France.

20 G.P. Allen (1973) *Étude des processes sédimentaires dans l'estuarine de la Gironde*, Mémoire 5. Institut Géologique Bassin d'Aquitaine.

21 K. Pye (1994) Properties of sediment particles. In: *Sediment Transport and Depositional Processes* (ed. K. Pye). Blackwell Scientific Publications, Oxford, pp. 1–24.

8 Tides and waves

One day I wrote her name upon the strand,
But came the waves and washèd it away:
Again I wrote it with a second hand,
But came the tide, and made my pains his prey.
Edmund Spenser (1552–99), *Amoretti* [1595], sonnet 75.

Chapter summary

Any casual observation of the sea from a headland or harbour reveals a range of periodicity of water motion, from the rapid oscillations associated with wind-generated waves, whose periodicity is between 1 and 20 s, to the twice-daily ebb and flow of the tide. The more astute observer will recognize that the twice-daily tides vary in height with a longer-term fortnightly periodicity related to the phases of the Moon and Sun, and with an even longer-term periodicity related to the passage of the seasons.

The tides are generated by the gravitational attraction of the Moon and, to a lesser extent, of the Sun. The magnitude of the tide-generating force on the surface of the Earth varies with position. Since the Earth is continuously rotating with respect to the Moon and Sun, the tide-generating force at a particular point on the Earth's surface is constantly changing. The ideal tide on an Earth with a continuous deep ocean is known as the equilibrium tide. The combination of solar and lunar gravitational attractions is weakest at half moon, causing neap tides, and strongest at new moon and full moon, giving spring tides. The tidal record can therefore be viewed as being made up of a number of harmonic constituents, of which the semi-diurnal, diurnal and fortnightly frequencies dominate.

Dynamical theories of real tides must incorporate a number of other factors, including the depths and configurations of the ocean basins, the rotation of the Earth resulting in a Coriolis force, and friction of ocean water on the sea bed and by wind acting on its surface. This causes tidal motions to be far more complex than in the equilibrium tide model. Ocean tidal waves (Kelvin waves) shallowing on the continental shelf steepen and shorten their wavelength, thereby increasing their amplitude and current speeds. In a partially enclosed sea the ocean Kelvin wave may resonate with the natural frequency of the gulf, causing a standing wave to develop which greatly amplifies tidal ranges. The Bay of Fundy in Nova Scotia is a fine example. Coriolis effects and friction on the sea bed and at the head of a gulf result in amphidromic systems in which tides rotate around a nodal point of no tidal range. Tidal flows may be strongly funnelled in estuaries, amplifying tidal heights, as in the St Lawrence estuary of eastern Canada.

Exceptionally long-wavelength (*c.* 200 km) ocean waves known as tsunamis may be generated by underwater earthquakes and volcanic eruptions. These waves behave as 'shallow water' waves on account of their very long wavelength, and travel with a celerity dependent on the water depth. They are responsible for considerable coastal damage, especially around the Pacific Ocean basin.

The sediment transport and preserved sedimentary deposits of tidal flows reflect their reversing nature. Deposits showing intermittent transport of sand separated by phases of deposition of fine sediment as drapes are typical of estuarine tidal sandwaves. The details of the internal structure of tidal sandwaves in straits and on the continental shelf are less well known, but their periodic surface forms and geometry have been well imaged by side-scan sonar and very high-resolution seismic reflection techniques.

Surface waves on a body of water are caused by gravity acting on surface disturbances caused by wind stress. In contrast to the situation in tidal waves, Coriolis effects are unimportant in the movement of progressive gravity waves. The fundamental relation between the period and the wavelength of a gravity wave is the wave dispersion equation. Waves can be categorized according to the ratio of their wavelength to water depth. In water which is deep compared to the wavelength, waves have a speed determined by their wavelength, that is, they are dispersive waves. In water which is shallow compared to the wavelength, the speed is only dependent on the water depth. Consequently, waves of different wavelength travel at different speeds, and individual waves are continuously catching up or being caught up by neighbouring waves. The wave energy of a sea is therefore related to the group velocity of a wave train.

The orbital motion of water particles under waves results in a to-and-fro fluid motion at the sea bed above wavebase. There is also a unidirectional mass transport at the bed, known as a Stokes drift, acting in the direction of wave propagation, which causes a build-up of water in the nearshore zone; this build-up must be compensated for by the action of seaward-flowing rip currents. Waves approaching the shoreline refract towards parallelism with the sea bed contours. However, obliquity of wave approach commonly induces an alongshore drift in the beach zone.

The threshold of sediment motion under waves for a given sediment grain size can be predicted from a knowledge of the maximum orbital velocity and amplitude of the near-bed water motion, and can be expressed in terms of a dimensionless stress and grain Reynolds number in a manner similar to the case for unidirectional currents. Wave ripples include steep trochoidal forms characterized by the formation of a vortex with each half cycle of the wave motion (vortex ripples) and low-steepness, undulating types where the near-bed wave motion passes over several ripple crests (post-vortex ripples). Vortex types have been used for wave hindcasting analyses. The internal lamination of wave ripples is commonly unidirectional in character. This may be due to the Stokes mass transport near the bed, or to the superimposition on the orbital flow of a current of some other origin.

8.1 Introduction to surface waves

It is a commonplace observation that the surface of the sea is not a flat plane, but that it is on all but the calmest of days moulded by the wind into waves with a period measured in seconds, and on most coasts rises and falls with an average period of a little over 12 hours. The motion of water on the continental shelf is dominated by the effects of tides and waves (which are considered in this chapter), and of intruding ocean currents and storms (Chapter 9). These water motions are the driving force for the erosion, transport and deposition of particulate sediment. They are also the conveyors of nutrients affecting biological productivity.

The range of surface waves can be shown in relation to the distribution of energy, known as an *energy density spectrum*. Waves are characterized by their frequency or period (Fig. 8.1). The shortest (less than 1 s) period waves (that is, highest frequency) are the tiny *capillary waves* on water surfaces controlled by surface tension. These are of no great importance to us since they are incapable of causing sediment transport. There is then a large concentration of energy in the surface waves generated by wind blowing over a water surface. *Wind waves* have periods between about 1 and 20 s and their form is largely determined by gravity acting on the disturbed water surface. Longer-period waves are also found, with periods of minutes to a few hours. These include the periodic oscillations of water in enclosed or partially enclosed bodies of water, such as lakes, known as *seiches*, and the surface waves on the ocean caused by subsea disturbances such as earthquake activity or volcanic eruptions, known as *tsunamis*. Finally, there are major contributions of energy from the waves operating at semi-diurnal (twice daily) and diurnal (once daily) frequencies due to the gravitational attraction of the Moon and Sun. These are the *tides*. The long-period waves such as tsunamis and tides are affected by both gravity and the Coriolis force due to the Earth's rotation (discussed in more detail in Chapter 9).

There are also internal waves developed at interfaces of various types, such as at density interfaces caused by salinity or temperature variations. Although these are of some importance (and were encountered in the case study of the Mississippi River outflow in Chapter 7), they are not dealt with further here.

8.2 Tidal observations

Inspection of records of sea level at ports from around

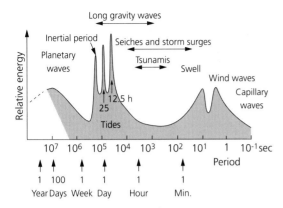

Fig. 8.1 Schematic energy spectrum for waves of various types. The area under the curve is roughly proportional to the contribution to the total energy.

the world shows a strong periodicity to be present. Although these tidal curves may differ substantially from one port to another, in general there is a dominant average period of 12.42 hours. Since the tide comes in and goes out twice a day, this is known as the *semi-diurnal tidal frequency*. Some stations show a diurnal frequency, and others show a mixture of semi-diurnal and diurnal components (Fig. 8.2). There are other periodicities seen in tidal records. High tides tend to be higher and low tides lower during particular phases of the Moon, within a few days of a new moon or full moon. These *spring tides* occur twice within a lunar cycle of 29.5 days. Spring tides alternate with periods when the range in height between high tide and low tide is at its minimum. These are known as *neap tides*. There are therefore 14.8 days in a spring–neap–spring cycle. Yet other periodicities are recognizable. Commonly, there is an inequality between the two semi-diurnal tides occurring in a lunar day (24 hours 50 minutes). On a longer time-scale, there are variations in the heights of spring and neap tides on a periodicity governed by the equinoxes, the greatest springs and lowest neaps occurring at the vernal and autumnal equinoxes.

The tides are a response to the combination of the gravitational forces of the Sun, Moon and Earth as they move relative to each other. Other planets can be neglected as being too small in mass and too distant to exert tidal forces on the Earth. Although the tide-producing force caused by the Sun and Moon is very small compared to the gravitational force of the Earth, it is sufficient when acting on the total mass of the ocean to cause significant elevation differences.

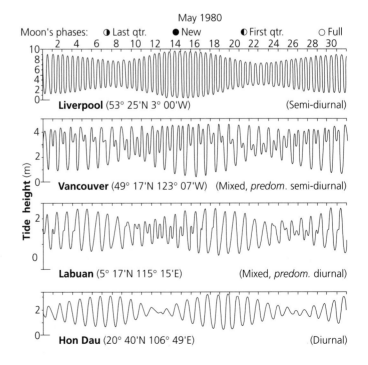

Fig. 8.2 Tidal curves for May 1980 (from Admiralty Tide Tables) showing semi-diurnal and diurnal frequencies.

The rise and fall of the tide observed at the coast must be associated with the flow of tidal currents. In an estuary or gulf there is a clear phase relationship between the elevation of the water surface and the tidal current, the current flowing on the rising tide being known as the *flood*, and that on the falling tide the *ebb*. Away from estuaries, inlets and gulfs, and further from the coast, however, the phase relationship is less simple and tidal currents commonly rotate, as well as change speed, during a tidal cycle.

In considering tidal currents on the shelf and at the coast it is important to emphasize that coastal and nearshore tides are driven essentially by the ocean tides, and not directly by the gravitational attraction of the Sun and Moon. That this must be the case can be illustrated by considering large water bodies such as the Mediterranean and Baltic Sea which have a restricted connection with the open ocean. Such water bodies have very low tides. The same principle applies to lakes. The prediction of tides at a coast therefore depends on a knowledge of the ocean tides, and their likely modification as they travel into shallower waters bordered by an irregular coastline.

8.3 Ocean tides

8.3.1 Tide-generating forces

Despite its relatively small mass, the effect of the Moon is stronger than that of the Sun because of its close proximity. Initially, we can consider the tidal forces as originating entirely from the relative position of the Earth and Moon. Having calculated the tide-generating force due to the Moon, it is a relatively simple extension of the problem to consider in addition the effect of the Sun.

The Earth and Moon comprise a single system mutually revolving around a centre of mass, denoted by Z (Fig. 8.3a). The Earth revolves eccentrically around this common centre of mass with a period of 27.3 days. Consequently, after a time t, the centre of mass of the Earth, C, has moved to a point shown as C_t. Any other point on the Earth's surface, such as A, will have moved in the same period of time to the point A_t. Each point on the Earth's surface has the same angular velocity, and the same centrifugal force as a result of this eccentric motion (we are here ignoring completely the spin of the Earth on its axis of rotation). As the Earth undergoes its eccentric motion, the Moon travels around the Earth as if attached to it by a string (Fig. 8.3b).

In order for the Earth–Moon system to be in equilibrium, the total centrifugal force must balance the gravitational attraction between the two bodies.

The centrifugal force must act parallel to a line joining the centres of the Earth and the Moon (Fig. 8.4a), and as we have already seen, it is the same for all points on the Earth's surface. However, the gravitational force exerted by the Moon is not the same everywhere on the Earth's surface. This is because, as shown by Newton in 1697, the gravitational force between two bodies is proportional to the product of the two masses, and varies inversely with the square of their separation distance. Consequently, the gravitational force is greater on the side of the Earth nearer to the Moon and less on the far side. In addition, the gravitational force acts from the point on the Earth's surface towards the centre of mass of the Moon.

The resultant of the centrifugal and gravitational forces is the tide-generating force (Fig. 8.4b). The vertical component of this force (or *acceleration*, since it is a force per unit mass) at point P results in a minute increase or reduction in g, the gravitational acceleration, and can be neglected. The horizontal component is the force causing tides. Its magnitude must vary with position on the Earth's surface. If θ is the angle between the line joining the centres of the Earth and Moon and a line from the centre of the Earth to the point P on its surface (Fig. 8.5a), the horizontal component, F, of the tide-generating force is given by

$$F = \frac{3}{2}\frac{M_m}{M_E}\left(\frac{a}{d}\right)^3 g \sin 2\theta \qquad (8.1)$$

where M_m and M_E are the masses of the Moon and Earth respectively, a is the radius vector from the centre of the Earth to the point P on its surface and d is the distance of the centre of the Earth from the centre of the Moon. This shows that the tide-generating forces act tangentially towards the points directly beneath the Moon, or *sublunar points*. The forces are at a maximum where $\theta = \pm 45°$ (Fig. 8.5b). Since the Earth is continuously rotating, the tide-generating forces at a point on the Earth's surface are constantly changing.

The equilibrium tide

So far, the distribution of tide-generating forces has been introduced, but what effect do these forces have on the tidal height around the Earth's surface? Let us first assume that the Earth is covered with a continuous ocean, so that the complicating effects of land masses can be ignored, and let us also assume that the water in the ocean can respond instantly to any

(a)

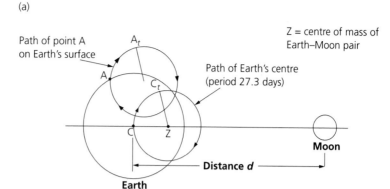

Path of point A
on Earth's surface

Z = centre of mass of
Earth–Moon pair

Path of Earth's centre
(period 27.3 days)

Moon

Distance *d*

Earth

(b)

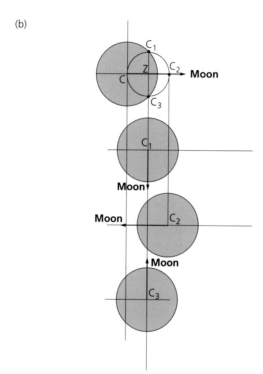

Fig. 8.3 Illustration of the eccentric orbit of the Earth around the common centre of mass of the Earth–Moon pair at Z. (a) Any point on the Earth's surface, such as A, travels with the same angular velocity due to the eccentric orbit, and with the same centrifugal force. (b) Position of the Moon in relation to the motion of the Earth around the common centre of mass at Z. The centre of mass of the Earth is shown at four times during its revolution.

changes in forces. The tide calculated using these assumptions is known as the *equilibrium tide*. The horizontal tide-generating force is balanced by a horizontal pressure gradient in the ocean water. The balance of the tide-generating force and the horizontal pressure gradient is achieved by the water surface assuming a slope. It can be seen by the triangle of forces in Fig. 8.6 that if the tide-generating force reduces, the slope likewise reduces. Since the slope is positive when the tide-generating force acts clockwise, there must be a maximum elevation of the equilibrium

tide at the sublunar points where $\theta = 0$ and $\theta = \pi$ and minimum elevations at angles of $\theta = \pm\pi/2$. Since the Earth is rotating, a given point on the Earth's surface should experience two high waters and two low waters per day (Fig. 8.6c). In addition, the Moon is not in the plane of the Earth's equator, having a declination north and south of it. Consequently, there should be a variation in the heights of the two tides per day, which is recognized in the *diurnal inequality* of tidal curves. The difference between the maximum and minimum elevations should be about half a metre in

(a)

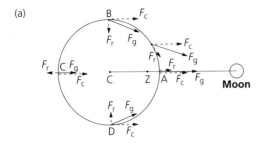

(b)

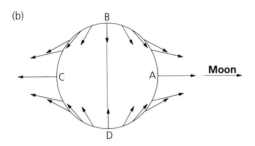

Fig. 8.4 (a) The tide-generating force is the resultant of the force due to the gravitational attraction of the Moon, and the centrifugal force due to the Earth's eccentric orbit around the common centre of mass. (b) The relative magnitude of this resultant force.

the equilibrium tide model. This is roughly similar to the tide measured at oceanic seamounts.

There should also be an equilibrium tide resulting from the attraction of the Sun. The solar tide-generating force, F', has the same form as that due to the Moon, F (equation (8.1)). Substituting the correct values for the relative masses of the Sun and Earth (3.3×10^5), and the distance of the centre of the Sun from the centre of the Earth $(1.496 \times 10^8 \text{ km})$, the ratio for the maximum values of the tide-generating forces is

$$\frac{F'_{max}}{F_{max}} = 0.46 \qquad (8.2)$$

The same ratio applies to the equilibrium tidal heights caused by the lunar and solar components. Consequently, the maximum tidal height caused by the solar component is about 0.24 m.

Since the Sun and Moon both have effects on the tide-generating forces and equilibrium tide heights, the relative positions of the Earth, Moon and Sun through the year should result in tidal variations. The distance from the Earth to the Sun is smallest at the winter solstice (perihelion) and greatest at the summer solstice (aphelion) (Fig. 8.7). The distance

(a)

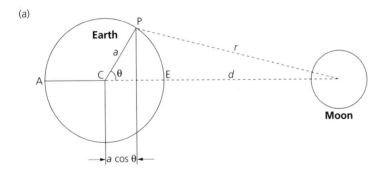

(b)

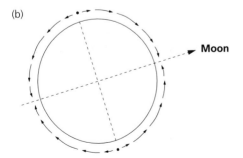

Fig. 8.5 The horizontal component of the resultant force shown in Fig. 8.4 varies with latitude, reaching a maximum at latitudes of 45° either side of the sublunar points. See text for explanation.

(a)

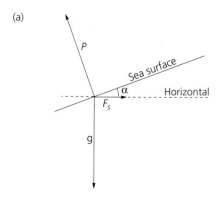

(b)

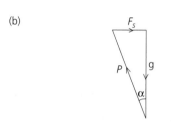

(c)

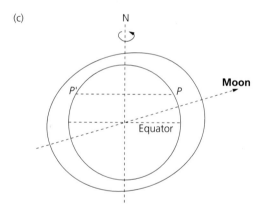

Fig. 8.6 (a) A horizontally acting tide-generating force F_s results in a sloping sea surface because of its vector addition with gravity (b) to produce a resultant pressure force P at right angles to the surface slope. (c) The equilibrium tide due to the Moon therefore involves two bulges and two depressions in the water surface around the Earth. A given point on the Earth's surface therefore experiences two high waters and two low waters each day as the Earth rotates on its axis.

from the Earth to the Moon has a monthly variation. Consequently, ignoring the effect of the declination of the Moon, the combination of solar and lunar

attractions is greatest at new moon or full moon, when spring tides result. Springs are highest during the vernal or autumnal equinoxes. The combination of solar and lunar attractions is at a minimum at half moon, causing neaps. The lowest neaps also occur at the equinoxes.

The combination of equilibrium tides for the Sun and Moon can be expressed in terms of the sum of a number of harmonic constituents. These constituents fall into three main classes:

1 semi-diurnal constituents, including the principal lunar constituent M_2, with a period of half a lunar day (12.42 hours), and the principal solar constituent S_2, with a period of half a solar day (12 hours);

2 diurnal constituents, with a period of approximately one day;

3 long-period constituents, with periods of two weeks or longer, such as the lunar fortnightly constituent M_f with a period of 13.66 days.

Although the equilibrium tide is a useful concept in visualizing the semi-diurnal lunar tide and the neap–spring cycle, the assumptions are clearly unrealistic. For instance, the ocean water has inertia and so does not respond instantly to changes in tide-generating forces. The actual response of the ocean water to tide-generating forces is rather different to the equilibrium tide predictions. Dynamical theories of real tides must incorporate the effects of a number of factors: the depths and configurations of the ocean basins; the rotation of the Earth resulting in a Coriolis force (explained in Chapter 9); the friction of ocean water on the sea bed, and friction by wind acting on the surface. This makes the solution of the flow equations complex. Another approach has been to describe tidal movements by collating many observations at the coast and plotting lines joining points experiencing high tide at the same time. These are known as *co-tidal lines*. In order to explain the tidal observations, it is necessary to radiate co-tidal lines from *nodal points* at which there is no tidal range, the tidal range increasing along co-tidal lines away from the nodal point. These tidal systems are known as *amphidromic systems*. Individual amphidromic systems vary greatly in size. The world-wide distribution of amphidromic systems showing co-tidal lines for the M_2 tide is shown in Fig. 8.8. It can be noted that the high-water crest rotates about the nodal point in an anticlockwise direction in the northern hemisphere and a clockwise direction in the southern. These senses of rotation can be understood once we have considered the motion of water in a gulf or semi-enclosed basin (Section 8.4.2).

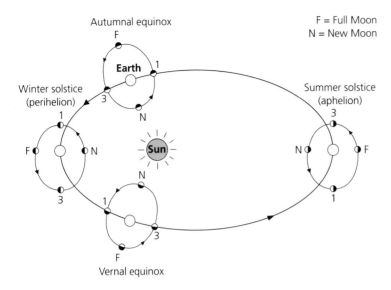

Fig. 8.7 The combination of lunar and solar components as the Earth moves through the seasons explains neap–spring variations in tidal heights.

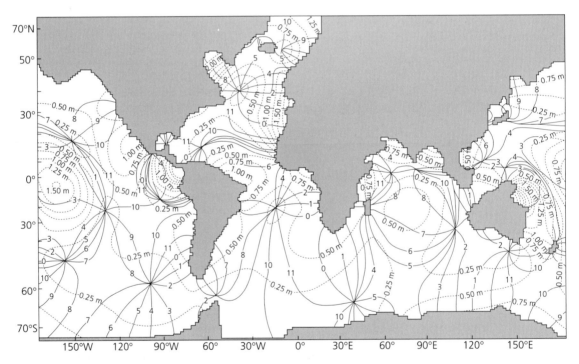

Fig. 8.8 The global lunar semi-diurnal tide (M_2), showing the world's major amphidromic systems. Full lines show the phase of the tide (co-tidal lines), dashed lines show the tidal range (co-range lines). After Accad & Pekeris (1978) [1].

8.3.2 Kelvin waves in the open ocean

Tides in the ocean can be thought of as due to the passage of waves of very long wavelength created by the tide-generating forces. A common type of progressive wave thought to approximate tidal motions is a *Kelvin wave*. Kelvin waves travel with a similar phase velocity (celerity) to a shallow water gravity wave (Section 8.6.1), that is

$$c = \sqrt{gh} \tag{8.3}$$

so that if the ocean depth, h, is 4 km and g is 9.8 m s^{-2}, the celerity of the Kelvin wave is almost 200 m s^{-1}. The period is $T = \lambda/c$, where T is the wave period and λ the wavelength. If the Kelvin wave has the period of the semi-diurnal M_2 tide (12.42 hours), the wavelength must be 8850 km. A characteristic of Kelvin waves is that they increase in amplitude parallel to the wavecrest because of the Coriolis acceleration (Fig. 8.9). In the northern hemisphere the amplitude increases to the right of the wavecrest by a factor $b = c/f$, where f is the Coriolis parameter (with units of reciprocal seconds (s^{-1}); the Coriolis acceleration is therefore fv, where v is the velocity normal to the propagation direction of the wave, giving the units of acceleration, m s^{-2}). Since the Coriolis parameter is given by

$$f = 2\Omega \sin \phi \tag{8.4}$$

where Ω is the angular rate of rotation of the Earth (7.29×10^{-5} rad s^{-1}) and ϕ is the latitude (positive north of the equator), giving a value of 11.2×10^{-5} s^{-1} at a latitude of 50°N, the amplitude of the Kelvin wave would increase by a factor e (2.78) in a distance of 1768 km to the right of the wave crest. Kelvin waves, therefore, are observed hugging the right-hand sides of oceans in the northern hemisphere, with amplitudes that decrease towards the open ocean.

Although the celerity of the Kelvin wave is very fast, the speed of the water particles under the wave is low. This can be obtained from

$$U = \frac{c}{h} A \tag{8.5}$$

where A is the amplitude of the Kelvin wave in the open ocean. Taking the same 4 km deep ocean as before, and letting $A = 0.5$ m as a typical amplitude, the current under the wave is 25 mm s^{-1}. This gives a good impression of the dimensions and speeds of Kelvin waves and their water motion. It should be noticed that Kelvin waves in the open ocean are singularly incapable of causing transport of any but the very finest suspended grains. The reason why tides are a powerful agent in the transport of particulate matter is that they transform in the shallow water of the continental shelf or are reinforced by the geometries of partially enclosed seas. These are the subjects of the following sections.

8.3.3 Tsunamis

Tsunamis (from the Japanese word for 'harbour wave')

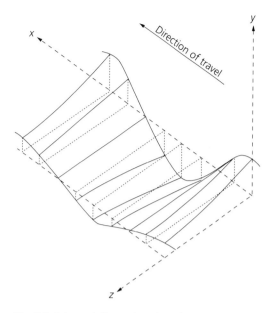

Fig. 8.9 Schematic illustration of a Kelvin wave banking up to the right in the northern hemisphere as a result of Coriolis effects.

are very long-wavelength water waves generated by shallow (focus at <40 km depth) submarine earthquakes in excess of 6.5 on the Richter scale, and to a lesser extent by submarine slides or submarine volcanic activity [2]. The eruptions of Santorini in 1628 BC and Krakatoa in AD 1883 were both accompanied by tsunamis. However, tsunamis are relatively rare, occurring perhaps with only one in a hundred submarine earthquakes. The source regions for recorded tsunamis are concentrated in the western Pacific along the subducting margin of the Pacific plate.

The typical wavelengths of tsunamis are in the region of 200 km. In an ocean of 4000 m depth, the ratio $h/L \approx 1/50$, so they behave as 'shallow water' waves, even though they occur in the deep ocean. Typical tsunami amplitudes in the ocean are only about 1 m. A 1 m rise and fall over a 200 km wavelength explains why an observer on a boat would most certainly not notice the passage of a tsunami wave. If the celerity of a shallow water wave is $\sqrt{gh}$, the wave should travel at about 200 m s^{-1} in our 4 km deep ocean, with a waveperiod of 1000 s. However, as the waves move into shallow water, they peak up, so that the run-up height far exceeds the deep water wave height. Run-up heights of over 10 m have commonly been measured. The Krakatoan eruption caused a tsunami with a run-up of as much as 40 m along

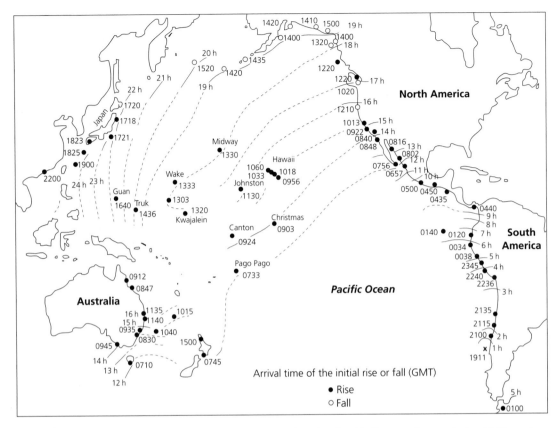

Fig. 8.10 The tsunami wavefront resulting from the Chilean earthquake of 1960. The wave front took about 24 hours to reach the coast of Japan. After Wadati *et al.* (1963) [3].

adjacent coasts. Clearly such waves can cause considerable damage and loss of life in coastal zones. Because tsunami waves are often far travelled (Fig. 8.10), there is usually good warning up to 24 hours in advance of the arrival of the wave.

8.4 Tides in shallow waters

8.4.1 Shoaling of ocean tides on the continental shelf

When progressive tidal waves move into the shallow water of the continental shelf they are forced to transform. The transformations taking place can be estimated by conserving the energy flux (rate of transmission of energy). A number of assumptions are made: first, the effects of a sloping sea bed are ignored; second, the energy reflected back from the sloping bottom is neglected; and third, friction on the sea bed in the shallow water depths of the shelf is, at least initially, ignored.

The rate of transmission of energy through a vertical plane of unit width parallel to the crestline and unit depth is the product of the excess of pressure above the hydrostatic pressure (in the absence of wave motion) ($= \rho g \varsigma$, where ς is the elevation at a time during the tidal cycle) and the velocity of the water particles passing through the plane (U). If the plane through which the energy is transmitted has unit width, but depth h to the sea bed, averaged over a tidal cycle, the mean energy flux is therefore

$$E = \frac{1}{2}\rho g A U h \qquad (8.6)$$

where A and U are the amplitudes of the elevation and tidal current velocity, respectively. Since $U = (c/h)A$ and $c = \sqrt{gh}$, the energy flux can be rewritten

$$E = \frac{1}{2}\rho g^{3/2} h^{1/2} A^2 \qquad (8.7)$$

Conserving energy flux between the open ocean and the shelf, the relative change in amplitude is proportional to the quartic root of the relative change in water depth:

$$\frac{h_1^{1/2}}{h_2^{1/2}} = \frac{A_2^2}{A_1^2}$$

$$\frac{h_1}{h_2} = \left(\frac{A_2}{A_1}\right)^4 \tag{8.8}$$

Consequently, if the Kelvin wave moves from 4 km water depth on to a continental shelf with depth 100 m, the amplitude increases from 0.5 m to 1.26 m. In a similar way,

$$\frac{U_2}{U_1} = \left(\frac{h_1}{h_2}\right)^{3/4} \tag{8.9}$$

If the current speed in the open ocean is 25 mm s^{-1}, as previously calculated, the velocity on the continental shelf will be 400 mm s^{-1}, which is vigorous enough to cause considerable sediment transport.

The shoaling effect of the continental shelf can be illustrated by the tides of the continental shelf bordering the eastern seaboard of USA [1] (Fig. 8.11). This represents an ideal case study because the M$_2$ tidal wave approaches the coastline and shelf edge more or less at right angles, and the shelf width varies sufficiently for any changes to the tidal wave to be attributed to its passage over the shelf. The shelf width is narrowest at Cape Hatteras, and widens both northwards to New England and south-westwards to Georgia. The tidal range increases with the width of the continental shelf, demonstrating the effect of shoaling of the ocean tide. The timing of high water, velocities and range can best be modelled as due to a shoaling ocean tide being damped by friction as it crosses the shelf co-oscillating with a reflection from the coast. These additional effects of sea bed friction and reflection are discussed below.

The transformation of a progressive tidal wave as it shoals may also be enhanced by a funnelling effect as it is constricted into a gradually narrowing cross-section (Section 8.4.4). Reflected waves and friction are likely to have very important effects in such situations.

8.4.2 Tidal co-oscillation in a partially enclosed sea
Oscillation of a standing wave
The tidal water motion, and therefore the tidal sediment transport patterns, in the shallow waters of

the continental shelf, gulfs, straits and estuaries are strongly affected by the configuration of the basin, Coriolis effects and friction. These effects can be explored with the following examples.

A body of water that is semi-enclosed has a natural mode of oscillation. This natural oscillation may resonate with the tidal periods of the ocean tide entering the gulf. If the gulf is rectangular in shape, sufficiently narrow for Coriolis effects to be ignored, and friction is initially neglected, there can be no water flux through the closed end of the gulf and the water motion approximates a *standing wave* (Fig. 8.12). Resonance occurs when the length of the gulf is one-quarter wavelength of the standing wave (and three-quarters of the wavelength, five-quarters and so on). Since the wavelength depends on the period and water depth according to

$$\lambda = T\sqrt{gh} \tag{8.10}$$

the first resonance occurs when

$$l = \frac{1}{4}T\sqrt{gh} \tag{8.11}$$

where l is the length of the gulf, T is the period and h is the water depth. Such a gulf is then said to be acting as a *quarter-wave resonator*. This may explain why some gulfs have exceptionally high tidal ranges at their heads. For example, the Bay of Fundy, eastern Canada, has tidal ranges at springs reaching over 15 m. The bay is about 250 km long and has an average depth of 70 m. The bay should therefore resonate with a period of $T = 10.6$ hours, which is close to the M$_2$ tidal period of 12.42 hours. If the assumptions are valid, this suggests that the Bay of Fundy should experience particularly high tides because of the constructive interference of the tidal and standing wave frequencies. In fact the Bay of Fundy is connected to the Atlantic Ocean by the shallow waters of the Gulf of Maine. Treating the Gulf of Maine and Bay of Fundy as a single system co-oscillating with the Atlantic Ocean tides modifies the estimate of the natural resonant period to even closer to the semi-diurnal period.

Coriolis effects
If the gulf is wider than in Practical Exercise 8.1 (see p. 278), Coriolis effects cause a motion of water across the gulf. The Coriolis acceleration varies according to latitude (equation (8.4)), causing a deflection of a water stream to the right in the northern hemisphere. A current entering the gulf (in the northern hemisphere) is banked up on the right-hand side, the transverse slope vanishing at high water when the

Practical exercise 8.1: Resonance in lakes, gulfs and fjords

The problem of tidal amplification in partially enclosed seas is of sufficient importance and, with certain assumptions, represents a simple enough physical system, to warrant further exploration. In all of the following calculations ignore the effects of friction and treat the water bodies as of constant width so that there are no funnelling effects.

1 We start by considering a body of water closed at both ends, approximating a steep-sided lake (Fig. 8.13). What will be the period of the simplest (fundamental) oscillation in the lake with depth h, length l, where c is the phase velocity (celerity) of the waveform? If the lake is 100 km long and 200 m deep, what is the period of the fundamental oscillation in hours?

2 If the body of water has just one closed end, approximating a shallow gulf with a length of 200 km and a depth of 50 m, what will be the period of the fundamental oscillation?

3 If the dimensions of a fjord are 500 m depth and 100 km length, values typical of Norwegian examples, what will be the wavelength of the fundamental oscillation compared to the length of the fjord? How does this explain the small tidal amplification at the head of the fjord?

Solution

The phenomenon of resonance refers to a body of water having a natural period of oscillation similar to the astronomical tide generated by the Moon and Sun. The first question on lakes allows the basic principles to be appreciated. The second question develops the ideas to the sedimentologically more important situation of a partially closed sea. The third question allows us to examine why long deep inlets like the Norwegian fjords do not experience the tidal amplification characteristic of, for example, the Bay of Fundy.

1 The lake acts like a bath-tub, the simplest oscillation being one in which the water alternately rises and falls at each end (out of phase) with a nodal line across the centre of the bath where

there is no vertical motion of the surface. Since the length of the lake is half the wavelength of the standing wave (or *seiche*), it is known as a *half-wave oscillator*. The water movements constitute standing waves. The time taken for the wave to travel from one end of the lake to the other and back again is the wave period T, equal to twice the length of the lake divided by the celerity. Since the celerity of a shallow water wave is $c = \sqrt{gh}$, the period of the standing wave is

$$T = \frac{2l}{\sqrt{gh}} \qquad (8.12)$$

It is possible for other oscillations to take place in the lake. For example, there may be two nodes in the lake, or three or more. The fundamental oscillation, however, is where there is one node. Using the dimensions for the lake given above, we have $T = 1.25$ hours.

2 In the case of the shallow gulf, the simplest oscillation involves the flowing in of water on the flood tide, and the flowing out on the ebb, with a node at the mouth of the gulf. The length of the gulf is now one-quarter of the wavelength of the standing wave, making it a quarter-wave oscillator. The fundamental period of the standing wave is therefore

$$T = \frac{4l}{\sqrt{gh}} \qquad (8.13)$$

This fundamental quarter-wave oscillation has a period of 10 hours. This is similar to the 12.42 hour period of the semi-diurnal M_2 tide. We should expect considerable tidal amplification at the head of the gulf.

3 The fundamental quarter-wave oscillation for the fjord of length 100 km and depth 500 m is 1.58 hours. This is much smaller than the semi-diurnal tidal frequency. The length of fjord required to resonate with the semi-diurnal astronomical tide is 790 km. This is far greater than the actual length of the fjord. The node is therefore situated far seawards of the mouth of the fjord. Since the fjord does not have a natural oscillation comparable with the semi-diurnal tide, tidal amplification is very small.

current is zero, and is banked up on the left-hand side when the current flows out of the gulf (Fig. 8.14). Note that this transverse oscillation is out of phase

with the longitudinal standing wave oscillation by a quarter of a period.

In the absence of Coriolis effects a nodal line crosses

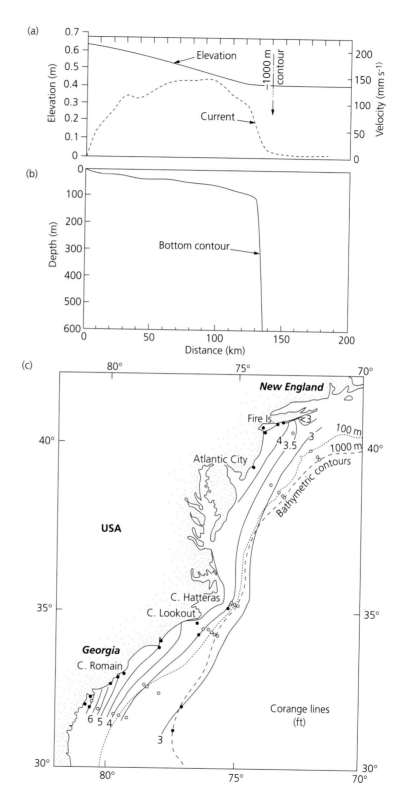

Fig. 8.11 Effects of shallowing of the continental shelf on tidal range and velocity, US eastern coast. (a) Estimated elevation of high water and maximum velocities of tidal currents across the shelf with the profile shown in (b) (off Atlantic City). (c) Estimated mean range of the tide (in feet), showing the amplification of the tide by up to a factor of 2. After Redfield (1958) [4].

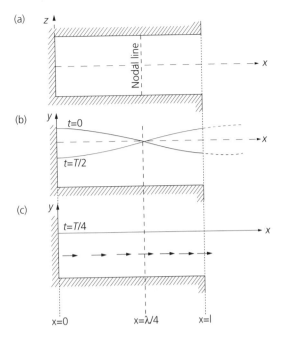

Fig. 8.12 Development of a standing wave due to tidal cooscillation in a partially enclosed sea. (a) Plan view of narrow gulf closed at one end. (b) Elevation at time $t = 0$ and $t = T/2$. (c) Elevation of the water surface and currents at time $t = T/4$.

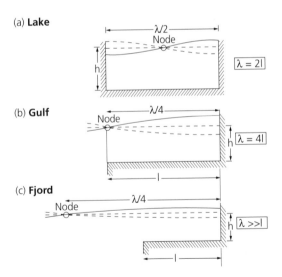

Fig. 8.13 Practical Exercise 8.1. (a) A lake (closed at both ends) has a fundamental oscillation where the length of the lake is half the wavelength (a half-wave oscillator). (b) The gulf acts as a quarter-wave oscillator. (c) In the fjord example the semi-diurnal oscillation requires a wavelength much longer than the length of the fjord. Consequently, there is unlikely to be tidal amplication at the head of the fjord.

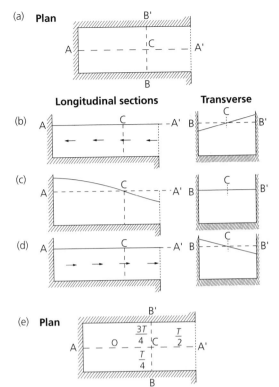

Fig. 8.14 The effects of the Coriolis acceleration on co-oscillation in a gulf. (a) Plan view of the gulf, closed at one end. (b), (c), (d) Transverse and longitudinal elevations at times $t = -T/4$, $t = 0$ and $t = T/4$. (e) The resulting amphidromic system with co-tidal lines radiating from a nodal point at C.

the gulf at a distance of a quarter of a wavelength from the head of the gulf (Fig. 8.12). When Coriolis effects are included, this nodal line becomes reduced to a nodal point by the transverse oscillation (Fig. 8.15). We can now appreciate why amphidromic systems develop, and why the high water crest moves anticlockwise around the amphidromic point in the northern hemisphere.

Frictional effects

Once again neglecting Coriolis effects for simplicity, how does friction affect the co-oscillation in the gulf? The standing wave oscillation can be treated as two progressive waves moving in opposite directions— first towards the head of the gulf, and then towards the entrance. When friction is incorporated, the progressive waves are damped so that the elevation decreases with distance in the wave propagation direction. This reduces the amplitude of the standing wave at the head of the gulf. The reflected wave may also be

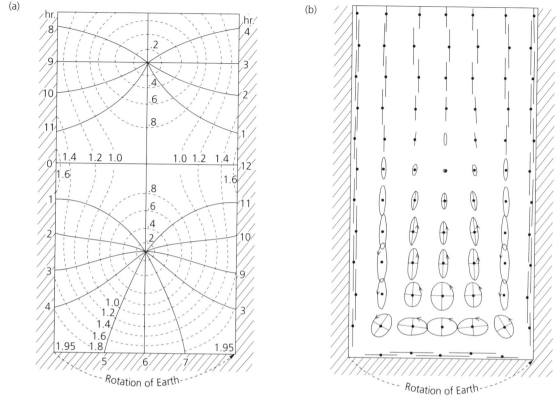

Fig. 8.15 (a) Co-tidal and co-range lines for the reflection of a Kelvin wave in a rectangular bay long enough to produce two amphidromic systems. (b) Current ellipses, showing the faster speeds along the edges of the bay and the transformation from rectilinear current patterns at the entrance to rotary patterns at the head of the gulf caused by reflection. After Taylor (1920) [5].

damped by an incomplete reflection from the head of the gulf by dissipation in shallow water in that region.

We can now introduce Coriolis effects into the problem. The incoming progressive (Kelvin) wave has a greater amplitude on the right-hand side of the gulf (in the northern hemisphere) and the reflected wave a greater amplitude on the left-hand side. However, since the reflected progressive wave is weaker than the incoming Kelvin wave because of being damped by friction and dissipation, the amphidromic point is displaced towards the left-hand side of the gulf from the centreline. Such a point is termed a *displaced amphidromic point* (Fig. 8.16). Frictional losses either on the bed of the gulf, or through dissipation at the head, may be extreme enough to push the amphidromic point beyond the edge of the gulf, causing a *degenerative amphidromic point* (Fig. 8.16). Both displaced and degenerative amphidromic systems

can be seen in the north-west European continental shelf. The North Sea contains excellent examples of amphidromic points displaced to the left of the northward-opening semi-enclosed sea. Degenerative patterns are seen in the English Channel and Irish Sea (Fig. 8.17). A qualitative explanation for the pattern observed, building on what we know from the previous discussion, is as follows.

The tide from the Atlantic approaches the NW European continental shelf as a northward-moving Kelvin wave. Part of this wave heads north-east generating co-oscillations in the English Channel, Irish Sea and Bristol Channel. The English Channel has a configuration which causes it to resonate (as a half-wave resonator) with maximum amplitudes (in opposite phase) at either end and potentially with a nodal line across the centre. However, the combination of friction and Coriolis effects causes the NE-moving Kelvin wave to bank up on the French coast, causing

Head of gulf

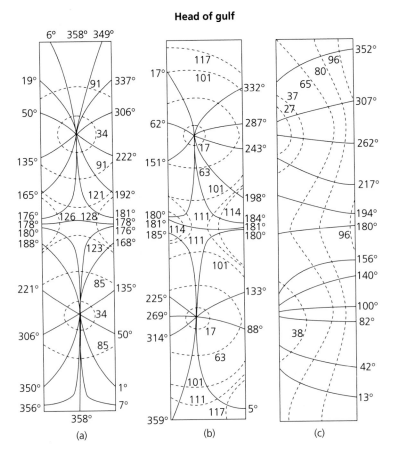

Fig. 8.16 Complete and partial reflection of a Kelvin way at the head of a rectangular gulf. (a) Complete reflection reproduces the pattern seen in Fig. 8.15 with two amphidromic systems. (b) and (c) Show increasing amounts of frictional losses causing partial reflection. This leads first to displaced amphidromic points in (b), and then to degenerative amphidromic points in (c). After Hendershott & Speranza (1971) [6].

high tidal amplitudes there, but the weaker, damped, reflected wave to travel along the southern English coast with lower amplitudes. The potential nodal line degenerates to a nodal point just inland of the southern English coast.

In the North Sea area, the tide enters as a Kelvin wave from the north. Much of the energy is dissipated in the shallow waters of the southern North Sea, so only a weak wave is reflected back northwards. The interference of the incoming Kelvin and reflected waves produces three amphidromic systems, two of them being markedly displaced towards the Danish and Norwegian coasts (Fig. 8.17).

8.4.3 Tidal currents in shallow waters

Description of tidal currents requires information on their speed and direction. If speed and direction are represented by vectors, the tips of the vector arrows would describe an ellipse through a tidal period

(Fig. 8.15b). These *tidal ellipses* may be substantially distorted where, for example, a mean flow is superimposed on the tidal currents, or where the tidal currents are forced up and down a narrow channel. The tidal current pattern may be further complicated by the current ellipses varying with depth in the water column.

We have already seen that tidal heights are the result of the summation of a large number of components of different frequency, and that they are strongly affected by co-oscillation, friction and Coriolis effects. However, tidal currents are more complex still, being affected by local bathymetry and topography such as headlands, constrictions and expansions. In general terms, there is a substantial difference in the tidal current resulting from a progressive Kelvin wave, on the one hand, and a reflected standing wave, on the other. Under a progressive tidal wave the maximum flow velocity occurs under the crest of the wave,

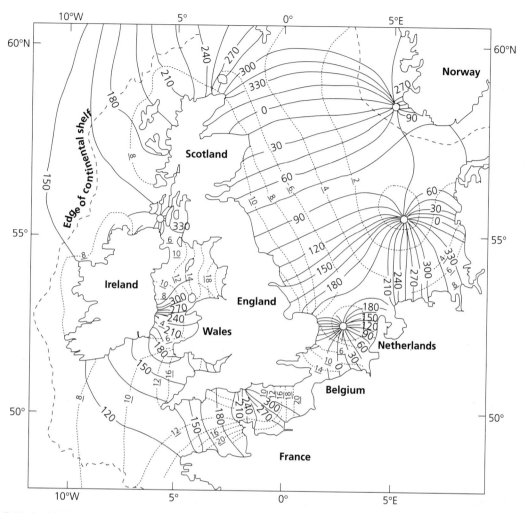

Fig. 8.17 Co-tidal (solid) and co-range (dashed) lines (in feet) for the M_2 tide on the north-west European continental shelf. Note the displaced amphidromic points in the North Sea, and the degenerative amphidromic points in the English Channel and Irish Sea. After Huntley (1980) [7].

corresponding to high tide, and under the trough, corresponding to low tide. Under a standing wave the flow velocity is at a minimum at high tide and low tide, and reaches a maximum mid-way through the period when the water level in the standing wave is horizontal. Consequently, flow velocities in partially enclosed seas are difficult to predict with accuracy.

The variation of flow velocity with depth for a unidirectional flow was discussed in Chapter 4 and developed further in Chapter 5. With tidal currents, unsteadiness is caused by changes of the current vector through a tidal period and, in very shallow waters, by changes in the flow depth. Friction on the sea bed is responsible for the reduction in flow velocity with depth. Friction also has the effect of causing the phase of the current velocity to occur slightly early compared to the phase of the surface elevation. For example, under a progressive wave, the maximum velocity would occur slightly earlier than the maximum elevation. Since frictional effects are concentrated near the sea bed, this should cause a turning (from ebb to flood or vice versa) of the current earlier near the bed than at the surface, a phenomenon recognized in tidal current observations. This is relevant to the problem of sediment transport because the bulk of the sediment transport takes place close to the bed. Seas shallower than 50 m are likely to be strongly affected by sea bed friction for the semi-diurnal M$_2$ tide.

8.4.4 Tides in estuaries

In Chapter 7 the phenomenon of salt wedge intrusion up river mouths was introduced. Tidal waves may extend up estuaries and the rivers feeding them for very long distances, as far as 800 km in the case of the Amazon River. The tidal wave must, of course, be strongly modified as it intrudes, losing energy by friction on the estuary and river bed, opposing the seaward river flow, and by constriction in the progressively narrowing and shallowing channel. The analysis of the funnelling effect is similar to that of shallowing on the continental shelf. We simply modify the analysis by allowing the width of the tidal flow to vary and consider the energy flux along a unit length of estuary. The energy transmitted across the estuary cross-section averaged over one tidal cycle is

$$E = \frac{1}{2}\rho g A U h w \qquad (8.14)$$

where w is the width of the estuary. Since $U = (c/h)A$ and $c = \sqrt{gh}$, this becomes

$$E = \frac{1}{2}\rho g A^2 (gh)^{1/2} w \qquad (8.15)$$

If we now make a major (and false, since we are ignoring friction) assumption that the energy flux is constant for any cross-section of the estuary, then it can be seen that the amplitude of the tidal wave varies inversely as the square root of the width, but the fourth root of the water depth. Funneling is therefore more important than shoaling in amplifying tidal heights. The St Lawrence estuary in north-eastern North America is an example of this effect [3]. The tidal height during springs varies from less than 3 m at Father Point near the mouth of the estuary, to 6 m at Quebec City, 350 km up-estuary (Fig. 8.18).

A further consequence of shallowing and funnelling is that the wavelength of the tidal wave decreases. Since we know already that the tidal height increases, this implies that the wave must become steeper. Eventually, the wave may form a steep front and move rapidly up the estuary in a manner similar to surf on a beach. Such tidal steep-fronted waves are called *bores*. Large bores have been reported in the mouth of the Amazon River. A smaller but better-known example is found in the Severn estuary, SW Britain.

In summary, tidal ranges and current velocities may be high in estuaries despite the fact that the amplitude of the tidal wave is low in offshore water depths. The configuration of the coastline and continental shelf therefore have a major impact on tidal processes and tidal sediment transport.

(a)

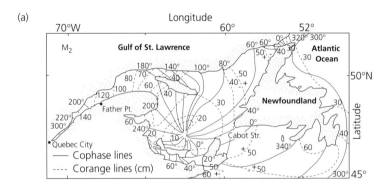

(b) **Tides in St. Lawrence estuary**

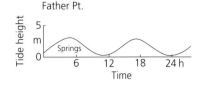

Fig. 8.18 (a) Co-tidal and co-range lines (in cm) for the St Lawrence estuary, Canada. (b) Variation in the tidal heights at springs between Quebec City and Father Point. After Godin (1980) [8].

8.4.5 Dissipation of tidal energy

The dispersal of river-derived sediment is controlled by a number of factors such as outflow dynamics and the energy of the receiving basin (Chapter 7). Clearly the tidal energy is of great importance in this context. It is important to know, therefore, something about the dissipation of tidal energy by friction as ocean tidal waves cross the shelf to the coast.

The rate of transmission of energy by a tidal wave has been previously introduced in studying the transformation of a tidal wave as it moves into the shallower water of the continental shelf. In the example in Section 8.4.1, we ignored friction. The total energy flux of a tidal wave entering a closed sea can be obtained by taking the same formulation for the energy flux (equation (8.6)) per unit width normal to the wave propagation direction, and integrating it across the entrance to the closed sea. This tidal energy flux does work in overcoming friction on the sea bed. This work done is the product of the bottom shear stress and the tidal velocity integrated over the entire area of the closed sea bed. Since the bottom shear stress is related to the friction coefficient, which is typically 2×10^{-3}, the dissipation of tidal energy depends on the surface area of sea bed over which the tidal energy is dissipated, the velocity (cubed) and the friction coefficient.

The continental shelves and shallow seas are a small proportion of the total surface area of the oceans. The Moon's (and Sun's) gravitational forces therefore do most work over the open oceans, whereas most tidal energy is dissipated in shallow seas. There is therefore a predominant transfer of tidal energy from the oceans to the coast. Some shallow seas are great consumers of tidal energy, such as the NW European shelf, the Sea of Okhotsk, Hudson Strait, and the Patagonian shelf of Argentina. We should expect high shear stresses and therefore high tidal sediment transport rates in these regions.

Fig. 8.19 (*Right.*) Velocity–time curves for tidal currents and their bedload transports (stippled). Time when current velocity exceeds the critical velocity for bedload transport is hatched. ΔU is the excess velocity. (a) Tidal currents are perfect sine waves, resulting in no net sediment transport over the tidal cycle. (b) Tidal wave distortion by shallow water effects, resulting in a short but vigorous flood and a slow but long duration ebb. Since the bedload sediment transport rate is proportional to the cube of the velocity, there is much greater sediment transport by the flood current, causing a net sediment transport over the tidal cycle. (c) A superimposed steady current U_{st} on a sinusoidal current pattern also results in a net sediment transport through a tidal cycle. After Allen (1985) [9].

8.5 Sediment transport under tidal flows

The distinctive character of tidal flows compared to unidirectional flows is that they are periodic on a time-scale of hours. The effect of this periodicity on

(a) Symmetrical oscillatory

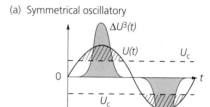

(b) Distorted

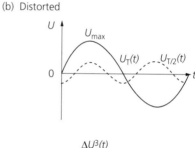

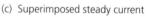

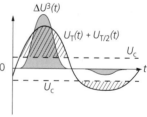

(c) Superimposed steady current

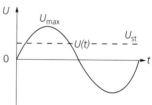

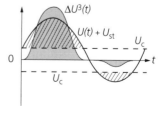

sediment transport can be approached by initially considering the tidal currents to vary according to a perfect sine wave (Fig. 8.19, see p. 285). Let us assume that the sediment transport rate of bedload is proportional to the cube of the excess velocity (as discussed in Chapter 5)

$$q_b \propto (U(t) - U_c)^3 \tag{8.16}$$

where $U(t)$ is the sinusoidal tidal current velocity as a function of time, and U_c is the critical velocity for the transport of bedload. Since the tidal current is symmetrical, the sediment transports on the flood and ebb currents are identical and there is no net sediment transport over the tidal cycle (Fig. 8.19a).

The perfect sinusoidal form of the tidal currents may, however, be distorted by shallow water effects. Conserving discharge, one of the currents may be higher in speed but shorter in duration, and the other current may be lower in speed but longer in duration (Fig. 8.19b). Since the sediment transport rate is proportional to the cube of the excess velocity, there will be a higher sediment transport on the short-duration but vigorous tidal current, resulting in a net sediment transport in one direction over the tidal cycle.

Furthermore, the tidal currents may be super-imposed by a steady unidirectional current. This might be a wind drift current (Chapter 9) super-imposed on tides, for example. The addition of the sinusoidal tidal currents and the steady current results in an imbalance in the discharge of water over a tidal cycle, and an imbalance in the sediment transport, with a net sediment transport in the direction of the steady current (Fig. 8.19c).

In these and other ways, tidal currents may cause a net drift of sediment down a tidal sediment transport path. The integration over time of this net drift can be recognized in the internal structure of tidal sandwaves (Section 8.5.1).

A distinctive characteristic of sedimentation in tidal environments is the preservation of fine-grained sediment deposited during periods of weak currents or slack water (see also Section 7.4). On the shelf where the tidal current ellipses are strongly rotational, there is no time during the tidal cycle when current speeds are almost zero. In addition, background wave activity is likely to keep the fine sediment grains suspended.

Consequently, *fine sediment drapes* are not expected to be well developed in these environments. However, where the tidal current ellipses are more rectilinear, so that there are distinct periods of slack water around high and low tide, and wave activity is restricted, muddy and silty sediment layers may drape the underlying bedform. Such drapes may be preserved beneath new accretions of sand transported by tidal currents, or may be partly consolidated and then eroded to form small chips of fine sediment. Tidal sediments with fine sediment drapes are particularly common in estuaries. They have been ubiquitously described from ancient tidal sediments such as the Lower Cretaceous of central and south-eastern England, the Jurassic of Luxembourg, the Miocene of Switzerland and France, the Eocene of the southern Pyrenees, and the Cretaceous of the Western Interior Seaway, USA [10].

8.5.1 Tidal sandwaves

Periodic, flow-transverse bedforms which persist over many tides, but which may have smaller bedforms such as ripples and dunes superimposed on their surfaces that are created and destroyed with every tide, are known as *sandwaves*. This is an important distinction, since sandwaves are more the equivalent of wave ripples under an oscillatory flow (Section 8.7.2) than dunes under a unidirectional flow (Section 5.4), which is one reason why tides and waves are discussed together in this chapter. Sandwaves vary greatly in size and geometry. They occur in estuaries, tidal straits and on the continental shelf. Their wavelengths are typically between 50 and 300 m and heights between 1 and 8 m, though larger examples are known. The sandwaves imaged on the sea bed of the English Channel are fine examples (Fig. 8.20), displaying superimposed dunes on their flatter limbs and a predominantly unidirectional internal stratification [11].

The likely internal architecture of sandwaves under a range of flow asymmetry has been explored by J.R.L. Allen [12]. He believed there to be two end-members (Fig. 8.21), one developed when the tidal currents were nearly symmetrical, and the other when there was a large inequality in the sediment-transporting capacity of the two tidal streams. Where the tidal currents are nearly symmetrical, the sandwaves migrate very slowly by an accretion of sediment on low, irregular sandwave surfaces. Each tide results in a migration of dunes over the bedform surface with much reworking of the previous tide's deposits. Under highly asymmetrical tides tall avalanche faces may be developed, with little modification by the reverse tide. These sandwaves should migrate more rapidly than the previous class. Both types of sand-wave have been recognized in ancient tidal deposits

(a)

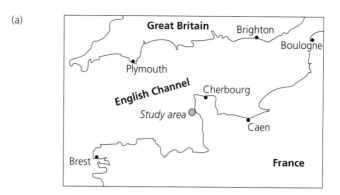

(b)

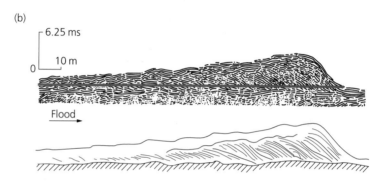

Fig. 8.20 (a) Location of the Surtainville area in the Golfe Normand-Breton of the English Channel. (b) An example of the internal structure and gross morphology of a tidal sandwave imaged in milliseconds and interpreted in the line drawing. After Berné *et al.* (1988) [11].

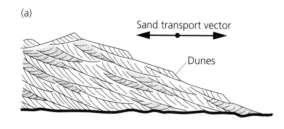

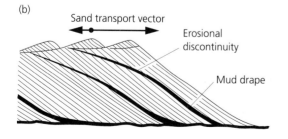

such as the Miocene sediments of the peri-Alpine seaway in France and Switzerland [13, 14].

The record of net sediment transport over ancient sandwaves has been documented by the recognition of sediment increments over tidal cycles. The sediment deposited under one tidal cycle of flood and ebb is known as a *tidal bundle*. Tidal bundles can be recognized and measured because they are bounded by the fine sediment deposited during slack water. In subtidal environments (below low tide) two mud or fine sediment drapes may be preserved in a tidal bundle—one at slack water at high tide and one at slack water at low tide—whereas in intertidal environments (between low and high tide) only one slack

Fig. 8.21 (*Left.*) Simplified and speculative end-member models of the internal structure of sandwaves in relation to the time–velocity asymmetry of the tidal regime. After Allen (1980) [12].

(a)

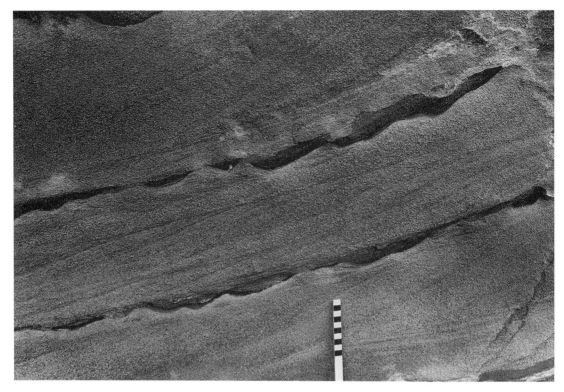

(b)

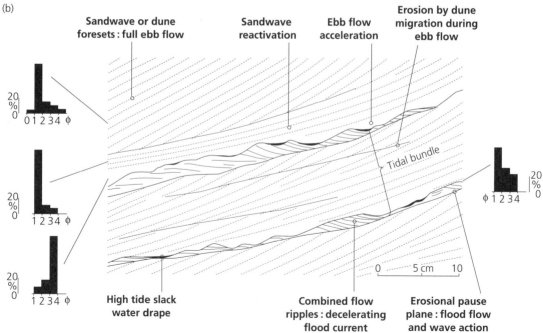

Fig. 8.22 (a) Tidal bundles preserved in a Miocene tidal sandwave, Switzerland; (b) interpretative sketch with grain size distributions of sediment samples. After Allen & Homewood (1984) [15].

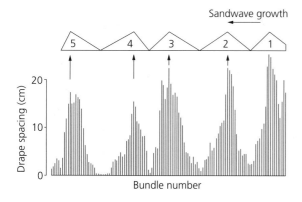

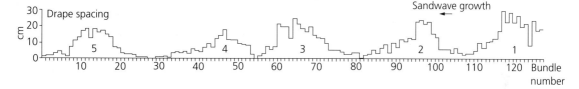

Fig. 8.23 Drape spacings (bundle thicknesses) from the sandwave illustrated in Fig. 8.22(a) and the same data replotted using a three-point moving average, showing the alternating asymmetry of the neap–spring variation in bundle thickness. After Allen & Homewood (1984) [15].

water drape is likely to be well preserved. The low-tide drape on the sandwave lee-side is commonly reworked by the flood tide, whereas the high tide drape is usually well preserved beneath new additions of ebb tide sediment being transported down the sandwave lee-side (Fig. 8.22). When measured roughly in the direction of the tidal flows the tidal bundles com-monly show a marked cyclicity in thickness that can be attributed to neap–spring variations [15–17] (Fig. 8.23).

8.6 Wind-generated waves

Surface waves on a pond, lake, sea or ocean are generated by the flow over the surface of the water of the wind (Table 8.1). On small ponds it is possible to see very small waves move rapidly across the water surface in response to short gusts of wind. On larger bodies of water one is familiar with a range of surface wave characteristics depending on the weather. On coasts fronting large seas and oceans widely separated, long-period waves may be seen even on a calm day. These commonplace observations give a clue as the factors influencing surface waves, which might include:

- the speed of the wind;
- the duration of the wind;

Table 8.1 Waves classified by period.

Period	Wavelength	Name
0–0.2 s	Centimetres	Ripples
0.2–9 s	Up to about 130 m	Wind waves
9–15 s	Hundreds of metres	Swell
15–30 s	Many hundreds of metres	Long swell or forerunners
0.5 min–hours	Up to thousands of kilometres	Long-period waves including tsunamis
12.5, 25 h, etc.	Thousands of kilometres	Tides

• the *fetch* of the wind, that is, the distance over which it blows uninterrupted by land.

We would further deduce that waves may travel long distances from their place of generation, and that waves of different dimensions travel at different speeds. But how does the wind impart energy on the water surface in such a way as to develop surface waves?

When the wind blows over a water surface we can treat the problem once again as one of instability (Section 5.4.6). The wind is compressed over upward undulations of the surface and expanded over downward undulations, which, from the Bernoulli equation, result in pressure differences, causing the instability to grow. Gravity balances the tendency for the instabilities to continue to grow, thereby establishing an equilibrium waviness to the water surface for the given wind conditions. The surface waves caused by the shear of the wind fall within a range of wave period from about 1 s to 15 s (Fig. 8.1). These wind waves are a type of *progressive gravity wave* since the waveform travels in one direction, and their form is the result of gravity acting on the surface disturbance. Unlike tidal waves, Coriolis effects are unimportant in the movement of progressive gravity waves.

8.6.1 Small-amplitude wave theory

There are a number of wave theories which vary in the complexity of their mathematics, and which apply to waves in different environmental settings. It is a generalization to say that waves in shallow water are considerably more complex than those occurring in open water where the water depth is not important. There are a number of excellent treatments of the common wave theories to which the interested reader is referred [18]. In this chapter it is sufficient to restrict our analysis to the simplest but most widely applicable wave theory.

The simplest wave theory makes the assumption that the amplitude of the wave is small compared to its wavelength. This is called *small-amplitude wave theory*. It is also called *linear* or *Airy wave theory* because the wave equations involve only the unit power of the wave height, rather than H^2 and so on. Consider waves with one frequency passing a stationary point in the water, such as a post. The crests of successive waves pass the post at regular intervals, the wave period. The wave period is therefore

$$T = \frac{L}{c} \tag{8.17}$$

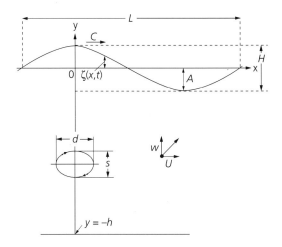

Fig. 8.24 Notation for wave theory. A surface wave of wavelength L, celerity C and height H (twice the amplitude A) produces a water surface elevation (relative to the still water level) ς which varies with horizontal distance x and time t. The wave is moving in a water depth h. The diameter of the orbital excursions of the water particles d and their maximum velocities U varies as a function of h.

where L is the wavelength and c is the celerity (Fig. 8.24). Two parameters are commonly used in physical oceanography which express the wavelength and wave period in a different form. These are the *wave number*, k, and the *angular* (or *radian*) *frequency*, σ:

$$k = \frac{2\pi}{L} \tag{8.18}$$

$$\sigma = \frac{2\pi}{T} \tag{8.19}$$

Consequently, $c = \sigma/k$. The simplest formulation for the elevation of the water surface is described by the sine function

$$\varsigma = A \sin(kx - \sigma t) \tag{8.20}$$

where ς is the vertical height of the water surface and x is the horizontal coordinate, A is the amplitude equal to half of the waveheight and t is time. If x is held constant, this function describes the variation of the elevation of the water surface at a fixed point. If t is held constant, the function describes the shape of the water surface, the wave profile.

The fundamental relation between the angular frequency and the wave number is known as the *wave dispersion relation*. It is given by

$$\sigma^2 = gk \tanh kh \qquad (8.21)$$

Since we know that $c = \sigma/k$, the general equation for the celerity, or phase velocity, is

$$c = \left[\frac{gL}{2\pi} \tanh\left(\frac{2\pi h}{L} \right) \right]^{1/2} = \left[\frac{g}{k} \tanh kh \right]^{1/2} \qquad (8.22)$$

where tanh is the hyperbolic tangent, h is the water depth and g is acceleration due to gravity. Now if the wavelength is small compared to the water depth, kh becomes large and $\tanh kh \to 1$. Consequently in 'deep water' the celerity is equal to $(g/k)^{1/2}$. Alternatively, if the wavelength is large compared to the water depth, $\tanh kh \to kh$. In 'shallow water', therefore, the celerity is equal to $(gh)^{1/2}$. This result has already been made use of in considering the speed of long-wavelength progressive tidal (Kelvin) waves in the ocean.

We can immediately categorize waves according to the ratio of their wavelength to water depth (Fig. 8.25). If $L < 2h$, the wave is in 'deep water' and if $L > 20h$ it is in 'shallow water'. 'Intermediate' water depths exist in between. As a deep water wave travels into shallow water, the retarding effect of the sea bed causes it to slow down. Whereas the speed of deep water waves depends on their wave number and therefore their wavelength—that is, they are dispersive waves—the speed of a shallow water wave is dependent only on the water depth. If there is a spectrum of waves generated at a certain location in the open sea, the longer-wavelength (longer-period) waves will travel faster than the shorter wavelength (shorter-period) waves, arriving at the coast earlier.

Since dispersive waves travel with different speeds according to their wavelength and wave period, individual waves are continuously catching up or being caught by neighbouring waves. The interference of wave trains of different frequency produces an amplification or damping of the wave amplitudes (Fig. 8.26), resulting in the creation of *wave groups*. Individual waves move through a group and die out at the front, while other waves join the group from the back. The wave group therefore moves with a speed that is slower than the celerity of individual waves. This is an important concept in relation to the energy expended by waves. Since individual waves grow and die, their energy is not of profound importance. The wave energy of a sea can, however, be estimated from the *group speed*.

The speed of the wave group is evaluated from the dispersion relation to give

$$c_g = \frac{c}{2} \left[1 + \frac{2kh}{\sinh(2kh)} \right] \qquad (8.23)$$

where c_g is the group speed, c is the phase speed of the individual wave and sinh is the hyperbolic sine. For 'shallow water' waves the group speed is the same as the celerity (phase speed) (they are not dispersive waves), whereas 'deep water' waves have a group speed that is half of the celerity of individual waves.

Orbital motion of water particles under waves

Although the wave form moves over the water surface with a phase speed or celerity, the water particles do not move with the same speed. Instead, they describe orbital motions under the wave, but the shape of the orbits varies between 'deep water' waves and 'shallow water' waves. Both the horizontal and vertical components of the velocity of the water particles, and the pressure due to the wave (in excess of or as a deficit from hydrostatic) vary as the crest, cross-over and trough pass over the surface (Fig. 8.25). Under the crest, the wave pressure is at a maximum, the horizontal velocity of the water particles is at a maximum and in the same direction as the wave propagation. Under the cross-over the wave pressure is zero (the pressure is hydrostatic), the vertical velocity is at a maximum (upwards) and the horizontal velocity is zero. Under the trough the wave pressure is at a minimum (it is less than hydrostatic), the horizontal velocity is at a maximum and in the direction opposite to the propagation direction. Under the second cross-over, the wave pressure is once again zero and the vertical velocity is at a maximum (downwards). If the sea bed is within reach of the waves, it therefore experiences a periodic variation in pressure due to the passage of surface waves. Under large waves, this *cyclic wave loading* may be great enough to cause failure of the sea bed sediment.

Under 'deep water' waves the water particles describe circular orbits which decrease in diameter downwards. The diameter is given by

$$d_y = H \exp(ky) \qquad (8.24)$$

where y is the height in the water column measured negative downwards from the still water level, and H is the wave height equal to twice the amplitude. The diameter at the surface is found by putting $y = 0$, showing that the orbit has a diameter equal to the wave height. At a depth of one wavelength below the surface, the orbital diameter is only 0.2% of its surface value.

(a)

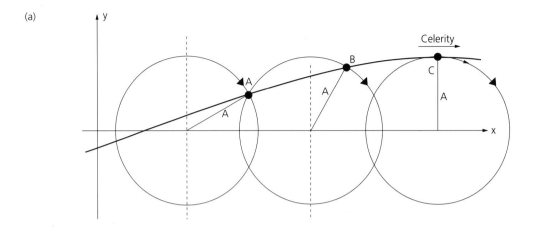

(b)

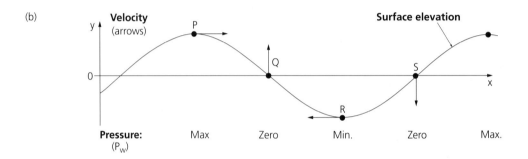

(c) (d)

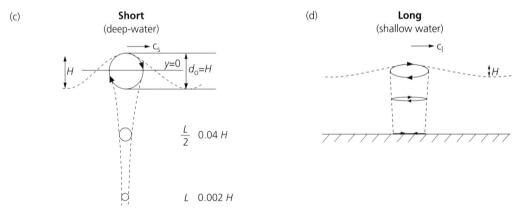

Fig. 8.25 (a) Water particles at the surface describe circular orbits of diameter equal to the wave height. The three water particles A, B and C each take part in circular orbits, with the maximum forward motion being when the particle is located at a wave crest, and the maximum reverse motion being when the particle is located at the trough. (b) The passage of waves causes a cyclic variation in pressure and velocity. (c) In water which is deep compared to the wavelength of the wave ($h/L > 0.5$) the orbits decrease in diameter exponentially with depth. At a water depth equal to the wavelength, the orbital diameter has reduced to $0.002H$. (d) In water which is shallow compared to the wavelength ($h/L < 0.05$), the water particle orbits do not decrease in diameter with depth, but instead take the form of progressively flattened ellipses.

(a)

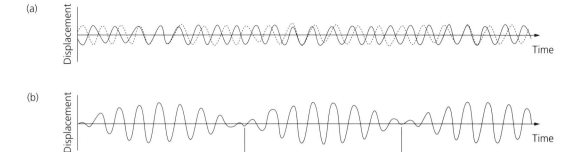

(b)

Fig. 8.26 The interference of waves of different frequency (different wavelengths but similar amplitudes) (a) causes wave groups to form (b). The wave groups travel with a speed which is different to the celerity of individual waves. After Open University (1989) [19].

Under 'shallow water' waves, the orbits of water particles are elliptical at the surface, with vertical diameters equal to the wave height, but horizontal diameters that are stretched. With increasing depth, the horizontal diameter decreases only very slightly, whereas the vertical diameter decreases linearly with depth. At the sea bed there is simply a to-and-fro movement with a diameter only slightly reduced from the surface value of $HL/2\pi h$.

Whereas 'deep water' waves do not 'feel' the sea bed, therefore, and so are of no great importance in terms of near-bed sediment transport, 'shallow water' and 'intermediate' waves have the potential to exert high shear stresses on the sea bed.

Since water velocities decrease with depth, the forward velocity at the top of an orbit must be greater by a small amount than the backward velocity at the bottom of the orbit. There must therefore be a net drift of water in the direction of the wave propagation (Fig. 8.27). Furthermore, the net drift must be greatest where the vertical diameters of the orbits are largest, at the surface. Consequently, the net drift under 'deep water' waves decreases with depth, being given by

$$U_{mt} = A^2 k^2 c \exp(2ky) \qquad (8.25)$$

where U_{mt} is the mass transport, known as the *Stokes drift*. Since the speed of orbital motion of water particles is $Akc \exp(ky)$, the Stokes drift relative to the orbital speed at a given depth can be easily calculated (Practical Exercise 8.2). It is generally rather small, but must result in an accumulation of water in the nearshore zone. This net accumulation due to the Stokes drift is compensated for by the occurrence of narrow jets of seaward-moving water close to the

(a)

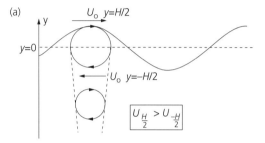

(b)

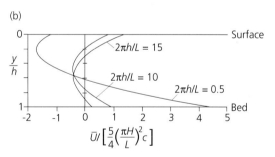

Fig. 8.27 (a) Conceptualization of the Stokes drift. Since the orbital velocities decrease with depth, there must be a small difference between the speed of the forward stroke and the speed of the reverse stroke for the same water particle. This results in a small net drift in the direction of wave propagation. (b) Theoretical wave drift velocities in a rectangular channel of finite length for three values of $2\pi h/L$. Negative values indicate drift up-channel, that is, in the opposite direction to wave advance. After Longuet-Higgins (1953) [20].

beach, known as *rip currents*. Since the magnitude of the Stokes drift is dependent on the wave height, rip currents are commonly located where the breaker height is smallest (Fig. 8.28).

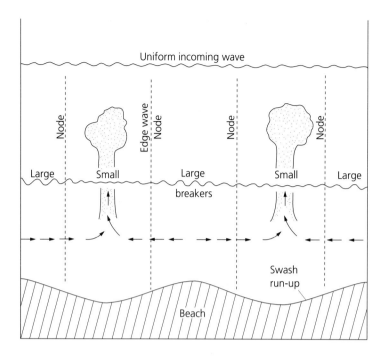

Fig. 8.28 Shoaling waves cause a build-up of water in the littoral zone which is compensated for by the action of seaward-flowing rip currents. These rip channels are commonly located where the breaker height is smallest.

8.6.2 Transformations in shallow water

A number of transformations of waves take place when they travel into shallow water, affecting both the surface wave form and the underlying water motion. These transformations eventually lead to breaking.

The transformations taking place are shown graphically in Fig. 8.29. Of particular interest are the orbital velocity and orbital diameter near the bed. These are given by (*see opposite*)

Practical exercise 8.2: The Stokes drift under progressive waves in 'deep water'

1 Waves of height 1 m and period 10 s are propagated in deep water. Calculate the wave number and radian frequency of these surface waves. What is the celerity (phase velocity), c?

2 Calculate the maximum orbital velocity and the Stokes drift under these waves (i) at the surface by setting $y = 0$, and (ii) at a depth of 10 m (where $y = -10$ m).

3 How large is the mass transport compared to the maximum orbital velocity as a percentage at the two depths?

Solution

1 The radian frequency can immediately be found from the wave period since $\sigma = 2\pi/T = 0.6284$. The wavelength in deep water can be found from the wave dispersion relation by allowing kh to become very large, in which case $\tanh kh \to 1$ and the deep water wavelength L_∞ is then given by

$$L = L_\infty = \frac{gT^2}{2\pi} \qquad (8.26)$$

The deep water wavelength is therefore 156 m and the wave number $k = 2\pi/L = 0.0402$. The celerity can be found either from $C = \sigma/k$ or directly from $C = L/T$, giving a value of 15.6 m s^{-1}.

2 The maximum orbital velocity is found from $U_{max} = Akc \exp(ky)$, giving 0.31 m s^{-1} at $y = 0$, and 0.21 m s^{-1} at $y = -10$ m. The Stokes drift can be found from $U_{mt} = A^2k^2c \exp(2ky)$, giving values of 6.3 mm s^{-1} at $y = 0$ and 2.8 mm s^{-1} at $y = -10$ m.

3 The mass transport velocity is 2% and 1.4% of the maximum orbital velocity at the surface and 10 m depth, respectively. This shows that the relative importance of the Stokes drift decreases with water depth in deep water. The mass transport velocity, however, increases strongly compared to the orbital velocity as the wave moves into shallow water (Section 8.6.2).

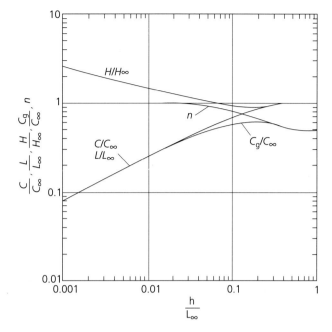

Fig. 8.29 The transformation of Airy waves in shallow water as functions of the ratio of water depth to deep water wavelength, h/L_∞. The subscript ∞ always refers to deep water values. C_g is the group velocity, n is a shoaling coefficient relating wave celerity to group velocity. After Wiegel (1964) [21].

$$U_{y=-h} = \frac{\sigma A}{\sinh kh} \quad \text{(8.27)}$$

$$d_{y=-h} = \frac{H}{\sinh kh} \quad \text{(8.28)}$$

$$\therefore U = \frac{\pi d}{T} \quad \text{(8.29)}$$

The mass transport at the bed can be found from

$$U_{mt} = \left\{ \frac{5}{4} \frac{kA}{\sinh kh} \right\} U_{max} \quad \text{(8.30)}$$

where U_{max} is the maximum orbital velocity near the bed. The factor in the curly brackets is clearly the proportionality between the orbital and mass transport velocities near the bed. For the wave conditions given in Practical Exercise 8.2 ($T = 10$ s, $H = 1$ m), this proportionality is 0.7% in 50 m water depth, 1% in 40 m, 1.7% in 30 m, 2.8% in 20 m and 6.1% in 10 m, showing that as the wave shoals the mass transport velocity becomes increasingly important relative to the maximum orbital velocity. This is of significance in understanding the migration and internal stratification of ripple-marks produced under waves. However, it should be recognized that as waves travel into shallow water they may be better described by a theory other than small-amplitude wave theory. These higher-order theories also involve a strong mass transport in the direction of wave propagation, the end-member in very shallow water depths being a *solitary wave* in which the motion is entirely translatory. Solitary waves characterize the surf zone.

Wave refraction

The speed of a small-amplitude 'shallow water' wave is determined only by the water depth, irrespective of the wavelength and wave period. As the wave moves into shallower water the phase speed slows down and therefore for the same wave period the wavelength decreases. This has the effect of increasing the steepness of the wave form. If the waves are shoaling on to a sea bottom which is aligned at an angle to the wave approach, the slowing down and shortening of wavelength will take place earlier at one point along a wave crest than at another (Fig. 8.30). Consequently, the wave crests close to shore will condense, steepen and orient themselves almost parallel to shore before breaking as surf. This is known as *wave refraction*. Oblique wave approach commonly results in the breaking wave running up the beach at a slight angle compared to the mean slope. The forward rush of the wave (*swash*) is therefore at an angle to the backward motion (*backwash*), resulting in a zigzag movement of sediment along the beach (Fig. 8.31). This is known as *longshore drift*.

Wave refraction will also take place where the sea bottom is irregular, perhaps associated with the presence of bays and headlands. If orthogonals are drawn at right angles to the wave crestlines, called *wave rays*,

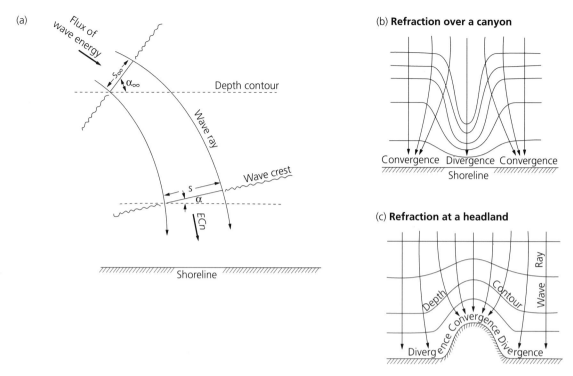

(a)

(b) Refraction over a canyon

(c) Refraction at a headland

Fig. 8.30 Wave refraction due to (a) oblique approach, where the wave energy flux *Ecn* is conserved; (b) divergence over a submarine canyon; (c) convergence on a headland.

they can be seen converging on the headlands and diverging into bays (Fig. 8.30). If the energy of the wave train is constant per unit width of wave crest in 'deep water', this implies a concentration of energy on the headlands. In general terms, this should result in erosion of headlands and deposition of sediment in bays. Waves may also be refracted around spits, headlands and islands.

Off major river mouths, the continental shelf is commonly incised deeply by submarine canyons. Submarine canyons may have large effects on the refraction of storm waves. The canyon found off the Hudson River in the north-eastern USA is an example (Fig. 8.32). The wave rays for storm waves of period *c.* 12 s approaching the shelf from the SSE diverge over the greater water depths over the canyon. This has the effect of concentrating wave energy on the mouth of the Hudson River (New York City) and the adjacent part of Long Island. The lowest wave energy during storms should be at the New Jersey coast known as Long Branch. This would be the best place for fishermen to shelter from a storm [22]. The focusing of wave energy by refraction also has implications for

delta morphology and coastal morphologies in general (Chapter 7).

Wave breaking

Waves may become unstable and break under two different sets of circumstances. In the open sea, waves may become too steep when the ratio of wave height to wavelength exceeds a critical value, although the ratio of H/L will normally be most strongly affected by waves steepening as they approach shore. In theory waves may exist up to steepnesses of $H/L = 1/7$, but real waves seldom exceed $H/L = 1/12$. As the steepness increases, the sinusoidal wave form changes to a series of peaked crests, which eventually break. Breaking may also take place at values of $H/L < 1/12$ if the water depth has become too shallow. In this case the parameter controlling breaking is the ratio of the wave height to the water depth, H/h. Waves will always break at about a value of $H/h = 0.8$ whatever their steepness.

As waves shoal, they become progressively more asymmetrical and the water motion becomes highly translatory towards shore. The wave motion then

(a)

(b)

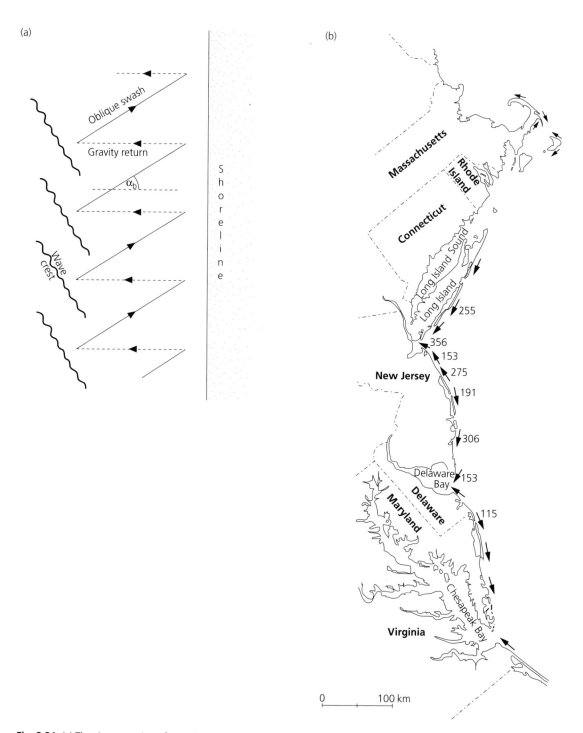

Fig. 8.31 (a) The zigzag motion of particles along a steep beach face caused by oblique approach of waves. (b) Drift directions in the littoral zone of the north-eastern USA, with magnitudes in thousands of cubic metres per year. After Johnson (1956) [23].

(a)

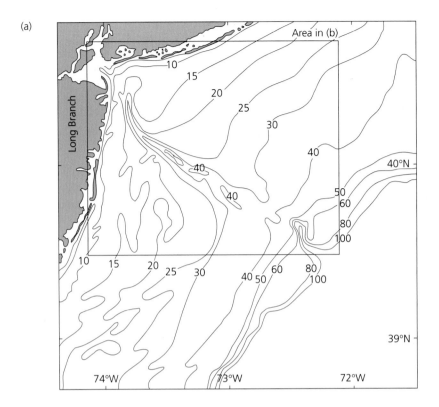

(b)

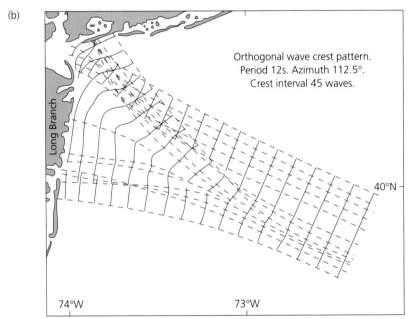

Orthogonal wave crest pattern.
Period 12s. Azimuth 112.5°.
Crest interval 45 waves.

Fig. 8.32 Effects of the Hudson Canyon on wave refraction. (a) Barthymetry off New York harbour at the mouth of the Hudson River. (b) Waverays (orthogonals to wave crestlines), showing the low wave energies at Long Branch beach. After Kinsman (1965) [22].

approximates a solitary wave, eventually breaking as surf. The tidal bores of estuaries (Section 8.4.4) are also solitary waves.

Wave energy and power

To the casual observer on a harbour wall or promenade on a stormy day there can be no doubt as to the large amount of energy transmitted by waves. The passage of surface waves is associated with kinetic energy of water particle motions, and gravitational potential energy due to the vertical displacement of the water surface. We can think of the kinetic energy being continuously transformed into potential energy, and back again into kinetic energy as the wave passes. Over a wave period the total energy per unit area of sea surface for a progressive sinusoidal wave is

$$E = \frac{1}{8}\rho g H^2 \qquad (8.31)$$

that is, it is proportional to the square of the wave height. This energy is transmitted by the waves at a certain rate. Since individual waves join wave groups at the rear and die at the front, the wave energy per unit surface area of the sea must be transmitted at the group speed, which for 'deep water' waves is half the celerity of individual waves. The rate at which energy is transmitted per unit width of wave crest is the *wave power*, given by

$$\omega = E c_g \qquad (8.32)$$

where c_g is the group speed.

Clearly there is a tremendous amount of wave energy, ultimately derived from the action of the wind over the sea surface, expended by waves shoaling and breaking on the world's coastlines. The power reaching the world's shorelines is estimated as about $2 \times 10^{12}\,\text{W}$. The efficient harnessing of this energy source is the goal of scientists hoping to replace the dwindling energy sources from the burning of fossil fuels.

A sea surface can be analysed by breaking the wave energy into its component frequencies. The energy is proportional to the square of the wave height. If H^2 is plotted against the frequency, it is seen that for fully developed seas in which the wind has been blowing for a sufficiently long time over a sufficiently long fetch for the effects of duration and fetch to be ignored, the peak in the energy occurs at different frequencies for different wind speeds (Fig. 8.33a). Where the fetch is limited, there is also a limit to the wave heights and therefore wave energies that can be produced under any wind speeds (Fig. 8.33b). This is the basis for

the understanding that small seas, lagoons and lakes have an upper bound on the wave periods and wave energies that they can support. This is an important result that has been made use of in hindcasting studies of ancient wave conditions and palaeogeographies (Section 8.7.2).

8.7 Sediment transport under waves

The essential difference between sediment transport under a steady unidirectional flow, as outlined in Chapter 5, and sediment transport under waves is that in the latter particle accelerations and decelerations are very important in the short-period reversals of flow near the bed. Whereas a turbulent boundary layer progressively develops when a unidirectional steady current moves over a boundary (such as a flat plate), under oscillatory flows the boundary is never able to develop fully. Consequently, it is always thinner than for the equivalent fluid speed in the unidirectional case. Since the boundary layer is thin, the velocity gradients close to the bed must be large, and the shear stresses on the bed must also be large. This is why waves are very efficient at stirring sediment up from the bed, but not in causing long-distance advection. The long-distance transport of sediment under a stormy sea is caused by unidirectional currents upon which the wave motion is superimposed (Chapter 9).

8.7.1 The threshold of sediment movement under waves

In the previous section the water motion under waves in 'deep' and 'shallow' water was introduced. Of particular importance in studying the threshold of sediment motion is the maximum horizontal speed of water particles close to the bed.

For movement of grains on a bed the maximum orbital velocity must exceed the threshold velocity of the grains. We should expect the nature of the wave boundary layer to have some influence on the threshold condition. It is found that for laminar boundary layers or turbulent boundary layers with a viscous sublayer—(that is, corresponding to hydraulically smooth flow (Chapter 5))—the threshold condition is given by

$$\frac{\rho U_{max}^2}{(\rho_s - \rho)gD} = 0.21\left(\frac{d_0}{D}\right)^{1/2} \qquad (8.33)$$

corresponding to grain sizes D of up to 0.5 mm, where d_0 is the near-bed orbital diameter, ρ_s is the sediment density and ρ is the density of water. In coarser grain

(a)

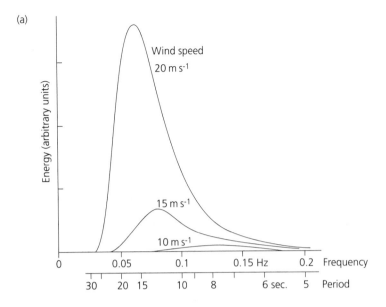

(b)

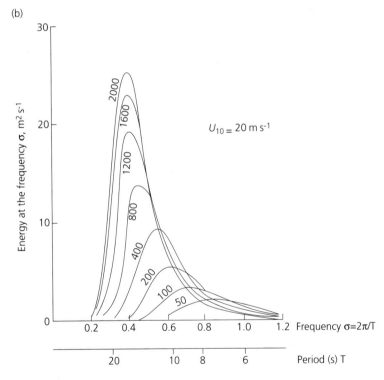

Fig. 8.33 (a) Idealized spectra of wind wave energy in relation to wave period (or frequency) for three wind speeds over a fully developed sea. (b) The effects of fetch for a given wind speed of 20 m s^{-1} measured at 10 m above the water surface, after Liu (1971) [24].

sizes the viscous sublayer disappears, corresponding to hydraulically rough flow:

$$\frac{\rho U_{max}^2}{(\rho_s - \rho)gD} = 0.46\pi \left(\frac{d_0}{D}\right)^{1/4}$$ (8.34)

The threshold condition can also be expressed in terms of a dimensionless stress o and a grain Reynolds number Re_* in order to make the analysis similar to that under a unidirectional current [25]. The overall shape of the threshold curve is similar to the unidirectional case (Fig. 8.34).

(a)

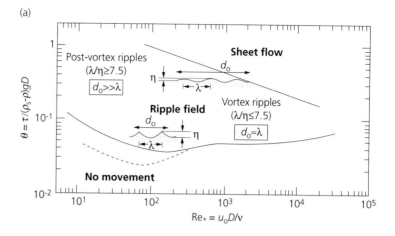

(b)

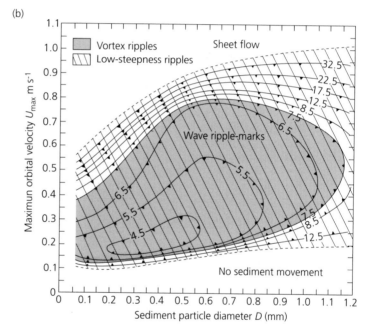

Fig. 8.34 The threshold under waves in the plane of dimensionless stress against grain Reynolds number (after Komar and Miller (1975) [25]), with inset showing ripple steepnesses, threshold and plane bed transition in the plane of orbital velocity against grain diameter (after Allen (1979) [26]).

8.7.2 Wave-generated bedforms
Wave ripple-marks

When the maximum orbital velocity near the bed exceeds the threshold velocity, grains begin to move. The first structures to form on the sandy bed would then be low ridges orientated at right angles to the wave propagation direction. These closely spaced low-amplitude ridges have been termed *rolling grain ripples* [27]. With continued wave motion the rolling grain ripples are replaced by much steeper bedforms which have a trochoidal profile. Flurries or clouds of grains are transported over the steep crest of the ripple

with every half-cycle of the wave motion, being thrown into a vortex on the lee side. Some of these grains settle on the bed, whereas some, still in suspension, together with newly eroded grains, are thrown back over the ripplecrest into another vortex. Bedforms of this type are known as *vortex ripples* [28] (also as *orbital ripples* [29]).

Vortex ripples occur in a wide variety of grain sizes from very fine sand to gravel. They invariably have profiles with steepnesses between about 4.5 and 7.5. Where the wave motion near the bed becomes significantly asymmetrical, as when waves shoal in shallower

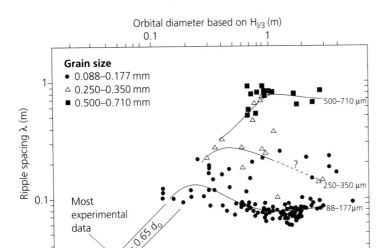

Fig. 8.35 Relation between ripple wavelength and orbital diameter based on a large number of laboratory experiments together with field data. Note the poor correlation between the orbital diameter and the ripple wavelength at wavelengths greater than 0.2 m. The relation for most experimental data is given in equation (8.35). Modified from Miller & Komar (1980) [30, 31].

water, the ripple profiles also become asymmetrical, as is typical in the wave ripples found on beaches just seawards of the swash–backwash zone.

With a further increase in the intensity of the oscillatory flow, the vortex ripples become flattened off, forming low-steepness ripples with rounded profiles, known as *post-vortex ripples* [30]. They are also known as *anorbital ripples* [29] or *decaying ripples*. They appear to form a transition between vortex ripples and a plane bed characterized by very high sediment transport rates. There therefore seems to be a bedform sequence conceptually similar to that found under unidirectional flows (Fig. 8.34).

The compilation of the results of a large number of experiments suggests that there is a relationship between the orbital diameter of the water particles close to the bed and the wavelength of vortex ripples (Fig. 8.35). This relationship is

$$\lambda = 0.65 d_0 \qquad (8.35)$$

If a wave-generated ripple-mark can be clearly identified as a vortex ripple and its wavelength and grain size determined, it is therefore possible to predict ancient wave conditions using small-amplitude wave theory. A number of studies, pioneered by Komar in 1974 [32], have followed this approach [33–37]. A practical exercise is provided below.

Experimental results also show that at a certain value of grain size, the linear relationship between

ripple wavelength and orbital diameter breaks down. Above this threshold the ripple wavelength does not increase despite further increases in orbital diameter. Instead the steepness decreases through the transition to a plane bed. It is hazardous to apply wave hindcasting methods to datasets containing ripple-marks other than those of the vortex type, where there is expected to be a clear relationship between the ripple wavelength and the orbital diameter near the bed [38].

The internal stratification of wave ripple-marks

Since the water motion at the bed is essentially a to-and-fro movement, one might think intuitively that wave ripple-marks do not migrate and that their internal laminations must record purely a symmetrical vertical build up of the ripple form. This turns out to not be the case. Both in the laboratory and in the natural environment [40] wave ripple-marks are observed to have a predominantly unidirectional internal lamination (Fig. 8.36). The internal lamination commonly dips in the direction of wave propagation, that is, towards shore in the nearshore zone. However, this is not always the case. Some studies show that the asymmetry points in a roughly offshore direction. To understand this, we need to return to the mass transports associated with waves, and to consider the other processes at work as waves approach shore

(a)

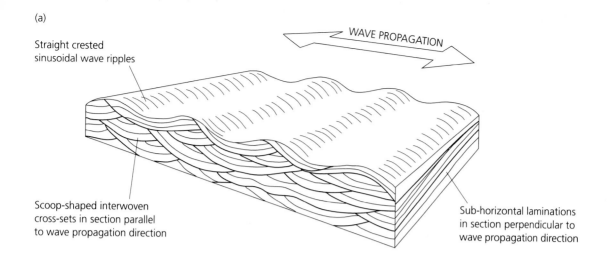

Straight crested
sinusoidal wave ripples

WAVE PROPAGATION

Scoop-shaped interwoven
cross-sets in section parallel
to wave propagation direction

Sub-horizontal laminations
in section perpendicular to
wave propagation direction

(b)

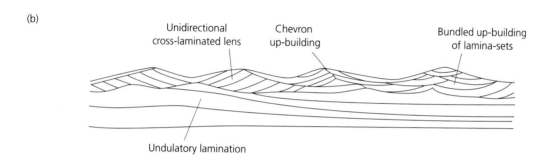

Unidirectional
cross-laminated lens

Chevron
up-building

Bundled up-building
of lamina-sets

Undulatory lamination

(c)

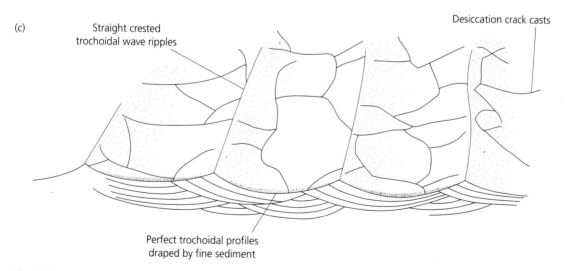

Desiccation crack casts

Straight crested
trochoidal wave ripples

Perfect trochoidal profiles
draped by fine sediment

Fig. 8.36 Wave ripple-marks. (a) Somewhat idealized three-dimensional structure of stratification produced by wave oscillation, after Boersma (1970) [40]. (b) Features characteristic of wave-generated cross-lamination, after Boersma (1970) [40]. (c) Sketch of a slab from the Rhaetic of the Penarth area, South Wales, showing trochoidal ripple profiles. See practical exercise 8.3.

Practical exercise 8.3: Wave hindcasting from preserved ripple-marks

Small amplitude wave theory can be used to calculate the period of waves responsible for fossilized ripple-marks. This is the task of this practical exercise. The combination of wave conditions and water depths can also be investigated. For guidance in how to do this, the reader is referred to [39].

1 The average wave ripple-mark wavelength on bedding planes from the Rhaetic of the Penarth area, South Wales, is 60 mm (Fig. 8.36a). The ripple-marks are symmetrical and trochoidal with an average ripple height of 12 mm. The rippled profile is commonly overlain by a layer of mudstone which thickens into rippled troughs. The mudstone drape contains polygonal, sand-filled desiccation cracks. The grain size of the rippled sandstone is 0.125 mm. What was the period of the waves that formed these Rhaetic ripple-marks?

2 The ubiquitous presence of desiccation cracks on bedding planes suggests that water depths were very shallow. If the water depth was 1 m at the time of formation of the ripple-marks, what would be the height of the waves responsible?

Solution

1 The preserved ripple-marks have a steepness of $60/12 = 5$, so the ripple-marks are clearly of the vortex ripple type. We can therefore use the relation between ripple wavelength and orbital diameter $\lambda = 0.65 d_0$, giving an orbital diameter of 92 mm. Since the grain size is 0.125 mm, the orbital velocity at the threshold of sediment motion is calculated from equation (8.33), giving 0.11 m s^{-1}. Orbital velocities must have exceeded this value in order to mould the bed into ripples. The maximum wave period is therefore given by $U_{max} > U_{cr} = \pi d_0/T_{max}$. The maximum wave period is therefore 2.6 s. This is quite small, and suggests locally generated waves over a small fetch.

2 In order to find out the wave height in 1 m water depth, we make use of the wave dispersion relation in order first to find the wavelength. The deep water wavelength is $gT^2/2\pi = 10.55$ m. The wavelength in 1 m water depth is given by $L = L_\infty \tanh kh$, which can be solved iteratively (by initially guessing L and converging on the solution) as $L = 7.33$ m. The wave height can be found from $d_0 = H/\sinh kh$, giving $H = 89$ mm.

These wave conditions can be checked against the limits for breaking. First, $L/h = 7.33$, so the wave is in 'intermediate' water depths. The steepness $H/L \approx 1/82$, so the wave is far from the limiting steepness of $1/12$. The relative water depth is $H/h \approx 0.09$, showing that the wave is perfectly stable in water depths of 1 m.

By a procedure of repeated calculations, the maximum water depth for stable waves could be found. For our purposes, it is sufficient to envisage locally generated short-period waves acting in shallow water depths. Since the rippled tops of most bedding planes have mudcracks, it is not unreasonable to infer a wave-rippled intertidal sand-mudflat origin for the Rhaetic sediments.

The existence of a slow mass transport in the direction of the wave propagation, the Stokes drift, has already been mentioned. This mass transport near the bed may introduce sufficient asymmetry in the orbital velocities to cause a net sediment transport in the wave propagation direction. This would explain the internal lamination dipping towards the shore. The drift of water towards the shore caused by the wind blowing over the water surface causes the water surface to be elevated at the shore. There is consequently a pressure gradient current flowing offshore near the bed (Chapter 9). This current may cause asymmetry in the orbital velocities due to waves at the sea bed. It may explain the internal laminations of ripple-marks that are directed offshore or at a high angle to it.

Further reading

K.F. Bowden (1983) *Physical Oceanography of Coastal Waters*. Ellis Horwood, Chichester.

P.D. Komar (1972) *Beach Processes and Sedimentation*. Prentice Hall, Englewood Cliffs, NJ.

Open University (1989) *Waves, Tides and Shallow-Water Processes*, course S330, Vol. 4. Pergamon Press, Oxford.

S. Pond & G.L. Pickard (1983) *Introductory Dynamical Oceanography*, 2nd edn. Pergamon Press, Oxford.

References

1 Y. Accad & C.L. Pekeris (1978) Solution of tidal equations for M$_2$ and S$_2$ tides in the world ocean form a knowledge of the tidal potential alone. *Philosophical*

Transactions of the Royal Society London, **290A**, 235–66.

2 S. Pond & G.L. Pickard (1983) *Introductory Dynamical Oceanography*, 2nd edn. Pergamon Press, Oxford.

3 K. Wadati, T. Hirono & S. Hisamoto (1963) On the tsunami warning service in Japan. *Proceedings of the Tsunami Meetings, 10th Pacific Science Congress*, Monograph 24. IUGG, pp. 138–45.

4 A.C. Redfield (1958) Influence of the continental shelf on tides of the Atlantic coast of the United States. *Journal of Marine Research* **17**, 432–48.

5 G.I. Taylor (1920) Tidal oscillations in gulfs and rectangular basins. *Proceedings of the London Mathematical Society* **20**, 144–81.

6 M.C. Hendershott & A. Speranza (1971) Co-oscillating tides in long, narrow bays: the Taylor problem revisited. *Deep Sea Research* **18**, 959–80.

7 D.A. Huntley (1980) Tides on the north-west European continental shelf. In: *The North-West European Shelf Seas: The Sea Bed and the Sea in Motion. II Physical and Chemical Oceanography and Physical Resources* (eds F.T. Banner, M.B. Collins & K.S. Massie). Elsevier, Amsterdam.

8 G. Godin (1980) *Cotidal charts for Canada*, MSS Report Series No. 55. Marine Sciences and Information Directorate, Department of Fisheries and Oceans, Ottawa.

9 J.R.L. Allen (1985) *Principles of Physical Sedimentology*. George Allen & Unwin, London.

10 K.F. Bowden (1983) *Physical Oceanography of Coastal Waters*. Ellis Horwood, Chichester.

11 S. Berné, J.-P. Auffret & P. Walker (1988) Internal structure of subtidal sandwaves revealed by high resolution seismic reflection. *Sedimentology* **35**, 5–20.

12 J.R.L. Allen (1980) Sand waves: a model of origin and internal structure. *Sedimentary Geology* **26**, 281–328.

13 P.A. Allen, P. Homewood, M.A. Mange-Rajetzky & A. Matter (1985) Dynamic palaeogeography of the open Burdigalian seaway, Swiss Molasse Basin. *Eclogae Geologicae Helvetiae* **86**, 121–72.

14 P.A. Allen & J.P. Bass (1992) Sedimentology of the Upper Marine Molasse of the Rhone–Alp region, eastern France; implications for basin evolution. *Eclogae Geologicae Helvetiae* **86**, 121–72.

15 P.A. Allen & P. Homewood (1984) Evolution and mechanics of a Miocene tidal sandwave. *Sedimentology* **31**, 63–81.

16 M.J. Visser (1980) Neap–spring cycles reflected in Holocene subtidal large-scale bedform deposits: a preliminary note. *Geology* **8**, 543–6.

17 J.R.L. Allen (1982) Mud-drapes in sandwave deposits: a physical model with application to the Folkestone Beds (early Cretaceous, southeast England). *Philosophical Transactions of the Royal Society, London* **306A**, 291–345.

18 P.D. Komar (1972) *Beach Processes and Sedimentation*. Prentice Hall, Englewood Cliffs, NJ.

19 Open University (1989) *Waves, Tides and Shallow-Water Processes*, course S330, Vol. 4. Pergamon Press, Oxford.

20 M.S. Longuet-Higgins (1953) Mass transport in water waves. *Philosophical Transactions of the Royal Society, London* **245A**, 535–81.

21 R.L. Wiegel (1964) *Oceanographical Engineering*. Prentice Hall, Englewood Cliffs, NJ.

22 B. Kinsman (1965) *Wind Waves: Their Generation and Propagation on the Ocean Surface*. Prentice Hall.

23 J.W. Johnson (1956) Dynamics of nearshore sediment movement. *Bulletin of the American Association of Petroleum Geologists* **40**, 2211–32.

24 P.C. Liu (1971) Normalized and equilibrium spectra of wind waves in Lake Michigan. *Journal of Physical Oceanography* **1**, 249–57.

25 P.D. Komar & M.C. Miller (1975) The threshold of sediment movement under oscillatory water waves. *Journal of Sedimentary Petrology* **43**, 1101–10.

26 J.R.L. Allen (1979) A model for the interpretation of wave ripple-marks using their wavelength, textural composition and shape. *Journal of the Geological Society of London* **136**, 673–82.

27 J.F.A. Sleath (1976) On rolling grain ripples. *Journal of Hydraulic Research* **14**, 69–80.

28 R.A. Bagnold (1946) Motion of waves in shallow water. Interactions between waves and sand bottoms. *Proceedings of the Royal Society of London* **187A**, 1–15.

29 H.E. Clifton (1976) Wave-formed sedimentary structures—a conceptual model. In: Beach and Nearshore Sedimentation (eds R.A. Davies & R.L. Ethington), Special Publication 24. Society of Economic Paleontologists and Mineralogists, Tulsa, OK, pp. 126–48.

30 M.C. Miller & P.D. Komar (1980) Oscillation sand ripples generated by laboratory apparatus. *Journal of Sedimentary Petrology* **50**, 173–82.

31 J. Dingler & D.L. Inman (1977) Wave-formed ripples in nearshore sands. *Proceedings of the 15th Conference on Coastal Engineering*, 2109–26.

32 P.D. Komar (1974) Oscillatory ripple marks and the evaluation of ancient wave conditions and environments. *Journal of Sedimentary Petrology* **44**, 169–80.

33 P.A. Allen (1981) Wave-generated structures in the Devonian lacustrine sediments of SE Shetland, and ancient wave conditions. *Sedimentology* **28**, 369–79.

34 P.A. Allen (1984) Reconstruction of ancient wave conditions, with an example from the Swiss Molasse. *Marine Geology* **60**, 455–73.

35 B. Diem (1984) Analytical method for estimating palaeoware climate and water depth from wave ripple marks, *Sedimentology* **32**, 705–20.

36 P.S. Moore (1982) Ripple-mark analysis of a fine-

grained epeiric sea deposit (Cambrian, South Australia). *Journal of the Geological Society of Australia* **27**, 71–81.

37 B. Sundquist (1982) Palaeobathymetric interpretation of wave ripple-marks in a Ludlovian grainstone of Gotland. *Geologiska Föreningen i Stockholm Förhandlingar* **104**, 157–66.

38 P.A. Allen (1981) Some guidelines in reconstructing ancient sea conditions from wave ripple-marks. *Marine Geology* **43**, M59–67.

40 J.R. Boersma (1970) Distinguishing features of wave-ripple cross-stratification and morphology. Doctoral thesis, University Utrecht, The Netherlands, 65pp.

39 See R.S. Newton (1968) Internal structures of wave-formed ripple marks in the nearshore zone. *Sedimentology* **11**, 275–92.

9 Ocean currents and storms

No coward soul is mine,

No trembler in the world's storm-troubled sphere:

I see Heaven's glories shine,

And faith shines equal, arming me from fear.

    Emily Brontë (1818–48) *No coward soul is mine* [1846]

Chapter summary

The motion of water in the oceans is profoundly affected by the rotation of the Earth. But ocean water is not homogeneous, differing in density from place to place as a result of variations in temperature and salinity. These variations set up horizontal pressure gradients causing flow. The thermohaline motions, and those due to the drag of the wind on the ocean surface, are acted upon by the Coriolis acceleration caused by Earth's rotation, and result in the great circulations of the ocean basins.

The blowing of wind over water imparts a tangential force on the top surface of the water in the direction of the wind. This causes a movement of water in the uppermost layer which is immediately deflected (to the right in the northern hemisphere) by the Coriolis force. The surface motion is transferred to lower layers by friction. Ekman quantified the resulting water motion with depth under a wind stress, which is in the form of a spiral with a characteristic depth of frictional influence known as the Ekman depth. The surface current velocity and mean Ekman transport can be calculated for deep, homogeneous water for a given wind stress as a function of latitude.

The currents caused by pressure gradients in the absence of friction and turned by the Coriolis force are termed geostrophic flows. The geostrophic equation expresses the balance between the pressure force and the Coriolis force. The geostrophic flows expected from known distributions of salinity and temperature in the ocean can therefore be computed.

The total or mean Ekman transport is at right angles to the surface wind stress. If the wind is blowing with a coastline on the right in the northern hemisphere, water is forced to pile up against the coast, causing a surface slope. This slope in turn drives a deeper geostrophic current moving offshore, a situation termed downwelling. If the wind is blowing with a coastline on the left in the northern hemisphere, water is removed from the nearshore region by the Ekman transport. The water lost is replaced at depth by an upwelling. Coastal upwelling commonly brings nutrient-rich waters to the surface, promoting plankton production and thereby supporting large fish populations. Upwelling characterizes the Pacific Ocean off Peru and the eastern Atlantic Ocean off South Africa.

In shallow water with a depth less than the Ekman depth, the Ekman layer is affected by sea bed friction, causing a partial cancelling out of the wind-driven spiral of water motion. This results in a net transport which is closer to the wind stress direction than in deep, homogeneous water.

The bottom currents in the ocean are those of the slow thermohaline circulation. These thermohaline currents intensify on the western sides of ocean basins, may accelerate as they pass though sea bed constrictions, and may mix with the edges of deep geostrophic flows where they develop large eddies. All of these processes affect sediment transport and deposition in the ocean. Bottom currents generate zones of very fine-grained suspended sediment called nepheloid layers. Nepheloid layers may detach from the sea bed at the continental edge and extend far into the ocean, providing a steady rain of fine material to the ocean floor. Bottom currents also deposit sediment drifts composed of contourites, commonly at the edges of current pathways.

The principles underlying the ocean circulation can be applied to the water motion associated with the passage of a storm on the continental shelf. The motion of water during the landfall of a storm is affected by: (i) elevation changes caused by lateral variations in the barometric pressure; (ii) a wind-driven surface current; (iii) the piling up of water in the nearshore zone, termed coastal set-up, which drives a return current known as a gradient or slope current; and (iv) a mass transport due to waves causing a wave set-up. Each of these components can be quantified. In sum, they may cause storm surge heights of several metres at the coast. The sediment transport during a storm is dominated by sea bed currents originating from: increased undertows from the beachface; a seaward-flowing, friction-dominated gradient current in the shallow waters of the nearshore zone; and a geostrophic flow in deeper water moving along the continental shelf. Superimposed on the storm currents are the oscillatory motions of waves. The sediment distribution resulting from the landfall of Hurricane Carla in the Gulf of Mexico in 1961 shows a sand bed extending to water depths of up to 50 m, stretching along the strike of the shelf for up to 200 km. This supports the idea of obliquely driven stormflows predicted by the geostrophic model outlined above.

The Atlantic shelf off the eastern seaboard of North America is one of the best documented pieces of sea bed in the world. For the typical winds from the north-east, a geostrophic core flow runs along the shelf to the south and south-west, with a surface current directed obliquely onshore and a bottom current obliquely offshore. The continental shelf is also crossed by long-period waves known as topographically trapped waves, caused by the relaxation of bulges of the sea surface overlying shallow waters when the wind dies down. Sand ridges attached to the shoreface and orientated with their long axes oblique to the coast are thought to be active under the present-day storm regime of the shelf. The Oregon–Washington shelf of western USA has a stronger wave regime than the Atlantic shelf. Waves and storms are crucial in the advection of fine sediment delivered by rivers. Large waves appear to efficiently stir up sediment from the sea bed, whereas steady storm-generated currents advect the sediment along the shelf.

The single most diagnostic bedform attributed to the passage of storms is hummocky cross-stratification, consisting of metre-scale, low-amplitude hummocks and swales of fine to very fine sand. Although perhaps most common on stormy shelves, the structure most likely forms wherever there is a combined flow field involving oscillatory and steady components.

9.1 Introduction

We have encountered the major circulation patterns of water in the ocean in Chapter 1 in the context of the global heat budget. In this chapter the physical principles underpinning this circulation pattern are introduced.

The motion observed (Figs 1.9 and 1.10) is profoundly influenced by the effects of the rotation of the Earth. This may not immediately be intuitively obvious. We ourselves are affected by the Coriolis force when we walk about on the Earth's surface, but it goes unnoticed. Why, then, is the water in the ocean so profoundly affected by the Earth's rotation? The answer is that water behaves as a Newtonian fluid with no yield strength. Consequently, although the Coriolis acceleration is a very weak force (per unit mass), in the ocean, where all the other horizontal forces are also very weak, the Coriolis component can be important. Another way of considering the problem is to consider the speed of rotation of the

Earth. At the equator the Earth is spinning at several thousand kilometers per hour, decreasing to zero at the poles. In the northern hemisphere, if one looks down the transport path of a water particle, the particle appears to veer to the right as the solid Earth beneath it spins to the left.

The ocean is far from a homogeneous mass of water. It differs in temperature and salinity from place to place as a result of processes such as variations in solar radiation, inputs of cold bottom water, and fresh water runoff from continents (Chapter 1). These variations in properties of ocean water, and the effects of the wind blowing over it, result in its large-scale dynamics. The dynamics of water on the continental shelves can be studied by the appropriate amendment of principles derived from the deep ocean. The hydraulic regime, and therefore the sediment transport within it, are extremely complex on the continental shelf. In the second half of this chapter, our main concern is the understanding of the response of the continental shelf to the passage of major atmospheric disturbances. Since 80% of the world's continental shelves are thought to be storm-dominated, and since the stratigraphic record of ancient shelf environments is so abundant, the problem of the atmosphere–ocean coupling and sea bed response during storms is an environmentally and geologically important one.

The subject of shelf and deep sea sedimentation is dealt with in some detail elsewhere [1, 2]. For those wishing to make further progress in the physical principles, a number of reviews and specialist texts are listed in the Further Reading section at the end of this chapter.

9.2 Currents in the ocean

The large-scale circulation of water in the ocean is caused by two main processes.

1 The shear of the wind over the ocean surface drives a surface flow which is transmitted to deep layers in the ocean. The belts of prevailing winds associated with atmospheric cells (Chapter 1) therefore have a major role in producing the large-scale surface layer circulation of the oceans, composed of clockwise (in the northern hemisphere) and anticlockwise (in the southern hemisphere) current systems.

2 The density differences set up by variations in temperature and salinity drive a very large global circulation that is slow in speed but very large in mass. This *thermohaline circulation* has a prime control on the distribution of heat over the Earth's surface.

It has been known for as long as navigators have sailed across the oceans that the surface currents flow in a direction at an angle to the prevailing winds, but it took until the beginning of the twentieth century before the motion was explained quantitatively by Ekman by invoking the Coriolis force. In Section 9.2.2 the solution proposed by Ekman for a deep, homogeneous and infinite water body subjected to a constant wind stress is introduced. The implications for the speed and direction of the surface current in relation to the wind stress, and the variation in current velocities with depth, are described and a practical exercise provided.

We continue our analysis of the currents in the ocean by considering currents set up by horizontal pressure gradients in the water mass, where friction and (tidal) gravitational forces can be ignored. These are termed *geostrophic flows*. Pressure gradients may result from differences in density caused by variations in temperature and/or salinity, as in the deep oceanic

circulation pattern observed in Chapter 1. They may also result from the water surface being elevated in one area compared to another, perhaps resulting from the action of a wind drift current. Geostrophic flows and wind drift currents therefore combine in a number of oceanographical situations, as in the case of upwelling and downwelling along continental margins.

Superimposed on these large-scale, long time-period motions which act in something approaching a steady state are short-term, transient effects related to individual weather systems, the focus of the second half of this chapter. However, the likely water motions and sediment transport patterns expected due to the passage of a storm are rooted in an understanding of the steady-state conditions characterizing the deep ocean.

9.2.1 An intuitive description of the Coriolis force

We cannot consider the circulation of water in the ocean, and the particulate material that the water carries, without knowing something about the effects of the rotation of the Earth. The crux of this issue is that we normally make observations relative to a coordinate system fixed to the Earth, but the Earth is itself rotating. This presents an immediate problem because the most fundamental equation of motion, which states that the force applied is equal to the mass times the acceleration, applies only when the acceleration is measured relative to axes which are fixed in space, that is, axes whose origin is not accelerating. Consequently, since we want to express motion relative to the Earth, the equation of motion must be adjusted to take account of the Earth's rotating frame of reference. Such an adjustment involves the Coriolis acceleration.

You can visualize the effect of the Earth's rotation by considering a person situated at the north pole who fires a projectile directly down a meridian (line of longitude) towards the equator. The projectile will move in a plane which is fixed relative to the 'fixed stars' (a way of stating that the origin of the co-ordinate system is truly fixed), whereas the distant target on the Earth's surface will swing away to the east during the flight of the projectile. Since the person is also fixed to the rotating Earth, the projectile will appear to swing to the west in his reference frame. It appears as if a force is causing the projectile to swing to the west, a 'force' we term the Coriolis force. Had the person fired a projectile from a point situated on the equator along the equatorial line, the projectile would not deviate left or right from its path.

But what happens if the person is situated between the equator and the pole? As one moves from the equator to the poles, a point on the surface experiences a smaller eastward velocity. But in addition to this linear velocity, there is an angular velocity that also varies between the equator and the poles (Fig. 9.1). The angular speed of rotation, Ω, is the angle that the Earth turns through each day, that is, 2π divided by the time taken to complete a revolution (relative to the 'fixed stars', this time is 23 hours 56 minutes, the *sidereal day*). The angular velocity about a vertical axis from the centre of the Earth to a point on the Earth's surface decreases from Ω at the poles to zero at the equator, with the general form $\Omega \sin \phi$ where ϕ is the latitude (Fig. 9.1). Consequently, a particle on the surface of the Earth (except exactly at the equator) and acted upon by no other forces, would rotate anticlockwise in the northern hemisphere and clockwise in the southern hemisphere by this effect alone. It is termed *planetary vorticity* and can be recognized in the characteristic spiralling motion of water in, for example, the Baltic Sea or the Gulf Stream.

Now imagine a projectile is fired along an initial line of latitude from A to B (Fig. 9.2). Let the average velocity of the projectile be V, so the distance travelled in time t is equal to Vt. However, in this time interval the target will have been displaced from B to B'. The projectile will appear to the observer to have veered to

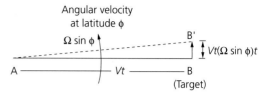

Fig. 9.2 Derivation of the apparent displacement of a target situated initially at B along a line of latitude from A. The second derivative of this apparent displacement gives the apparent (Coriolis) acceleration.

the right in this northern hemisphere illustration. Since the angular velocity of the Earth at latitude ϕ (in the northern hemisphere) is $\Omega \sin \phi$, the apparent displacement of B in the flight time t is

$$Vt(\Omega \sin \phi)t \qquad (9.1)$$

The apparent acceleration caused by the rotation of the Earth is then the second differential of the apparent displacement:

$$\frac{d^2}{dt^2}\Omega(\sin\phi)Vt^2 = 2\Omega(\sin\phi)V \qquad (9.2)$$

This is often given in shorthand as fV, where f is the Coriolis parameter, and the Coriolis force would therefore be mfV.

9.2.2 The wind-driven circulation

Imagine a cube of water with a top surface over which the wind blows. The wind imparts a tangential force F_t over the top surface in the direction of the wind direction. As soon as the water in the surface layer begins to move, it is deflected to the right (in the northern hemisphere) by the Coriolis force F_c, so the resultant motion is in the direction between the two forces (Fig. 9.3). Since the water in the cube is moving along this resultant direction, there is a retarding fluid frictional force F_f along the base of the cube which is oriented opposite to the resultant motion in the cube. Because the tangential, Coriolis and frictional forces are in balance (Fig. 9.3 inset), the water in the cube moves with a steady velocity, V_0. Ekman quantified this tendency for the water to flow at an angle to the wind stress by considering the equations of motion.

Ekman first made a number of assumptions in order to simplify the solution. He assumed that the water was homogeneous and that there was no surface slope

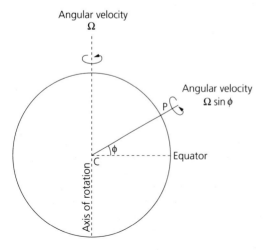

Fig. 9.1 The dependence of the angular velocity about a vertical axis at any point P on the Earth's surface.

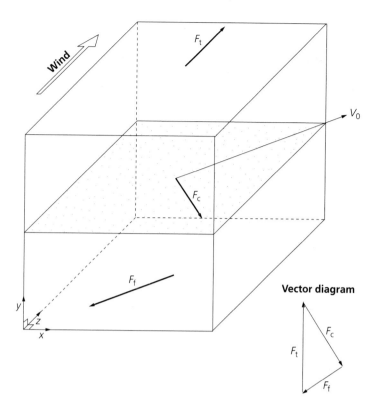

Fig. 9.3 Forces on a cube of water in the surface layer of the ocean with a tangential wind stress on the surface F_t, a Coriolis force F_c acting to the right (in the northern hemisphere) of the resultant water motion V_0, and a frictional force F_f acting opposite to the water motion. The inset shows the force balance between wind stress, friction and Coriolis forces. Since the vector triangle is closed, the water has no acceleration and moves with a steady velocity.

to the water body, in which case pressure gradients in the horizontal plane could be ignored. He further assumed that the water was very deep and of infinite extent, so there was no friction at the sea bed, only fluid friction between layers of water moving at different velocities. Finally, he assumed a steady wind blowing over the surface for a long time, so that transient effects could be ignored. (He also assumed that the eddy viscosity was constant with depth, an assumption we examine later in this section). Ekman's equations are sufficiently important to be repeated here, though the reader is referred to texts in physical oceanography for their derivation. If u_E and v_E are the components of the Ekman velocity in the horizontal plane at any depth y, and V_0 is the *Ekman surface current* (at $y = 0$), the Ekman velocities have the form of cosine or sine waves damped exponentially with increasing depth (Fig. 9.4):

$$u_E = \pm V_0 \cos\left(\frac{\pi}{4} + \frac{\pi}{D_E} y\right) \exp\left(\frac{\pi}{D_E} y\right) \qquad (9.3)$$

$$v_E = V_0 \sin\left(\frac{\pi}{4} + \frac{\pi}{D_E} y\right) \exp\left(\frac{\pi}{D_E} y\right) \qquad (9.4)$$

where the sign is positive for the northern hemisphere and negative for the southern and you are reminded that y is negative downwards. The parameter D_E is a measure of the depth of frictional influence known as the *Ekman depth*. The Ekman surface current and the Ekman depth can be found from

$$V_0 = \left(\tau_w \pi \sqrt{2}\right)\left(D_E \rho |f|\right) \qquad (9.5)$$

$$D_E = \pi\left(\frac{2\eta}{|f|}\right)^{1/2} \qquad (9.6)$$

where τ_w is the wind stress on the surface, acting in the direction of the wind and roughly equal to the square of the wind speed, $|f|$ is the absolute magnitude of the Coriolis parameter, and η is the eddy viscosity.

The Ekman spiral results in a mean mass transport representing the current velocities integrated over depth, known as the *total Ekman transport*. The total Ekman transport, T, can be expressed in terms of the wind stress on the sea surface.

$$T = \frac{\tau_w}{\rho_w f} \qquad (9.7)$$

(a)

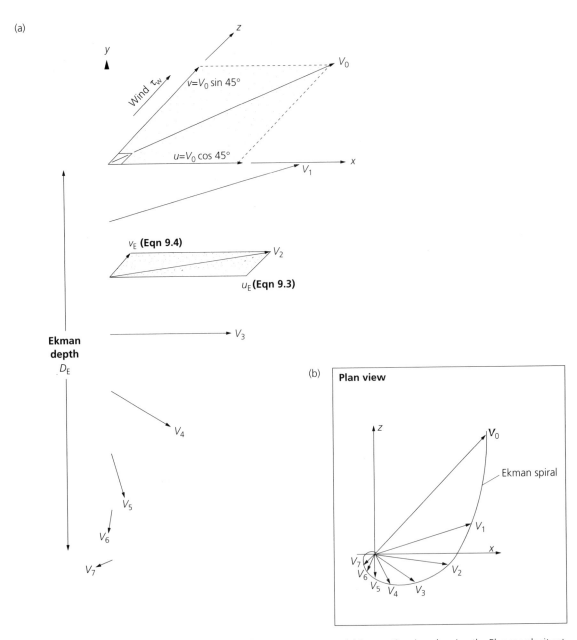

Fig. 9.4 The Ekman spiral and Ekman depth in deep homogeneous water. (a) Perspective view showing the Ekman velocity at seven depths below the surface. (b) The characteristic Ekman spiral shown in plan view.

where τ_w is the wind stress on a unit area of the sea surface, ρ_w is the density of the water and f is the Coriolis parameter (Fig. 9.5).

Although the Ekman spiral is thought to be a rough approximation of the wind-driven currents dominated by friction in the upper layer of the ocean, it is ex-

tremely difficult to make measurements of the Ekman velocities with depth. In particular, the precise value of the eddy viscosity is uncertain.

The driving force for the Ekman spiral is wind stress on the sea surface. The effective tangential wind stress on the sea surface is thought to be related to the

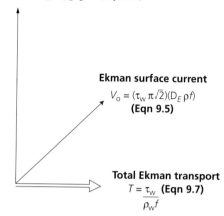

Wind stress $\tau_{\mathrm{w}} = \rho_{\mathrm{a}} C_{\mathrm{D}} W^2$ **(Eqn 9.8)**

Ekman surface current
$V_0 = (\tau_{\mathrm{w}} \pi \sqrt{2})(D_E \rho f)$
(Eqn 9.5)

Total Ekman transport
$T = \tau_{\mathrm{w}}$ **(Eqn 9.7)**
$\dfrac{}{\rho_{\mathrm{w}} f}$

Fig. 9.5 Relation of the total Ekman transport to the Ekman surface current and the tangential wind stress in deep, homogeneous water.

square of the wind velocity

$$\tau_{\mathrm{w}} = \rho_{\mathrm{a}} C_{\mathrm{D}} W^2 \tag{9.8}$$

where ρ_{a} is the density of the air, C_{D} is the drag coefficient and W is the wind speed, usually measured 10 m above the sea surface. It is perhaps instructive to pause to consider what this drag coefficient represents and what it depends on. The energy of the wind is used in generating surface waves which thereby determine the surface roughness of the sea. The value of the drag coefficient therefore depends on the velocity of the wind itself, as well as on the stability of the lowermost few metres of the atmosphere. If the wind above the sea surface has a logarithmic velocity profile, with a wind shear velocity given as $w_* = \sqrt{\tau_{\mathrm{w}}/\rho_{\mathrm{a}}}$, the wind shear stress can be calculated and therefore the drag coefficient can be found [3]. Under neutral atmospheric conditions, the drag coefficient takes values of about 1.1×10^{-3} for a wind speed of $5\,\mathrm{m\,s^{-1}}$ (at $y = 10\,\mathrm{m}$), 1.4×10^{-3} at $10\,\mathrm{m\,s^{-1}}$, 2.0×10^{-3} at $20\,\mathrm{m\,s^{-1}}$ and 3.2×10^{-3} at $40\,\mathrm{m\,s^{-1}}$. A related parameter is the *roughness length* (again referring back to the logarithmic velocity law for unidirectional flows). The value of this parameter is determined by the geometrical roughness of the boundary, and so should in some way be related to the spectrum of waves and their heights on the sea surface. Observations suggest that it is related to the square of the wind shear velocity:

$$y_0 = \frac{a w_*^2}{g} \tag{9.9}$$

where a is a constant equal to 0.0144. There are therefore two coefficients characterizing the resistance of the sea surface, the drag coefficient and the roughness length, and these parameters might be expected to vary according to the state of the sea surface. It has been found that they are higher in the period of a storm before a fully developed sea, that is, in the early stages when the wave spectrum is of higher frequency.

Observations on wind drift currents

The relationship between the wind speed and the surface current velocity is a central concern of oceanographers. Such information would, for example, help in the prediction of the movement of an oil slick under known wind conditions. This relationship is commonly known as the *wind factor*. The results of many observations using different techniques suggest that the wind drift current is roughly 3% of the wind velocity measured 10 m above the sea surface [4]. Although the Ekman theory predicts a substantial angle between the direction of the prevailing wind and the direction of the surface current (equations (9.3) and (9.4), also Practical Exercise 9.1), observations suggest that this angle is very small, seldom exceeding 10° (to the right in the northern hemisphere). These observations on the surface drift of markers are in fact measuring both the Stokes drift, caused by the mass transport due to waves in the direction of wave propagation (Section 8.6.1) and the true wind drift current. The magnitude of the Stokes drift can be found from equation (8.25). For waves in deep water of height 3 m and period 8 s, the average surface Stokes drift is $0.11\,\mathrm{m\,s^{-1}}$. Observations from fully developed seas suggest that it is about 1.6% of the wind speed and increases linearly with it. It seems highly likely that between one-third and one-half of the observed surface current velocity is made up of the wave-induced Stokes drift.

Actual observations of wind velocities and surface wind drift current velocities allow the relationship to be simplified, and for the Ekman depth and the eddy viscosity to be determined [5]:

$$D_{\mathrm{E}} = \frac{4.3W}{(\sin\phi)^{1/2}} \tag{9.10}$$

$$\eta = 1.37 W^2 \tag{9.11}$$

for latitudes greater than 10° either side of the equator and for wind speeds of greater than $6\,\mathrm{m\,s^{-1}}$. However, since little is known of the variation of eddy viscosity with depth (if any), it is informative to consider the impact of a variable eddy viscosity.

Measurements in the atmospheric boundary layer suggest that the effective eddy viscosity may be very small at the boundary and increase linearly away from it. Using a depth-dependent eddy viscosity and including Coriolis terms, Madsen [6] found that the wind-driven current in deep water had a surface current at only about 10° from the wind direction, in harmony with actual measurements of surface currents, instead of the 45° in Ekman's theory, and that the magnitude of the wind drift current decreased rapidly with depth. Figure 9.6 compares the two theories. Madsen showed the surface current to have a magnitude directly proportional to the shear velocity (or wind stress), and the thickness (L) of the Ekman layer also to be related to the shear velocity

$$V_0 \approx 25 u_* \tag{9.12}$$

$$L_E = 0.4 \frac{u_*}{f} \tag{9.13}$$

Now since the water shear velocity is directly proportional to the wind shear velocity ($u_* = (\rho_a/\rho_w)w_* = 0.035\,w_*$), the depth of the Ekman layer using a variable eddy viscosity can be expressed in terms of the wind shear velocity, and is thus immediately comparable with the characteristic depth in Ekman's treatment (equation (9.10)):

$$L_E = 96 \frac{w_*}{\sin \phi} = 96 \frac{WC_D^{1/2}}{\sin \phi} \tag{9.14}$$

With a wind speed of $10\,\mathrm{m\,s^{-1}}$, measured 10 m above sea level, and a drag coefficient of 1.4×10^{-3}, the estimates of the Ekman depth from the two methods (equation (9.10) and equation (9.14)) are almost identical (51 m) at a latitude of 45°, but diverge strongly at lower latitudes and less strongly at higher

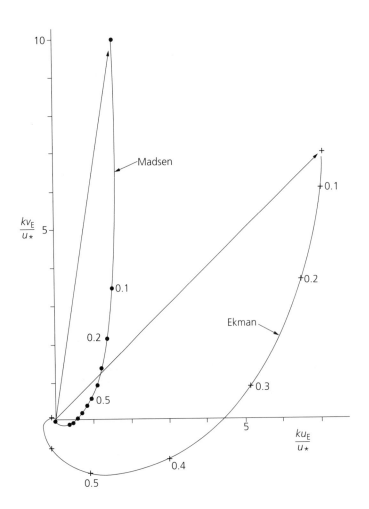

Fig. 9.6 Comparison of the Ekman and Madsen theories for the wind-driven current in deep water, in the plane of the non-dimensional velocities ku_E/u_* and kv_E/u_*, where k is von Karman's constant, equal to 0.4 for clear water, and u_* is the shear velocity. The numbers on the curves are values of $\sqrt{y/D_E}$, where D_E is the Ekman depth. After Madsen (1977) [6].

Practical exercise 9.1: The Ekman layer

1 What are the Ekman velocities in the horizontal plane at the surface of the ocean for a surface current velocity of V_0? What does this imply about the direction of the surface current? The wind direction is in the direction of the v component of the Ekman transport.

2 Calculate the current speed as a function of depth by selecting values of y at regular increments. If you are doing this using a speadsheet program, choose closely spaced increments; otherwise, select water depths at 10 m intervals. Let $D_E = 100$ m.

3 At what depth does the current flow become opposite to that at the surface? How much has the current decreased relative to the surface current speed at this depth?

4 Field observations of the Ekman surface current compared to the wind velocity indicate that it depends only on the latitude, ϕ, outside of a band of 10° latitude north and south of the equator:

$$\frac{V_0}{W} = \frac{0.0127}{\left(\sin|\phi| \right)^{1/2}} \tag{9.15}$$

What is the magnitude of the surface Ekman current for a wind of 10 m s^{-1} blowing over the ocean surface at 45°N? What is the corresponding Ekman depth?

Solution

1 Putting $y = 0$, we have

$$u_E = \pm V_0 \cos 45°$$
$$v_E = V_0 \sin 45°$$

showing that the Ekman surface current flows at 45° to the right of the wind direction in the northern hemisphere and at 45° to the left in the southern.

2 The current speed as a function of depth is given in Fig. 9.7.

3 The current speed decreases below the surface by the factor $\exp(\pi y / D_E)$. Remember that with increasing depth y becomes more negative since y is measured positive upwards from the sea surface. The current direction will be opposite that at the surface when $u_E = \cos(3\pi/4)$, $\cos(5\pi/4)$,... and $v_E = \sin(5\pi/4)$, $\sin(7\pi/4)$,... in the northern hemisphere. Consequently, the Ekman current is opposite in direction to the surface current at the depth corresponding to when

$$\left(\frac{\pi}{4} + \frac{\pi}{D_E} y \right) = \frac{5\pi}{4} \tag{9.16}$$

$$y = D_E$$

showing that the current reverses in direction at the Ekman depth. When y is the Ekman depth, the speed has reduced to $\exp(-\pi) = 0.04$ of the surface Ekman current. The spiralling pattern of current velocity above this depth is known as the *Ekman layer*.

4 The surface Ekman current has a speed of 0.15 m s^{-1}. The Ekman depth for a wind of 10 m s^{-1} at 45°N (from equation (9.11)) is 51 m. The Ekman depth decreases only very slowly with an increase in latitude (it is 43 m at 80°N) but increases more rapidly towards the equator (it is 103 m at 10°N).

latitudes. The main feature of the Madsen spiral is that there is a much more rapid change in velocity with depth in the first few metres than in the Ekman solution.

9.2.3 Currents without friction: geostrophic flows

Let us now take the situation where the equations of motion for ocean water are simplified by the assumption that there are no frictional forces (Section 9.2.1), no forces set up by the gravitational attraction of the Sun and Moon (Chapter 8) and no slope to the water surface. A *geopotential surface* can be described as the surface upon which the gravitational force acting on a mass is everywhere the same, so the geoid (Chapter 1) is an example of a geopotential surface. A plumb line would be exactly perpendicular to a geopotential surface. One example of a geopotential surface is the surface of a lake where there are no currents and no waves (including seiches). Another is a perfectly set-up snooker table. A geopotential surface is different to an *isobaric surface*, upon which the pressure is everywhere the same. Take the example of an isobaric surface that is sloping relative to the geopotential surface (Fig. 9.8). For a unit mass of water situated on the isobaric surface at A, there is a pressure force $\alpha \Delta p_n$ acting normally (hence the subscript) to the isobaric surface. The vertical component of this force, $\alpha \Delta p_n \cos i$, is balanced by gravity acting on the unit mass of water. But there is also a small force $\alpha \Delta p_n \sin i$ acting horizontally in the direction of the water surface slope which is unbalanced. The horizontal force (to the left) can be written (see Eqn 9.17, p. 317)

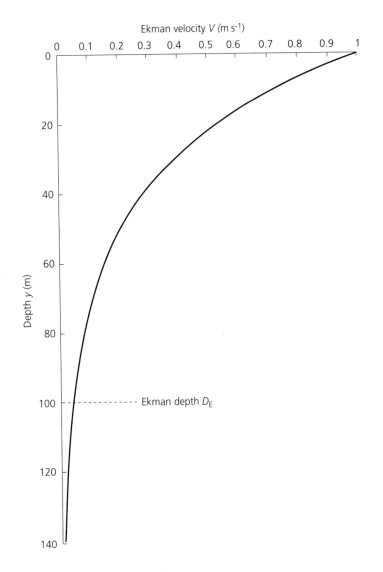

Fig. 9.7 Magnitude of the Ekman current versus depth for Practical Exercise 9.1.

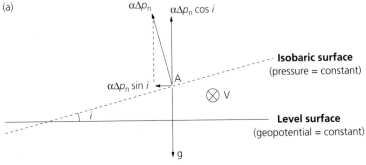

Fig. 9.8 (a) Pressure forces on a unit mass of water at A on an isobaric surface sloping at an angle i to the geopotential surface. The pressure force is proportional (through α) to the pressure gradient normal to the isobaric surface (Δp_n). The horizontal component of this force is balanced by the Coriolis force, resulting in a flow into the page (in the northern hemisphere) with velocity V.

$$\alpha \Delta p_n \sin i = (\alpha \Delta p_n \cos i) \frac{\sin i}{\cos i} = g \tan i \qquad (9.17)$$

which simply makes use of the fact that $\alpha \Delta p_n \cos i = g$. To balance this, a force is required in the opposite direction. The Coriolis force fulfils this function. Since it acts to the right in the northern hemisphere, the particle should move into the page with a velocity V. This can be expressed by the *geostrophic equation*, geostrophic meaning 'Earth-turned', which expresses the balance between the pressure force and the Coriolis force,

$$2\Omega(\sin \phi)V = g \tan i \qquad (9.18)$$

where the left-hand side is the Coriolis force per unit mass (Ω is the angular rate of rotation of the Earth, ϕ is the latitude), and the right-hand side is the force per unit mass caused by the isobaric surface sloping at an angle i, and g is gravitational acceleration. The slope required to balance a surface current moving with a given velocity at a given latitude can be found from the geostrophic equation. If the current is moving at $1\,\mathrm{m\,s^{-1}}$ at 45°N, and the angular rate of rotation is $7.29 \times 10^{-5}\,\mathrm{rad\,s^{-1}}$, $\tan i$ is about 10^{-5}. The Gulf Stream in the North Atlantic (Chapter 1) is about 100 km in width, so the south-eastern edge should be elevated by about 1 m compared to the north-western. Clearly, the water slopes are very small, even for a large, strong current such as the Gulf Stream. The sea surface slope can now be measured accurately using satellite altimetry. Traditionally, surface currents are estimated from knowledge of the density distribution in the ocean (measurement of salinity and temperature) using the geostrophic method.

The simplest way to visualize the pressure gradient, Coriolis force and current direction is as follows. An unspecified pressure gradient causes fluid to flow down the pressure gradient. However, it is deflected to the right by the Coriolis force in the northern hemisphere, causing the fluid eventually to move along the isobars, with the pressure force down the pressure gradient to the left and the Coriolis force up the pressure gradient to the right. Alternatively, if there is no pressure gradient initially, the flow of a fluid, deflected to the right by the Coriolis force, will eventually cause a build-up of water there, creating a surface slope and pressure gradient to the left (in the northern hemisphere). In both these situations a balance is achieved between the Coriolis and pressure forces, irrespective of how they came about.

The geostrophic method allows ocean currents to be estimated at different levels as long as friction can be ignored. Each ocean has a different salinity and temperature distribution, and therefore a different pattern of geostrophic flows. Taking two examples schematically (Fig. 9.9):

• in the west Pacific, the deep water is uniform in properties below a depth of about 1000 m, so there is essentially no motion below this depth;

• in the west Atlantic (Gulf Stream region), a level of no motion separates the surface Gulf Stream and the North Atlantic Deep Water.

It is theoretically possible that isobaric surfaces would slope relative to geopotential surfaces at a constant angle with depth in the ocean. In this case there would be no depth of negligible movement, a *slope or gradient current* existing throughout the oceanic water column (Fig. 9.10).

Isobaric and isopycnal surfaces

We have already defined isobaric surfaces as those on

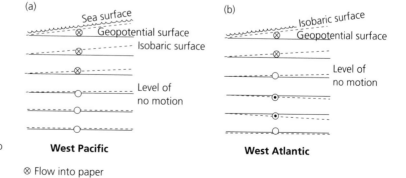

Fig. 9.9 Relationship between isobaric and geopotential surfaces to show schematically the situations in (a) the west Pacific Ocean and (b) the west Atlantic Ocean in the region of the Gulf Stream.

⊗ Flow into paper
⊙ Flow out of paper
○ No flow

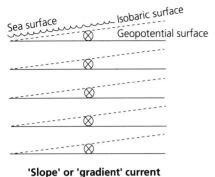

'Slope' or 'gradient' current

Fig. 9.10 Isobaric surfaces slope uniformly with depth relative to geopotential surfaces, giving rise to a slope or gradient current.

which the hydrostatic pressure is constant. An *isopycnal surface* is one on which the density of the fluid is constant. If the density of the fluid is entirely due to the ambient pressure, the isobaric and isopycnal surfaces would parallel each other, known as a *barotropic field* (Fig. 9.11a). But if the density is a function of other parameters such as temperature (applicable, for example, to a fresh water lake) or temperature and salinity (applicable to the sea), the two sets of surfaces may be inclined to each other—a *baroclinic field* (Fig. 9.11b). In real oceans, the upper 1000 m or so are typically baroclinic, and the deep water barotropic.

In mid- and low latitudes, the baroclinic field is caused primarily by variations in temperature. Looking down the flow direction in the northern hemisphere,

we should have light (warm) water on the right, the isotherms also dipping in this direction.

9.2.4 Coastal upwelling and downwelling

The interaction of a wind drift current and a geostrophic flow is classically displayed in the oceanographic situations of upwelling and downwelling. This is based on the fact that the net total transport of the wind-driven Ekman current (equation (9.5)) must be directed further to the right (in the northern hemisphere) than the surface current because of the progressive rotation with depth. In fact the net current acts at right angles to the wind direction. If there is an outflow to the right from a control volume of the ocean then from continuity there must be an inflow from the left to the surface layer. In an infinite ocean this inflow is accomplished easily. However, if the wind is blowing along a coast which is situated to the left of the wind direction in the northern hemisphere, there is a lack of surface water to replace that lost in the Ekman transport. Consequently, water replaces the Ekman transport from depth (Fig. 9.12), causing upwelling. This occurs when a wind blows equatorwards along the eastern boundary of an ocean, or polewards along a western boundary, the latter situation being less common.

Coastal upwelling may bring deep nutrient-rich water to the surface, promoting plankton production. The world's great fisheries are therefore in areas of coastal upwelling. For example, winds blowing equatorwards along the eastern side of the Pacific Ocean off Peru (southern hemisphere) and the western USA (northern hemisphere) cause upwelling. The same is

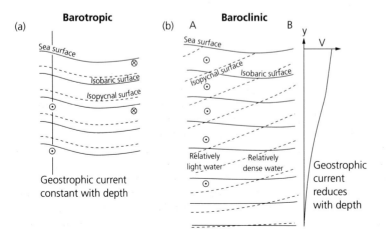

Fig. 9.11 (a) Barotropic and (b) baroclinic fields with associated geostrophic flow directions. In (a) the ocean water is homogeneous, whereas in (b) the water is relatively lighter under A than under B. At shallow depths the isobaric surfaces are parallel to the sea surface, but they flatten with depth because of the lateral variation in water density. Where the isobaric and isopycnic surfaces are approximately parallel, the geostrophic current reduces to zero.

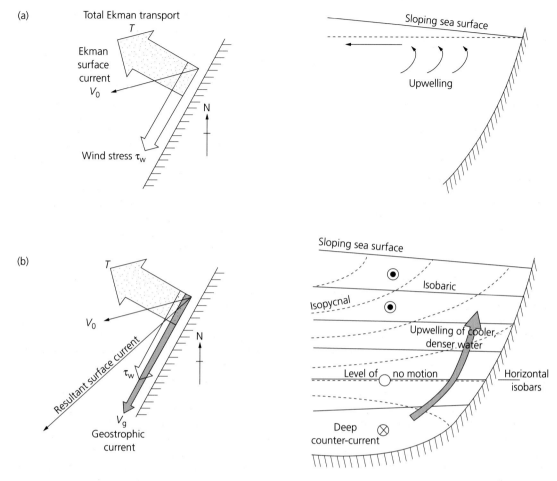

Fig. 9.12 Upwelling forced by the Ekman effect along the eastern side of an ocean in the northern hemisphere under a prevailing wind direction parallel to the coast. (a) The total Ekman transport pulls water away from the coast, causing upwelling and a lowering of the water surface towards the coast. (b) The upwelling of cold, dense water causes a baroclinic field, resulting in a geostrophic flow to the south in the surface layer (out of the page), a level of no motion (zero geostrophic velocity) and a deep countercurrent to the north (into the page). The resultant surface current continues to encourage upwelling. After Open University (1989) [7].

true of the western coast of South Africa along the eastern margin of the Atlantic Ocean.

When the Ekman transport is towards the coast, the water level may rise there, causing a surface slope. The surface slope cannot build up indefinitely, so there must be a return flow of water moving offshore. This return flow is driven by the pressure gradient set up by the superelevation of the water surface at the coast, producing a *geostrophic current*. This is the situation of downwelling. Vigorous downwelling may prevent the water near the coast from becoming stratified.

9.2.5 Effects of sea bed friction on a geostrophic current

Consider a current caused by horizontal pressure gradients within the water body—that is, a geostrophic current. A current flowing over the sea bottom will experience friction. The effects of the friction on the sea bed are to produce a current pattern that is essentially an inverted Ekman spiral (Fig. 9.13). Let us call this the *Ekman bottom layer* to distinguish it from the wind-driven surface Ekman spiral. This can be understood qualitatively as follows. With no friction,

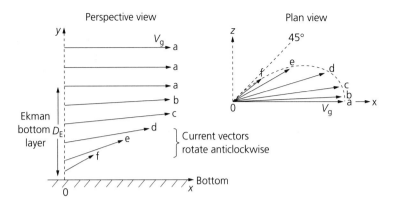

Fig. 9.13 The bottom Ekman layer as an inverted Ekman spiral due to the effects of friction on a geostrophic flow of velocity V_g in the northern hemisphere. After Pond & Pickard (1983) [5].

there is a geostrophic flow, with the Coriolis force acting to the right and a pressure gradient force acting to the left. Friction reduces the speed of the geostrophic flow, so the Coriolis force reduces since it is proportional to speed, which leaves the pressure gradient force unbalanced. Since the force acting to the right has reduced, the flow will deflect to the left so as to achieve a balance between the Coriolis and frictional forces and the pressure gradient force. The bottom Ekman layer therefore spirals to the left upwards.

In shallow water with a depth less than the Ekman depth, the wind-driven Ekman spiral may interfere with the bottom Ekman layer, causing a certain cancelling out. This causes the net transport to be close to the wind direction rather than at right angles. In very shallow water depths of about $D_E/10$ the net transport will be almost exactly in the wind direction because the frictional effects have completely dominated the Coriolis effects.

Although the results of Ekman may be criticized on the basis of the reality of the assumptions, they provide a good framework for thinking about the wind-driven flow of ocean water and the occurrence of upwelling and downwelling. The very large-scale circulation patterns of the oceans, including the intensification of ocean currents on the western sides of oceans, are explained by physical oceanographers by including pressure gradient terms and the effects of vorticity (the tendency of portions of the fluid to rotate). By including pressure gradient terms, the circulation due to the Ekman flow and the geostrophic flow can be integrated. These more complex models are successful at explaining the oceanic gyres shown in Figs 1.9 and 1.10, particularly the equatorial current systems. However, their formulation is well beyond the scope of this text.

9.2.6 Interaction between ocean currents and coastal waters

In Chapter 8 we saw that the energy for tidal processes on the shelf was provided by the ocean tides. Wind waves generated in the ocean may also travel long distances as swell before breaking at the coast. But the interaction between ocean currents and coastal circulation is less clear. There are a number of situations where ocean currents interact with coastal waters.

• A common occurrence is the formation of a shallow water countercurrent to the offshore deep ocean current. An example is the western Atlantic where, north of Cape Hatteras, a nearshore current flows to the SW in the opposite direction to the Gulf Stream. If such current pairs involve a transverse circulation, oceanic water can be exchanged with shelf water.

• Another form of interaction is the upwelling along the eastern sides of oceans, such as the Peru Current in the Pacific and the Benguela Current in the south Atlantic. This has previously been discussed in Section 9.2.4.

• Oceanic water is also driven in strong transient currents during cyclones, moving roughly parallel with the coast. We shall study this phenomenon in some detail below, since it has a major impact of sedimentary processes on the shelf.

• In general, ocean currents do not intrude significantly on to the continental shelf. The percentage of continental shelves where intruding ocean currents are important is very small (less than 3%). The intrusion of the Agulhas Current on the south-east African coast is a good example (Fig. 9.14). The narrowness of the shelf, between 10 km and 40 km, and the steepness of the continental slope (c. 12°) allows the powerful western boundary Agulhas Current to dominate fluid and sediment movement on the

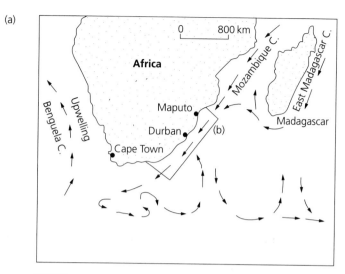

(a)

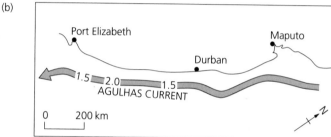

(b)

Fig. 9.14 (a) The ocean current systems around South Africa, showing the intrusion of the Agulhas Current on to the south-east African continental shelf. (b) Current velocities (in metres per second) associated with the Agulhas Current. After Flemming (1981) [8], pp. 259–77, with kind permission of Elsevier Science-NL, Sara Burgerhartstraat 25, 1055 KV Amsterdam, The Netherlands.

shelf, producing NE-SW currents of up to 2.5 m s^{-1} at approximately the shelf edge. The inner shelf (up to 50–60 m depth) is consequently dominated by surface wind waves producing a well-sorted prism of quartz sand, whereas the outer shelf is dominated by the ocean current, driving large dunes of carbonate sand over a pavement of gravels deposited during a relative sea level lowstand earlier in the Holocene.

9.3 Deep water sediment drifts

With the exception of the currents set up by internal waves acting on density gradients in the oceanic water, and the currents moving up and down submarine canyons, the bottom currents in the deep sea are those of the slow thermohaline circulation and to a lesser extent the larger wind-driven circulations where they impinge on the sea bed, such as the Gulf Stream gyres of the North Atlantic. The global circulation pattern has tremendous importance in the distribution of heat through the Earth surface system (Chapter 1), but it is also responsible for the transport and accumulation of very fine particulate sediment. Three main effects are of interest in controlling the sediment transporting capabilities of these deep ocean currents: first, currents are intensified on the western sides of ocean basins, the so-called *western intensification*; second, current speeds are increased as they flow through bathymetric constrictions and reduced where they expand; and third, large-scale eddies may develop at the edges of deep geostrophic flows, resulting in enhanced mixing and loss of ability to maintain sediment in suspension. Bottom current speeds in the absence of these intensifying effects are very low, generally less than 20 mm s^{-1}. Deep western boundary currents may reach 0.1–0.2 m s^{-1}, and constriction by sea bed irregularities may cause further enhancement.

The more vigorous bottom currents may be associated with concentrations of very fine-grained particulate matter (about 12 μm mean size). These concentrations are very low, but are nevertheless substantially above that of the ambient clear seawater [9, 10]. They are called *nepheloid layers*. The highest concentrations of perhaps 0.01 ppm (0.2 mg l^{-1}) are found near the sea bed, but some nepheloid layers extend from the continental shelf edge far into the ocean as zones of anomalously high suspended matter concentrations, 1–2 km thick. These detached nepheloid layers provide a steady rain of very fine particulate matter to the deep sea far from the shelf edge.

Bottom currents may locally erode the sea floor, but they are also responsible for the formation of large sediment drifts. These large expanses of sea bed sediment are particularly common at the edges of deep bottom water pathways (Figs 9.15 and 9.16). They may occur where secondary gyres are shed from the main current, or where the bottom current runs along the contours of the continental slope. The fine-grained beds within deep water sediment drifts are commonly termed *contourites*[11]. Along the base of the continental slope, it is particularly common to find contourites interstratified with the deposits of density-driven currents (turbidites) (Chapter 6).

The large wind-driven and thermohaline currents also affect the slow settling of the primarily biogenic and organic debris of siliceous and calcareous planktonic organisms living in the photic zone of the ocean. These materials form the pelagic deposits of the deep sea. When admixtures of very fine terrigenous, aeolian, volcanic and cosmogenic sediment are found together with pelagic components, the sediment is called *hemipelagic*.

9.4 Passage of a storm/cyclone

Having considered the large-scale, steady-state processes of wind-driven and geostrophic circulations, we can now turn our attention to the processes taking place during a disturbance of short time-scale and small spatial scale—a storm (hurricane or cyclone) crossing the continental shelf and approaching a coast.

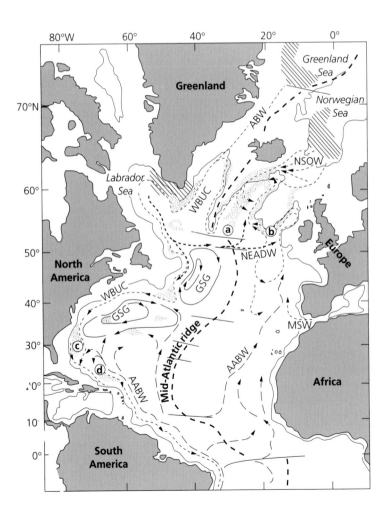

Fig. 9.15 Contourite drifts in the central and north Atlantic Ocean (stipple), deep water circulation (arrows) and areas of deep water formation (hatched). AABW, Antarctic Bottom Water; ABW, Arctic Bottom Water; MSW, Mediterranean seawater; NSOW, Norwegian Sea outflow water; WBUC, western boundary undercurrent; GSG, Gulf Stream gyre; NEADW, north-east Atlantic Deep Water. Bathymetric contour shown at 2000 m. After Stow & Holbrook (1984) [12]. Letters a, b, c and d refer to sediment drifts shown in Fig. 9.16.

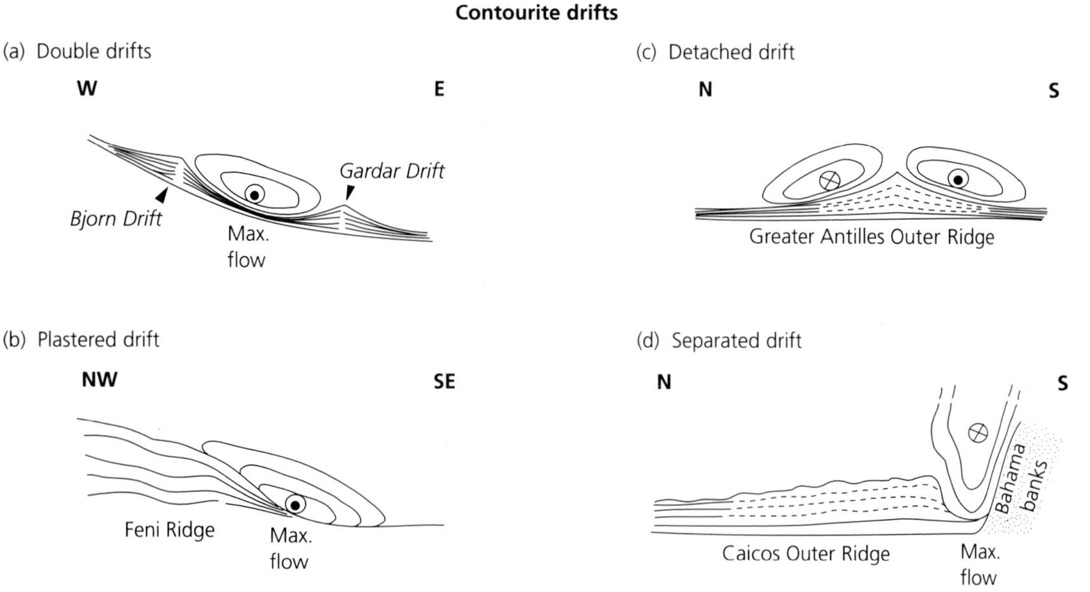

Contourite drifts

(a) Double drifts

(c) Detached drift

Fig. 9.16 Possible relationships between deep sediment drift deposition and deep ocean currents. Examples are (a) Gardar and Bjorn Drifts; (b) Hatton and North Feni Drifts; (c) Greater Antilles and Blake Drifts; (d) Caicos Drift. After McCave & Tucholke (1986) [13].

There are a number of reasons for being particularly concerned about this situation: first, cyclones represent the single most important natural hazard in terms of loss of human life and damage to human infrastructure (Section 9.5); second, storms may cause considerable environmental change at the coastline; and third, storms are thought responsible for the transportation of large amounts of particulate matter on to the continental shelf from the nearshore zone—storm deposits have been widely described from the Earth's sedimentary record.

The fluid and sediment movements during a storm can be thought of as a resultant of a number of different processes, some of which we have considered earlier in this chapter. (i) storms are associated with deep atmospheric depressions—the spatial variations in atmospheric pressure have a barometric effect on the elevation of the water surface; (ii) storms are associated with strong winds, which must therefore cause a wind-driven surface current (along the lines described by Ekman); (iii) the piling up of water at the coast during a storm by the wind drift current, known as *wind set-up*, and by the mass transport associated with progressive gravity waves, called *wave set-up*, results in a pressure gradient which must drive water seawards as a gradient current (meaning pressure gradient current).

There may be yet other important processes taking place during a storm that should be incorporated into any model for fluid and sediment transport. These would include the surface wind wave action, the outflow of rivers on to the shelf, and the transient effects when the wind stops blowing, the depression weakens, or the cyclone passes through a coastal locality. These effects include the propagation of storm surge water as long waves along the shelf (*shelf waves*), or where the coastal zone allows water to be stored in lagoons, the post-storm flow back on to the shelf of coastal water, known as *storm surge ebb*.

9.4.1 The barometric effect

The lowering of atmospheric pressure in the centre of a cyclone should result in an upward bulging of the water surface. This can be verified by viewing the cyclone as moving very slowly compared to the speed of response of the water to atmospheric pressure changes, in which case the static condition adequately describes the situation under a moving weather system. Consider a water body with a horizontal plane at some depth in the water (Fig. 9.17). Now let the atmospheric pressure be different at points A and B on this submerged plane. The total pressures at A and B are therefore the sum of the hydrostatic pressures due to the overlying water and the atmospheric pressures.

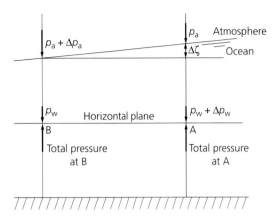

Fig. 9.17 The barometric effect on the water surface elevation in a storm. After Allen (1985) [14].

If there is no force gradient on the submerged horizontal plane, which would cause motion, the total forces at A and B can be equated, from which it can be derived that

$$\Delta p_a = \Delta p_w \qquad (9.19)$$

where Δp_a and Δp_w are the differences in pressure between A and B due to the atmosphere and water, respectively. If the difference in elevation of the water surface between A and B is responsible for the difference in the hydrostatic pressure, then

$$\rho g \Delta \varsigma = \Delta p_w \qquad (9.20)$$

which allows the elevation change of the water surface to be expressed in terms of the change in atmospheric pressure

$$\Delta \varsigma = \frac{\Delta p_a}{\rho g} \qquad (9.21)$$

The atmospheric pressures in the centre of a cyclone are typically 50 millibars $(500\,\mathrm{kg/m^{-1}\,s^{-2}})$ less than normal atmospheric pressure. Consequently, we should expect the sea surface to bulge upwards by about 0.5 m under the centre of the cyclone. This is a significant addition to the effects caused by coastal set-up (see below). The maximum barometric effect at the coast should be felt at the landfall of the cyclone.

9.4.2 Wind set-up

The fluid motion in a deep, homogeneous water column in response to the shear of the wind over the surface was treated in Section 9.2.2 of this chapter, where it was called the Ekman transport. Here, we wish

to know the wind drift current where there is a landward boundary and a shallow sea bed. This complicates the dynamical theory of Ekman considerably.

There are three questions that we must address if we are to evaluate the importance of the Ekman effects for the movement of water during storms in the shallow waters of the nearshore zone. These can be summarized as follows:

1 The effect of decreasing water depth relative to the Ekman depth is to reduce the angle between the Ekman surface current and the wind direction, the current becoming almost parallel to the wind direction, with little rotation with increasing depth, as water depth shallows.

2 The water motion in a storm is in a transient state. If a wind stress is suddenly applied to a still body of water, the surface current flows in a very shallow layer in the direction of the wind, the typical steady-state Ekman spiral only developing after a period of the order of one day.

3 If the water is stratified, as may occur particularly in the summer months near the coast, the vertical eddy viscosity is reduced, causing the surface Ekman spiral to be confined to very shallow water depths. Since the total Ekman transport remains the same as in the homogeneous case, the speed of the surface current increases.

The elevation of the water surface due to the wind stress driving water towards a closed boundary such as a coastline or the head of a gulf results in a shorewards surface layer drift and a compensatory flow offshore in the lower layer. These currents are associated with shear stresses at the surface and at the sea bed τ_s and τ_b. The depth-integrated equation for the wind stress effect only is

$$\frac{\partial \varsigma}{\partial x} = \frac{\tau_s - \tau_b}{g \rho (h + \varsigma)} \qquad (9.22)$$

where ς is the elevation of the water level in a rectangular gulf of depth h (Fig. 9.18). If the elevation is small compared to the water depth in an idealized rectangular cross-section (an invalid assumption in the very shallow water depths at the coast), this can be simplified to

$$\frac{\partial \varsigma}{\partial x} = \frac{C \tau_s}{g \rho h} \qquad (9.23)$$

where C is a factor accounting for the relation between the surface shear stress and the sea bed shear stress, ranging between 1 and 1.5. The slope gradient can therefore be found for appropriate values of wind stress and water depth. If the wind has a velocity of

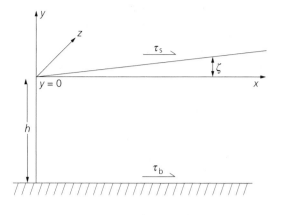

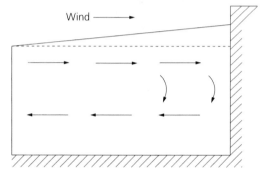

Fig. 9.18 Notation for the water surface slope and current flow caused by the wind stress in a rectangular gulf.

20 m s^{-1} (at 10 m above the sea surface), the stress is about 1 N m^{-2}. In a water depth of 20 m, the sea surface gradient should be 5×10^{-6} (taking $g = 9.81$ m s^{-2}, $\rho = 1025$ kg m^{-3}, $C = 1$). If the water surface slope occurs over a lateral distance of 400 km, the elevation at the coast or head of a gulf should be 2 m.

9.4.3 Wave set-up

In Chapter 8 it was found that the propagation of progressive gravity waves over the Earth's surface results in a small but significant mass transport in the wave propagation direction. Where the waves are propagating towards shore, this build-up of water in the nearshore zone is compensated by the development of rip currents. Shorewards of the breaker zone there is a particularly strong mass transport towards shore.

We know that the diminishing water depth causes waves to break as they approach shore, a value of $H/h = 0.4$ being a good approximation for breaking

(Section 8.6.2). If the amplitude A ($\equiv H/2$) of waves inshore of the breaker zone is proportional to the water depth h,

$$A = \alpha h \tag{9.24}$$

where the proportionality coefficient is related to the beach slope and varies between 0.3 and 0.6, and is equal to 0.4 for the criterion for breaking given above. The gradient of the sea surface normal to the beach is then proportional to the gradient of the water depth (equivalent to beach slope) in the same direction:

$$\frac{\partial \bar{\zeta}}{\partial x} = -\frac{3}{2}\alpha^2 \frac{\partial h}{\partial x} \tag{9.25}$$

where $\bar{\zeta}$ is the mean elevation of the water surface above the level found in the absence of waves, or wave set-up. If a is taken to be 0.4, the horizontal shore-normal gradient in the wave set-up is $-0.24\partial h/\partial x$. If the gradient in the wave set-up occurs over a transverse distance of 100 m, and the waves have an amplitude of 1 m at the point of breaking, the beach gradient is 0.025 and from equation (9.25) the set-up at the highest point on the beach reached by waves is 0.6 m. Large amounts of wave set-up are clearly favoured by high-amplitude storm waves with long surf run-up distances. Wave set-up is a potent process of elevating mean sea levels on long beaches fronting oceans, and contributes to the breaching and washing over of barriers.

Wave set-up is balanced by a seaward flow caused by the surface slope. This return flow or *undertow* has an average velocity U given by

$$U = -\frac{1}{2}\alpha^2 \sqrt{gh} \tag{9.26}$$

where h is the depth of water inshore of the breaking point. This undertow is unlikely to have speeds in excess of about 0.1–0.2 m s^{-1}, but the seaward flow is more often concentrated in rip channels where it may be very vigorous. The velocities in rip channels where the discharge is concentrated may be up to 5 m s^{-1} in severe storm conditions. Rip currents therefore have great potential for the seaward transport of coarse beach material during storms. The location of these rip channels is controlled by low points in the wave set-up. This is because longshore variations in the wave set-up due to longshore variations in wave-height result in longshore pressure differences. The pressure gradients act from high points of set-up to low points of set-up, thereby explaining the nearshore circulation.

9.4.4 The pressure gradient current

The piling up of water at the coast by wind set-up, wave set-up and barometric effects, termed *coastal set-up*, causes a slope on the water surface. To simplify our study, we assume that the slope occurs under a uniform atmospheric pressure and is in a steady state (Figs 9.18 and 9.19).

A surface slope produces a horizontal pressure gradient in the underlying water. In homogeneous water, this horizontal pressure gradient will be uniform with depth (a barotropic field), and will generate a current.

In the absence of friction, this is a geostrophic current flowing at right angles to the horizontal pressure gradient. If the sea surface is elevated above the horizontal by an amount ς, so that there are water surface gradients in the x and z directions denoted by $\partial\varsigma/\partial x$ and $\partial\varsigma/\partial z$, the components u and v of the geostrophic current are

$$u = -\frac{g}{f}\frac{\partial\varsigma}{\partial z} \tag{9.27}$$

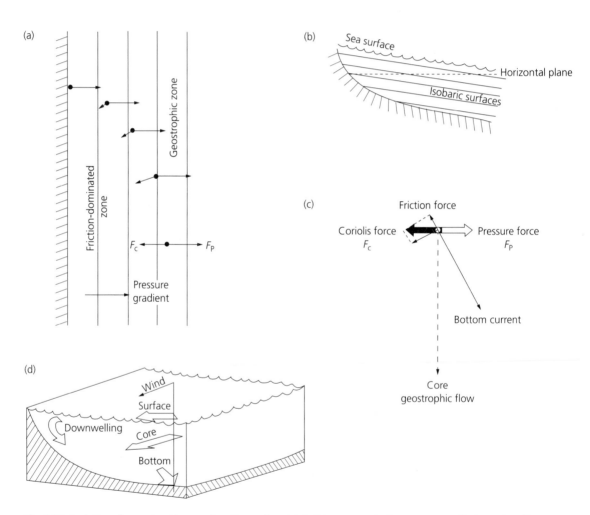

Fig. 9.19 Evolution of a geostrophic current on the continental shelf in response to the pressure gradients produced by coastal set-up. (a) Pressure gradient and Coriolis forces on a particle subjected to smaller frictional forces as it travels into deeper water. (b) Cross-section of the continental shelf showing the inclined isobaric surfaces and sloping water surface. (c) Balance of forces on a unit mass of water near the bed. The bottom flow is oblique offshore, whereas the core geostrophic flow acts along shelf. (d) 3D view of the differing orientations of the wind stress, surface Ekman current, bottom current and core geostrophic flow. After Swift *et al.* (1986) [15].

Practical exercise 9.2: Storm surge height

The actual height of the water surface at the coast above (or below) the tidal prediction is the resultant of the barometric effect, wind set-up and wave set-up. Consider a storm approaching land as follows. The atmospheric pressure in the centre of a deep tropical cyclone is 60 millibars (601×10^2 kg m^{-1}s^{-2} below the normal atmospheric pressure. It produces a water surface slope over a lateral distance of 250 km where the average water depth is 30 m. Winds are 25 m s^{-1}, measured at 10 m above the sea surface. An additional set-up is caused by waves of height 2.6 m just before breaking with a surf run-up distance of 160 m. Calculate the three different set-ups and sum them to calculate the resultant storm surge height.

Solution

We have previously seen that the barometric effect is related to the pressure difference in the atmosphere (equation (9.21)). The elevation change at the coast during landfall of the cyclone should be 0.6 m for the pressure difference of 60 hPa.

The wind stress is given by equation (9.9). For a density of air of 1.3 kg m^{-3}, a drag coefficient of 2.3×10^{-3} at 25 m s^{-1}, and a wind speed measured at 10 m above the sea surface, the wind stress is 1.87 N m^{-2}. The water surface slope associated with wind set-up is inversely proportional to the water depth (equation (9.23)). For an average water depth of 30 m, the water surface slope is 6.2×10^{-5}, giving a wind set-up of 1.55 m over a lateral distance of 250 km.

From equations (9.24) and (9.25), the gradient in the water surface in the surf zone due to wave set-up is 0.0052, giving a set-up of 0.84 m over a lateral distance of 160 m.

The total set-up at the coast is therefore the sum of the barometric effect, the wind set-up and the wave set-up, equal to 3 m.

$$v = \frac{g}{f} \frac{\partial \varsigma}{\partial x} \qquad \textbf{(9.28)}$$

We know, however, that in shallow water there is likely to be friction at the sea bed. This produces a bottom Ekman layer (an inverted Ekman spiral) which counteracts the surface Ekman spiral. The result is a surface current that flows more or less down the pressure gradient. The velocity profile between the sea surface and the sea bed will depend on the vertical variation in the effective viscosity of the sea water. If the flow is laminar, the velocity profile will be parabolic [14]. If the gradient current is turbulent, its velocity profile should approximate a logarithmic law.

The relative contribution of the various effects causing a storm surge must vary according to whether the disturbance is a large mid-latitude storm, in which case the wind set-up will dominate, or a smaller hurricane where the barometric effect may be of greater importance. Wave set-ups are largest on shoaling ocean-fronted beaches.

The storm surge heights actually measured during hurricanes, typhoons and cyclones support the range of values calculated in Practical Exercise 9.2, with heights commonly between one and a few metres. The 1970 cyclone in the Bay of Bengal produced an exceptional surge height of 7 m (see also Section 9.5). Once developed, the storm surge may propagate as a free long wave along the shelf, in much the same manner as a Kelvin wave (Chapter 8). This phenomenon will be further discussed in relation to the hydraulic regime of the Middle Atlantic shelf of North America.

9.4.5 Sediment transport and bedforms under storms

The adequate prediction of sediment transport during a storm relies on knowing something about the water velocity at the bed. It can be appreciated from the above discussion that this is by no means an easy exercise because of the complexity of the water motion and added problems caused by irregular sea bed topography. Storm activity should be associated with:

1 increased undertow/rip current discharges from the beachface due to wave set-up;

2 a seaward-flowing gradient current where friction dominates Coriolis effects in the shallow waters of the nearshore zone;

3 a geostrophic flow in deeper water where the Coriolis force balances the pressure gradient and the near-bed current moves approximately at right angles to the surface slope, that is, along the continental shelf;

4 and, superimposed on these steady or quasi-steady currents, the oscillatory motions of storm waves.

The discussion above, and the equations provided, allow the water depth at which the seaward-moving flow transforms to a longshore geostrophic flow (Fig. 9.19) to be approximated. This water depth is relative to the Ekman depth. As a guide, if the Ekman depth is 100 m in deep homogeneous water, friction should completely dominate at water depths of less than about 10 m. In water depths less than the Ekman depth, the surface Ekman spiral and the bottom Ekman layer interfere. Consequently, in water depths between 10 and 100 m (where D_E is 100 m) we should expect to see the transition from a friction-dominated to geostrophic flow. This is also a range of water depth where the effects of surface wind waves are likely to be felt (Chapter 8). It is highly likely, therefore, that in the shallower parts of the continental shelf the flow near the bed will be a combined wave and current field, the waves dominating in progressively shallower water, and the storm currents dominating in progressively deeper water.

The zone of friction is found in the shallow water depths where wave action is intense, corresponding to the *shoreface*. On the continental shelf, however, an along-shelf flow should dominate storm sedimentation [16]. The inverted Ekman spiral, or Ekman bottom layer, causes a deflection to the left (in the northern hemisphere) of the geostrophic core flow, so where the geostrophic flow is in water depths shallow enough for a bottom Ekman layer to develop, there should be a net sediment transport obliquely offshore to the strike of the shelf [17]. These concepts can be tested by considering the processes and deposits of modern continental shelves.

Hydraulic regimes and sediments of modern continental shelves

Hurricane Carla, Gulf of Mexico The concepts of storm-driven flows can be tested by studying the sediment distribution following Hurricane Carla, one of a number of historical tropical cyclones to affect the Gulf of Mexico region (Fig. 9.20a). Although there were no near-bed flow measurements taken during the passage of Hurricane Carla across the Texas shelf in September 1961, there is information on the wind field, the coastal set-up and the surface gravity waves during the storm, and the distribution and thickness of a sand bed attributed to the storm are also well documented (Fig. 9.20b).

A dynamical model used to reconstruct the storm surge, and validated by more recent, smaller hurricanes in the same area, indicates that a substantial cross-shelf sea surface slope of about 10^{-5} was built up about a day before the landfall of the hurricane when the eye of the storm was still beyond the shelf edge, the water surface having a superelevation of over 2 m at the coast close to Houston (Fig. 9.21). A NW surface wind drift, SW geostrophic core flow and S or SSW bottom flow, oblique to the continental shelf

contours, are expected from our dynamic model. We can estimate the speed of the geostrophic current resulting from this sea surface slope from the geostrophic equation (equation (9.18)). Taking the latitude of Houston as 30°N, the speed of the geostrophic current is $1.35\,\mathrm{m\,s^{-1}}$. Six hours before the landfall when the eye of the storm was less than 50 km from the coast between Corpus Christi and Houston, the sea surface slope had increased to 4×10^{-5}, with a coastal set-up of a maximum of 2.5 m. The corresponding geostrophic speed is over $5\,\mathrm{m\,s^{-1}}$ using the geostrophic equation. However, this high figure ignores friction on the sea bed and the transient state of the flow, making it an unrealistic maximum. Nevertheless, the magnitudes of these flows driven by the surface slope were clearly able to cause substantial along-shelf sediment transport in the day before the landfall of the hurricane.

Storm waves are likely to have modified the near-bed flow field during the passage of the hurricane. Knowledge of the surface wave conditions allows the orbital velocities near the bed to be calculated. Such orbital velocities appear to have been similar in magnitude to the maximum steady geostrophic flows on the inner and middle shelf, but directed roughly orthogonally to the steady current. In shallower waters, the wave orbital velocities would have dominated the steady flow, producing a more shore-normal flow field.

These predictions can now be compared with the distribution of the Carla sand bed, and with the sedimentary structures found from shallow coring of the Carla sea bed sediment (Fig. 9.20b). The observed geometry of the Carla sand bed shows it to occur in water depths up to 50 m, and to thicken to up to 9 cm in water depths of between 20 and 30 m, but, importantly, to extend continuously along the strike of the shelf for about 200 km. Information landwards of the 20 m isobath is lacking because of biological reworking and truncation by younger events. Associated with the seaward thinning is a seaward reduction in grain size. These observations are in harmony with the proposed obliquely offshore storm-driven flows deriving their sand and silt from the extensive shorefaces of this barrier-island coast.

Box cores through the Hurricane Carla deposit on the Gulf of Mexico shelf generally fine upwards in grain size from a sharp, basal surface overlain by upper stage planar laminations, to low-angle inclined laminations, and a gradational upper contact. In shallow waters close to the shoreface, the Carla sand has a truncated, sharp upper contact and rests within an amalgamated sequence of beds.

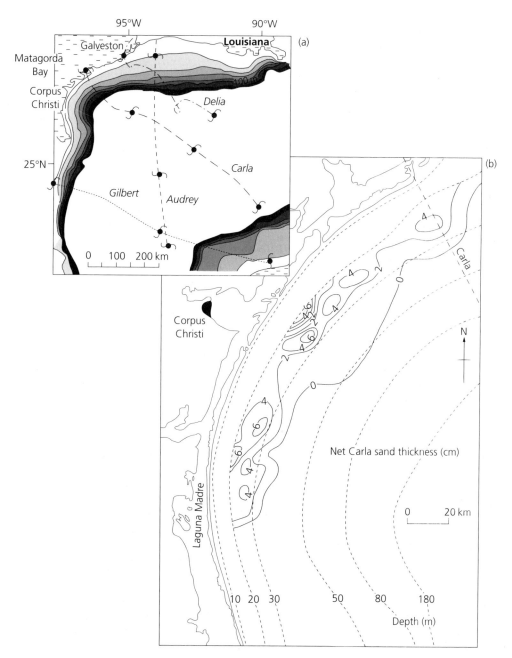

Fig. 9.20 (a) The bathymetry of the Texas–Louisiana shelf of the Gulf of Mexico with the paths of four recent tropical cyclones: Tropical Storm Delia (1973) and Hurricanes Audrey (1957), Carla (1961) and Gilbert (1988). The symbols show the location of the eye of the cyclone every 24 hours except for Delia. After Keen & Slingerland (1993) [18]. (b) Distribution and thickness of the Carla sand bed on the Gulf of Mexico shelf. After Snedden & Nummedal (1991) [19]. The loss of data landwards of the 20 m isobath is because of the pervasive bioturbation which churns up the sea bed sediment, making recognition of the Carla sand bed difficult.

(a)

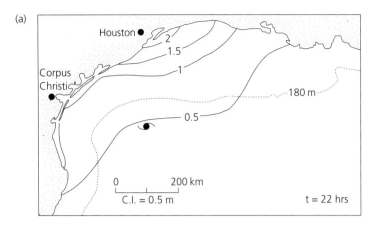

(b)

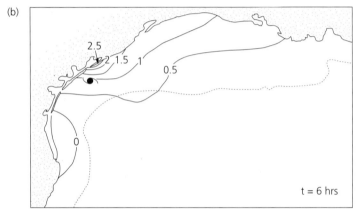

Fig. 9.21 Modelled storm surge heights 22 hours and 6 hours before the landfall of Hurricane Carla in 1961. Modified from Miyazaki (1965) [20]. Set-ups have also been modelled by Keen & Slingerland (1993) [18].

The adequacy of the combined flow field due to the pressure gradient steady current and the orbital motion due to storm waves to cause sediment transport from the shoreface to the shelf, and the along-strike geometry of the Carla sand body, make a previous model of transport and deposition by a storm surge ebb draining rapidly from coastal lagoons in a shore-normal direction unsatisfactory [21].

Middle Atlantic shelf of North America The storm regime on the Middle Atlantic shelf is probably better known than that on any other of the world's storm-dominated shelves. Storms on the shelf are mid-latitude low-pressure systems known as north-easters, which generate sustained southwesterly flows on the shelf over distances in excess of 1000 km. The tropical, intense, but smaller hurricanes that typify the Gulf of Mexico are less important in affecting the hydraulic regime of the shelf. The predominantly along-shelf fluid movement results in a predominantly along-shelf sediment transport (Fig. 9.22), and offshore directed

transport is minimal.

The response of water on the Middle Atlantic shelf of North America is very well described by the Ekman theory previously developed. For a wind from the north-east, there is a surface wind-driven current directed obliquely onshore (to the south and west) by the Coriolis force, a geostrophic core flow running approximately along the shelf typically with speeds of about $0.4 \, \mathrm{m \, s^{-1}}$, and a bottom current directed obliquely offshore. The pile-up of water at the coast causes a set-up resulting in a pressure gradient sloping offshore, and *coastal downwelling*. For a wind blowing from the south-west, also common on this shelf, the surface current is directed offshore, resulting in *coastal set-down*, and consequently *coastal upwelling*. Current meters deployed on the shelf have recorded much stronger current velocities in the winter (mean of $0.15 \, \mathrm{m \, s^{-1}}$) than in the summer (mean of less than $0.08 \, \mathrm{m \, s^{-1}}$).

There are additional interesting phenomena that occur on the shelf and have significance for the dis-

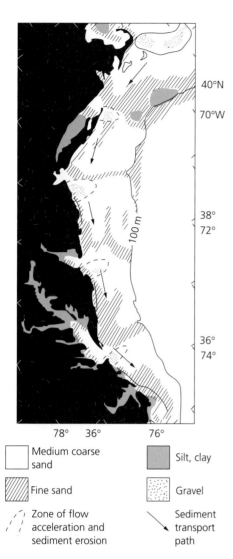

Fig. 9.22 The Middle Atlantic shelf of the USA. Observations on the distribution of grain size, and interpreted along-shelf net sediment transport paths (arrows). Areas of sea bed gravel are thought to represent regions of flow acceleration and erosion. After Swift *et al.* (1981) [22].

tribution of particulate sediment, pollutants and the dispersal of planktonic material such as larvae. First, during the summer months the water on the shelf becomes stratified, with light, brackish water at the surface overlying denser, cooler, more saline water. The piling up of fresher surface water at the coast by the wind may in these conditions cause the pycnocline

to slope downwards towards the coast, generating a baroclinic flow along the coast known as a *coastal jet*. This is in contrast to the barotropic flow caused by the pressure gradient during coastal set-up. Baroclinic flows may be particularly important during periods of high fresh water runoff from the North American continent, through arteries such as Chesapeake Bay. Tongues of brackish water may stream along the coast for considerable distances as baroclinic jets, as in 1972 during Hurricane Agnes off Chesapeake Bay.

Second, there have been a number of well-documented storms on the Middle Atlantic shelf where the build-up of water either at the coast or over shallower sea bed has resulted in the propagation of a long-wavelength wave on the shelf (with a celerity equal to $\sqrt{gh}$) when the wind forcing dies down. This is sometimes known as a *relaxation current*, or a *topographically trapped wave*, because they are trapped by the coast to which they attempt to veer because of the Coriolis force. (Where gravity is the restoring force for a wave of this type it is identical to the Kelvin waves described in Chapter 8. Where the Coriolis force is the restoring force, they are known as *Rossby waves*.) The currents under these shelf waves may be very vigorous, and move in the northern hemisphere to the south along the western boundary of an ocean, and to the north along the eastern. Because they are damped by friction, there is a net fluid transport and net sediment transport in the direction of wave propagation.

The dissipation of the coastal set-up as the storm wanes is achieved by the propagation of the shelf wave, which moves along the shelf losing energy by friction and causing a progressively smaller coastal set-up to occur. Consequently, there is no sudden storm surge ebb of water back on to the shelf, but rather a gradual fall in the coastal set-up by an along-shelf movement of the shelf wave. It is not expected therefore to find strong cross-shelf post-storm sediment transporting currents in settings analogous to the Middle Atlantic shelf of North America.

Current meter records from 10 m depth on the Long Island coast (Fig. 9.23) show that reversing tidal currents (less than $0.11\ \mathrm{m\,s^{-1}}$) are always below the threshold of sediment movement for the fine to medium sand on the sea bed. The passage of typical north-easters causes storm-driven currents of 0.05–$0.4\ \mathrm{m\,s^{-1}}$ over a period of about 2 days. Oscillatory wave action is only important during this storm period, and contributes to very high sediment concentrations of $100\ \mathrm{mg\,l^{-1}}$. The storm flow field is therefore a strongly combined wave–current field, resulting in

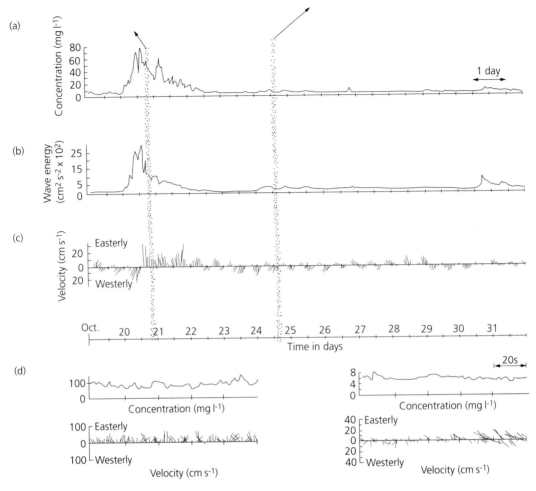

Fig. 9.23 Water motion and sediment concentrations over a two-week period on the inner Long Island shelf at the 10 m isobath: (a) sediment concentration; (b) wave energy; (c) current velocity. Details in (d) show the wave orbital velocity and sediment concentrations sampled once per second. Note that the vertical scales for orbital velocity and concentration are different between the storm event on 20 October (left-hand panel) and during fair weather (24 October, right-hand panel). The near-bed motion during fair weather is a typical tidal sinusoid (c) whose velocities are insufficient to entrain sea bed sediment. However, during the storm there is a strong combined flow field of a steady along-coast flow superimposed by high wave energy, causing a tenfold increase in the sediment concentrations measured 1 m off the sea bottom. After Swift *et al.* (1986) [15].

high suspended sediment concentrations and advection in the direction of the steady component.

The release of tracers, and sediment transport calculations based on fluid measurement, both suggest that the great bulk of the sediment transport on the shelf is caused in three to five major winter storms each year, the sediment transport lasting a period of a few days in each case. This results in a general, uniform sediment transport direction to the south-west along the shelf but slightly offshore, in the same direction

as the geostrophic flows. The net south-westerly geostrophic transport is aided by the relative set-down at the shelf edge caused by the presence of the Gulf Stream (Section 9.2.3) and, perhaps more importantly, by the anticlockwise propagation of shelf waves.

Although the hydraulic regime and sediment transport paths are complex, the Middle Atlantic 'model' clearly provides an important guide to the interpretation of ancient shelf sediments preserved in the geological record [15].

On the Middle Atlantic continental shelf, storm flows appear to fashion sediment accumulations that are larger than the deposits of individual storm events such as that of Hurricane Carla. These storm sand ridges are particularly well developed and well documented on the continental shelf of the eastern USA (Fig. 9.24). The inner shelf is characterized by a sand sheet which is being actively reworked by the modern hydraulic regime following Holocene transgression, but little material is currently being supplied to the shelf by rivers. A highly conspicuous feature of the inner shelf are the swarms of longitudinal ridges, typically 3–9 m high, 9–15 km long and up to 3 km wide, oriented with their long axes at angles of about 20° to both the shoreline and the maximum current directions (Fig. 9.24).

The origin of the sand ridges is complicated by the fact that the shelf has undergone major sea level change over the late Pleistocene and Holocene. Consequently, there is much debate as to whether the ridges picked out in the sea bed bathymetry are currently active, or represent degraded and inactive coastal and nearshore ridges drowned during the Holocene transgression. Some ridges are clearly coast-parallel, whereas others are orientated at about 20–30° to the coast. The obliquely oriented ridges are attached to the shoreface at their landward ends and are interpreted as a response to winter storms [24]. The origin of the mid-shelf ridges subparallel to shore is more controversial. There are different schools of thought: one that the ridges are degraded and inactive former nearshore-coastal barriers [25]; another that they are drowned (relict) shoreface-attached ridges [26]; and a third, more dynamic hypothesis that they are currently highly active in mid-shelf setting [27] rather than being relict.

The Oregon–Washington shelf of western North America
The Oregon–Washington shelf of western North America has a stronger wave regime than the eastern seaboard, and has the added interest of a significant sediment supply by rivers. Consequently, much can be learnt about shelf processes by the dispersal of particulate sediment from distinct point sources such as the Columbia River (Fig. 9.25). The Washington shelf north-west of the Columbia River mouth has been studied in detail [28]. This area is exposed to high-energy waves and currents generated by North Pacific storms originating mainly in the Gulf of Alaska. The combined action of waves and storm-driven steady flows is crucial to an understanding of the sediment transport on the shelf.

In the summer the northerly winds cause an offshore Ekman transport, causing coastal upwelling and a coastal set-down. This in turn produces a geostrophic current flowing to the south. However,

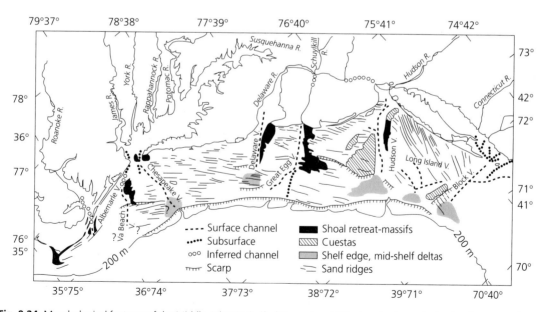

Fig. 9.24 Morphological features of the Middle Atlantic shelf of the eastern USA, showing in particular the alignment of sand ridges at an oblique angle to the coast. After Swift *et al.* (1973) [23].

(a)

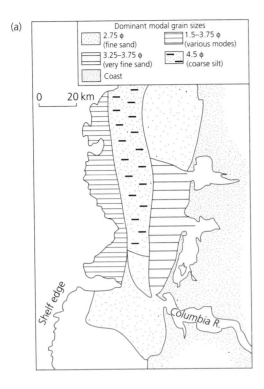

(b)

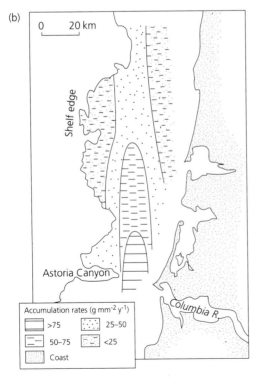

the upwelling causes the pycnocline to dip seawards, causing a baroclinic rather than barotropic current because of the presence of a density stratification. This baroclinic flow opposes the barotropic flow (due purely to the pressure gradient caused by the slope of the water surface) at the bed, and when the stratification is intense results in a net northward flow, but the current velocities are too low to affect sediment transport substantially.

During winter storms, the winds are from the south, causing coastal set-up and a northward barotropic current. Coastal downwelling and nearshore mixing break down the density stratification, so that the geostrophic current is essentially barotropic. Friction on the sea bed causes the bottom flow of the geostrophic current to turn offshore relative to the core flow. The combination of a high barotropic flow and the lack of an opposing baroclinic flow because of the lack of stratification results in large amounts of sediment transport during winter storms.

Nevertheless, the sediment transport taking place must also be affected by the oscillatory motion of storm waves. These waves typically have periods of 12 s and deep water wave heights of 4–6 m. The small-amplitude wave theory presented in Chapter 8 (equation (8.27)) allows the likely orbital velocities to be calculated. The radian frequency is $\sigma = 2\pi/T = 0.524$, the deep water wavelength is $L_\infty = gT^2/2\pi = 225\,\text{m}$, so the wave number is $k = 2\pi/L = 0.028$. In 100 m of water, the maximum orbital velocity under surface waves of height 6 m, or amplitude 3 m, is therefore $U \approx 0.2\,\text{m s}^{-1}$. This shows that these surface waves have a major impact on sea bed pro-cesses out to 100 m water depth at least, and probably as far as the shelf break. Waves capable of producing near-bed orbital velocities in excess of $0.3\,\text{m s}^{-1}$ at water depths of 100 m or more for more than 12 hours are estimated to occur on average nine times over the winter months of November to February each year. The probability of a such periods of storm wave activity coinciding with periods of mid-depth geostrophic flow velocities exceeding $0.5\,\text{m s}^{-1}$ for 12 hours is one event every 4 years. Most of the sediment transport on the Oregon–Washington shelf

Fig. 9.25 (*Left.*) The Oregon–Washington shelf. (a) Distribution of dominant modal grain sizes. Most sediment entering the shelf is derived from the Columbia River. After Nittrouer & Sternberg (1981) [29]. (b) Accumulation rates show a northward and lateral decrease away from a mid-shelf axis starting near the mouth of the Columbia River. After Nittrouer et al. (1979) [30].

should therefore take place in these combined wave–current events roughly every 4 years.

Modelling of storm currents, storm waves and swell waves [31] suggests that large waves are responsible for mobilization of sediment by erosion and disturbance of the bed, whereas steady currents cause diffusion of sediment upwards and transport of the finer grain size components along the shelf. Bedload was modelled to be transported more obliquely offshore because of the frictional effects in the bottom Ekman layer, while the finer suspended load was modelled as travelling more nearly parallel to the isobaths due to being transported higher in the geostrophic flow (Fig. 9.26). This is supported by maps of the distribution of sea bed sediment, which show a mid-shelf deposit trending NNW from the mouth of the Columbia River, representing the typical transport path of fine suspended material (Fig. 9.25). The sand-grade sediment, on the other hand, is distributed along a trend at about 30° to the isobaths. The predicted typical storm sediment bed on the shelf is a graded layer from very fine sand to silt up to 50 mm thick. However, the hydraulic separation of different grain size fractions noted above is a method by which thicker, clean (clay-free) sands may be produced by repeated sorting events on the shelf.

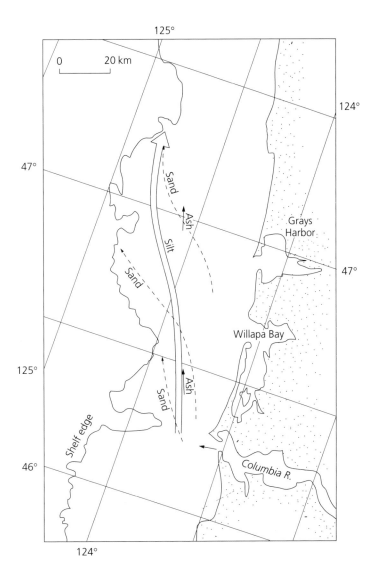

Fig. 9.26 Predicted transport path for fine sediment (silt) carried in suspension by storm events on the Washington continental shelf (solid arrows), showing a direction about 10° west of the isobaths, for sand travelling as bedload (broken arrows) directed 20–25° west of the isobaths, and the movement of ash from the Mount St Helens eruption of 1980 (short arrows). Data in Kachel & Smith (1986) [31] and Healy-Ridge & Carson (1987) [32].

In summary, the Gulf of Mexico is an example of a shelf affected by small, intense, but short-lived hurricanes. The Middle Atlantic shelf is a type example of a winter storm shelf dominated by massive geostrophic discharges reworking older sediment. The Oregon–Washington shelf is a good example of a storm-dominated shelf of high wave energy with significant riverine input.

The hummocky cross-stratification controversy

The single most diagnostic bedform thought to

be attributable to the passage of storms over the continental shelf is a structure characterized by wavy low-angle laminations and occasionally accretionary hummocks (Fig. 9.27). These hummocky bedforms generally have wavelengths of the order of 1 m or more, though smaller-scale varieties on the decimetre scale have also been described. Hummocky cross-stratified beds of generally very fine to fine sand have been widely described from sedimentary sequences thought to have been deposited on the continental shelf.

Unfortunately, unambiguously identifiable hummocky cross-stratification has never been observed

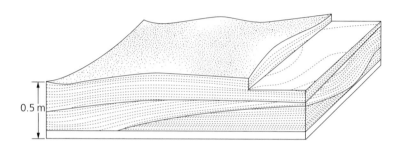

Fig. 9.27 Schematic cross-sectional view of hummocky cross-stratification. After Harms *et al.* (1975) [33].

Shallow-marine storm beds

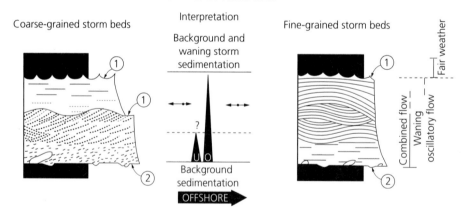

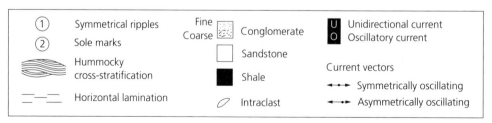

Fig. 9.28 Schematic comparison of the vertical profile of coarse-grained and fine-grained storm-generated beds. The succession of sedimentary structures suggest that strong combined flow fields give way to waning oscillatory flow before fair weather conditions are resumed. After Cheel & Leckie (1993) [34].

forming on the continental shelf or recovered in box core samples. The debate on the origin of hummocky styles of stratification has therefore been led by geologists attempting to explain preserved structures in ancient sedimentary rocks. Many workers now believe that it is most likely that the structure forms under a combined flow field generated during a storm (Fig. 9.28). In the early stages of a storm the unidirectional (geostrophic) current dominates, whereas as the storm recedes the remaining orbital motions of waves dominate over the unidirectional current. This gives rise to a bed with an erosional, sharp base, overlain by upper-stage plane bed laminations, hummocky laminations, then oscillatory vortex ripples and a capping of fine-grained sediment deposited by fallout from suspension (Fig. 9.28). The structure has also been identified in palaeoenvironments other than the continental shelf. For example, it has been interpreted to have originated close to an estuary mouth in the Jurassic of southern England, and smaller-scale hummocky styles of lamination have been described from ancient turbiditic sequences,

and the shallow nearshore of modern lakes. Clearly, there is a need to be circumspect about the precise environment of deposition of hummocky cross-stratification. The essential conditions appear to be a combination of a steady current and an oscillatory component. The steady current might be a turbiditic underflow, a geostrophic flow or a littoral current. The unsteady component might be due to wind waves, internal waves or Kelvin-Helmholtz instabilities.

9.5 Storms as hazards

Tropical cyclones are intense low-pressure weather systems generated over warm tropical seas with sea temperatures in excess of 26°C. They develop, therefore, in the northern hemisphere between June and October, and in the southern hemisphere between December and May, especially on the western sides of oceans where warm waters accumulate as a result of the wind-driven circulation (Fig. 9.29). Convergence causing upward motion of air, and the release of latent heat through evaporation from the sea surface and condensation in the atmosphere, lead to extreme

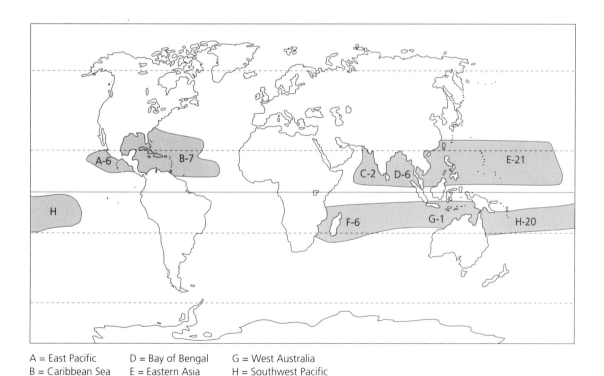

A = East Pacific   D = Bay of Bengal   G = West Australia
B = Caribbean Sea   E = Eastern Asia   H = Southwest Pacific
C = Arabian Sea   F = Madagascar

Fig. 9.29 Occurrence of tropical cyclones (average number per year). Most are generated above the warm waters of the western Pacific. After Gray (1975) [35].

atmospheric instability. Air drawn into the zone of convective instability rotates under Coriolis forces, developing a major vortex. As the convective zone grows, a central vertical core of air subsides, with violent winds along its margins, giving the cyclone its typical 'eye' structure.

Tropical cyclones are very common, the western Pacific typically experiencing about 20 per year (Fig. 9.29). Some cyclones penetrate very large distances inland. In Australia, at least one cyclone every 20 years has penetrated the continent as far as Alice Springs. The number of reported cyclones has increased in the last decade or two, certainly in the south-west Pacific, coincident with more frequent El Niño southern oscillation (ENSO) events. Globally, perhaps 100 or more tropical cyclones develop every year. These events have considerable potential for geomorphological change in drainage basins and at the coast, and for the transport of substantial volumes of sediment on the continental shelf. Coastal barrier systems may be strongly dissected during cyclone storm surge events.

Tropical cyclones are, in terms of loss of life, the most important natural hazard. The main threats to life come from: flooding caused by high rainfalls, leading to contaminated water supplies, the spread of disease and drowning; high winds causing destruction to buildings, transportation, and crops, and the spread of fires; and storm surge at the coast causing flooding and erosion, damage to coastal structures, salt damage

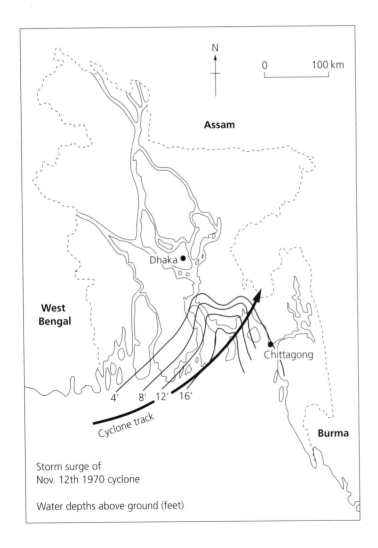

Fig. 9.30 Flooding in Bangladesh caused by storm surge set-up associated with the tropical cyclone of November 1970, showing storm surge heights (in feet). Coastal flooding is exacerbated by river flooding caused by intense rainfall.

to soils and intrusion of saltwater into aquifers. Bangladesh, in particular, has suffered devastating tropical cyclones in the recent past. The enormous loss of life in the Ganges–Brahmaputra delta area at the head of the Bay of Bengal has been caused by tropical cyclone storm surges (Fig. 9.30) backed up by river floods resulting from the associated high rainfall. Catastrophic flooding by the Yangtze and Huanghe rivers in China resulting from the landfall of tropical cyclones has claimed tens of millions of lives.

Extra-tropical cyclones develop along the polar front and in regions overlying warm water outside the tropics. Polar front cyclones result from the instabilities caused by the meeting of cold polar and warm westerly air masses at about 40–50°N and S. Polar front cyclones are much larger than their tropical counterparts (2000 km versus 400 km diameter) and develop more frequently. Some cyclones, known as *east coast lows*, appear to develop where easterly moving low-pressure systems pass over a mountainous coastline and then a warm, offshore, poleward-moving ocean current. This situation is found, for example, on the east coasts of Japan, where air travels over the mountainous spine of Japan and then the Kirushio Current, and on the eastern seaboard of the USA, where air crossing the Appalachians passes over the warm Gulf Stream.

Further reading

J.R.L. Allen (1982) Physical oceanography of continental shelves. *Reviews of Geophysics and Space Physics* **21**, 1149–81.

K.F. Bowden (1983) *Physical Oceanography of Coastal Waters.* Ellis Horwood, Chichester.

S. Pond & G.L. Pickard (1983) *Introductory Dynamical Oceanography*, 2nd edn. Pergamon Press, Oxford.

C.P. Summerhayes & S.A. Thorpe (1996) *Oceanography: An Illustrated Guide.* Manson Publishing, London.

References

1 H.G. Reading (ed.) (1996) *Sedimentary Environments and Facies*, 3rd edn.

2 K.T. Pickering, R.N. Hiscott & F.J. Hein (1989) *Deep Marine Environments: Clastic Sedimentation and Tectonics.* Unwin Hyman, London, 416pp.

3 J.R. Garratt (1977) Review of drag coefficients over oceans and continents. *Monthly Weather Review* **105**, 915–29.

4 K.F. Bowden (1983) *Physical Oceanography of Coastal Waters.* Ellis Horwood Ltd., Chichester, England, p. 125.

5 S. Pond & G.L. Pickard (1983) *Introductory Dynamical Oceanography*, 2nd edn. Pergamon Press, Oxford.

6 O.S. Madsen (1977) A realistic model of the wind-induced Ekman boundary layer. *Journal of Physical Oceanography* **7**, 248–55.

7 Open University (1989) *Ocean Circulation*, course S330, Vol. 3. Pergamon Press, Oxford.

8 B.W. Flemming (1981) Factors controlling shelf sediment dispersal along the southeast African continental margin. *Marine Geology* **42**, 259–77.

9 I.N. McCave & D.J.P. Swift (1976) A physical model for the rate of deposition of fine-grained sediments in the deep sea. *Bulletin of the Geological Society of America* **87**, 541–6.

10 S. Eittreim & M. Ewing (1972) Suspended particulate matter in the deep waters of the North American basin. In: *Studies in Physical Oceanography*, Vol. 2 (ed. A.L. Gordon). Gordon & Breach, New York, pp. 123–68.

11 D.A.V. Stow & J.P.B. Lovell (1979) Contourites: their recognition in modern and ancient sediments. *Earth Science Reviews* **14**, 251–91.

12 D.A.V. Stow & J.A. Holbrook (1984) North Atlantic contourites: an overview. In: *Fine Grained Sediments: Deep-Water Processes and Facies* (ed. D.A.V. Stow and D.J.W. Piper), Geological Society of London Special Publication 15. Blackwell Scientific Publications, Oxford, pp. 245–56.

13 I.N. McCave & B.E. Tucholke (1986) Deep current-controlled sedimentation in the western North Atlantic. In: *The Geology of North America, Volume M, The Western North Atlantic Region* (eds P.R. Vogt and B.E. Tucholke). Geological Society of America, Boulder, CO, pp. 451–68.

14 J.R.L. Allen (1985) *Principles of Physical Sedimentology.* George Allen & Unwin, London.

15 D.J.P. Swift, G. Han & C.E. Vincent (1986) Fluid processes and sea floor response on a modern storm-dominated shelf: the Middle Atlantic shelf of North America. Part 1: The storm-current regime. In: *Shelf Sands and Sandstones* (ed. R.J. Knight & J.R. McLean), Memoir 11. Canadian Society of Petroleum Geologists, Calgary, pp. 99–119. [An excellent summary of the hydraulic regime, with an eye constantly on the applicability to ancient shelf sequences.]

16 G.T. Csanady (1976) Mean circulation in shallow seas. *Journal of Geophysical Research* **81**, 5389–99.

17 D.J.P. Swift & A.W. Niedoroda (1985) Fluid and sediment dynamics on continental shelves. In: *Shelf Sands and Sandstone Reservoirs* (eds R.W. Tillman, R.G. Walker & D.J.P. Swift), Short Course Notes 13. Society of Economic Paleontologists and Mineralogists, Tulsa, OK, pp. 47–135.

18 T.R. Keen & R.L. Slingerland (1993) Four storm-event beds and the tropical cyclones that produced them: a numerical hindcast. *Journal of Sedimentary Petrology* **63**, 218–32.

19 J.W. Snedden & D. Nummedal (1991) Origin and geometry of storm-deposited sand beds in modern

sediments of the Texas continental shelf. In: *Shelf Sand and Sandstone Bodies: Geometry, Facies and Sequence Stratigraphy* (eds D.J.P. Swift, G.F. Oertel, R.W. Tillman & J.A. Thorne), Special Publication of the International Association of Sedimentologists 14. Blackwell Scientific Publications, Oxford, pp. 283–308.

20 M. Miyazaki (1965) A numerical computation of the storm surge of Hurricane Carla 1961 in the Gulf of Mexico. *Oceanographical Magazine* **17**, 109–40.

21 M.O. Hayes (1967) *Hurricanes as geologic agents: case-studies of Hurricane Carla, 1961, and Cindy, 1963* Report of Investigation 61. Texas University Bureau of Economic Geology, Austin, TX.

22 D.J.P. Swift, R.A. Young, T. Clarke & C.E. Vincent (1981) Sediment transport in the Middle Atlantic Bight of North America: synopsis of recent observations. In: *Holocene Marine Sedimentation in the North Sea Basin* (eds S.D. Nio, R.T.E. Schuttenhelm & T.C.E. van Weering), Special Publication of the International Association of Sedimentologists 5. Blackwell Scientific Publications, Oxford, pp. 361–83.

23 D.J.P. Swift, D.B. Duane & T.F. McKinney (1973) Ridge and swale topography of the Middle Atlantic Bight, North America: secular response to the Holocene hydraulic regime. *Marine Geology* **15**, 227–47.

24 D.J.P. Swift (1976) Continental shelf sedimentation. In: *Marine Sediment Transport and Environmental Management* (eds D.J. Stanley & D.J.P. Swift). Wiley, New York, pp. 311–50.

25 W.L. Stubblefield, D.W. McGrail & D.G. Kersey (1983) Development of middle continental shelf sand ridges: New Jersey. *Bulletin of the American Association of Petroleum Geologists* **67**, 817–30.

26 D.J.P. Swift, T.F. McKinney & L. Stahl (1984) Recognition of transgressive and post-transgressive sand ridges on the New Jersey continental shelf: Discussion. In: *Siliclastic Shelf Sediments* (eds R.W. Tillman & C.T. Siemers), Special Publication 34. Society of Economic Paleontologists and Mineralogists, Tulsa, OK, pp. 25–36.

27 J.M. Rine, R.W. Tillman, S.J. Culver & D.J.P. Swift (1991) Generation of late Holocene sand ridges on the middle continental shelf of New Jersey, USA—evidence for formation in a mid-shelf setting based on comparisons with a nearshore ridge. In: *Shelf Sand and Sandstone Bodies* (eds D.J.P. Swift, G.F. Oertel, R.W. Tillman & J.A. Thorne), Special Publication of the International Association of Sedimentologists 14. Blackwell Scientific Publications, Oxford, pp. 395–423.

28 J.D. Smith & T.S. Hopkins (1972) Sediment transport on the continental shelf off Washington and Oregon in light of recent current meter measurements. In: *Shelf Sediment Transport* (eds D.J.P. Swift, D.B. Duane & O.H. Pilkey). Dowden, Hutchinson & Ross, Stroudsburg, PA, pp. 143–80.

29 C.A. Nittrouer & R.W. Sternberg (1981) The formation of sedimentary strata in an allochthonous shelf environment: the Washington continental shelf. *Marine Geology* **42**, 201–32.

30 C.A. Nittrouer, R.W. Sternberg, R. Carpenter & J.J. Bennett (1979) The use of ^{210}Pb geochronology as a sedimentological tool: application to the Washington continental shelf. *Marine Geology* **31**, 297–316.

31 N.B. Kachel & J.D. Smith (1986) Geological impact of sediment transporting events on the Washington continental shelf. In: *Shelf Sands and Sandstones* (ed. R.J. Knight & J.R. McLean), Memoir 11. Canadian Society of Petroleum Geologists, Calgary, pp. 145–62.

32 M.J. Healy-Ridge & B. Carson (1987) Sediment transport on the Washington shelf: estimates of dispersal rates of the Mt. St. Helens ash. *Continental Shelf Research* **7**, 759–72.

33 J.C. Harms, J.B. Southard, D.R. Spearing & R.G. Walker (1975) *Depositional Environments as Interpreted from Primary Sedimentary Structures and Stratification Sequences*, Short Course Notes 2. Society Economic Paleontologists and Mineralogists, Tulsa, OK.

34 R.J. Cheel & D.A. Leckie (1993) Hummocky cross-stratification. *Sedimentology Review* **1**, 103–22.

35 W.M. Gray (1975) *Tropical Cyclone Genesis*, Atmospheric Science Paper 234. Department of Atmospheric Science, Colorado State University, Fort Collins, CO.

10 Wind

The answer, my friend, is blowin' in the wind,
The answer is blowin' in the wind.

> Bob Dylan (b. 1941), *Blowin' in the Wind* [1962]

Chapter summary

The atmosphere is dominated by a small number of gases, but also contains particles which are concentrated in the lower boundary layer affected by the frictional resistance of the surface of the Earth. This boundary layer, approximately 1 km thick, is the lowermost layer of a well-mixed, turbulent troposphere separated from the layered stratosphere by the tropopause at altitudes of 10–16 km.

The stability of the atmosphere depends on its temperature variation with height, known as the atmospheric lapse rate, in relation to the variation of temperature caused by changes in pressure, the dry adiabatic lapse rate. The release of latent heat by the formation of clouds results in a wet adiabatic lapse rate.

The global circulation of the atmosphere takes the form of three cells per hemisphere, producing a zonal wind system. In addition, wind systems respond to seasonal variations of heating and cooling, such as the Asian monsoon, and to zones of atmospheric instability, as in the mid-latitude westerlies. The geostrophic wind is caused by lateral pressure gradients, and shares the same dynamics as geostrophic flows of water in the ocean. The geostrophic wind is affected by the Coriolis force and flows parallel to isobars with a speed determined by the pressure gradient and latitude. The force balance between the pressure gradient, Coriolis and frictional components adequately explains large wind systems such as the jet streams and global surface winds such as the trade winds and westerlies. The observed planetary wind field of three zonal cells per hemisphere can be explained by the effects of conservation of angular momentum on geostrophic winds.

The velocity profile of the wind in the atmospheric boundary layer can be estimated from the same principles as used for unidirectional flows of water. The profile in the lower 40% of the boundary layer can be expressed as a logarithmic function with height, depending on the roughness of the boundary and the shear velocity of the wind. The datum from which heights are measured is referred to as the zero plane displacement.

The threshold of sediment motion under the wind and the dynamics of aeolian sediment transport are slightly different to the subaqueous case for reasons emanating from the very low viscosity and density of air. Grains entrained by the wind are subject to large lift forces, and they are transported downwind with high kinetic energies. Impact of entrained grains on the bed like ballistic projectiles is highly effective in dislodging other grains. There is consequently a fluid threshold related to fluid lift and drag forces, and a lower impact or dynamic threshold incorporating the effects of ballistics. The entrainment threshold is also affected by surface moisture and boundary roughness.

Transport by wind takes place in a number of different modes. The process of short trajectory jumps of grains is known as saltation, whereas grains moving much closer to the bed are said to move by surface creep or reptation. Finer grains move as suspended load. A consideration of the dynamics of saltation allows the typical trajectory of a saltating grain to be predicted, consisting of a steep ascent to a height of several millimetres, and a longer low-angle descent to the bed. Mean saltation lengths are typically 12 times mean saltation heights.

Oversteepened accumulations of sand on a slope collapse by generating sandflows (or grainflows) which move as highly concentrated dispersions. This process typifies the lee sides of aeolian dunes where the slope is kept steep by the sediment transport rate over the windward side of the dune.

Dust is transported in suspension over long distances, producing deposits known as loess. Field measurements suggest that general diffusion theory explains the concentration of dust in the boundary layer. Fine grains are capable of being lofted high (hundreds of metres) above the ground. Their downwind distribution depends on the trapping effects of vegetation interrupting their dispersal. For reasonable diffusivities, we should expect an exponential decline in the thickness of aeolian dust deposits for hundreds and even thousands of kilometres away from the source, the rapidity of the decline depending on factors such as vegetation cover.

Aeolian sand seas cover large parts of the subtropical deserts today and were more extensive during glacial maxima in the Pleistocene. The deposits of aeolian sand dunes consist of a distinctive set of lamination styles produced by wind-ripple lamination, grain fallout from suspension over a dune slipface, and grain flows down the lee side. The wavelength of impact ripples is thought to depend on the mean saltation length, but other workers believe that the dynamics of reptation is more important. The fastest-growing wavelength from stability theory of the fluid-like layer composed of the reptating population and saltating population would represent the wind-ripple wavelength. Dunes take on a myriad of forms, varying from flow-transverse barchan dunes to the flow-longitudinal linear and seif dunes. The barchanoid types can be explained in the same way as the ripples and dunes under aqueous flows. The longitudinal dunes, however, are different. They may be due to a helical, corkscrew-like motion of air caused by convective heating of the atmosphere, but an alternative model is that they form under highly oblique winds which set up secondary flows blowing along the linear dune's lee flank.

10.1 Introduction

Earth's atmosphere is a complex fluid system of gases and suspended particles, dominated by just five gases: nitrogen (78.09% by volume of dry air), oxygen (20.95%), argon (0.93%), carbon dioxide (0.03%) and water vapour (which varies widely). Suspended particles include water droplets, sea spray, dust and industrial pollutants such as soot. These main components occur in distinct vertical distributions. In the bottom *c.* 11 km the air is very well mixed because of turbulence, but above this height there is a banding of constituents (Fig. 10.1). Water vapour, for example, which is derived by diffusion from liquid water at the Earth's surface, is concentrated in the lowermost 6–8 km of the atmosphere. Most suspended particles are concentrated in even lower altitudes since gravity acts on their weight to cause settling. Fine dust may be kept aloft at heights of 10 km or more for several years, as in the case of major volcanic eruptions such as Mount Pinatubo in 1991, but most particles are concentrated in the lowermost 1 km of the atmosphere within the boundary layer dominated by the effects of friction of the land and sea surface.

The principles of atmospheric chemistry are beyond the scope of this book, but the distribution and behaviour of oxygen are of particular interest. Although it is distributed throughout the lowest 120 km of the atmosphere, it occurs mainly in its dissociated atomic form (O) above about 60 km because of the breakdown of oxygen molecules (O_2) by cosmic radiation above this level. The dissociation of the oxygen molecule during the absorption of short-wave radiation produces atomic oxygen, which recombines with molecular oxygen to produce ozone (O_3). This leads to a concentration of ozone at about 30 km, where there is a suitable combination of large quantities of molecular oxygen and short-wave radiation, and low rates of destruction by reaction with oxides of nitrogen. The absorption of radiant energy by ozone leads to a second temperature maximum in the atmosphere, separating two layers undergoing convectional mixing. This is the basis for the zonation of the atmosphere into the lower *troposphere*, and the upper *mesosphere*, separated by a stratified layer known as the *stratosphere*. The boundary between the troposphere and the layered

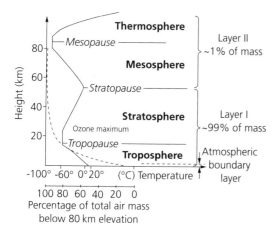

Fig. 10.1 Vertical zonation of the atmosphere showing temperature (solid line) and the percentage of the total mass of the atmosphere below the 80 km elevation (dashed line). The troposphere encloses not only the atmospheric boundary layer but also all of the Earth's weather systems.

stratosphere (the *tropopause*) marks the upper limit to the well-mixed lower atmosphere, and varies in height between about 16 km above the tropics and 10 km above the polar regions. One can think of the atmosphere as consisting of two concentric shells—a lower shell consisting of the troposphere and stratosphere, up to a height of about 50 km, and an outer shell above it. All of the global circulation discussed in this chapter and summarized in Chapter 1 occurs within this inner shell, where 99% of the mass of the atmosphere is concentrated.

The broad features of the general circulation of the atmosphere were introduced in Chapter 1 in the context of the heat budget of the Earth. This chapter investigates a little more deeply the underlying physics of this cellular circulation pattern, and then focuses on the particularities of sediment transport under the wind in what may be termed the *atmospheric boundary layer*. The principles of sediment transport by the wind are not fundamentally different to those introduced in Chapter 5, which dealt with the general mechanics of sediment transport under unidirectional fluid flows. However, the very low viscosity and density of the atmosphere, and the great thickness of the atmospheric boundary layer (*c.* 1 km), result in distinctive characteristics to both the mechanics of sediment transport by wind and the typical bedforms in a cohesionless substrate fashioned by the wind. An understanding of the mechanics involved is therefore of prime

importance in the management of contemporary environments susceptible to wind erosion and deposition, and to the correct identification and interpretation of wind-deposited or *aeolian* sediment in the geological record (from *Aeolus*, the Greek god of the winds).

10.2 Atmospheric circulation

As we have seen in Chapter 1, the global circulation of the atmosphere is a response to two major phenomena. First, the uneven distribution of solar radiation over the Earth's surface results in temperature gradients, causing heat to be transported polewards in both hemispheres, in the form of a coupled ocean–atmosphere system of convection and advection of heat. Second, the rotation of the Earth results in a geostrophic relation for the forces on parcels of air in this atmospheric circulation.

The three atmospheric cells per hemisphere noted in Chapter 1 produce prevailing zonal (that is, latitudinally zoned) wind systems. In the northern hemisphere, the trade winds blow at low latitudes from the north-east and east towards the equator, the moisture-laden westerlies blow from the south-west and west in middle latitudes, and the polar easterlies occupy the highest latitudinal zone. These prevailing winds belong to the Hadley, Ferrel and Polar cells, respectively.

At a smaller, intermediate scale, winds associated with transient weather systems are responsible for a substantial advection of heat. These weather systems may owe their origin to regional heating or cooling of the atmosphere and therefore density differences set up between different air masses. The monsoon of southern Asia is an example. Other weather systems have a dynamical origin for atmospheric instability along zones of rapid horizontal temperature gradient. These cyclones or depressions are common in the belt of mid-latitude westerlies. They commonly have scales of 1000 km.

Still smaller wind patterns are associated with the flow behind solid obstacles such as mountain ranges known as *lee waves*, as waves between two air masses of different characteristics in relative motion (they are therefore Kelvin–Helmholtz instabilities), or as *thermals*, representing small (100–1000 m across) and relatively light masses of air that rise buoyantly through the ambient, cooler atmosphere.

10.2.1 Atmospheric stability

Since the atmosphere is heated principally by long-wavelength radiation from the Earth's surface rather than direct incident solar radiation, there is a general fall in temperature with height in the atmosphere, known as the *atmospheric lapse rate*, averaging 6.6 K km⁻¹.

Air changes its density in proportion to pressure and in inverse proportion to temperature. If a parcel of air changes its temperature purely as a result of changing its pressure or volume, that is, no external heat source has been applied to it, the change is said to be *adiabatic*. Since the pressure decreases vertically upwards in the atmosphere, there is a corresponding *dry adiabatic lapse rate*, averaging 9.8 K km⁻¹. It is the combination of the atmospheric lapse rate and the dry adiabatic lapse rate that determines the stability of the atmosphere.

- If the atmospheric lapse rate is smaller than the dry adiabatic lapse rate, the parcel of air rising through the atmosphere will be cooler than its surroundings, causing it to be relatively denser and to sink. This is a *stable atmosphere*.
- If the two lapse rates are equal, the parcel of air will have the same temperature and density as its surroundings, so will remain in position. This is referred to as a *neutral atmosphere*.
- If the atmospheric lapse rate is greater than the dry adiabatic lapse rate, the parcel of air becomes warmer than its surroundings and continues to rise because of its smaller density. This is known as an *unstable atmosphere*.

The rising of air does not always result in an adiabatic cooling. The release of latent heat during cloud formation may offset the adiabatic cooling, causing the air to cool at a lower rate, the *wet adiabatic lapse rate*, of 5–6 K km⁻¹. This makes the air mass less dense than its surroundings, promoting atmospheric instability, even though the atmospheric lapse rate is lower than the dry adiabatic lapse rate. The presence of water vapour and the rapid release of latent heat during condensation are responsible for the atmospheric instability leading to thunderstorms.

10.2.2 The geostrophic wind

Although vertical motion of air masses is very important, most winds are caused by horizontal pressure gradients, just as ocean currents are driven by horizontal variations in temperature and salinity (Chapter 9). Much of what follows is therefore directly analogous to the dynamics of the ocean.

The force acting on a volume of air acts down a pressure gradient. As is now familiar, it is deflected to the right in the northern hemisphere by the Coriolis force, a balance being reached in the absence of friction between the pressure gradient force and the Coriolis force

$$-\frac{1}{\rho}\frac{\mathrm{d}p}{\mathrm{d}x} = 2V\Omega\sin\phi \qquad (10.1)$$

where ρ is the density of the air, $\mathrm{d}p/\mathrm{d}x$ is the horizontal pressure gradient measured normal to straight isobars (with a minus sign since the air flows from high to low pressure), V is the wind velocity, Ω is the angular rate of rotation of the Earth $(7.29 \times 10^{-5}\,\mathrm{rad\,s^{-1}})$, and ϕ is the latitude. This force balance describes the *geostrophic wind*, which runs parallel to the isobars. It has a speed

$$V = \frac{-\mathrm{d}p/\mathrm{d}x}{2\Omega\rho\sin\phi} \qquad (10.2)$$

which increases with an increase in the horizontal pressure gradient and with a decrease in latitude.

Just as in the ocean, friction at the boundary causes a reduction in the wind speed close to the boundary, producing an Ekman bottom layer. Since the frictional effects of the land and sea surfaces are different (they have different drag coefficients), the thickness of the Ekman layers and the rotation of the surface wind from the geostrophic direction towards the lower pressure will differ over the land (30–$40°$) and sea (10–$20°$) boundaries. Above a height of 0.5–1 km the geostrophic wind is not affected by the frictional resistance of the boundary. The turbulence generated by the presence of the boundary causes thorough mixing in this bottom Ekman layer.

The force balance between the pressure gradient, Coriolis and frictional components adequately explains very large-scale systems such as the jet streams and global surface winds. Smaller weather systems such as *cyclones* are characterized by strongly curved isobars, however, so that geostrophic winds flowing parallel to isobars experience a centrifugal (cyclone) or centripetal (anticyclone) force (Fig. 10.2). If R is the radius of curvature of the flow path of the wind, the force balance becomes

$$-\frac{1}{\rho}\frac{\mathrm{d}p}{\mathrm{d}x} = 2\Omega V\sin\phi + \left(\frac{V^2}{R}\right) \qquad (10.3)$$

for a cyclone, and

$$-\frac{1}{\rho}\frac{\mathrm{d}p}{\mathrm{d}x} = 2\Omega V\sin\phi - \left(\frac{V^2}{R}\right) \qquad (10.4)$$

for an anticyclone. The resulting winds are termed *gradient winds* [1]. An interesting result is that near the equator, where the Coriolis force is small, the

Northern Hemisphere

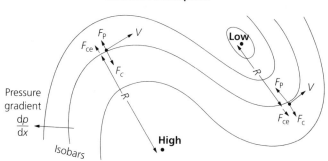

Fig. 10.2 Isobars and force balance on air in cyclonic and anticyclonic systems in the northern hemisphere. Force balances involve the horizontal pressure gradient force F_p, the centrifugal force F_{ce} and the Coriolis force F_c and result in a gradient wind with velocity V. After Pye & Tsoar (1990) [2].

centrifugal force is of great importance, as in tropical hurricanes. However, in anticyclones, the negative centrifugal force causes a reduction in the pressure gradient, which becomes smaller and smaller as the radius of curvature decreases. This results in progressively lighter winds as the centre of the anticyclone is approached. These anticyclonic weather systems typify the subtropical deserts. In contrast, as the centre of a cyclone is approached, the pressure gradient becomes higher as the Coriolis force is reinforced by the centrifugal force, causing winds to become more violent towards the centre of the cyclone.

10.2.3 The planetary wind field

We have previously considered the planetary wind field in Chapter 1. The purpose here is to place it on a firmer physical footing. The development of convectional cells of air movement can be viewed

in terms of the pattern of isobaric surfaces in the atmosphere. These surfaces become uplifted and more widely separated over regions of excess heating such as the tropics. This causes horizontal pressure gradients resulting in flow divergence above the heated region, and therefore convergence and sinking at its flanks (Fig. 10.3). This is the simple cellular circulation suggested by Hadley in 1735. However, it does not fully consider the effects of the Earth's rotation. On a rotating planet, there is a need to conserve *angular momentum* (Fig. 10.4). A parcel of air which is stationary with respect to the Earth's surface has the angular velocity of the Earth, which varies with latitude according to $\Omega r \cos \phi$, where Ω is the angular velocity of the Earth's rotation, r is the radius of the Earth and ϕ is the latitude, acting eastwards in the northern hemisphere. The same parcel of air at a more poleward latitude has a considerably smaller angular velocity

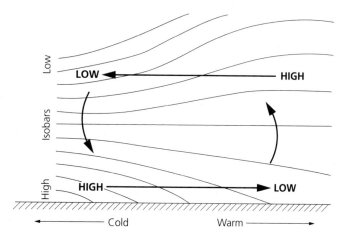

Fig. 10.3 Simple circulation in response to differential heating of the Earth's surface, causing uplift and expansion of isobars above the heated region.

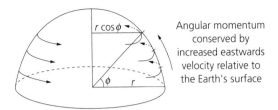

Fig. 10.4 Conservation of angular momentum of a parcel of air moving to higher latitudes.

because of the effect of latitude. Consequently, the speed of the parcel of air relative to the Earth' surface must increase as it moves polewards to compensate for the loss of momentum derived from the Earth's rotation. If air flowed polewards for long distances, it would need to achieve exceptionally high velocities in order to conserve angular momentum, which would cause it to disintegrate into a fully turbulent stream of eddies. Consequently, the Hadley cell is restricted to latitudes below 30°. This conclusion can be verified by carrying out Practical Exercise 10.1.

Finally, the global circulation of air is affected by the distribution of continental and oceanic areas. This strongly affects the pattern of temperature and thereby atmospheric pressure, and introduces strong seasonality into the flow of air. These variations in temperature might result, for example, from differences in the reflectivity, or albedo, of the Earth's surface. Above a warm area the isobars are uplifted and expanded, causing horizontal pressure gradients towards atmosphere overlying cold air. At the surface,

air flows from the high pressures over the cold area towards the low pressure over the warm area. The variations in temperature therefore induce circulation patterns. On a large scale these circulations can be seen in the seasonal winds blowing into the warm continental interiors in the summer, and from the cold continental interiors in the winter (see below). On a much smaller scale are the *sea breezes* which take place due to the heating of the land surface relative to the sea surface, causing a landward horizontal pressure gradient.

How this works in practice can be seen by considering the seasonal variations in air flow related to the distribution of land and sea. The sea level pressure during July is low throughout the equatorial regions. North and south of this belt are the subtropical high-pressure areas, but they differ markedly in their continuity. In the southern hemisphere, the subtropical high pressures form a more or less continuous belt across a large expanse of ocean. But in the northern hemisphere, the large land masses of North America and particularly Asia cause continental low pressure systems to develop because of intense heating. These low pressures cause a vary large-scale sucking in of air towards the continental interiors.

In January, there is a wholesale southward shift of the pressure distribution. Whereas the pattern in the ocean-dominated southern hemisphere remains essentially zonal (that is, a predominantly latitudinal variation), in the northern hemisphere continental high-pressure systems develop over North America and Asia because of surface cooling, contrasting strongly with the low pressures over the adjacent

---

Practical exercise 10.1: Conservation of angular momentum

A wind at the equator has an eastward component of velocity of 10 m s^{-1}. Conserving angular momentum, what in theory is the eastward component of velocity of the wind at (a) 30°N, and (b) 45°N? Take the radius of the Earth as 6.37×10^6 m and the angular rate of rotation of the Earth as $7.29 \times 10^{-5} \text{ rad s}^{-1}$.

Solution

The angular rate of rotation of the solid Earth surface at the equator is Ωr. The angular rate of rotation of the solid Earth surface at any latitude is $\Omega r \cos \phi$. At the equator the absolute speed of

rotation of the wind with eastward velocity U_1 is therefore $\Omega r + U_1$. At latitude ϕ the absolute speed of rotation of the wind is $\Omega r \cos \phi + U_2$. Assuming that the mass of the air stream does not change, and conserving angular momentum,

$$\Omega r + U_1 = \Omega r \cos \phi + U_2$$
$$\therefore \quad U_2 = U_1 + \Omega r (1 - \cos \phi)$$

Substituting the appropriate numerical values, the eastward component of the speed of the wind at 30°N relative to the surface of the Earth is 72 m s^{-1}, and at 45°N is 146 m s^{-1}.

It is unlikely that the airstream could maintain its integrity at such high speeds. This explains why the Hadley cell does not extend north and south of about 30° latitude.

oceans. Consequently, the warmer air over the oceans is blocked from the continental interiors.

Lateral temperature variations may also be caused by the wind drift along the eastern sides of oceans at latitudes of 20–30°N and S, giving rise to coastal upwelling (Chapter 9). This brings cold water to the surface, cooling the atmosphere and contributing to stable conditions. Deserts are consequently found along the western sides of continents in subtropical latitudes, such as the Namibian and Western Sahara deserts in Africa, and the Atacama and southern California deserts of America.

The seasonal movement of the zonal pressure distribution also has an effect on the generation of smaller weather systems. When the intertropical convergence zone (ITCZ) is located over the equator, the Coriolis force is very weak and no cyclones can develop. When the ITCZ is displaced relatively far (10–15°) from the equator, cyclones may bring much-needed rain.

Although the distribution of land and sea strongly affects surface pressures and therefore low-level winds, the effect is lost above a few kilometres in the atmosphere. The pressure distribution at a level corresponding to half of the average surface pressure (500 millibars) shows a clear pattern of low pressure at the poles and high pressure in the subtropics, demonstrating that the global circulation pattern is not substantially disrupted by the effects of land and sea. The jet streams flow more or less parallel to these isobaric surfaces.

A map of the global distribution of wind energy (Fig. 10.5) shows it to be concentrated around coasts and at the poleward extremes of continents. The lower frictional resistance of the sea surface compared to the land results in higher surface wind velocities over the ocean. These oceanic winds expend their energy on meeting the rougher surface of the land. The high wind energies of polar regions may be due to the strong temperature gradients between areas of ice and permafrost and fringing oceans. Wind erosion and deposition may be extremely important therefore in polar regions where large amounts of

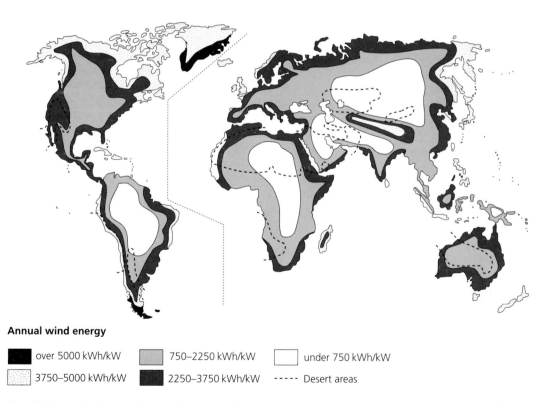

Annual wind energy

- ■ over 5000 kWh/kW
- ▨ 3750–5000 kWh/kW
- ▨ 750–2250 kWh/kW
- ■ 2250–3750 kWh/kW
- ☐ under 750 kWh/kW
- ----- Desert areas

Fig. 10.5 Map of estimated global wind energy, in kilowatt-hours per year, relative to the power output (in KW) of a wind machine operating at a constant velocity of 11 m s^{-1}. The main desert areas are superimposed. After Eldridge (1980) [3].

unconsolidated glacial outwash (Chapter 11) or coastal sand are available.

The wind energy map also shows clearly that the world's great deserts are not located in regions of particularly high wind energy. Their location in the subtropics of Africa and the Middle East, and in the heart of Asia, is a direct function of the arid rainfall regime, and nothing to do with wind velocities. Coastal sand dunes, such as those on the Port Elizabeth coast of SE Africa, have much higher sediment transport rates than the dunes of mid-continent deserts.

10.3 The atmospheric boundary layer

The frictional effects of the land and sea surface cause a reduction of wind speed above the Earth's surface. The zone in which the retarding effects of the surface are felt is the atmospheric boundary layer. Much of the boundary layer theory and elementary fluid mechanics introduced in Chapter 4 is applicable to the aerodynamic problems of the atmospheric boundary layer. The critical Reynolds number at which the atmospheric boundary layer becomes turbulent is about 6000 [4]. Since the kinematic viscosity of dry air at 15°C at sea level is approximately $1.5 \times 10^{-5} \, \text{m}^2 \text{s}^{-1}$, the flow Reynolds number must be turbulent even for very small length-scales and velocities. In practice, therefore, we can treat the atmosphere as always turbulent. Consequently, the relation between the shear stress and velocity gradient (rate of shear) in the turbulent boundary layer is given by the *eddy viscosity*.

10.3.1 The velocity profile in the wind

The velocity profile in the wind is treated in exactly the same manner as that in an aqueous turbulent flow. It can be expressed as a logarithmic function with depth, depending on the roughness of the boundary and the shear velocity at the bed, having the familiar form

$$\overline{U}_y = \frac{u_*}{k} \ln\left(\frac{y}{y_0}\right) \tag{10.5}$$

where $\overline{U}_y$ is the time-averaged velocity at height y above the boundary, u_* is the shear velocity, k is von Karman's constant, generally taken as equal to 0.4, and y_0 is the roughness length, dependent on the frictional characteristics of the boundary. The roughness length varies according to the nature of the boundary. Where the roughness of the boundary is simply due to the individual roughnesses of sand grains (so-called *skin friction*), the roughness length

is taken as a fraction of the grain size ($D/30$), but this is strictly applicable to a fixed sand roughness such as that due to uniform sand grains glued to the inside of a pipe, and is of questionable relevance to the flow of air over natural boundaries. The roughness length may also reflect the frictional effects of larger roughness elements, such as ripples, dunes, vegetation, boulders, fences or houses on land, or wave conditions at sea. This is known as *form roughness*, as opposed to skin friction. The roughness length can be found by plotting velocity measurements as a function of depth (see Practical Exercise 4.3).

There is a second difficulty in the applicability of the velocity law, and that is the choice of the elevation of the boundary. This is not a problem where the wind is blowing over an expanse of bare sand, but becomes acute where the wind is blowing over a forest with a tall canopy (Fig. 10.6). If the canopy height is h, the level at which y equals zero is taken at an elevation of between $0.6h$ and $0.8h$. The argument of the natural logarithm in equation (10.5) therefore becomes between $(y-0.6h)/y_0$ and $(y-0.8h)/y_0$. The new datum for measuring wind speeds is known as the *zero plane displacement height*.

The logarithmic velocity law applies to the lower part of the boundary layer under neutral, unstratified atmospheric conditions. When modified to account for the zero plane displacement height it may give a good indication of the air velocities within the bottom 40% of the boundary layer, but the entrainment and transport of sediment affects the velocity profile. This is because the wind must expend energy in keeping grains aloft. The wind therefore appears to feel additional roughness when the bed is mobile. This topic will be developed below.

10.3.2 The effect of topography

The flow of air over hills, large dunes, or other forms of complex topography is of considerable interest to the problem of sediment transport and deposition. The underlying principles are: the Bernoulli equation, which describes the streamwise variations in pressure and shear stress as streamtubes are either compressed or expanded (Section 4.2); the phenomenon of flow separation due to the build-up of pressure gradients over the lee of an obstacle (Section 4.4); and the sediment continuity equation, which relates the changes in bed elevation to lateral changes in the sediment transport rate (Section 5.1). These principles can be applied to the 'classical' case of flow over an asymmetrical bedform such as a large dune (Section 5.4) and need not be repeated here.

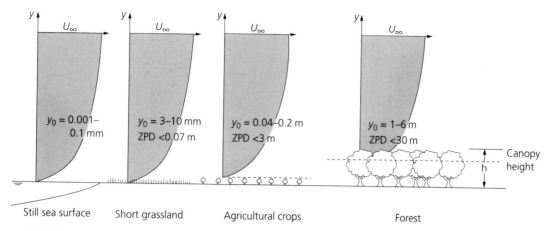

Fig. 10.6 Concept of zero plane of displacement f (ZPD) or different surfaces. Roughness lengths (y_0) and the height of the zero plane displacement increase from a still water surface to short grassland to areas of agricultural crops to tall forests. Data from Oke (1978) [5].

A particular problem of interest is the flow of air over hills which, because of the thickness of the atmospheric boundary layer are entirely enclosed within the flow. The Bernoulli equation indicates that where the flow is compressed over the upstream-facing flank of a hill, the velocity should increase and the pressure reduce. The effect of the compression of streamtubes should be felt for a certain distance above the hill. If $\bar{U}_1$ is the mean velocity at height y over flat ground upwind of the hill, and $\bar{U}_2$ is the mean velocity above the hill at the same height y, the ratio of the two gives the *amplification factor* [6]. The value of the amplification factor (A_x) tells us something about the impact of the shape of the hill or escarpment on the flow of air. There is commonly a reduction in velocity immediately upwind of the hill ($A_x < 1$), then an increase in velocity as the crest is approached ($A_x > 1$)—see Fig. 10.7. In the following section we

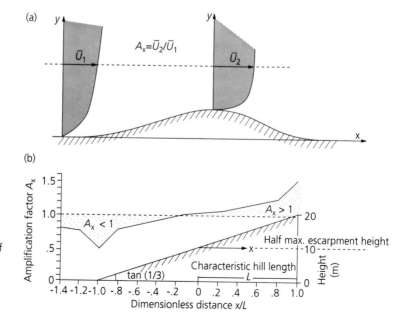

Fig. 10.7 (a) Derivation of the amplification factor.
(b) Amplification factor (A_x) based on Jackson (1977) [7] for the case of a steady slope facing the flow with the horizontal axis scaled by a characteristic hill length. After Tsoar (1986) [8].

shall see the implications of this for the transport and deposition of sand and silt.

10.4 Sediment transport by the wind

The threshold of sediment motion under a unidirectional flow was introduced in Chapter 5. Is there anything special about the threshold under the wind? The answer is 'yes'. Grains entrained by the wind are subject to relatively large lift forces compared to drag, and are transported violently downwind to bombard the bed with high kinetic energies. This high-energy bed bombardment process results in further dislodgement of grains and the dispensation of large amounts of energy among the top layers of sediment in the bed. The problem of sediment transport by wind is therefore one of ballistics. It is not a topic restricted to the prediction of sand transport across the Earth's desert floors or coastal beaches. Much interest has been shown recently in the question of transport by wind on planets such as Mars. We shall return to the planetary dimension in considering the threshold of sediment motion and the saltation process on the Earth and Mars in Practical Exercises 10.2 and 10.3.

10.4.1 Threshold of sediment motion

The forces arising from the flow of air acting on a stationary spherical grain resting on a loose granular bed include fluid drag (F_D), fluid lift (F_L) and a rotational force or moment (M). These forces are opposed by gravity acting on the mass of the grain, and frictional and cohesive forces. The drag, lift and moment forces can be expressed as follows:

$$\begin{aligned} F_D &= K_D \rho_s u_*^2 D^2 \\ F_L &= K_L \rho_s u_*^2 D^2 \\ M &= K_M \rho_s u_*^2 D^3 \end{aligned} \tag{10.6}$$

where the drag, lift and moment coefficients are denoted by K_D, K_L and K_M, respectively [9]. The experimentally determined values of the drag and lift coefficients in air (at high grain Reynolds numbers in the range 1000–4000) are 3–4.7 and 2.2–5 respectively, but the moment coefficient can only be estimated from theory (0.74). The important point to recognize is that the fluid forces tending to promote grain movement vary with the square of the shear velocity, and hence linearly with the shear stress on the bed since $u_* = \sqrt{\tau_0/\rho_a}$ (where ρ_a is the density of the air), and with the square or cube of the grain diameter. The theoretical use of a critical shear velocity for grain entrainment under the wind was pioneered by Bagnold [10], who suggested that:

$$u_{*c} = A\sqrt{\frac{\rho_s - \rho_a}{\rho_a} gD} \tag{10.7}$$

where the coefficient A was determined from experiment, and found to be about 0.1 for grain Reynolds numbers greater than 3.5. This equation shows that the critical shear velocity for grain entrainment varies with the square root of the grain diameter (Fig. 10.8). Let us call this the *fluid threshold*. Below the critical grain Reynolds number, the grains are fully enclosed within a viscous sublayer, the flow is aerodynamically smooth, and the value of A increases markedly. The critical threshold shear velocity is then more dependent on A than it is on grain diameter.

The process at entrainment is principally a steep lift-off from the bed, followed by a low-trajectory downwind path back to the bed, the process being termed *saltation* (see Section 10.4.2). Once grains are set in motion, their bombardment of the bed makes it easier for further grains to be entrained. This is achieved in a number or ways. The descending grain may simply ricochet off a bed grain and continue

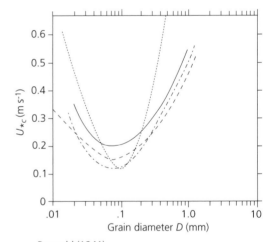

----- Bagnold (1941)
——— Chepil (1951)
- - - Horikawa and Shen (1960)
-·-·- Iversen and White (1982)

Fig. 10.8 The threshold of sediment motion under wind expressed by a plot of the threshold shear velocity against the grain diameter for quartz density grains, using data from Bagnold (1941) [10], Chepil (1951) [11], Horikawa & Shen (1960) [12] and Iversen & White (1982) [13]. Iversen & White's data include the effects of cohesion. Chepil's data relate to an equivalent quartz diameter of soil containing different size fractions. After Pye & Tsoar (1990) [2].

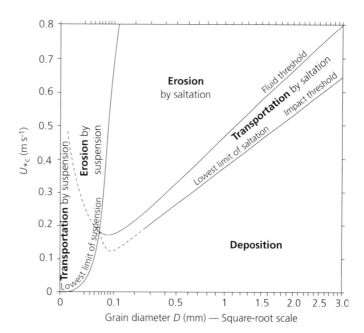

Fig. 10.9 Variation of the fluid threshold velocity and the dynamic impact velocity versus grain size. Data from Bagnold (1941) [10].

its high-energy path downwind. In addition, the projectile grain may dislodge a bed grain (or grains) by the kinetic energy of impact. There is consequently an *impact* or *dynamic threshold* which is lower than the fluid threshold (Fig. 10.9). Bagnold's experiments suggested that the impact threshold had the same dependence on the square root of the grain diameter, but that the coefficient was equal to 0.08.

The approach of Bagnold, in particular the use of his coefficient *A*, has been questioned by a number of workers who have developed the theoretical understanding and carried out detailed experiments to investigate the dependence of *A*. Experiments in wind tunnels using a range of grain sizes and densities show that *A* is indeed more or less constant over a large range of grain sizes and densities (Fig. 10.10), the value of *A* being the gradient of the line for the experimental results when u_{*c} is plotted against $\sqrt{\Delta \rho g D / \rho_a}$. However, at small values of $\sqrt{\Delta \rho g D / \rho_a}$ the slope reverses in sign after a transition region. The variation of *A* at small grain sizes is complex, and does not appear to be simply dependent on the grain Reynolds number. In such small grain sizes ($<80\,\mu m$) interparticle forces arising from electrostatic charges and moisture films leading to cohesion are important. A generally applicable threshold equation therefore must involve the calculation of *A* as a function of both the cohesionless and cohesive properties of the

bed material. The specific effects of surface moisture are discussed below.

Other factors affecting entrainment

The presence of *surface moisture* has a major effect on the entrainment and sediment transport rate by the wind. Wind tunnel experiments demonstrate that even very small moisture contents can cause dramatic increases in the critical threshold velocity. A theoretical model can be developed which attributes the effect of moisture to capillary forces. Such capillary forces depend on the geometrical properties of the grain contacts, the surface tension of the water and the moisture tension developed in the wedges of water between grains [14]. The Bagnold threshold equation is modified accordingly:

$$u_{*c} = A \left(\frac{\Delta \rho}{\rho_a} \right)^{1/2} (\alpha F_c + 1)^{1/2} \qquad (10.8)$$

where $\Delta \rho$ is the density difference between air and sediment, α is a parameter incorporating the various effects of particle weight and resting angle and F_c is the capillary force. The threshold velocity increases rapidly above the dry threshold value with increases in moisture content. Similar effects result from the presence of *bonding agents* in the surface soil or sediment, such as fine particles (clays), salts and organic

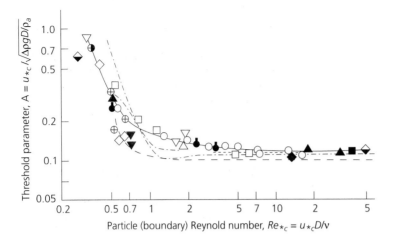

Material	Density (g cm⁻³)	Diameter (μm)
▼ Instant tea	0.21	719
▽ Silica gel	0.89	17; 169
☐ Nut shell	1.10	40 to 359
■ Clover seed	1.30	1290
☆ Sugar	1.59	393
⊕ Glass	2.42	31 to 48
○ Glass	2.50	38 to 586
◆ Sand	2.65	526
◓ Aluminium	2.70	36 to 204
△ Glass	3.99	55 to 519
◑ Copper oxide	6.00	10
▲ Bronze	7.80	616
◇ Copper	8.94	12, 37
◈ Lead	11.35	8, 720

————— Bagnold (1941)
– – – Chepil (1945, 1959)
–·–·– Zingg (1953)

Fig. 10.10 Variation of Bagnold's threshold parameter A against the boundary Reynolds number for a number of data sets—Bagnold (1941) [10], Chepil (1945) [15], Chepil (1959) [16], Zingg (1953) [17]—compiled by Greeley & Iversen (1985) [9].

matter. *Crust formation* due to the presence of algae and fungi or salts is also effective at severely reducing entrainment.

The entrainment threshold is also affected by the roughness of the boundary. This can immediately be appreciated by considering the basic friction equation for a unidirectional flow in an open channel (Chapter 5) illustrated by:

$$\tau_0 = \frac{f}{8} \rho U^2 \qquad (10.9)$$

where ρ is the fluid density, so that an increase in the friction factor (f) must reduce the shear stress (τ_0) on the bed for the same flow velocity (U). One might qualitatively think of the bulk of the shear stress being expended on the larger, immovable grains on the substrate, so that little is available to move the finer, relatively 'sheltered' grains. One of the key factors is the spacing or 'density' of the roughness elements on the substrate. Widely spaced roughness elements (sand or gravel particles) are more easily entrained than closely spaced identical particles, suggesting that the formation of pressure gradients associated with flow separation behind exposed grains is greater than over closely packed grains. There is therefore a certain concentration of particles on the bed above which activation of erosion changes to protection, known as the *inversion point* [18]. The inversion point depends on grain size and shape. The concept is an important one, since as erosion proceeds and finer particles are winnowed from a mixed grain size substrate, the larger grains will become more concentrated. The influence of both increased bed roughness and concentration of larger grains leads to a self-limiting reduction in erosion and the development of a stable gravelly surface. This erosion of a substrate under the wind is known as *deflation*. The stabilized, gravelly surface is consequently termed a *deflation lag*.

10.4.2 Modes of sediment transport

The various modes of sediment transport under a unidirectional flow of water were introduced in Chapter 5, where a fundamental distinction was made between bedload and suspended load. We saw that a useful way of considering the type of sediment transport was the transport stage, a simple ratio between the shear velocity (or stress) on the bed

---

Practical exercise 10.2: Threshold condition on Earth and Mars

With the growth of interest in aeolian sediment transport on Earth and on other planets such as Mars and Venus, the physics of the threshold of motion by the wind, and the modes of sediment transport, has come under close scrutiny. Part of the interest is because sand in transport can cause considerable damage to surface probes and landing craft on planets such as Mars. *Mariner 9* is thought to have landed on the Martian surface during a planet-wide duststorm in which the dust was whipped up by saltation of sand. Another interesting perspective is that seasonal darkening of the Martian surface has in the past been attributed to biological activity and fuelled the debate about life on Mars. The effect can be better explained by duststorms removing the highly reflective (bright) fine particles from the surface, leading to darkening.

1 Bagnold's threshold equation (10.7) can be used with $A = 0.1$ above grain Reynolds numbers of about 7. Calculate the threshold shear velocity in the following two cases:

(a) a 0.1 mm diameter sand grain on the Earth, density 2650 kg m^{-3}, with an air density of 1.2 kg m^{-3}, and gravitational acceleration 9.81 m s^{-2}.

(b) the grain size estimated from *Mariner 9* photometric data is 0.5 mm, density 2950 kg m^{-3} (basalt), Martian atmosphere density 0.018 kg m^{-3}, gravitational acceleration 3.75 m s^{-2}.

2 Assuming that the velocity profile is logarithmic, and using a roughness length that best fits a range of desert conditions of 0.2–0.3 mm, calculate the wind speed at 1 m above the surface for the grain sizes and shear velocities at the threshold condition. In the Martian case, what would be the wind velocity at a height of 1 km above the surface? Geostrophic winds on Mars have been estimated to be about 100 m s^{-1}. How does your result compare?

Solution

1 Using Bagnold's expression, and putting $A = 0.1$, we have

(a) Earth: $u_* = 0.147$ m s^{-1} for a 0.1 mm grain;

(b) Mars: $u_* = 1.75$ m s^{-1} for a 0.5 mm grain.

As a check on the validity of the estimate of A, the grain Reynolds number $Re_* = \rho_a u_* D / \mu$ can be calculated for the Earth. It is equal to 1. Although this is rather low, it is sufficiently high to assume that A is not far from 0.1. If A were slightly higher than 0.1 at this rather low grain Reynolds number, it would mean that the estimate of the critical shear velocity would be too low, by perhaps 10%.

Comparing the Earth and Mars, note that the shear velocity required to initiate sediment movement by wind is an order of magnitude higher on Mars than on Earth. This is not simply due to the larger grain size used for the Martian case, as can be found by redoing the calculations with identical grain sizes. For example, if a 0.5 mm grain was used for Earth, the critical shear velocity would be 0.33 m s^{-1}. By contrast, the threshold shear velocities would be an order of magnitude lower on Venus than on the Earth. The difference between Earth and Mars is that the Martian atmosphere is less dense than the Earth's, so the weight of grains is relatively much greater. This effect outweighs the opposing effect of the lower gravitational acceleration.

2 Using a roughness length of 0.25 mm and a logarithmic velocity law, the wind velocities at 1 m above the bed are

(a) Earth: $U = 3.0$ m s^{-1};

(b) Mars: $U = 36$ m s^{-1}.

At a height of 1 km above the Martian surface the wind velocity corresponding to the fluid threshold condition at the surface is 66.5 m s^{-1}. Since geostrophic winds commonly exceed this value, we can conclude that surface winds should frequently exceed the threshold condition and that sand- and duststorms should be common.

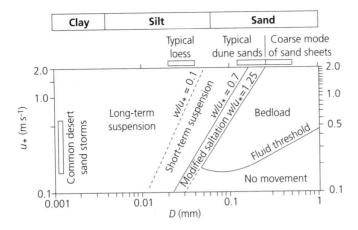

Fig. 10.11 Plot of shear velocity against grain size for the various types of sediment transport by wind. After Tsoar & Pye (1987) [19].

driving sediment transport, and the settling velocity causing resedimentation, that is, u_*/w. In the wind, the bedload sediment transport process is dominated by the short trajectory hops known as saltation (from the Latin word for 'jump'), and by a surface creep or reptation (from the Latin word for 'crawl'). As in aqueous unidirectional flows, there is no sharp boundary between pure suspension (dominated by upward-directed turbulent velocity fluctuations) and pure saltation (dominated by ballistics). This hybrid transport mode is called *modified saltation*, and is analogous to the intermittent suspension of aqueous flows. The mechanics of suspension is particularly relevant to the transport of dust (silt and clay) by wind. Bedload transport, which includes saltation, is more applicable to sands (Fig. 10.11).

Transport of sand

Saltation In order to understand the saltation process, it is necessary to consider the forces on a saltating particle. The steep ascent (50–90°) of a sand grain from the bed supports the view that fluid lift is an important force. But fluid lift vanishes quickly above the bed. This is because the lift force depends on an unequal pressure distribution on a grain which can only be produced while the grain is resting in a bed or is very close to it [20]. The saltation trajectory must therefore result from the combined action of gravity and fluid drag after this initial impulse from fluid lift.

Considering gravity alone, a grain impelled upwards from the bed with a velocity V should reach a height H given by

$$H = \frac{V^2}{2g'} \tag{10.10}$$

where $g' = (\Delta\rho/\rho_s)g$ is the 'reduced' gravity (modified by buoyancy). If the density of the fluid is very small compared to the grain density, as in the case of quartz grains in air, the reduced gravity is replaced by g since there are negligible buoyancy effects. Consequently, a grain should be expected to rise to a greater height in water than in air because of buoyancy. But observations show this not to be true. We should suspect that the reason is something to do with the second force on the particle, the fluid drag force.

The fluid drag force on a grain of largest cross-sectional area A, moving at velocity V, is given by

$$F_D = C_D \frac{\rho V^2}{2} A = C_D \frac{\rho V^2}{2} \frac{\pi D^2}{4} \tag{10.11}$$

where C_D is the drag coefficient. The gravitational force on the spherical grain of diameter D and volume $\pi D^3/6$ is

$$F_G = \Delta\rho g \frac{\pi D^3}{6} \tag{10.12}$$

Comparing the ratio of the two forces in air and water, the fluid drag is much more important in water, and gravity much more important in air. For example, taking the density of air as $1.3\ \mathrm{kg\,m^{-3}}$, of water as $1000\ \mathrm{kg\,m^{-3}}$, and of a 0.5 mm quartz sphere as $2650\ \mathrm{kg\,m^{-3}}$, with an initial velocity of $100\ \mathrm{mm\,s^{-1}}$, we obtain for water

$$\frac{F_G}{F_D} = \frac{4\Delta\rho g D}{3\rho C_D V^2} = \frac{3.7}{C_D} \tag{10.13}$$

and for air

$$\frac{F_G}{F_D} = \frac{1334}{C_D} \qquad (10.14)$$

If we assume that the drag coefficient does not vary between air and water, it can be seen that the gravitational force is much more important in air than in water, and vice versa for the fluid drag. In water therefore, a particle lifted off the bed is very quickly transported downstream by the drag force, only leaving the bed by a few grain diameters. In air, saltating grains rise up to about 0.1 m above the bed.

The relative importance of saltation in air is, however, not simply due to fluid drag effects. Of crucial importance is the dispensation of kinetic energy from saltating grains impacting the bed in air, and the relative ease of suspension in water.

• The settling velocity of sand grains is greater in air than in water, making it difficult to suspend sand grains in all but the most powerful winds. Sand grains therefore must travel close to the bed in wind.

• The kinetic energy of sand grains in air is far in excess of their counterparts in water [21]. The impact process is therefore particularly important in air, acting as a positive feedback to further sediment transport. Impacting sand grains in air have 10–20 times their kinetic energy at lift-off. This kinetic energy is dissipated in deformation and jostling of the sand bed, and in the expulsion of grains into the flow of air. The impacting grains themselves may rebound into the air stream (Fig. 10.12).

The height to which sand grains saltate is of considerable importance in the prediction of wind erosion, or wind sand blast damage. Experiments suggest that the sediment transport rate declines exponentially with height above the bed. If we define

a height h below which the sediment discharge is half of the total, so it is the mean height to which sand grains rise, we find that h depends primarily on the shear velocity and grain size, but also on the grain shape, increasing with increased sphericity. However, since the initial impulse causing the grains to rise from the bed is due not only to fluid lift but also to the kinetics of grain impacts, we should not expect the prediction of the height of saltating grains to be a simple matter.

In the absence of fluid forces the height of a saltating grain would depend on its initial velocity and gravitational acceleration ($h = V^2/2g$), where all of the kinetic energy of the grain is converted to potential energy at the top of its ascent. If the grain is dislodged by an impacting grain, which itself has an impact velocity proportional to the shear velocity, its initial vertical velocity should also be proportional to the shear velocity. The coefficient of proportionality is termed the *impact coefficient*, denoted by

$$V = \beta u_* \qquad (10.15)$$

where β may vary between about 0.8 and 2. Consequently, the mean saltation height is given by [23]

$$h = C \frac{u_*^2}{g} \qquad (10.16)$$

where C is between 0.32 and 2.0 for the range of values of β given above, and is given as 0.82 by Owen [23]. However, the grain size, forward velocity of the grain as it is impelled by the flow of air, and the angle of ascent all cause the actual saltation height to be smaller than the theoretical result in equation (10.16) [24]. The smaller the grain, the larger the effect of

Fig. 10.12 Pattern of particle ejections resulting from the impact of a 4 mm steel pellet on a bed of similar grains. Note the single high-energy rebound which is usually the impacting grain itself, and a number (about ten) of low-energy ejecta in short trajectories. After Anderson (1987) [22].

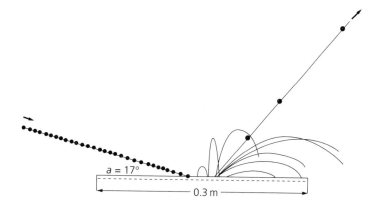

$a = 17°$

0.3 m

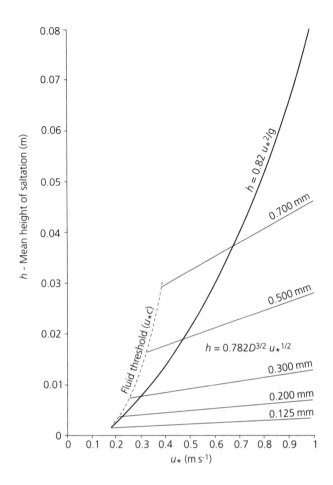

Fig. 10.13 Mean saltation height for various grain sizes and shear velocities, with the theoretical maximum (Owen (1980) [25]) and empirical (Zingg (1953) [17]) results for uniform sand with the grain sizes given. After Pye & Tsoar (1990) [2].

fluid drag in accelerating the grain downstream and preventing it from reaching its theoretical maximum height (Fig. 10.13). Observations in sandstorms confirm that larger sand grains saltate to greater heights than smaller grains. Empirical studies suggest that the mean saltation height is approximated by

$$h = \alpha D^{3/2} u_*^{1/2} \qquad (10.17)$$

where α is 2.47×10^3. If the shear velocity is very close to the fluid threshold value, say $0.2\,\text{m s}^{-1}$, and the grain diameter is $0.2\,\text{mm}$, the saltation height is therefore $3.1\,\text{mm}$, compared to the theoretical maximum of $3.3\,\text{mm}$. However, under more vigorous wind conditions where the shear velocity is $0.6\,\text{m s}^{-1}$, the same grain of diameter $0.2\,\text{mm}$ has saltation heights of $5.4\,\text{mm}$ and $30\,\text{mm}$ according to the empirical and theoretical methods, respectively.

The grain size dependence given above indicates that larger grains should jump to greater heights than small grains. However, when the size of saltating grains is measured in wind tunnel experiments, it is found that the grains at the top of the saltating layer are small. These grains probably have a component of upward-directed fluid turbulence responsible for their large heights above the bed, demonstrating the difficulty of strictly defining pure saltation. As a rule of thumb, the maximum thickness of the saltating layer is roughly ten times the mean saltation height. The mean saltation distance is also related to the mean saltation height, the theoretical mean saltation path being given by

$$l = 10.3 \frac{u_*^2}{g} \qquad (10.18)$$

Comparing this with equation (10.16) for the mean saltation height shows that the aspect of the saltation trajectory is about 1:12 (Fig. 10.14). There is

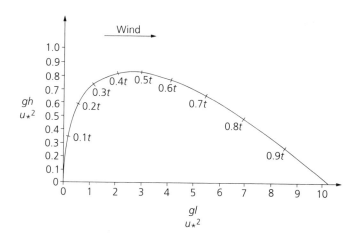

Fig. 10.14 The mean saltation trajectory based on the theory of Owen (1980) [25], where h is the maximum jump height, l is the jump length, t is the time taken to complete the jump, and u_* is the shear velocity of the wind. The particle has a steep ascent path, but a low-angle descent. Vertical scale exaggerated by a factor of 5 compared to horizontal.

considerable variation about this theoretical saltation length since it is also affected by grain size and shape. The saltation length is of particular importance in explaining the occurrence and scale of aeolian wind ripples (Section 10.5).

Observations using high-speed film have shown that the actual trajectories of saltating grains are higher than theory predicts. This enhanced lift has been attributed to the very rapid spinning (in excess of 1000 revolutions per second) of the airborne grains, the *Magnus effect*.

The presence of a large number of saltating grains close to the bed affects the wind velocity gradient in this region. This is because the grains extract energy/momentum from the wind to keep them in transport. The effect can be seen dramatically by comparing the velocity profiles over a damp, flat sand surface over which no movement takes place, and a similar but dry surface over which aeolian saltation is occurring (Fig. 10.15). When sediment transport is taking place, the roughness length increases, and the velocity profile becomes non-logarithmic, presumably due to the discontinuities caused by the saltation process. Empirical results suggest that the new roughness length is about ten times the mean grain diameter, though there may not be one well-defined value. The saltating layer can therefore be thought of an an added source of resistance or roughness. The logarithmic velocity profile can therefore be modified to

$$U_y = \frac{u_*}{k} \ln\left(\frac{y}{\delta}\right) + \overline{U}_\delta \qquad (10.19)$$

where δ is the thickness of the saltating layer and $\overline{U}_\delta$ is

the mean velocity of the wind at the top of the saltating layer.

The wind velocity commonly falls below the fluid threshold at heights above the bed roughly corresponding to the roughness length when active sediment transport is taking place (Fig. 10.15). Consequently, it must be assumed that sediment transport is being achieved through the grain impact process, not by fluid lift and drag. This is an interesting phenomenon, suggesting that once grain motion is initiated at a certain rate it is self-sustaining and self-limiting. If the concentration of saltating grains falls below a critical level, the velocity profile close to the bed will change towards the static profile, increasing the shear velocity and causing more entrainment by fluid forces. This equilibrium is known as *steady-state saltation*.

Surface creep or reptation Most studies of bedload sediment transport by wind have concentrated on saltation. When saltation is fully developed the fluid stress on the bed is insufficient to transport grains along it in contact with the bed, but the impact of saltating grains can cause a slow downwind movement of grains which Bagnold termed surface creep. Particles in the size range 0.5–2 mm are particularly prone to this mode of transport. Just as there is no sharp division between saltation and suspension, there is a transition between surface creep, where grains do not leave contact with the bed, and saltation. Some grains make very short trajectory paths where they barely leave the substrate. This process is termed reptation [22]. It is generally thought that surface creep and reptation represent a small part of the total sediment transport. It is determined, however, by

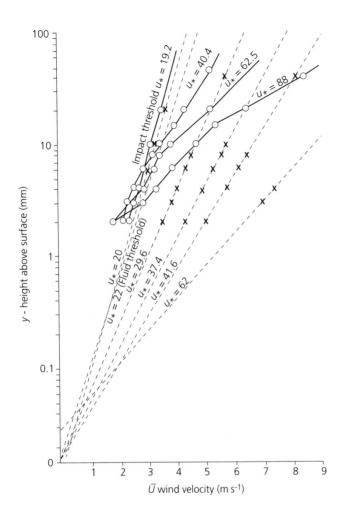

Fig. 10.15 Wind velocity profiles over active (solid lines) and inactive (dashed lines) beds of uniformly sized 0.25 mm quartz sand. After Bagnold (1936) [26].

the grain size distribution of the sediment comprising the bed, increasing in importance in coarser bed stocks.

The bedload sediment transport rate The theoretical and empirical basis for most bedload sediment transport rate formulae was provided by Bagnold, who found that the mass flux as a result of saltation and surface creep depended on both bed material characteristics and the cube of the shear velocity:

$$q = C\left(\frac{D}{D_{ref}}\right)^{0.5}\frac{\rho}{g}u_*^3 \qquad (10.20)$$

where the term in parentheses is the mean grain diameter of the sand compared to a reference diameter of 0.25 mm, and C is a coefficient introduced to account for the sorting of the bed material, varying from 1.5

for uniformly sized sand, to 2.8 for poorly sorted sand and 3.5 for a pebbly surface. It is immediately apparent that Bagnold's bedload sediment transport formula does not contain a threshold term; consequently, it makes the erroneous prediction that finite bedload sediment transport rates will occur for all values of shear velocity, even those below the threshold. A modification of Bagnold's equation to account for the threshold has the form

$$q = C_1\left(\frac{D}{D_{ref}}\right)^n\frac{\rho}{g}u_*\left(u_* - u_{*_c}\right) \qquad (10.21)$$

where $u_* - u_{*_c}$ is the excess velocity above the threshold, and the coefficient C_1 has a value of 4.2.

The problem with both of these formulae is that they predict that the sediment transport rate should continue increasing indefinitely, whereas we know

Practical exercise 10.3: Saltation on Earth and Mars

If the mean height of saltating grains is given by equation (10.17), calculate the mean height of saltation for the conditions on Earth and Mars given in Practical Exercise 10.2. Compare this with the theoretical mean saltation height based on no fluid drag, given in equation (10.16). Convert the fluid threshold shear velocity to the impact shear velocity before carrying out the calculations.

Solution

For the impact threshold, we use $A = 0.08$ instead of 0.1. In other words, the impact threshold is about 80% of the fluid threshold. The impact threshold velocities and mean saltation heights are then as follows.
(a) Earth: $u_* = 0.12 \, m \, s^{-1}$; $h = 0.86 \, mm$, compared to the theoretical 'no-drag' result of 1.2 mm.
(b) Mars: $U_* = 1.4 \, m \, s^{-1}$; the theoretical 'no-drag' result is 0.43 m. It is not possible to use the empirical result in equation (10.7) on Mars because of the different gravitational acceleration.

It is more sensible to compare the mean saltation height for typical sandstorm conditions rather than at the threshold. On Earth, considerable transport of sand takes place under winds where the shear velocity is $0.6 \, m \, s^{-1}$. This would produce a no-drag result of 30 mm for the mean saltation height. On Mars the shear velocity at the surface under a typical geostrophic wind with a velocity of $100 \, m \, s^{-1}$ at a height of 1 km is $2.7 \, m \, s^{-1}$. Remembering that Mars has a gravitational acceleration of $3.75 \, m \, s^{-2}$, this would produce a mean saltation height of 1.6 m. These results confirm that sand grains saltate to much greater heights on Mars than on the Earth, and pose a major hazard to landing craft and equipment left on the Martian surface.

It is also interesting to consider the size of impact ripples on Mars. If their wavelength is related to the mean saltation length, impact ripples on Mars must have wavelengths of tens of metres. They would be the same size, relatively speaking, as an impact ripple appears on Earth to an ant!

that saltation reaches a steady state. The transport stage u_*/u_{*_c} can be used to estimate the concentration of sand in the saltation layer. When the shear velocity is approximately three times the threshold, the saltation layer becomes saturated with grains. If the mean grain size is 0.18 mm, saturation is achieved at a shear velocity of $0.6 \, m \, s^{-1}$ [27].

As in bedload transport under water, the various sediment transport formulae give widely differing results. Some of the variation may be caused by factors such as the effect of vegetation, moisture, bed slope and grain shape.

Flow of sand on a slope The flow of sand down a subaerial slope has been discussed in Chapter 6 as a problem of the mechanics of highly concentrated dispersions. This mode of transport is particularly common in aeolian environments where sand avalanches down the slip-face of a dune. The topic will be returned to in Section 10.5 when considering lamination types associated with wind-generated bedforms. The onset of sliding of a mass of grains on a slope occurs when the friction coefficient equals the tangent of the angle of internal friction. In a moving mass of grains there is a stress directed away from the boundary which is transmitted by grain collisions and near misses which Bagnold termed *dispersive pressure*. When the dispersive pressure is unable to maintain grains in motion, the dispersion freezes at a characteristic angle which is somewhat smaller than the angle of internal friction. This angle, at which a pile of loose sediment will accumulate, is known as the *angle of repose*. On the upper part of a dune slip-face, sediment accumulates in an oversteepened bulge which periodically fails, generating flows of sand down the slip-face until a new lower slope is achieved. These *grainflows* are extremely commonplace on the lee-side slip-faces of dunes. The pivot point between an upper slip-face dominated by oversteepening, and a lower part dominated by accumulation from grainflows is about 2 m downslope from the brink of a dune of height 5 m [28].

Transport of dust

The transport of dust is essentially a problem of long-term suspension. Some of the fundamental considerations are presented in Chapter 5. The transport of dust by the wind can be extremely widespread. Thick dust clouds reach kilometres in height. The extent and thickness of deposits of wind-blown dust (*loess*) (Chapter 2) preserved in the recent geological record leave no doubt that aeolian dust transport is an extremely important surface process. Dust is produced from a large number of environmental settings, arid and semi-arid deserts being only one. Glacial outwash plains, dried up river and lake beds, coastal zones and agricultural soils all provide substantial quantities of dust in addition to deserts. Dust from soil erosion alone accounts for about 500 Mt of particulates in the atmosphere per year.

Measurements in duststorms indicate that the suspended sediment concentration decreases with distance from the land surface, often exponentially. Fine sediment grains are commonly liberated from the bed by bombardment during saltation, and then swept up into the air by turbulent eddies. Field measurements suggest that the general diffusive theory of suspended sediment concentrations is applicable (equation (5.36)). In the lower atmosphere the height at which the sediment concentration is zero always greatly exceeds the depth at which information on sediment concentrations on required, so the general equation simplifies to

$$\frac{C}{C_{a}} = \left(\frac{a}{y}\right)^{z} \tag{10.22}$$

where $z = 2.5 w/u_*$ is the Rouse number, w being the settling velocity of suspended grains. In the Stokes range the settling velocity is given by

$$w = \frac{1}{18} \frac{\Delta \rho g D^2}{\mu} \tag{10.23}$$

where $\Delta \rho$ is the difference in density between the sediment and the enclosing fluid, in this case air, D is the grain diameter, μ is the viscosity of the fluid, and g is gravitational acceleration. If a substrate is made of spherical particles with diameter 20 µm, and a strong wind has a shear velocity $0.7\,\mathrm{m\,s^{-1}}$, the particle concentration at a height of 10 m relative to that at 0.5 m from the ground will be about 60% (the viscosity of air at sea level is approximately $1.8 \times 10^{-5}\,\mathrm{N\,s\,m^{-2}}$), whereas at 100 m above the ground it is about 40%. These grain sizes are therefore capable of being transported very long distances high in a dust cloud. But if the grain size is increased to 50 µm, the relative concentrations fall to 5% and 0.5% at heights of 10 m and 100 m, respectively. These grain sizes would be transported in suspension at low levels above the ground and would most likely be trapped by vegetation after a short transport distance. A practical exercise on the significance of this diffusive theory for the deposition of loess is provided below.

If the vertical flux of suspended sediment was known, it would be possible to estimate the transport distance of grains of given size and settling velocity in winds of given shear velocity. The vertical particle flux is given by Fick's first law of diffusion (equation (5.22))

$$Q = -\kappa \frac{\partial C}{\partial y} \tag{10.24}$$

where κ is the turbulent diffusivity. Let us now assume that suspended sediment diffuses upwards at the same rate as the diffusion of momentum, in which case the diffusivity can be replaced by the *kinematic eddy viscosity*, ε. The kinematic eddy viscosity varies with shear velocity and height above the boundary. Experiments suggest it varies between about 10^{-1} and $10^{3}\,\mathrm{m^2\,s^{-1}}$. The maximum likely transport distance, L, for particles obeying Stokes law of settling is inversely proportional to the square of the settling velocity, w, and therefore to the fourth power of the grain size [19]:

$$L = \frac{\hat{U}^2 \varepsilon}{w^2} \tag{10.25}$$

where $\hat{U}$ is the average speed of the wind.

The thickness of loess deposits decreases exponentially with distance from the source (Fig. 10.16), the rate of exponential decline depending on the vegetation cover over which the dust-laden winds move. Forest cover causes a more rapid change in deposit thickness than grassland or bare land.

10.5 Aeolian bedforms and deposits

Although the sand seas or *ergs* characterized by large dunes occupy a relatively modest proportion (about 25%) of the world's desert areas, they are particularly important from a geological perspective because deposits of ancient *ergs* have been unambiguously identified in the geological record. They are also the sources of dust for long-distance transport to form loess deposits on desert margins, and as a contaminant to the pelagic deposits recovered in deep sea cores. The high sediment transport rates of sand and silt in ergs also pose environmental problems for development.

In the final section of this chapter, the physical principles underlying the various types of aeolian bedform are introduced, and the characteristic stratification patterns produced by their migration discussed.

Practical exercise 10.4: Dust transport by wind

1 Assuming that suspended sediment in the wind can be approximated by the diffusive model of Rouse (equation (10.22)), calculate the concentration, relative to a reference height of 0.5 m, at heights of 1 m, 10 m and 100 m for grain sizes of 10, 30 and 60 μm Dust transporting winds generally have shear velocities that fall in the range 0.2–0.6 m s⁻¹. Plot two sets of curves of relative concentration against height for the limits of this range. Take the viscosity of air as $1.8 \times 10^{-5}\,N\,m^2\,s^{-1}$.

2 Using equation (10.25), calculate the maximum transport distance of grains of diameter 10, 30 and 60 μm and density 2650 kg m⁻³, in air of viscosity $1.8 \times 10^{-5}\,N\,m^2\,s^{-1}$ with an average velocity of 15 m s⁻¹. Take kinematic eddy viscosities to vary between 1 and $10^2\,m^2\,s^{-1}$.

3 Loess deposits consist of predominantly medium and coarse silt grains, that is, in the region of 30–60 μm. Using the higher value of the kinematic eddy viscosity given, how far from the silt source would you expect the material to be dispersed by winds of average velocity 15 m s⁻¹?

Solution

1 The values of C/C_{ref} for the three heights and range of shear velocity are given in Table P10.1. The results plotted on a graph of height against relative concentration are found in Fig. 10.17.

Table P10.1 Values of C/C_{ref}.

Height: (m)	Shear velocity: (m s⁻¹)	Grain size (μm)		
		10	30	60
1	0.2	0.933	0.536	0.082
1	0.6	0.977	0.812	0.435
10	0.2	0.741	0.067	2.01×10^{-5}
10	0.6	0.906	0.406	0.027
100	0.2	0.589	8.49×10^{-3}	4.93×10^{-9}
100	0.6	0.840	0.203	1.73×10^{-3}

Table P10.2 Maximum transport distance.

Kinematic eddy viscosity, ε: (m² s⁻¹)	Grain size (μm)		
	10	30	60
1	3 498	43.2	2.7
100	349 830	4 320	270

2 The maximum transport distance (in kilometres) for the different grain sizes and kinematic eddy viscosities are given in Table P10.2.

3 Using the higher kinematic viscosity, the medium to coarse silt found in loess should be transported distances of hundreds to a few thousands of kilometres. These large distances of transport are in agreement with the very wide extent and distribution of loess deposits (Chapter 2).

10.5.1 Lamination styles

It is now recognized that there is a distinct suite of lamination types which can be attributed to sediment transport and deposition by the wind [30]. The recognition of distinctive lamination styles has enormously aided the recognition of ancient aeolian

Fig. 10.16 Measured thicknesses of loess deposits as a function of distance from source for relatively small transport distances, with a curve for a diffusion model. After Frazee *et al.* (1970) [29].

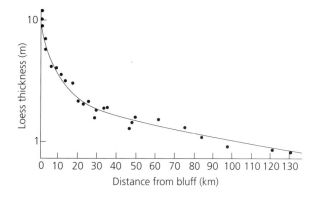

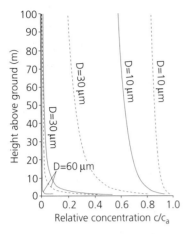

Fig. 10.17 Practical exercise results on vertical profiles of suspended dust concentrations for grain sizes of 60, 30 and 10 μm.

deposits in the geological record. These lamination types can be summarized as follows (Fig. 10.18).
• The avalanching of grainflows (concentrated dispersions) down the slip-faces of dunes produces well-sorted, massive or coarsening-up strata whose thickness depends on the volume of sediment involved in the massflow. Larger slip-faces tend to produce thicker grainflow strata. Grainflows produce laterally impersistent strata with planform terminations wedging out in the downslope direction (Fig. 10.19).

• The fallout of suspended particles on to the lee side of a dune produces less well-sorted, finer-grained, *grainfall* laminae. Grainfall laminae are coarser as the lee face is climbed, but are destroyed by repeated avalanching high on the slip-face. Lower on the slip-face, grainflows and grainfall laminae may be interbedded.
• The migration of wind ripples produces a millimetre-scale parallel lamination (Fig. 10.20), where individual laminae coarsen upwards. The angle of the parallel *wind-ripple lamination* depends on the slope of the depositional surface over which the wind is blowing. Wind-rippled aprons are very common on the lower flanks of dunes. Consequently, wind-ripple lamination typically is interstratified with the downslope parts of grainflows.

A wide range of aeolian bedforms are found in nature, commonly superimposed one on the other. The smallest centimetre-scale bedforms are ripples. They are commonly superimposed on metre-scale dunes which may be flow-transverse, longitudinal or complex. Dunes may make up larger sand accumulations with heights of hundreds of metres known as *draa*. Dunes and draa strongly interact with the airflow creating them, producing secondary airflows.

10.5.2 Ripples
The origin and wavelength of the most common variety of wind ripple (Fig. 10.21) has been attributed to the saltation process. They are consequently termed *impact ripples*. They have asymmetrical profiles with

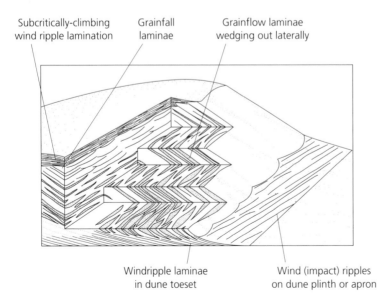

Subcritically-climbing wind ripple lamination

Grainfall laminae

Grainflow laminae wedging out laterally

Windripple laminae in dune toeset

Wind (impact) ripples on dune plinth or apron

Fig. 10.18 Lamination styles on an idealized dune with a well-developed slip-face and wind-rippled lee-side apron. Modified from Hunter (1977) [30].

Fig. 10.19 Downdip terminations of grainflow laminae intercalated with wind-ripple lamination. Lower Jurassic Navajo Sandstone, Zion National Park, Utah, USA.

low angle (8–10°) upwind flanks and steeper lee slopes (20–30°). Cross-sections of ripples commonly show an accumulation of coarser sand overlying a finer-grained plinth exposed in the troughs between ripplecrests, the coarsest grains being found near the ripplecrest. The migration of wind ripples therefore produces an upward-coarsening layer whose thickness depends on the angle of climb (see Section 5.4.8).

It has been proposed that the wavelength of impact ripples approximates the saltation length [10]. For well-sorted sands and steady winds the grains would move with a constant saltation length and impact angle. Once saltation is well developed, parts of the bed facing the downward trajectories of saltating grains should suffer erosion due to grain impacts, whereas parts of the bed facing away from the downward trajectories would be sheltered and experience deposition (Fig. 10.21). The ripple form is therefore translated downwind through time. Coarser grains move downwind, however, by surface creep.

They are concentrated near the ripplecrest because immediately downwind from the change in surface slope the rate of bombardment drops markedly. This model therefore explains both the occurrence of ripples and the concentration of coarse grains in the crestal region. A simple test is to compare the theoretical or empirical saltation length with the wavelengths of wind ripples. The theoretical mean saltation length is given in equation (10.18). For a wind (on Earth) with a shear velocity of 0.2 m s^{-1}, the mean saltation length is 42 mm. As the shear velocity increases, so does the mean saltation length. When the shear velocity is 0.4 m s^{-1}, the mean saltation length is 168 mm. These figures are certainly of the right order to explain the typical wavelengths of impact ripples of 70–140 mm.

Other workers believe that the short-trajectory reptation process is fundamentally important in impact-ripple migration. If the mass flux of the reptating population and the high-trajectory saltating population is considered in terms of the sediment

Fig. 10.20 Well developed wind-ripple lamination preserved in toeset region of aeolian dune, Lower Jurassic Navajo Sandstone, Arches National Park, Utah, USA.

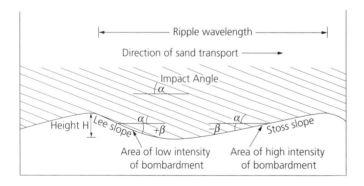

Fig. 10.21 Bagnold (1941) [10] model of the impact-ripple wavelength determined by the saltation length, showing active impact and sheltered (shadow) zones.

continuity equation and stability theory, the fastest-growing wavelength is about six times the mean reptation distance, rather than being scaled on the saltation length [22].

Other ripples may form under the wind that are more closely analogous to the ripples under aqueous flows. They have longer wavelengths (0.2–2 m) than impact ripples and heights of only 2–50 mm. They are thought to be due to fluid drag at high wind velocities causing high amounts of modified saltation and suspension transport. They are termed *fluid drag ripples* or *aerodynamics ripples*.

Some decimetre- to metre-scale ripples contain very coarse grains and are dominated by surface creep.

They are termed *granule ripples* and *megaripples*.

10.5.3 Dunes

Dunes exhibit an extraordinarily wide range of forms (see Fig. 10.22). Many classification systems are available, each with its own special terminology. A widely used scheme derived from a global analysis of dune form using satellite imagery [31] is given in Table 10.1. The main distinction is between dunes with axes oriented transverse to flow, with axes oriented parallel to flow, with multiple axes (star dunes), and with none at all (dome dunes).

A common transverse dune is an isolated crescentic form with its horns extending downwind, known as a *barchan*. Barchans are generally between 0.3 and 10 m high, and form where winds are relatively uniform and sand supply limited. They have well-developed slip-faces at the angle of repose of dry sand, and migrate by the successive avalanching of grain-flows down the lee side in response to oversteepening of the upper part of the slip-face from grainfall over the brink point. Barchans may amalgamate into *barchanoid ridges* where the sand supply is greater, and into *transverse ridges* where the characteristic crescentic form is replaced by a simple roughly linear flow-transverse ridge.

Dunes with axes parallel to the prevailing wind direction are termed *linear*. They commonly occur as low (5–20 m) sandy ridges separated by wide (0.3–2.7 km) pavements of sandy or gravelly material. Linear dunes are the most common dune type in ergs. They may retain their linear form along the wind direction for over 100 km. Their surface slopes average 10–20°, which is well below the angle of repose, although small avalanche slip-faces superimposed on a

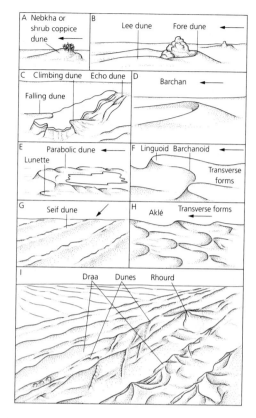

Fig. 10.22 Main types of dune and draa and their relationship to dominant wind directions where appropriate. After Cooke & Warren (1973) [32].

low-angle ridge are also found. Linear dunes with slip-faces which alternate from one side to the other are common in the Libyan desert, where they are called

Table 10.1 Global analysis of dune forms from satellite imagery. After McKee (1979) [31] and Greeley & Iversen (1985) [9].

Name	Form	Type*	Slip-face(s)	Wind†
Transverse				
Barchan	Crescent in plan view	S,C	1	Transverse
Barchanoid ridge	Rows of connected crescents in plan view	S,C,CX	1	Transverse
Transverse ridge	Asymmetric ridge in cross-section	S,C,CX	1	Transverse
Longitudinal	Symmetric ridge in cross-section	S,C,CX	2	Parallel
Parabolic	U-shaped in plan view	S,C	1 or more	Parallel
Dome	Circular or elliptical mound	C,CX	none or poorly defined	–
Star	Central peak with three or more arms	S,C,CX	3 or more	Multiple

* S, simple; C, compound; CX, complex.

† Orientation of dune axis with respect to wind direction or the vector of more than one wind direction.

seif dunes. Linear dunes may form larger complexes up to 150 m high composed of a set of linear ridges and star-shaped forms.

The origin of linear dune forms has attracted considerable attention. One theory is that convective heating of the atmosphere leads to the formation of helical or corkscrew vortices with horizontal axes (Fig. 10.23a), causing a sweeping up of sediment from the core towards the flanks of the vortices. Shear stresses are greatest where the velocity gradients are highest in zones of flow divergence. This would correspond to the centres of the interdune corridors, keeping them swept clean of loose sand. Once a linear fabric has been produced, the different roughnesses of the dune and interdune areas serves to perpetuate the helical motion. However, the transverse spacing of the convectional cells is somewhat greater (4–12 km) than the observed linear dune spacing (from less than 1 km up to about 3 km).

Other ideas include the modification of other dune types by a change in wind regime. For example, the horns of barchan dunes might be extended downwind while the central core was degraded. It seems unlikely that this model would explain the extremely extensive and regular pattern of simple linear dunes observed in some deserts such as the Simpson desert of Australia or the compound linear dunes of the Namibian erg.

Critical evidence comes from the wind and sediment transport directions associated with a linear dune topography. There is little evidence that the sediment transport direction is symmetrically away from the interdune corridor towards the two adjacent ridge crests. Instead, when the wind flows obliquely over a

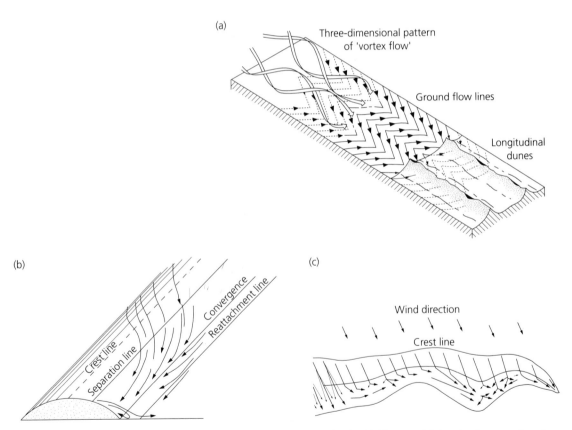

Fig. 10.23 Models for the origin of linear, longitudinal seif-type dunes. (a) Helicoidal flow model showing the sweeping of the interdune corridors by divergent near-ground flows. After Cooke & Warren (1973) [32]. (b) Flow visualization over a symmetrical rounded model for an oblique approach of the airflow. After Tsoar *et al.* (1985) [33]. (c) Plan view of model involving secondary circulation over a slightly sinuous crestline oblique to the main wind direction. Arrows on lee side show airflow vectors in separated and reattached flows. After Tsoar (1983) [35].

brink line of a dune, a secondary flow pattern is set up in the separated flow to the lee of the dune which has a strong along-crest component (Fig. 10.23b,c). The magnitude of the along-crest flow is determined by two factors. One is dune shape, particularly the dune height compared to the length of the windward slope. The other is the angle of wind approach relative to the dune crest. When this angle is high the dune behaves as a transverse element, with deposition under a sluggish separated flow. When the angle is between 35° and 50°, the lee slope is characterized by strong along-slope secondary winds which transport sand parallel to the crest and prevent tall slip-faces from being well developed.

If the crest is sinuous, there should be along-crest variations in the secondary flow and therefore in sediment transport rates, causing peaks and saddles to develop. This model [34] therefore is dependent on the existence of oblique winds for a significant part of the year.

10.6 Wind as a hazard

Wind poses a hazard in essentially two ways which are not mutually exclusive: one is through its strength alone; and the other is through its ability to erode, transport and deposit particulate material. Whereas hazards in the first category relate directly to atmospheric weather systems, those in the second are concentrated in the world's drylands (Fig. 10.24). We focus on the second category in the remainder of this chapter.

10.6.1 Strong winds and drought

Strong winds accompany the cyclones described in Chapter 6, but may also occur in smaller air volumes as tornadoes, and as winds enhanced by topographic effects. Strong winds pose a natural hazard in their own right, but they are also responsible for considerable sediment transport in duststorms.

Tornadoes are vertical vortices that initially develop from larger horizontal vortices in clouds. Where these vertical funnel-like vortices touch the ground they have immense destructive power, but their tracks are generally less than a kilometre in diameter and their life-spans very short (generally less than an hour). About 80% of tornadoes occur on the Great Plains of USA, where about 600 occur each year. This may be because cold Arctic air meets the warm air of the Gulf of Mexico, creating atmospheric instability. The maximum temperature difference between the two air masses is in the late spring, which is when most tornadoes occur. The most serious threat to life in a

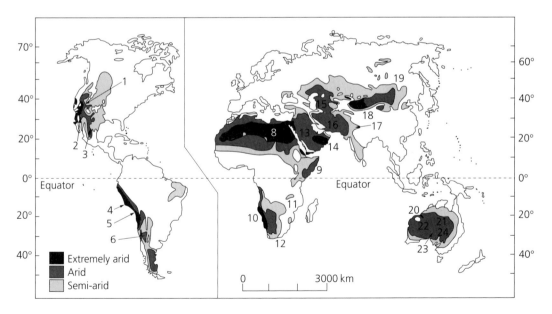

Fig. 10.24 Distribution of the world's drylands. The main deserts are: (1) Great Basin; (2) Sonoran; (3) Chihuahuan; (4) Peruvian; (5) Atacama; (6) Monte; (7) Patagonian; (8) Sahara; (9) Somali-Chabli; (10) Namib; (11) Kalahari; (12) Karroo; (13) Arabian; (14) Rub al Khali; (15) Turkistan; (16) Iranian; (17) Thar; (18) Taklimakan; (19) Gobi; (20) Great Sandy; (21) Simpson; (22) Gibson; (23) Great Victoria; (24) Sturt. After Cooke & Warren (1973) [32] and Greeley & Iversen (1985) [9].

tornado is not the damage caused by airborne debris, but the results of the dramatic pressure change. The pressure in the centre of the tornado is so low, and the pressure change so rapid, that air trapped in buildings during the passage of a tornado may explode, causing destruction of the building and burning of people inside.

Duststorms are responsible for a large amount of transport of fine particulate sediment, and deep sea cores and sea bed samples indicate that they are the primary mechanism for distribution of terrestrial material to huge parts of the deep ocean. Saharan dust is found in the Atlantic Ocean thousands of kilometres away as far as the Caribbean islands, and dust from China has been reported in Hawaii and Alaska 10 000 km away. The direct hazard posed by duststorms is negligible, but the secondary effects of soil loss and soil deterioration are extremely important. The winds responsible for duststorms originate from depressions tied to the westerlies, as in the northern Saharan region, from the seasonal movement of the monsoon,

as in north-east Africa and Arabia, from convection linked to the approach of the intertropical convergence, as in central Asia and north-east Africa, or from katabatic winds in mountainous regions, as in California, the Argentinian foothills of the Andes and the south-west margin of the Himalayan–Hindu Kush chain. Duststorms are most frequent in semi-arid climatic zones with annual rainfall of 100–200 mm, and occur most commonly in late spring and early summer. They are extremely common, with tens to hundreds occurring every year in some vulnerable areas such as the Gobi desert in the dry heart of Asia (Fig. 10.25).

Although human activities have important effects on duststorm vulnerability, the classic example being the introduction of the plough and land rush of the Great Plains in the 1930s (the so-called *dust bowl*) (Section 10.6.2), duststorms appear to be cyclical in occurrence, suggesting a climatic link. *Sunspots*, which are regions on the Sun's surface of particularly high magnetic field, are the location of *solar flares* of

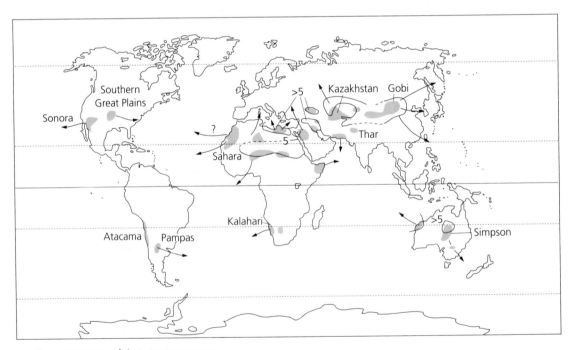

———5——— Frequency of dust storms per year
? Frequency of dust storms unknown
● Source areas of dust
⌒ Main dust trajectories

Fig. 10.25 Frequency of duststorms, major source areas of dust, and main dust transport pathways. After Middleton *et al.* (1986) [35], in Bryant (1991) [36].

electromagnetic radiation which reach the Earth. The arrival of the enhanced solar wind distorts the Earth's magnetosphere, leading to changes in the upper atmosphere and consequently to climatic effects. Sunspot activity operates on an 11 y cycle. Another astronomical mechanism is the variation of the orbit of the Moon. The Moon orbits Earth in a plane that is related to the equator of the Sun, and varies by 5° from this equatorial plane over a 18.6 y cycle. Some workers believe that the 18.6 y *lunar precession cycle* also affects climate through its tidal effect on the atmosphere. Records of drought and heavy rainfall from the Great Plains of America, southern Africa, and the great river floodplains of the Nile, China and India show evidence of an approximately 10 and 20 y cyclicity. Analysis of such records has led some workers [37] to conclude that the 18.6 y lunar cycle dominates, but that the effect of lunar precession 'flip-flops' over a 100–300 y cycle, lunar phase sometimes being associated with drought and sometimes with excessive rainfall.

Periods of drought, leaving the land susceptible to wind erosion, are also closely linked to the El Niño southern oscillation (ENSO) (Section 1.3.3). The effects of ENSO events can be detected in the rainfall measurements and river discharges in widely separated areas spanning the continents of South America, Africa, Asia and Australasia. In Indonesia, historical records going back to the mid-nineteenth century show a very strong correlation between ENSO events and the failure of the monsoon. It is increasingly recognized that ENSO events have implications for meteorological change and occurrence of natural hazards on a global scale.

10.6.2 Soil erosion by wind

Soil erosion by wind causes damage to crops, reduces soil productivity by removing humic and fine-grained material and reducing water-retention capacities. The products of wind erosion, exemplified by the duststorms of the 1930s in the Great Plains of the USA, represent additional hazards by endangering land use downwind, and by the transfer of airborne disease. Excessive soil loss by wind is commonly associated with inappropriate land-use practices, such as overgrazing or intensified farming on 'marginal' land. Soil erosion by wind is a particular problem in arid and semi-arid areas, but also where soils lend themselves to transport, such as the loessic soils of central Europe and China, and in areas of coastal sand dunes and sandflats.

The factors affecting wind erosion have already been discussed from the point of view of the basic physics earlier in this chapter. As with soil erosion by water, the variables controlling soil loss can be grouped into those affecting the *erosivity* of the wind (principally climatic variables) and the *erodibility* of the land surface (variables relating to surface sediment properties). One practical formulation is the wind erosion equation, which was developed for the conditions prevailing on the American Great Plains. It is similar in many respects to the universal soil loss equation (Chapter 3), having the form

$$A = f(I', K', C', L', V')$$

where A is the annual rate of erosion; I' is the soil erodibility compared to that of an unsheltered, non-crusted, wide, bare, smooth field; K' is a roughness factor; C' is a climatic factor incorporating the effects of wind velocity and soil moisture; L' is the length over which the wind blows in the wind erosion direction (fetch); and V' is an index of the vegetation cover, including its effect on roughness. It suffers from the same problems as the universal soil loss equation discussed in Section 3.3.2, that is, it lacks any real physical significance.

A number of techniques have been used to predict soil loss by wind erosion. Wind tunnel experiments, with appropriate scaling, give information on erodibility and sediment transport rates. *In situ* field experiments using soils of documented technical properties and monitored climatic conditions allow soil losses to be measured. Soil loss predictions can also be made from meteorological data, based essentially on the duration and velocity of winds from all directions in excess of the threshold for dust and sand entrainment. The potential erosion by wind can also be estimated by a regression of the climate parameter in the wind erosion equation, C, or of other climatic variables, on measured frequency and intensity of duststorms or periods of sand transport.

The dust bowl of the Great Plains

The continuing problem of soil erosion is less to do with a lack of understanding of the physics involved in soil loss and conservation, and more to do with failed agricultural programmes and policies, many of which are applied without due appreciation of local cultural and pastoral practices [38]. The success of soil conservation programmes must take account of the social, political and economic dimensions of the proposed changes to land use. The wind erosion on the Great Plains during the 1930s, known as the dust bowl, is a classic example of the link between changes in land-use practices, climatic deterioration and natural hazard [39]. During the 1930s severe wind erosion caused enormous crop and livestock damage, social hardship, and generated spectacular duststorms (Fig. 10.26). The cause of the problem was primarily the stripping of the land of its natural vegetation cover of grassland in order to plough it for wheat as a cash crop. A number of events contributed to motivate this: the development of mechanized agriculture; the growing market for grain after the First World War; the capitalist ethos which led to speculative and irresponsible commercialism; and the 'get rich quick' culture which distorts rational perceptions of risk. It is true that the wind erosion problems of the Great Plains were exacerbated by drought. The climatic effects simply pushed an unsustainable system over the edge.

The conservation practices developed for the High Plains provide a case study for management of wind erosion in general. Conservation measures involve a reduction of the power of the wind, and the improve-

ment of ground conditions. Four key principles help to achieve these conservation goals [41]:

1 the establishment and maintenance of vegetation and vegetative residues;

2 the production of non-erodible soil aggregates (clods);

3 the reduction of the fetch by reducing field width along the wind erosion direction;

4 and the increase in the roughness of the land surface.

Field lengths can be shortened and roughness increased by the planting or construction of *wind-breaks*. As a rule of thumb, wind-breaks offer shelter to a region downwind by 10–30 barrier heights and 5–10 barrier heights upwind, these figures depending on the nature of the barrier, particularly its permeability, and the characteristics of the wind. The planting of 'cover crops' reduces the exposure of the ground surface at critical times when erosion is likely, notably in the period prior to spring planting. *Strip cropping* of erosion-resistant and erosion-vulnerable crops is

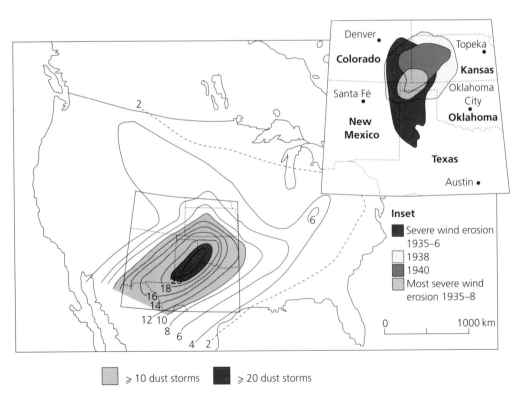

Fig. 10.26 (a) The extent and number of duststorms in USA in March 1936. After Lockeretz (1978) [40]. (b) Detail of the Great Plains with extent of severe wind erosion during selected time intervals in the 1930s. After Worster (1979) [39].

also practised. The leaving of *stubble* and other crop residues provides organic matter for the soil as well as protecting the soil from erosion. Finally, ploughing methods may help by creating a rough, cloddy surface rather than a smooth, tilled surface, and soil conditioning may improve moisture and organic contents and clod formation.

Further reading

R.A. Bagnold (1941) *The Physics of Blown Sands and Desert Dunes*. Methuen, London.

R.U. Cooke & A. Warren (1973) *Geomorphology in Deserts*. Batsford, London.

A.S. Greeley & J.D. Iversen (1985) *Wind as a Geological Process*. Cambridge University Press, Cambridge.

J.R. Holton (1979) *An Introduction to Dynamical Meteorology*, Academic Press, New York.

K. Pye & H. Tsoar (1990) *Aeolian Sand and Sand Dunes*. Unwin Hyman, London.

References

1 D. Brunt (1939) *Physical and Dynamical Meteorology*. Cambridge University Press, Cambridge.

2 K. Pye & H. Tsoar (1990) *Aeolian Sand and Sand Dunes*. Unwin Hyman, London.

3 F.R. Eldridge (1980) *Wind Machines*, 2nd edn. Van Nostrand Reinhold, New York.

4 J.T. Houghton (1986) *The Physics of Atmospheres*, 2nd edn. Cambridge University Press, Cambridge.

5 T.R. Oke (1978) *Boundary Layer Climates*. Methuen, London.

6 A.J. Bowen & D. Lindley (1977) A wind tunnel investigation of the wind speed and turbulence characteristics close to the ground over various escarpment shapes. *Boundary-Layer Meteorology* **12**, 315–38.

7 P.S. Jackson (1977) Aspects of surface wind behaviour. *Wind Engineering* **1**, 1–14.

8 H. Tsoar (1986) Two-dimensional analysis of dune profile and the effect of grain size on sand dune morphology. In: *Physics of Desertification* (eds F. El-Baz & M.H.A. Hassan). Martinus Nijhoff, Dordrecht, pp. 94–108.

9 R. Greeley & J.D. Iversen (1985) *Wind as a Geological Process*. Cambridge University Press, Cambridge.

10 R.A. Bagnold (1941) *The Physics of Blown Sands and Desert Dunes*. Methuen, London.

11 W.S. Chepil (1951) Properties of soil which influence wind erosion, IV. State of dry aggregate structure. *Soil Science* **72**, 387–401.

12 K. Horikawa & H.W. Shen (1960) *Sand Movement by Wind Action*, Technical Memo 119. US Army Corps of Engineers, Beach Erosion Board, Washington, DC.

13 J.D. Iversen & White (1982) Saltahan threshold on Earth, Mars and Venus. *Seimentology* **29**, 111–19.

14 C. McKenna-Neuman & W.G. Nickling (1989) A theoretical and wind tunnel investigation of the effect of capillary water on the entrainment of sediment by wind. *Canadian Journal of Soil Science* **69**, 79–96.

15 W.S. Chepil (1945) Dynamics of wind erosion, II. Initiation of soil movement. *Soil Science* **60**, 397–411.

16 W.S. Chepil (1959) Equilibrium of soil grains at the threshold of movement by wind. *Proceedings of the Soil Science Society of America* **23**, 422–8.

17 A.W. Zingg (1953) Wind tunnel studies of the movement of sedimentary material. *Proceedings of the 5th Hydraulic Conference Bulletin* **34**, 111–135. Institute of Hydraulics, Iowa City.

18 M. Logie (1982) Influence of roughness elements and soil moisture of sand to wind erosion. *Catena* (Supplement) **1**, 161–73.

19 H. Tsoar & K. Pye (1987) Dust transport and the question of desert loess formation. *Sedimentology* **34**, 139–53.

20 W.S. Chepil (1961) The use of spheres to measure lift and drag on wind-eroded soil grains. *Soil Science Society of America Proceedings* **25**, 343–5.

21 J.D. Iversen, R. Greeley, J.R. Marshall & J.B. Pollack (1987) Aeolian saltation threshold: the effect of density ratio. *Sedimentology* **34**, 699–706.

22 R.S. Anderson (1987) A theoretical model for aeolian impact ripples. *Sedimentology* **34**, 943–56.

23 P.R. Owen (1964) Saltation of uniform grains in air. *Journal of Fluid Mechanics* **20**, 225–42.

24 R.S. Anderson & B. Hallet (1986) Sediment transport by wind; towards a general model. *Bulletin of the Geological Society of America* **97**, 523–35.

25 P.R. Owen (1980) *The Physics of Sand Movement*. Lecture Notes, Workshop on Physics of Flow in Deserts. International Centre for Theoretical Physics, Trieste.

26 R.A. Bagnold (1936) The movement of desert sand. *Proceedings of the Royal Society of London* **157A**, 594–620.

27 K.M. Gerety & R. Slingerland (1983) Nature of the saltating population in wind tunnel experiments with heterogeneous size-density sands. In: *Eolian Sediments and Processes* (eds M.E. Brookfield & T.S. Ahlbrandt). Elsevier, Amsterdam, pp. 115–31.

28 R.S. Anderson (1988) The pattern of grainfall deposition in the lee of aeolian dunes. *Sedimentology* **34**, 943–56.

29 C.J. Frazee, J.B. Fehrenbacher & W.C. Krumbein (1970) Loess deposition from a source. *Proceedings of the Soil Science Society of America* **34**, 296–301.

30 R.E. Hunter (1977) Basic types of stratification in small eolian dunes. *Sedimentology* **24**, 361–87.

31 E.D. McKee (1979) A study of global sand seas. Professional Paper 1052. US Geological Survey, Washington DC.

32 R.U. Cooke & A. Warren (1973) *Geomorphology in Deserts*. University of California Press, Berkeley.

33 H. Tsoar, K.R. Rasmussen, M. Sørensen & B.B. Willetts (1985) Laboratory studies of flow over dunes. In: *Proceedings of the International Workshop on the Physics of Blown Sand* (eds O.E. Barndorff-Nielsen, J.T. Møller, K.R. Rasmussen & B.B. Willetts), Memoir 8. Department of Theoretical Statistics, Institute of Mathematics, University of Aarhus, pp. 327–49.

34 H. Tsoar (1983) Dynamic processes acting on a longitudinal (seif) dune. *Sedimentology* **30**, 567–578.

35 N.J. Middleton, A.S. Goudie & G.L. Wells (1986) The frequency and source areas of dust storms. In: *Aeolian Geomorphology* (ed. W.G. Nickling). Allen & Unwin, London, pp. 237–60.

36 E.A. Bryant (1991) *Natural Hazards*. Cambridge University Press, Cambridge.

37 R.G. Currie (1984) Periodic (18–6 years) and cyclic (11 year) induced drought and flood in western North America. *Journal of Geophysical Research* **89**, 7215–30.

38 P. Blaikie (1985) *The Political Economy of Soil Erosion in Developing Countries*. Longman, London.

39 D. Worster (1979) *Dust Bowl*. Oxford University Press, New York.

40 W. Lockeretz (1978) The lessons of the dust bowl. In: *Use and Misuse of Earth's Surface* (ed. B.J. Skinner). Kaufmann, Los Altos, CA, 140–9.

41 R.U. Cooke & J.C. Doornkamp (1990) *Geomorphology in Environmental Management*. Oxford University Press, Oxford, 410pp.

11 Glaciers

Wherever glaciers have passed over, during one of the ages of the
Earth's existence, the aspect of the country has been
transformed by their action. As do avalanches, they carry the rubbish of
the crumbling mountains into the plains, not by
violence, but by the patient labour of every moment.

> E. Reclus [1]

Chapter summary

Snow and ice cover about 10% of the Earth's land surface. Of this the bulk is found in the topographically unconstrained Greenland and Antarctic ice sheets. Valley glaciers, such as those studied by early glaciologists in the Alps of Switzerland, comprise a minutely small amount of ice in the Earth's cryosphere. Glaciers act as an important component in regional denudational systems.

The heat budget of a glacier depends on the balance of solar radiation on its surface, geothermal and frictional heat sources at its base, and latent heat from refreezing of meltwater, concentrated near its surface. Ice close to its pressure melting point is termed warm ice; ice well below its pressure melting point is termed cold ice. Heat generation at the base of a glacier may result in it being warm-based, which facilitates basal sliding. Cold-based glaciers are frozen to bedrock, restricting basal sliding. Ice deformation is dominated by creep. Laboratory experiments on ice and measurements of the velocities of natural glaciers suggest that the creep is dependent on both temperature and stress. Ice appears to behave as a fluid, since it deforms at even low levels of stress, but to have an apparent viscosity that changes with the strain rate. That is, it appears to obey a non-Newtonian fluid constitutive law.

The flow of a Newtonian fluid in a channel can be modified to account for the flow of a fluid with a nonlinear power law relation between the stress and the strain rate (or velocity gradient). The resulting velocity profile shows the velocity gradient to be concentrated in a thin zone close to the boundary of the channel, with a lower strain rate in the core of the flow. The vertical velocity profile can be similarly obtained by a change of coordinate system. The theoretical horizontal (cross-channel) and vertical velocity profiles match reasonably well the observed velocities of natural glaciers, with an exponent in the power law of $n = 3$.

In some channel-like flows, such as the flow in the aesthenosphere beneath the lithospheric plates, the flow is highly temperature-dependent. Although the power law flow of ice is temperature-dependent, it is necessary to know how important this is for the temperature regimes common in natural glaciers. The temperature difference between the top and base of a glacier is small compared to the surface temperature. However, the activation energy for ice is such that for a given stress, the strain rate is about 30% lower at a temperature of 249 K (−24°C) than at 260 K (−13°C), the temperatures of the surface and base of the Greenland ice sheet at Camp Century. This would cause the deformation of the ice to be focused into a zone close to the base of the glacier. However, in glaciers with small temperature differences, such as most valley glaciers, the effect is likely to be small.

Ice sheets can be modelled as an ideal plastic material resting on a flat substrate. The thickness of the overlying ice, which is maintained by snow accumulation, produces a shear stress at the base of the ice which may exceed its shear strength, thereby causing it to flow. In order to keep the basal ice at about its shear strength, the ice sheet must take up a surface profile which steepens in gradient towards the margin. Comparison of theory with the profile of the Antarctic ice sheet suggests that in outline ice sheets can be treated as ideal plastic materials in this way.

The mass balance of a glacier involves gains by accumulation and losses by ablation, which over a balance year are separated by the equilibrium line. Discharges of ice vary along the glacier, with a component of descent in the accumulation zone and a component of upward flow in the ablation zone. Longitudinal variations in the discharge result in extending and compressive flow. Changes in mass balance are propagated through the glacier, but the response time to a change in forcing mechanism depends on a number of factors which make its prediction difficult. This is extremely relevant to the response of the cryosphere to anthropogenic climate change.

Glaciers erode their bases and sides by abrasion and by plucking of blocks, producing a range of erosional landforms. The debris carried by ice may be transported on the glacier surface, within it, or at its base. The deposits are termed moraine and take on a great variety of forms depending on the precise mode of transport and deposition. The action of meltwater is very important in transferring glacial sediment to the fluvioglacial system. Meltwater discharges are typically highly concentrated in sediment, and spread their sediment load over braided and anastomosing plains known by the Icelandic term *sandar*. There is conclusive evidence from the Pleistocene of catastrophic meltwater drainages of glacial lakes caused by dam bursts which have essentially instantaneously reshaped the downstream terrain by erosion and deposition. Deposition at marine ice sheet margins where icebergs calve from the ice sheet includes a distinctive sedimentary facies of ice-rafted diamictite, containing large pebbles and boulders which have been dropped from floating ice into muddy sea bed sediment.

A case study of the Himalayas of north-western India illustrates the importance of glaciers in the sediment routing system. Geomorphic activity is zoned according to altitude and controlled to a large extent by a moisture stream from high- to low-altitudinal zones rather than by local climates. Consequently, hydrological activity in the semi-arid upper Indus valley is determined by glacial and paraglacial processes high in the Karakorum. Secular changes in climate are recognized by altitudinal migration of the different geomorphic zones.

11.1 Introduction: the cryosphere

Snow and ice cover large proportions of polar regions and terrains at high elevations in all latitudes, comprising $14.9 \times 10^6 \, \text{km}^2$, about 10% of the Earth's land surface. Of this area of ice, the vast majority is found in the Antarctic ice sheet (Plate 11.1, facing p. 204), and to a lesser extent in the Greenland ice sheet. As a percentage, very little ice makes up the valley glaciers that characterize a number of the Earth's mountainous regions such as the European Alps, Alaska, North America, or the Himalayas of Asia (Table 11.1). We know, too, that during the relatively recent past in the Pleistocene, glaciers and ice sheets were far more extensive (Chapter 2). At 18 000 BP, the area covered by ice is thought to have been as high as 30% of the Earth's land surface. This has profound significance for today's landscapes, since landforms in many regions are a legacy from a glacial past rather than reflecting present-day conditions.

The glacial system receives inputs from precipitation in the form of snow, and mineral debris from the erosion of the glacier bed and sides and from erosion of surrounding hillslopes. The glacial system suffers losses caused by solar radiation inducing melting, and by the deposition and efflux of sediment and solutes. The development and fluctuations of glacial systems are therefore controlled primarily by factors determining the rate of precipitation on the glacier and the solar radiation received, together with their seasonal distribution. These factors include latitude, distance from ocean, atmospheric circulation, topography and aspect. The causes of gains and losses in the glacial system are therefore varied and responses complex. For example, it does not strictly follow that global warming will cause glaciers to shrink. If global warming is associated with increased storminess and precipitation as snow over polar ice caps, they may expand. In addition, the response of a glacial system to a forcing mechanism must take account of the characteristic response time of the system to a perturbation. In the case of a small Alpine glacier, this may be very short, measured in tens of years, but for a large ice sheet such as that of Antarctica, it may be measured in thousands of years.

Glaciers fall into two basic types.

1 Ice sheets and ice caps are unconstrained by topography, with typically convex-up profiles. They

Table 11.1 The present distribution of ice. After Flint (1971) [2].

Region	Approximate area km²	% of total
North polar area		
Greenland ice sheet	1 726 400	11.59
Queen Elizabeth Islands	106 988	0.72
Other Greenland glaciers	76 200	0.51
Spitsbergen and Nordaustlandet	58 016	0.39
Other	114 012	0.77
Subtotal	2 081 616	13.97
Continental North America		
Alaska	51 476	0.35
Other	25 404	0.17
Subtotal	76 880	0.52
South American Cordillera	26 500	0.18
Europe	9 276	0.06
Asian continent	115 021	0.77
African continent	12	0.08×10^{-3}
Australasia	1 015	0.007
South polar area		
Antarctic ice sheet (excluding ice shelves)	12 535 000	84.14
Other glaciers on Antarctic continent	50 000	0.34
Sub-Antarctic islands	3 000	0.02
Subtotal	12 588 000	84.49
Total	14 898 320	100

flow radially from their centres and commonly generate outlet glaciers where the ice flows out through basement depressions, or ice shelves where a floating ice cap forms at the interface with the ocean.

2 Glaciers constrained by topography include ice fields and cirque glaciers ponded in bedrock hollows, and valley glaciers occupying pronounced bedrock valleys with fringing hillslopes.

The efficacy of glacial processes in modifying the landscape has been debated vigorously for over a hundred years. Travellers in the Swiss Alps had long ago noted the milky white meltwater discharges from glaciers clouded by the presence of suspended fine sediment. Data on the sediment loads of streams flowing from mountain glaciers initially suggested that rates of glacial erosion were higher than in equivalent non-glaciated areas [3]. Others [4,5] have challenged this assertion of enhanced rates of erosion associated with glaciers, urging caution in the use of sediment load data. A comparison of sediment yields from glacierized and non-glacierized catchments of equivalent climate, bedrock geology and tectonic

setting in New Zealand suggested instead that the erosion rates were broadly comparable. These rates may be several millimetres per year.

The erosive power of ice sheets past and present has also been a subject of debate. By estimating the volume of sediment deposited in the sea adjacent to the late Cenozoic Laurentide ice sheet of North America from seismic studies and deep sea cores, it was proposed [6] that the Laurentian shield was eroded at about 0.04–0.07 mm y^{-1} averaged over a 3 My time-span, representing 120–200 m of erosion over the glaciated Canadian shield. Others have suggested that considerably lower amounts of erosion would occur under sluggish ice sheets frozen to their beds and moving over low-gradient terrains [7].

Although descriptions of glaciers date back to eleventh century Icelandic literature, little attention was focused on their dynamics until the eighteenth century when gravitational forces were correctly identified as causing the flow of ice over its bed. Systematic measurements of glaciers in the Swiss Alps in the nineteenth century allowed the elaboration of the surface velocity structure of moving glacier ice. But more recent advances in the last 50 years which allow the flow of ice to be mathematically modelled stem from the treatment of ice as a polycrystalline solid which deforms like many other crystalline solids close to their melting point, such as metals. According to W.S.B. Paterson [8],

> a mere handful of mathematical physicists, who may seldom set foot on a glacier, have contributed far more to an understanding of the subject than have a hundred measurers of ablation stakes or recorders of advances and retreats of glacier termini.

With this in mind, this chapter introduces the fundamental dynamics of ice, before considering the erosional processes, depositional products and role of glaciers in the regional sediment routeing system.

11.2 The dynamics of ice

11.2.1 Introduction

We know that ice is capable of flow from the now commonplace observation of valley glaciers, and perhaps the less widely appreciated movement in ice sheets in Greenland and Antarctica. The relatively slow velocities and accessibility of flowing ice have made it an ideal natural laboratory for the testing of theoretical models of the behaviour of viscous materials. Consequently, a number of detailed treatises exist on the subject [9, 10]. In common with the rest of Part Two, the aim here is to provide enough theoretical background that the essential features of the flow of ice can be explained and simulated. The implications of the flow of ice for the transfer of sediment in cold regions, and for the hydrology of environments fed by the glacial system, will then be examined.

In Chapters 4, 5 and 6 the constitutive laws of Newtonian, non-Newtonian and Bingham plastic materials were introduced and applied to natural flows. Lava flowing down the slopes of a volcano can be approximated by a laminar Newtonian fluid. Most rivers behave as turbulent Newtonian fluids. Highly concentrated dispersions of grains such as grainflows of dry sand down the slip-face of an aeolian dune appear to behave as non-Newtonian fluids. Masses of rock and mud moving down a slope, such as debris flows, are better approximated by a Bingham plastic model. But what about ice? Sections 11.2.2 and 11.2.3 focus on this question.

Ice is not a homogeneous mass with spatially constant properties. It varies in its density and temperature within the glacier.

- Ice forms by the progressive compaction and recrystallization of snow. During this process voids between crystals are progressively removed. The density of ice may vary from 50–70 kg m^{-3} for virgin snow, through about 400 kg m^{-3} in an intermediate stage known as *firn*, to 800–900 kg m^{-3} in fully compacted ice (Fig. 11.1).
- Heat is supplied to the glacier at its surface by solar radiation and by the release of latent heat on refreezing of water. Latent heat production is most important in small glaciers in which percolation of meltwater causes refreezing. The glacier or ice sheet may receive heat at its base by friction and as a basal heat flow from the underlying lithosphere. The temperature structure of the ice body depends on the relative contribution of these different sources.
- The melting or freezing point of ice varies as a function of pressure, decreasing with increasing pressure. Consequently, under very thick ice sheets the ice may melt at temperatures below 0°C. The *pressure melting point*, the temperature at which ice melts at ambient pressure, reduces with an increase in pressure at about 1°C per 14 MPa. The effect is therefore minimal for most valley glaciers, but may be significant under thick ice sheets. Since the pressure under an ice load can be assumed to be $\rho g h$ (where ρ is the density of ice, which increases with depth

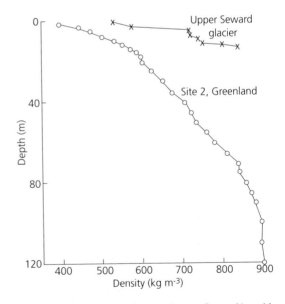

Fig. 11.1 Variation of the density of snow, firn and ice with depth in a temperate glacier (Seward) and the Greenland ice sheet. After Sharp (1951) [11] and Langway (1967) [12].

to about $900 \, kg \, m^{-3}$ when fully compacted; g is $9.81 \, m \, s^{-2}$; and h is the thickness of the column of ice above the point of interest), a thickness of fully compacted ice of 1585 m would be required to cause a 1°C change in the pressure melting point.

Geothermal and frictional heat sources may be capable of causing partial melting of the base of the glacier, producing a mixture of ice and liquid water. This is known as *warm ice*, and the glacier is termed *warm-based*. *Cold ice*, on the other hand, is at a temperature below its melting point and characterizes *cold-based* glaciers. In practice, a glacier may have both warm and cold bases at different places and at different times depending on the heat balance and pressure distribution.

Glaciers may move by two distinct mechanisms.

• *Basal sliding* accounts for almost all of the ice movement in warm-based glaciers. In fluid-mechanical terms, we have a finite velocity at zero height, invalidating the no-slip condition. The precise mechanisms of basal sliding include: slip over a lubricated thin (millimetre-scale) sole of water; melting and re-freezing at high- and low-pressure locations related to small (metre-scale) irregularities along the base of a warm-based glacier, known as *regelation creep*; flow around larger irregularities caused by the high stresses on their up-glacier flanks; and deformation of a water-saturated, sediment-covered bed close to its melting temperature.

• Internal deformation dominates the flow of cold-based glaciers, in which case the no-slip condition applies. In Section 11.2.2 it will be seen that the flow of the ice can be approximated by a flow law which is dependent on both temperature and stress in their effect on effective viscosity. Internal deformation takes place within and between ice crystals by the mechanism of creep.

In addition to the effect on the strain rate of the stress, it is important to consider the temperature dependence of ice if the flow of ice in real glaciers is to be modelled successfully. The essential problem here is that one of the boundaries of the glacier is heated by friction and geothermal supply. We need to know the effect this has on the effective viscosity of the ice and therefore its velocity profile.

Early measurements of the flow of a stream of ice in a valley glacier showed that the velocity profile departed strongly from that predicted by the flow equations for a viscous, incompressible fluid in a channel. It did so in two main ways.

1 The velocity did not become zero at the edge of the flow—that is, the no-slip condition was violated.

Clearly, there was a finite rate of slip between the glacier ice and the containing boundary.

2 The velocity profile a long way from the boundary was less steep than for viscous flow models. In other words, there was more deformation of the ice very close to the boundary and less deformation far from the boundary than predicted from the viscous flow model.

Together with the results of laboratory measurements of the deformation of ice, it became clear that although ice behaved like a fluid (it deforms even at very low levels of stress), it was non-Newtonian, with an apparent viscosity depending on the strain rate. With increasing strain rate, the apparent viscosity of ice decreases. The following pages develop the theory for two constitutive models of the deformation of ice. First, in Section 11.2.2, a rheology where the strain rate (velocity gradient) is dependent on the stress is developed from the simple case of a Newtonian fluid and extended to the non-Newtonian case where the strain rate depends on a power of the stress. The temperature dependence of the flow of a fluid in a channel is then investigated. The second constitutive model of an ideal plastic is applied to the motion of an ice sheet resting on a flat substrate in Section 11.2.3.

11.2.2 Non-Newtonian fluid model
Stress dependence

The viscosity of ice is dependent on its temperature, but also on the stress. It is important to see how this temperature and stress dependence of ice affects its deformation, as shown by its velocity profile across a channel cross-section, and its velocity profile with depth. The simplest starting point is to consider a channel flow of a viscous fluid with a stress-dependent viscosity. We first do this to solve for the velocity across a channel cross-section, and then modify the coordinate system to find the velocity profile with depth. Having satisfied ourselves that the solutions give close approximations to the measured horizontal and vertical velocity profiles in real glaciers, we can proceed to consider the more complex case of a strongly temperature-dependent viscosity.

Consider a channel flow of width h with a centreline at $z=0$ and lateral boundaries at $z=\pm h/2$ (Fig. 11.2). The flow in the channel is driven by a pressure difference p_1-p_0 along the channel of length L. The shear stress in the fluid varies with distance from the confining boundaries dependent on the applied pressure gradient:

$$\frac{d\tau}{dz} = \frac{dp}{dx} = -\frac{p_1 - p_0}{L} \tag{11.1}$$

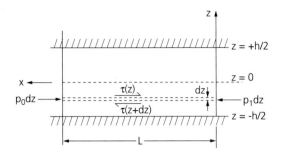

Fig. 11.2 Notation for the flow of a viscous fluid in a channel under a pressure gradient.

For a Newtonian fluid we know that

$$\tau = \mu \frac{du}{dz} \tag{11.2}$$

Consequently, an equation for the velocity can be obtained by combining equations (11.1) and (11.2):

$$\mu \frac{d^2 u}{dz^2} = \frac{dp}{dx} \tag{11.3}$$

which on integration gives

$$\frac{du}{dz} = \frac{1}{\mu} \frac{dp}{dx} z + c_1 \tag{11.4}$$

and integrating again

$$u = \frac{1}{2\mu} \frac{dp}{dx} z^2 + c_1 z + c_2 \tag{11.5}$$

where c_1 and c_2 are constants of integration which can be evaluated by satisfying boundary conditions. The boundary condition $du/dz = 0$ at $z = 0$ can be applied to the first integration, from which it can be seen that $c_1 = 0$. The no-slip condition which states that $u = 0$ at $z = \pm h/2$ can be applied to the second integration, giving $c_2 = (1/2\mu)(dp/dx)(h/2)^2$, and hence

$$u = \frac{1}{2\mu} \frac{dp}{dx} \left\{ z^2 + \left(\frac{h}{2} \right)^2 \right\} \tag{11.6}$$

showing that the velocity profile is a parabola symmetric about the centreline of the channel (Fig. 11.3). This is the solution for a Newtonian fluid. For a power law fluid the velocity gradient or strain rate is related to the shear stress as follows:

$$\frac{du}{dz} = a\tau^n \tag{11.7}$$

where, for the moment, we shall take the coefficient a to be a constant (and investigate its temperature dependence later). Rearranging and substituting this into equation (11.1) gives

$$\frac{1}{a^{1/n}} \frac{d}{dz} \left\{ \frac{du^{1/n}}{dz} \right\} = -\frac{p_1 - p_0}{L} \tag{11.8}$$

Assuming symmetry about the centreline gives the integration

$$\frac{du}{dz} = -a \left\{ \frac{p_1 - p_0}{L} \right\}^n z^n \tag{11.9}$$

Integrating again and then applying the no-slip condition gives

$$u = \frac{a}{n+1} \left\{ \frac{p_1 - p_0}{L} \right\}^n \left\{ \left(\frac{h}{2} \right)^{n+1} - z^{n+1} \right\} \tag{11.10}$$

The mean velocity is

$$\bar{u} = \frac{2}{h} \int_0^{h/2} u \, dz = \frac{a}{n+2} \left\{ \frac{p_1 - p_0}{L} \right\}^n \left(\frac{h}{2} \right)^{n+1} \tag{11.11}$$

and the ratio between the velocity at any point and the mean velocity in the channel is

$$\frac{u}{\bar{u}} = \left(\frac{n+1}{n+2} \right) \left\{ 1 - \left(\frac{2z}{h} \right)^{n+1} \right\} \tag{11.12}$$

Field measurements and laboratory results suggest that for polycrystalline ice the exponent varies between 1.9 and 4.5, with a mean value of $n \approx 3$, and that at temperatures close to the melting point $a \approx 5 \times 10^{-15} \, \text{kPa}^{-3} \text{s}^{-1}$. The value of a is, however, sensitive to temperature, decreasing as the temperature falls below 0°C [13]. This temperature dependence is investigated in Practical exercise 11.1, opposite.

It is now a simple matter to change our coordinate system to investigate the velocity profile with depth and to substitute a downslope component of the weight of ice for the lateral pressure gradient (Fig. 11.4). So let y be measured vertically from the bed to the surface of a body of ice of thickness h on a uniform slope of angle α.

(a)

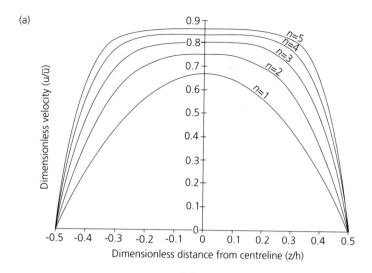

(b)

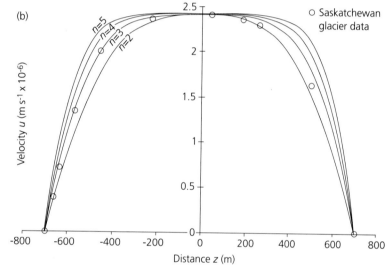

Fig. 11.3 (a) Plan-view dimensionless velocity profiles for a Newtonian fluid ($n = 1$) and for non-Newtonian (power law) fluids ($n = 2$ to 5). (b) Saskatchewan glacier velocity data (after Paterson (1981) [8]), compared with power law predictions where all curves are constrained by the maximum velocity.

Practical exercise 11.1: Flow of non-Newtonian fluids in open channels

1 Plot the velocity of a Newtonian fluid ($n = 1$) compared to its mean velocity in a channel of width h assuming the flow to be symmetrical about the centreline and to be zero at the lateral boundaries. Superimpose on the chart the velocity profiles for non-Newtonian fluids with $n = 2$ to $n = 5$.

2 The Saskatchewan glacier, 1400 m wide, has a horizontal distribution of surface velocity [14] as shown in Table P11.1. Plot the velocity profile for the Saskatchewan data. What value of n best fits the data?

Solution

1 The chart of non-dimensional velocity ($u/\bar{u}$) against non-dimensional distance (z/h) is given in Fig. 11.3a.

2 Figure 11.3b shows velocity profiles for non-Newtonian fluids constrained by the boundary condition that the maximum velocity is 2.41 µm s^{-1}. The best fit for the Saskatchewan glacier data is for $n \approx 3$ for negative values of z and between 2 and 3 for positive values of z. Taking the entire profile, a power of $n = 3$ fits the field data well.

Continued on p. 380.

Practical exercise 11.1: *Continued*

Table P11.1

Distance from centreline (m)	Velocity	
	m y^{-1}	(m 10^{-6}s^{-1})
−660	12	(0.38)
−640	22	(0.70)
−570	42	(1.33)
−460	63	(2.00)
−220	74	(2.35)
40	76	(2.41)
180	74	(2.35)
260	72	(2.28)
500	51	(1.62)

We know that the shear stress in a flow of fluid down a slope is due to the downslope component of the weight of the overlying fluid

$$\tau = \rho g h \sin \alpha \qquad (11.13)$$

or along any plane distance y above the bed

$$\tau = \rho g (h - y)\sin \alpha = \gamma_x (h - y) \qquad (11.14)$$

where $\gamma_x = \rho g \sin \alpha$ is the component of the gravity force acting downslope. Consequently, combining equations (11.2) and (11.14), the constitutive equation for a Newtonian fluid becomes

$$\frac{\gamma_x (h - y)}{\mu} = \frac{du}{dy} \qquad (11.15)$$

and integrating

$$u = \frac{\gamma_x}{\mu}\left(yh - \frac{y^2}{2} \right) + c$$

where $c = 0$ from the boundary condition that $u = 0$ at $y = 0$, which can be expressed

$$u = \frac{1}{2}\left(\frac{\gamma_x}{\mu} \right)\left\{ h^2 - (h - y)^2 \right\} \qquad (11.16)$$

The expression for shear stress in a fluid flow down an inclined plane in equation (11.14) can be modified for the case of a non-Newtonian fluid, where the strain rate, or velocity gradient, varies as a power of the shear stress (equation (11.7)), giving

$$a\left[\gamma_x (h - y) \right]^n = \frac{du}{dy} \qquad (11.17)$$

where for a Newtonian fluid $a = 1/\mu$ and $n = 1$.

To obtain the velocity profile, we need to integrate equation (11.17). Abandoning the no-slip condition and setting the velocity at the bed as u_b, we have

$$u = u_b + \frac{a}{n+1}\gamma_x^n\left[h^{n+1} - (h - y)^{n+1} \right] \qquad (11.18)$$

which for ice becomes

$$u = u_b + \frac{a}{4}\gamma_x^3\left[h^4 - (h - y)^4 \right] \qquad (11.19)$$

Measurements of velocities in glaciers vary greatly.

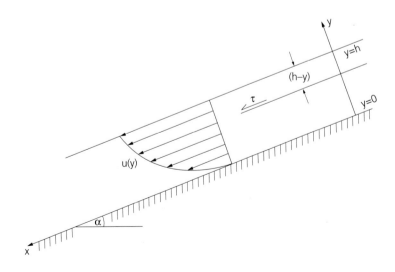

Fig. 11.4 Notation for the flow of a viscous fluid down an inclined plane.

Along a single glacier, the surface velocity has been shown to vary as a function of the surface slope. Most valley glaciers have velocities in the range 10–200 m y^{-1}, whereas the glaciers moving off the Antarctic ice sheet range from 300 to 1400 m y^{-1}. Glacier surges may involve velocities of several kilometres in a few months. The velocity profile, obtained from a borehole through the Athabasca glacier, Canada [8, 15, 16] (Fig. 11.5), is provided in Practical exercise 11.2, p. 382.

Temperature dependent flow of ice

If the deformation of ice has a temperature dependence, it is important to know what the temperature regime of a glacier is. If T_0 is the temperature of the surface of the glacier, and T_1 is a higher temperature of the base because of frictional heating or a geothermal contribution, what are the likely values of the ratio $(T_1 - T_0)/T_0$? In a situation such as the temperature-dependent flow in the aesthenosphere the temperature ratio is likely to be in the region of 0.5. For a glacier, it is likely to be very small, suggesting initially that the flow of ice in natural glaciers is likely to be only mildly affected by the thermal regime. However, this view needs to be quantified.

We have previously noted the different sources of heat in a glacier or ice sheet. The steady-state solution where the temperature along a given fixed vertical plane in the ice does not change with time as the ice continues to flow past is sketched in Fig. 11.6. The temperature profile is seen to be sensitive to the net mass balance b of the ice expressed as a thickness of ice accumulated/added per year. This drives the descent of ice from the surface into the interior of the ice sheet. It is known as *Robin's solution*. At Byrd Station, Antarctica, the surface net mass balance is 0.15 m, the surface temperature –28°C (245 K) and the thickness of the ice sheet about 2500 m. We should expect a temperature difference of 25.5 K between the top and base of the ice sheet. The temperature of the ice at the base should therefore be –2.5°C (270.5 K), which is very close to its melting temperature at the pressure at this depth (pressure melting temperature) of –1.6°C (271.4 K). It is thought that about half of the Antarctic ice sheet is at its pressure melting point at its base. This has important implications for the rheology of the ice sheet.

The temperature gradient at the base of the ice sheet is controlled by the geothermal heat flux, and the temperature itself is determined by the ice sheet thickness. Melting of ice along a thin layer at the base may take place when the surface temperature is

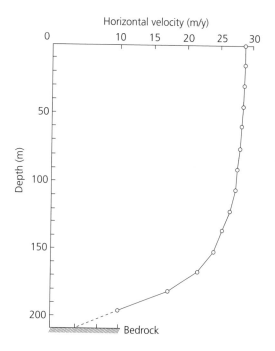

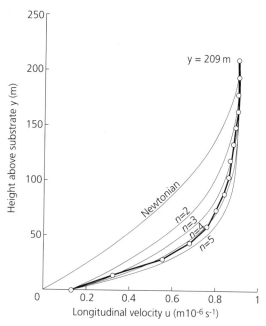

Fig. 11.5 Velocity profile of the Athabasca glacier, Canada, with theoretical velocity profiles for a Newtonian fluid, and for non-Newtonian fluids from $n = 2$ to $n = 5$. After Savage and Paterson (1963) [15].

Practical exercise 11.2: Flow of ice in the Athabasca glacier

The vertical velocity profile measured in the Athabasca glacier is as shown in Table P11.2

1 Plot the Athabasca glacier velocity profile on a graph of distance above the bed against velocity.

2 Calculate and plot the velocity profile for a Newtonian fluid on the same graph. As a reminder (Section 4.2.2), the velocity profile for a laminar Newtonian flow down an inclined plane with the no-slip condition of $u = 0$ at $y = 0$ is

$$u = \frac{\gamma_x}{\mu}\left(hy - \frac{y^2}{2}\right) \qquad (11.20)$$

Constrain this parabolic profile to satisfy the boundary condition that the maximum velocity of $u = 0.90 \times 10^{-6}\,\mathrm{m\,s^{-1}}$ is at $y = h = 209\,\mathrm{m}$.

3 Fit a velocity profile for a non-Newtonian fluid to the Athabasca glacier data with $a = 5 \times 10^{-15}\,\mathrm{kPa^{-3}\,s^{-1}}$. Constrain the velocity profiles to pass through the maximum velocity of $0.90 \times 10^{-6}\,\mathrm{m\,s^{-1}}$. What is the value of the power exponent n for the best-fitting curve?

Solution

1 The measured velocity profile is given in Fig. 11.5.

2 Applying the boundary condition that $u = 0.90 \times 10^{-6}\,\mathrm{m\,s^{-1}}$ at $y = 209\,\mathrm{m}$ gives, by use of equation (11.20), $\gamma_x/u = 41.2 \times 10^{-12}\,\mathrm{m^{-1}\,s^{-1}}$. The

Table P11.2

Depth from surface (m)	Height from base y (m)	Horizontal velocity $\mathrm{m\,y^{-1}}\,(10^{-6}\,\mathrm{m\,s^{-1}})$
0	209	28.6 (0.91)
15	194	28.5 (0.90)
30	179	28.5 (0.90)
45	164	28.4 (0.90)
60	149	28.2 (0.89)
75	134	28.0 (0.89)
90	119	27.7 (0.88)
105	104	27.2 (0.86)
120	89	26.5 (0.84)
135	74	25.5 (0.81)
150	59	24.0 (0.76)
165	44	21.5 (0.68)
180	29	17.5 (0.55)
195	14	10.0 (0.32)
209	0	Sliding velocity of approx. 4 (0.13)

parabolic velocity profile characteristic of a laminar Newtonian fluid is given in Fig. 11.5.

3 To calculate the velocity profiles we use equation (11.18). To solve it for the velocity we need to know the value of γ_x. This can be found once again from the boundary condition for the maximum velocity. The observational data for the Athabasca glacier are best fitted by curves with $n = 3$ to $n = 4$.

insufficiently low to offset the geothermal contribution at the base. Melting may be promoted, for example, by an increase in the thickness of the ice sheet as it flows into a hollow or bedrock valley.

The surface layer of a glacier responds to the seasonal variations in the overlying atmosphere by a conductive heat flow. For an annual temperature change of 10 K, the depth of the upper layer experiencing temperature fluctuations is roughly 15 m. Longer-period temperature changes, such as those due to long-term climate change, would be propagated to greater depths in the ice. In addition, melting near the surface and downward percolation of meltwater to depths at which refreezing takes place releases latent heat which causes significant warming. This can make the temperature of the firn warmer than the mean annual air temperature.

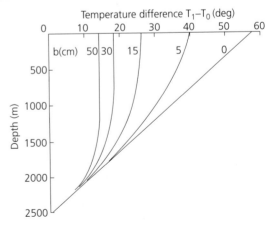

Fig. 11.6 Steady-state temperature distribution in an ice sheet 2500 m thick. The relationship between temperature and depth is shown for various values of the net mass balance (in centimetres of ice). After Paterson (1981) [8].

Measurements of the temperature variation through the Greenland ice sheet revealed the initially surprising result that temperatures decrease with depth from the surface. This negative temperature gradient near the surface has since been found to be a general feature of the Greenland and Antarctic ice sheets. This is not consistent with Robin's steady-state solution. It is most likely that these negative temperature gradients reaching depths of up to several hundred metres are due to the advection of ice from the cold surface layer to depths, so that the advection dominates the effects of heat conduction. Another possibility is that the climate has changed so that surface layers are warmer than older, deeper layers.

The temperature dependence of a (introduced first in equation (11.7)) is given by

$$\exp\left(\frac{-E_a}{RT}\right) \qquad (11.21)$$

where T is the absolute temperature, E_a is the activation energy for creep 133 kJ mol^{-1} [17] or 58.6 kJ mol^{-1} [18] according to different authors), and R is the molar gas constant (8.3144 J K^{-1} mol^{-1}). This formulation of temperature dependence is familiar to chemists as of the same form as the reaction rate given by the *Arrhenius equation*.

The temperature profile of the Greenland ice sheet at Camp Century, located 100 km from the western margin of the ice sheet, is given in Fig. 11.7. The ice is 1387 m thick at Camp Century. The temperature changes from –24°C (249 K) at the surface to –13°C (260 K) at the base. If we take the lower activation energy of 58.6 kJ mol^{-1}, the value of a at 249 K is 30% of its value at 260 K. In other words, the strain rate for a given stress is 30% at the lower temperature of the rate at the higher. This figure is very sensitive to the choice of activation energy. At the higher activation energy of 133 kJ mol^{-1}, the strain rate drops to 7% at the lower temperature for the same stress. It is predicted that, dependent on the temperature gradient in the glacier, the deformation in the ice will be more or less concentrated into a zone close to the base where the temperatures are highest.

Temperature effects on viscosity may be significant in warm ice glaciers with relatively high surface temperatures.

11.2.3 Ideal plastic model

Other authors have suggested that an ice sheet

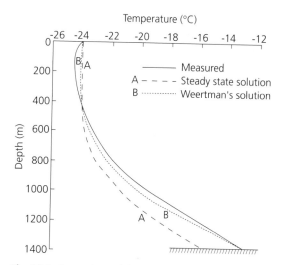

Fig. 11.7 Comparison of the observed and theoretical temperature profiles for the Greenland ice sheet at Camp Century. After Weertman (1968) [19].

may move like an ideal plastic [20] rather than a non-Newtonian fluid. In ideal plastics the velocity distribution is uniform, with the entire flow occupying a rigid plug and shear taking place in an infinitesimally thin layer next to the boundary. Ideal plastics have a finite shear strength but zero viscosity. Comparison with observations on ice sheets and valley glaciers suggest that this shear strength is about 10^5 Pa. Although the idea that an ice sheet moves as a rigid plug with no internal deformation seems quite daft, the approximation to an ideal plastic rheology would hold as long as the flow rate at the base of the ice sheet is very fast compared to the rate at which material (snow) is added to the top.

The thickness of an ice sheet required for the shear strength to be exceeded at its base can be found from the following analysis [16]. We consider a sheet of ice moving with an almost constant velocity over a horizontal surface, so that accelerations can be ignored. The ice sheet has a surface slope which varies along the x direction from its maximum elevation to its margin. The weight of the ice exerts a shear stress on its base τ_0. Across a vertical slice of the ice sheet of width ∂x and unit breadth (into or out of the page, Fig. 11.8), the horizontal force on the bed is simply $\tau_0 \partial x$. This horizontal force must result from the horizontal pressure difference across this vertical slice of ice. The pressure force along one side of the slice is the integration over the thickness of the slice of the pressure forces due to the overlying column of ice. At

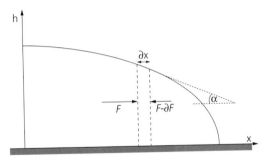

Fig. 11.8 Notation for an ice sheet acting as an ideal plastic on a flat substrate with a surface slope.

the position x, this can be denoted by

$$F = \int_0^h \gamma(h-y)\mathrm{d}y \tag{11.22}$$

which on integration becomes

$$F = \frac{\gamma h^2}{2} \tag{11.23}$$

At the position $x + \partial x$, the pressure force on the vertical plane of the slice must be less than at x because the ice sheet is thinner. The force at $x + \partial x$ is

$$F - \partial F = \frac{\gamma(h - \partial h)^2}{2} \tag{11.24}$$

Since the distance ∂x is very small, the quantity ∂h is also very small and ∂h^2 vanishes. Taking this into account, we have

$$-\tau_0 \partial x = \partial F = \gamma h \partial h \tag{11.25}$$

and dividing by ∂x gives the shear stress on the base of the ice sheet in terms of the local slope and elevation:

$$\tau_0 = -\gamma h \frac{\partial h}{\partial x} = -\gamma h \tan \alpha \tag{11.26}$$

We can think of the deformation of the ice as follows. When the shear stress at the base of the ice sheet is less than the shear strength of the ice, no flow will take place. Snow which builds up on the surface of the ice sheet, however, causes it to thicken. At a critical basal shear stress equal to the shear strength the ice starts to flow. Flow causes the stress to be relaxed, so that in an ideal plastic material the stress is nowhere significantly above the shear strength (yield stress). Continued flow at the base is driven by continuing accumulation on the surface.

Equation (11.26) can be rearranged to give

$$h\,\mathrm{d}h = -\frac{\tau_0}{\gamma}\,\mathrm{d}x \tag{11.27}$$

The quotient τ_0/γ has the units of length, so denoting it by h_0, and integrating

$$\frac{h^2}{2} = -h_0 x + C \tag{11.28}$$

We have the boundary condition that $x = L$ when $h = 0$, so $C = h_0 L$. Consequently, the thickness (elevation) of the ice sheet is given by the simple relationship

$$h = \sqrt{2h_0(L - x)} \tag{11.29}$$

11.2.4 Discharge variations in glaciers

A glacier experiences gains and losses so that we can think in terms of an ice budget or mass balance. The gains result from accumulation of snow on the surface of the glacier. The losses result largely from melting, but calving of icebergs into lakes and seas may also be important. Losses are termed *ablation*. There is therefore a balance between accumulation and ablation over time. This also varies across the surface of the glacier. Over a balance year, the upper part of the glacier appears to experience a net accumulation, whereas the lower part suffers net ablation, with an *equilibrium line* in between where there is a balance. This necessitates a flow of ice from the rear of the glacier to the front. The flow patterns in the glacier must reflect the longitudinal variations in discharge of ice. These come about partly through variations in bedrock topography, and partly by the gross trend down the glacier.

Rates of flow vary greatly between different glaciers. Discharges also vary within a single glacier (Fig. 11.10). In the *accumulation area*, the velocity also increases with distance down-glacier, whereas the reverse trend occurs in the *ablation zone* below the equilibrium line. The characteristic pattern (Fig. 11.10), therefore, is of maximum discharge at the equilibrium line, where the ice is also thickest.

Detailed studies of individual glaciers suggest that surface velocity maxima occur at maxima in surface slope. We also know that there are longitudinal variations in the velocity along a centreline of the glacier, giving a longitudinal strain rate or velocity gradient which can be denoted by $\partial u/\partial x$. In most parts of the accumulation zone this longitudinal velocity gradient is positive, whereas most of the ablation zone has negative velocity gradients. The velocity, in general, is at a maximum at the equilibrium line. We can therefore define an area where $\partial u/\partial x > 0$ as *extending*

Practical exercise 11.3: Dynamics of an ice sheet

1 The Antarctic ice sheet comprises 84.5% of the present-day extent of glaciated areas. From a peak to the north-east of the south pole, its surface descends to the coast in the Norwegian and Australian dependencies. The profile of the Antarctic ice sheet between Vostok and Mirny, 850 km away on the coast, is given in Table P11.3 [21]. Plot the profile of the Antarctic ice sheet, assuming it to be situated on a flat substrate. Assume that it deforms as an ideal plastic. For the ice to flow under Vostok where the maximum thickness of ice (H) occurs, $\frac{1}{2}\rho\,gH^2 = \tau_0 L$, so $h_0 = H^2/2L$. Fit a theoretical curve to this profile.

2 The Scandinavian ice sheet had its centre over the position of the present Gulf of Bothnia, northern Baltic Sea, during the Last Glacial Maximum. At its maximum extent it was approximately 2000 km across in a north–south direction between the North Cape of Norway and the position of Warsaw, Poland. Geomorphological studies suggest that it occupied nearly 7×10^6 km^2 at its maximum extent. What would have been the maximum elevation of the active ice sheet, assuming it to have behaved as an ideal plastic?

Solution

1 The profile is given by equation (11.29) and is sketched in Fig. 11.9. Note that the slope of the surface increases towards the margin of the ice sheet.

The ideal plastic rheology is a satisfactory approximation for an ice sheet.

2 The maximum elevation is at $x = 0$ where it is equal to $(2h_0 L)^{0.5}$. For the Scandinavian ice sheet this gives a maximum height of 4.5 km, not dissimilar to the maximum height of the present-day Antarctic ice sheet.

Table P11.3

Elevation (m)	Distance from Vostok (km)
3550	0
3500	50
3460	100
3400	150
3300	200
3210	250
3070	300
2950	350
2870	400
2730	450
2600	500
2420	550
2300	600
2100	650
1850	700
1650	750
750	800
0	850

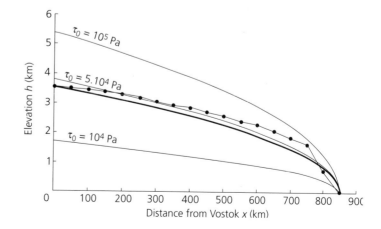

Fig. 11.9 Profile of Antarctic ice sheet between Vostok and Mirny (black dots), together with approximation from ideal plastic theory using the boundary condition at Vostock (heavy curve), and three curves for an ideal plastic with different yield strengths (Practical Exercise 11.3). Data from Vialov (1958) [21].

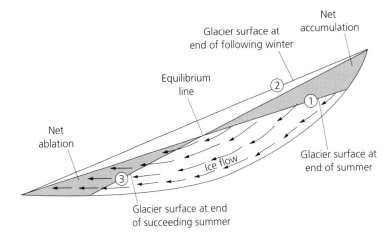

Fig. 11.10 Net annual losses (ablation) and gains (accumulation) over an idealized glacier. Glacier surfaces (1), (2) and (3) show the evolution of the glacier from the end of one summer to the end of the following winter and then to the end of the succeeding summer. The zones of net ablation and net accumulation are separated by the equilibrium line. The net accumulation above this line drives an ice movement down-glacier (arrows) which has a maximum velocity beneath the thickest ice near the equilibrium line. Modified after Summerfield (1991) [23].

flow, and a region where $\partial u/\partial x < 0$ as *compressive flow* [22]. Since ice is incompressible, extending and compressive flow must result in changes in the height or width of the ice. If lateral valley walls prevent changes in width, the ice must change its thickness in response to extending and compressive flow. For compressive flow, there should therefore be an upward component to the velocity vector of the ice. Consequently, below the equilibrium line there is an emergence velocity which is counteracted by ablation. On the other hand, above the equilibrium line in the area of extending flow, there must be a downward component to velocity vectors. There is thus a net downward movement of ice in the net accumulation zone, and a net upward movement (relative to the surface of the ice) in the area of net ablation. This is clear from the emergence from the ice in the lower parts of glaciers of debris previously carried at deeper levels.

The mass balance of a glacier changes with different rates of accumulation and ablation. This is reflected in variations in the extent of glaciers and ice sheets, but the processes involved and their response times are complex. There are two reasons why the response of the glacial system to a forcing mechanism, such as climate change, is difficult to judge. First, the dynamics of a moving stream of ice determines the magnitude and response time to a change in a forcing mechanism. Second, individual glaciers might be expected to react differently to the same change in forcing mechanism because of their particular size, steepness and velocity distribution.

A change in a forcing mechanism such as climate can be related to a change in the mass balance of a glacier. Climate change may also affect ice temperature and amount of meltwater, both of which may affect ice velocity. If we restrict the analysis to temperate valley glaciers, we can safely assume that changes in mass balance dominate. If mass is conserved, and changes in the width of the glacier are negligible, a climatically driven increase in the mass balance $b(x,t)$ must result in an increase in the thickness $h(x,t)$ of the ice and an increase in discharge $Q(x,t)$. Such mass balance changes are propagated down the glacier in the form of a *kinematic wave*. That is, a given property of the glacier, in this case a given discharge, travels down the glacier at a different speed than (faster than) the velocity of the ice itself. The bulge of high discharge associated with a kinematic wave tends to flatten because the increased ice thickness causes longitudinal variations in the discharge which diminish the amplitude of the wave. This is termed *diffusion*. It substantially slows the response times of glaciers to perturbations in mass balance.

The response times of valley glaciers to a change in mass balance are of the order of hundreds of years, and the advance and retreat of a glacier front may lag behind a periodic variation in mass balance (with a period of a few hundred years or less) by roughly a quarter of a period. These response times are largely determined by the speed and diffusivity of kinematic waves, steep, fast-flowing glaciers reacting most quickly to a perturbation. It is important to emphasize that present-day rates of advance and retreat of glaciers and ice sheets are very little to do with current climatic variations. Our grandchildren will experience first-hand the variations in glacial extent driven by today's global climate changes. The response time for large ice sheets such as those of Antarctica and Greenland is of the order of thousands of years.

Some glaciers sporadically or periodically increase their flow rate by a factor of 10 or 100 in so-called *glacier surges*. Such increases may be associated with rapid glacial advance, or ice thickening in the snout region. Advance rates of metres per hour have been measured during glacier surges. The triggering mechanism for glacier surges is poorly understood. It may be due to the build-up of high pore pressures at the base of the glacier, assisted by the presence of easily deformable subglacial sediments. Higher than normal rates of outflow of meltwater during surges support a mechanism such as this. What causes the sudden or periodic nature of the phenomenon is currently being debated.

Fast-flowing narrow zones of ice movement have been recognized in the Antarctic ice sheet. These ice streams move much faster than the adjacent ice mass and commonly feed outlet glaciers. Their accelerated movement may be related to variations in the nature of the base of the ice sheet. In west Antarctica icestream B appears to be moving over a saturated and easily deformable substrate [24].

11.3 Sediment transport by ice

What role do glaciers play in the sediment routeing system and what implications do glacial processes have for more downstream components of the sediment routing system? The best way of appreciating this is to look at a sediment routing system fed by a glaciated source region—the upper Indus of northern Pakistan. This section outlines the main processes of glacial erosion and deposition, before presenting a case study where the impact of glaciers on a sediment routing system can be assessed.

11.3.1 Glacial erosion

Sediment is provided to glaciers by erosion. This erosion may take place at its base and sides in direct contact with ice by a number of processes, the most important of which is the *abrasion* of bedrock by debris carried along by the ice. Factors important in determining rates of abrasion are the concentration and hardness of debris in the basal ice, the effective normal pressure at the base of the glacier, and the rate of basal sliding. The effective normal pressure is that due to the weight of the overlying overburden less, in the case of warm-based glaciers, the basal water pressure. The rate of erosion first increases with

effective normal pressure, then decreases as the high normal pressures cause debris to be pressed on to the bed rather than dragged along it. Consequently, maximum abrasion rates are expected to be found at intermediate effective normal pressures, with the maximum rates occurring at progressively higher pressures as the velocity of ice flow increases (Fig. 11.11). Abrasion under cold-based glaciers is likely to be minimal because of the negligible rate of basal sliding.

Glacial erosion also takes place by a plucking of blocks from the bedrock surface, often along pre-existing joint planes, or along fractures caused by frost weathering under warm-based glaciers.

Landscapes eroded by glacial action possess a wide range of distinctive features attributable to the movement of ice and the interaction with periglacial processes. Such a large geomorphological subject cannot be dealt with in any detail here, and the reader is referred to the large number of physical geography texts dealing with glacial and periglacial landforms.

Unconfined ice flow in ice sheets produces erosional landforms oriented in the direction of the ice movement. Examples are the asymmetrical streamlined forms known as *roches moutonnées*, or the similar but larger erosive forms, up to a few hundred metres in length, known as *whalebacks*. The markedly asymmetrical nature of *roches moutonnées* suggests that the long upstream flank is smoothed by subglacial abrasion, whereas the shorter, steeper and irregular downstream flank is formed by frost wedging and block removal by glacial plucking. Flow of ice within a channel results in steep-sided glacial troughs, truncating the spurs of side valleys and commonly leaving

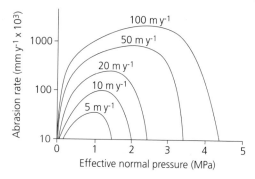

Fig. 11.11 Relationship between the theoretical abrasion rate and effective basal pressure for various values of ice velocity. Actual abrasion rates will be strongly affected by other factors such as the hardness of the bed of the glacier and the concentration of subglacial debris. After Boulton (1974) [25].

tributary valleys stranded high above the floor of the glacial trough, known as *hanging valleys*. Bowl-shaped depressions known as *cirques* are commonly found at the heads of deep glacial valleys. They are formed by the accumulation of firn in a depression which is then gradually enlarged by the action of frost weathering, creep caused by freeze–thaw processes (*gelifluction*) and meltwater action at the base of the snow patch, a combination of processes termed *nivation* [26]. Eventually the accumulating firn turns to ice which then flows from the nivation hollow, abrading its base. Such deepening steepens the headwall at the back of the cirque, causing it to erode by mass wasting.

11.3.2 Glacial deposition
Subglacial, englacial and supraglacial debris
The debris transported by ice is typically very poorly sorted, ranging in size from silt to large boulders. It travels on the surface of the glacier, within it, and at its base (Fig. 11.12).

Supraglacial debris accumulates on the surface of the glacier by falling, sliding or flowing from adjacent hillslopes. Supraglacial debris is therefore common in valley glaciers, but almost absent from ice sheets. Supraglacial debris deposited in the accumulation area above the equilibrium line descends into the ice mass to become *englacial debris*. Its fate is then varied. It may be carried by ice downslope and reappear in the ablation zone below the equilibrium line. It may travel all the way to the snout of the glacier. Or it may descend to the base of the glacier to become a portion of the *subglacial load*. Other subglacial material is derived from plucking and abrasion of bedrock. Subglacial debris is commonly dumped at the snout, but may be returned to the main body of the glacier as englacial material in zones with a component of

upward flow, as in zones of compressive flow (Section 11.2.5).

Moraine
Sediment carried by the glacier can be deposited in a number of settings (Fig. 11.13), some directly under the ice or from the surface of the ice, and some at the ice margin. Processes within the meltwater system are discussed in brief below. The mechanics of sediment transport and deposition in rivers in general, and the estimation of palaeodischarges from meltwater streams, are discussed in more detail in Chapter 5.

- Subglacial deposition results from melting of the overlying ice in active or inactive, warm-based glaciers; by the smearing along the base of active, predominantly warm-based, generally thick glaciers of *basal lodgement*; or by the flow of subglacial sediment into basal irregularities under warm-based glaciers.

- Supraglacial deposition takes place primarily through melting of the glacier surface, causing any debris to be dropped to the base. This is most marked near the snout where ablation rates are highest. Supraglacial sediment may also flow down the surface of the melting glacier as an ice–sediment mixture. This is also concentrated near the snout where most melting takes place.

- Deposition at the ice margin occurs by dumping of supraglacial and englacial material which has been transported as on a conveyor belt to the terminus of active glaciers; and by the pushing of debris by an actively advancing cold- or warm-based glacier and then dropping of the material when the advance stops.

The three-dimensional deposits of ice are known as *moraines*, and are composed of *till*. For more information, the reader is referred to the texts listed in the Further Reading section at the end of this chapter,

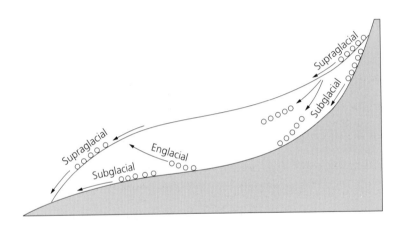

Fig. 11.12 Typical routes of supraglacial, englacial and subglacial loads in valley glaciers. After Summerfield (1991) [23].

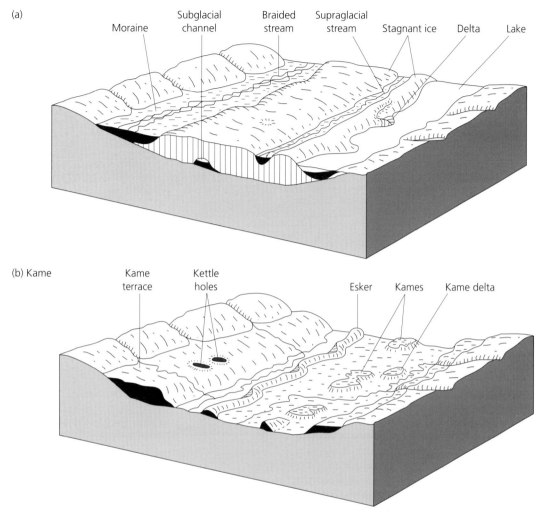

Fig. 11.13 Types of moraine and fluvioglacial environments and deposits (a) at a late stage of glacial activity and (b) after deglaciation. After Flint (1971) [2].

which have been selected from a rich literature on glacial erosion and deposition.

Sheets of subglacial till known as *ground moraine* are fashioned by both erosive and depositional processes, with a fabric determined by the flow direction at the base of the ice. *Drumlins* represent the most characteristic and most intensively studied depositional feature of ground moraine. These elongate features are bluntly rounded on the upstream margins and taper to a point in the downstream direction. They are probably formed by local patterns of ice movement associated with an initial bedrock irregularity.

Some moraines represent deposition on the ice surface or at its margin. *Lateral moraines* accumulate from frost-weathered debris which falls from hillslopes on to the ice margin. *Medial moraines* are formed by the confluence of tributary glaciers each carrying lateral moraines. *End moraines* formed at the glacier or ice sheet margin may be complex mixtures of supraglacial and englacial debris, lake sediments and fluvioglacial outwash. End moraines may be particularly thick where ice margins are relatively static, and can be used to map the maximum extent of glaciation. Some moraine at the ice margin may originate from bulldozing of material by active ice producing ridges oriented at right angles to the mean ice flow direction.

Fluvioglacial deposits

The action of glacial meltwater is very important in the transport of sediment through the glacial system and into the fluvial system beyond. We have also seen that meltwater is important in glacial dynamics, especially in its role in the rate of basal sliding. The movement of meltwater depends to a large extent on the thermal regime of the glacier (Fig. 11.14).

• Meltwater movement in cold-ice glaciers is restricted to the surface of the ice, increasing is discharge below the equilibrium line (Fig. 11.14a). Surface runoff may take place in extensive drainage networks.

• Warm-ice glaciers have both surface drainage and internal routes of meltwater movement. Water commonly permeates the warm ice at crystal boundaries, and drains down through tunnels and cracks, to reach a water table within the glacier. Subglacial and englacial lakes and water pockets form in this zone and discharge meltwater to the glacier margin (Fig. 11.14b). High-altitude glaciers in equatorial latitudes may be composed of warm ice entirely above the water table, in which case meltwater percolates through the ice mass but is unable to accumulate as englacial water pockets or lakes.

• Glaciers with cold ice in the ablation zone and warm ice in the accumulation zone (Fig. 11.14c) may be dominated by internal drainage in the zone of warm ice, and by surface runoff in the zone of cold ice.

The discharge of meltwater from the glacier varies on a number of frequencies. There are high-frequency variations reflecting daily temperature variations and short-term meteorological effects. At somewhat longer time-scales, there are seasonal variations between winter and summer, with meltwater outputs peaking in the early summer. At longer time-scales high meltwater discharges are associated with glacier surges and with the catastrophic drainage of glacial lakes following dam bursts, known as *jökulhlaups*. Finally, there are variations in meltwater discharges through periods of glacial advance and retreat driven by climatic factors, up to the frequency of glacial–interglacial cycles (of the order of 10^5y) in the Pleistocene.

Meltwater drainage is capable of both erosion and deposition. Meltwater commonly transports high quantities of bedload, suspended load and solute load, as attested by their typical greyish-white colour. Much of the sediment load is derived by glacial abrasion. The swift-flowing, sediment-charged waters

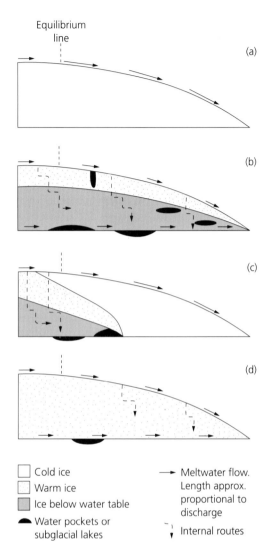

Fig. 11.14 Meltwater routes in glaciers depending on their thermal regime. (a) Cold ice. (b) Warm ice throughout with an elevated water table. (c) Cold ice in the ablation zone and warm ice in the accumulation zone. (d) A high-altitude glacier in equatorial latitudes. After Sugden & John (1976) [27].

are potent agents of channel erosion. Their high proportions of bedload and steep slopes encourage braided and anastomosing channel networks to be developed in the proglacial environment. Extensive proglacial depositional plains are known as *sandar* (singular *sandur*).

Subglacial (and englacial) channels are commonly restricted in their lateral movement by the ice tunnels in which they occur. They give rise to ridges of sediment elongated along the axis of the glacier, known as *eskers*. Supraglacial stream deposits and englacial sediments formed in pockets may be gradually lowered during melting, producing mounds and terraces known as *kames*.

There is convincing evidence for the repeated catastrophic drainage of large proglacial lakes during the Pleistocene. Although these events are increasingly recognized from around the world, the best known are those of the north-west USA [29]. Lake Missoula was a large proglacial lake (volume up to 2000 km^3) which occupied a large region in the north-west of Montana at the margin of the Laurentian (Rocky Mountain) ice sheet. Outbursts of the lake took place repeatedly by the failure of a natural dam in the north-west of the lake (Fig. 11.15), causing peak catastrophic discharges of over $20 \times 10^6 \text{m}^3\text{s}^{-1}$ between 16 and 12 ka. To place this in context, the annual discharge of water of the world's highest discharge river, the Amazon [30], is

6300 km^3, or $0.2 \times 10^6 \text{m}^3\text{s}^{-1}$. The short-lived terror of the Lake Missoula outbursts is almost unimaginable. The floods carved deep canyons, rock pools and rock basins, and deposited enormous ripples whose dimensions and orientation record the ferocity and direction of the flood waters. The ripples are typically about 5 m high, of the order of 100 m in wavelength, and are composed of coarse gravel. The impact of the essentially instantaneous discharge of tens of cubic kilometres of fresh water into the global ocean reservoir is also worth contemplating.

Marine ice sheet margins

When ice bodies reach the sea they may take on two forms.

1 The ice terminus may be grounded below sea level, the line along which icebergs calve off (*calving line*) coinciding with the seaward limit of grounded ice. This applies to tidewater valley glaciers and marine ice sheets. Coarse sediment accumulates close to the *grounding line*, with some finer material being

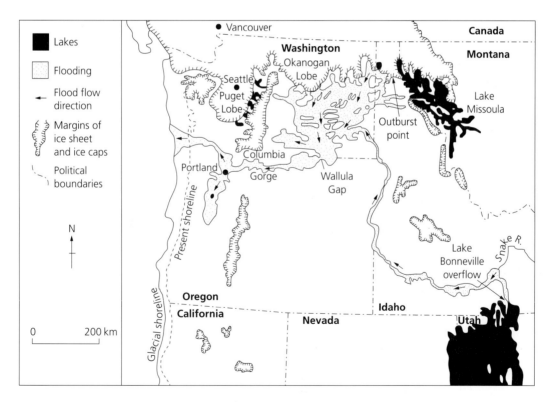

Fig. 11.15 Location of the catastrophic outburst of glacial Lake Missoula, Montana and Washington, and of the catastrophic overspills from Lake Bonneville, Utah and Idaho. After Baker & Bunker (1985) [29].

transported further from the ice margin in meltwater plumes, and as slumps and debris flows down the proximal subaqueous slope.

2 The ice terminus may take the form of a floating *ice shelf* which extends far beyond the grounding line. Icebergs calve off from the outer edge of the floating ice shelf. Melting of the base of the ice in the floating shelf causes the progressive dropping of debris to the sea bed in front of the grounding line so that at the ice terminus only supraglacial debris is present.

A number of additional factors control glaciomarine environments and processes. One is the amount of meltwater generated, which is high in temperate glaciers, moderate in subpolar settings, and low in arid polar settings. High amounts of meltwater favour the far-field distribution of fine sediment in meltwater plumes, as is seen in glacier-fed fjords. A second factor is the sea bottom relief; narrow fjords are typified by low wave and tidal energies and the occlusion of intruding ocean currents which limits the amount of reworking of glacial material. The steep glacial troughs act as traps for sediment, fed by meltwater flows and by mass transport from flanking slopes. A third factor is the tidal, wave and ocean current regime of the receiving water body; in contrast to protected, deep glacial troughs, shallow marine, exposed margins may be subject to a great deal of wave reworking, and scour by tides and ocean currents. These currents control iceberg tracks as well as surface and bottom current activity.

The poorly sorted and generally poorly stratified marine sediments formed by deposition of debris directly carried by ice are called *ice-rafted diamictites.* They commonly have large pebbles or boulders which have been dropped (*dropstones*) into finer-grained, muddy marine sediments. Blankets of ice-rafted diamictites tens of metres (and up to 200 m) thick occur in the Late Cenozoic of Alaska, Spitsbergen and Russia, and comprise 90% of the glaciomarine sediments of the Ross Sea shelf, Antarctica.

11.3.3 Altitudinal zonation of surface processes in glaciated landscapes

Glacial processes are part of a range of erosional and depositional processes which act in an altitudinally organized system in mountainous regions such as the Himalayas [31]. To understand the role of glaciers in the sediment routing system, we need to understand its context within this broader altitudinally zoned picture. Rather than attempting a comprehensive treatment of glacial and paraglacial surface processes, this chapter concludes with case studies of regions where denudational processes are accelerated and altitudinal zonation particularly marked. The example of the upper Indus valley and fringing peaks of the Karakoram and Nanga Parbat provides information on the context of glaciation within the regional geomorphological setting, and of the dynamics of individual steep, erosive glaciers such as the Raikot.

The Karakoram Himalayas lie across the upper Indus valley to the north of the Nanga Parbat (Fig. 11.16). Glacial ablation zones and adjacent steep valley walls comprise a large percentage of the central Karakoram. The high and steep topography and deep valley incision combine to make it a region of exceptional erosional activity. The hypsometry of the Karakoram (Fig. 11.17) shows it to comprise:
• a lower arid to semi-arid region (80–350 mm precipitation) below 3000 m, which coincides with a zone of predominantly fluvial processes characterized by deep valley incision in gorges, making up less than 25% of the region;
• a subhumid to humid zone (350–800 mm precipitation) between 3000 and 6000 m, which contains most of the mountain ridges and glaciated surfaces, occupies the bulk of the remainder of the region;
• a very small (less than 2%) humid (1000–1800 mm) zone composed of the very highest ridges and peaks above 6000 m.
The variation of geomorphological processes in the Karakoram is fundamentally controlled by the passage of a 'moisture stream' from the humid highlands to the arid lowlands through the altitudinal zones (Fig. 11.17).

The bulk of precipitation in the Karakoram is snowfall. Solar radiation is the main control on snow and ice ablation, with little runoff where temperatures remain below freezing for most or all of the day. However, an altitudinal band of diurnal freeze–thaw cycles moves through the Karakoram with the seasons, reaching heights greater than 6000 m in the summer, and occupying the lower fluvial zone below 3000 m in the winter. This causes temporary melting which can result in considerable geomorphic activity.

The moisture stream has a far greater impact on geomorphic processes than direct precipitation. This moisture stream has the following internal dynamics.
• Above about 5500 m the landscape is dominated by all-season avalanching of snow and ice and by the accumulation zones of glaciers. Both avalanche-fed glaciers ('Turkistan type') and firn-fed glaciers ('Alpine type') occur in the Karakoram. This is an extremely effective mechanism of fluxing moisture to the altitudinal zone below 5000 m.

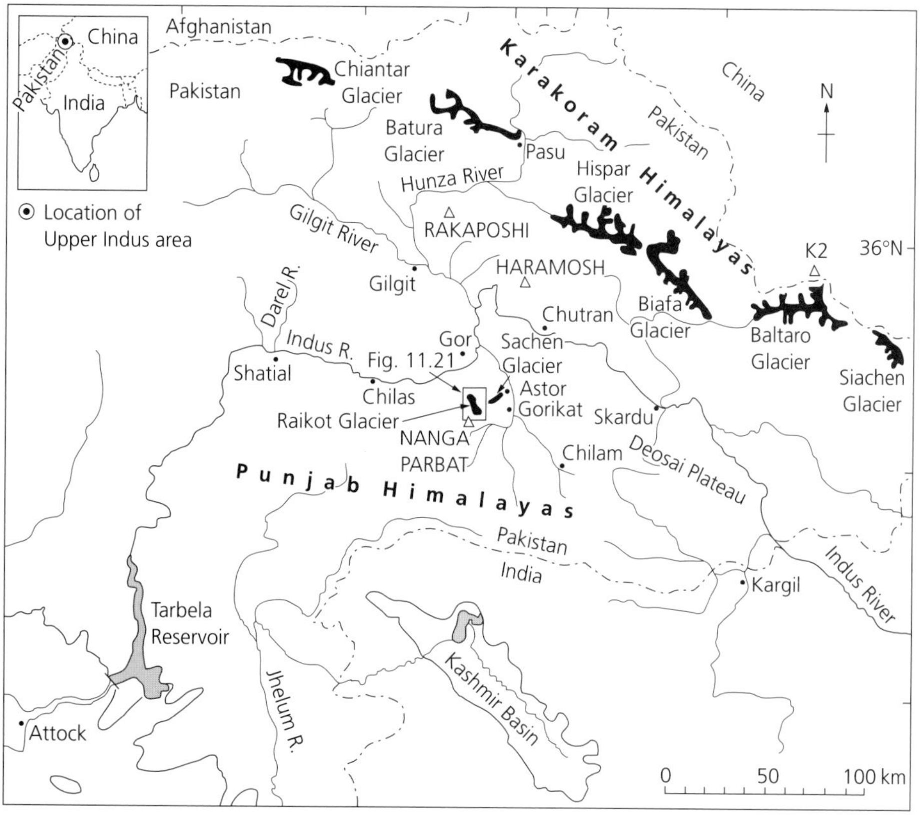

Fig. 11.16 Location map of the upper Indus between the Nanga Parbat and Karakoram in northern Pakistan. The largest modern glaciers and major rivers are shown.

- In the zone between 3500 and 5000 m precipitation is only moderate, but the ablation of avalanched and glacier ice provides a major source of runoff, concentrated into a period of intense melting during 6–10 weeks of the summer. Meltwater derived from avalanched snow is responsible for widespread debris-flow and mudflow activity below about 4000 m.

- The valleys below 3000 m are subjected to prolonged periods of drought. Few glaciers descend far into this zone, and nearly all of the region is in moisture deficit. Apart from aeolian activity, geomorphic processes are almost entirely dependent on snowmelt streams and mass movements from higher altitudes. Indeed, most of the tributaries of the upper Indus emanate from glacier basins. The dynamics of this 'paraglacial' zone is therefore controlled to a very large extent by processes in the higher-altitude zones of the Karakoram. Pleistocene and Holocene sedimentary deposits are currently

being reworked, particularly during high-magnitude dam burst floods.

The most substantial depositional forms in the heavily glacierized Karakoram between 3000 and 5500 m comprise a thick depositional wedge along the margins of glaciers in their ablation zones, consisting mostly of avalanche and massflow deposits, kame terraces, meltwater stream and lacustrine deposits and minor lateral moraines. These complex depositional forms have been termed *ablation valleys* (Fig. 11.18). This zone has several preserved major landslides, some of which must have completely buried the glacier, and a link between catastrophic landslide activity and glacier surging has been proposed. Near the glacier termini between 3000 and 4000 m lodgement and lateral moraines dominate and are subject to frequent landsliding. The lower precipitation rates here, however, restrict hillslope erosion to minor mudflows at springs, minor rockfalls and hillslope wash.

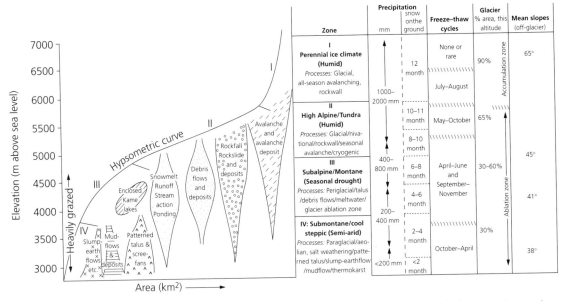

Fig. 11.17 Geomorphic processes in the central Karakoram shown in relation to altitudinal zones. The hypsometric curve is for the Barpu Bualtar basin. After Hewitt (1993) [31].

Secular changes in climate would cause these altitudinal zones to migrate in concert with changing temperature and precipitation patterns. Evidence of such shifts is found today in geomorphic forms which belong to another altitudinal zone. For example, in the semi-arid fluvial zone below 3000 m river terraces are cut into thick glaciofluvial and glaciolacustrine deposits which record a former more extensive glaciated zone. But even in this zone today, the geomorphic activity is largely determined by the processes taking place in the glaciated higher terrains of the Karakoram.

11.3.4 The sediment budget of the Raikot glacier

Raikot glacier is one of the largest of a number of glaciers on the north slope of the Nanga Parbat Massif (8125 m) in the upper Indus basin of northern Pakistan (Fig. 11.19). Vigorous glaciers such as the Raikot are driven by high accumulation rates, steep topographic gradients and high rates of water loss in the ablation zone, all of which contribute to very high sediment concentrations in glacier meltwaters. This glacier is therefore an important component in the regional denudation-sediment routing system.

Raikot glacier is at present just 14 km in length, but from head to terminus drops a staggering 4600 m (Fig. 11.19). In the late Pleistocene (it was at a maximum at 140 ka) it provided sediments directly to the Indus Valley [32]. Now its terminus is situated 15 km from and 2000 m above the Indus. High accumulation rates feed a high discharge of ice across the equilibrium line to the ablation zone ($70 \times 10^6 \, m^3 \, y^{-1}$ water equivalent). High ice velocities ($900 \, m \, y^{-1}$) are found in the glacier at the base of a high-gradient icefall (4100–5500 m) during midsummer, decreasing strongly down-glacier, suggesting compressive flow in the lower section of the glacier.

The sediment load of the glacier has been estimated from the distribution and thickness of supraglacial debris, and from concentrations of englacial debris. Within 1.6 km of the terminus the glacier is completely covered with debris up to 1.6 m in thickness, with greatest accumulations along the glacier margins. Supraglacial debris extends up to the icefall region as lateral moraines. The debris cover retards ablation and thereby allows ice movement to lower altitudes. Over $2 \times 10^6 \, m^3$ of surface debris are calculated to be present [33]. Estimation of the englacial load is made difficult by the strong banding of sediment concentrations, but an estimate of nearly $2.5 \times 10^6 \, m^3$ for the total englacial load suggests that it is comparable to the supraglacial load. This combined sediment load originates in different ways.

- Basal erosion may be important in the cold-based sections in the accumulation and upper ablation areas.
- A contribution through the admixture of debris with snow and ice avalanching from the north face

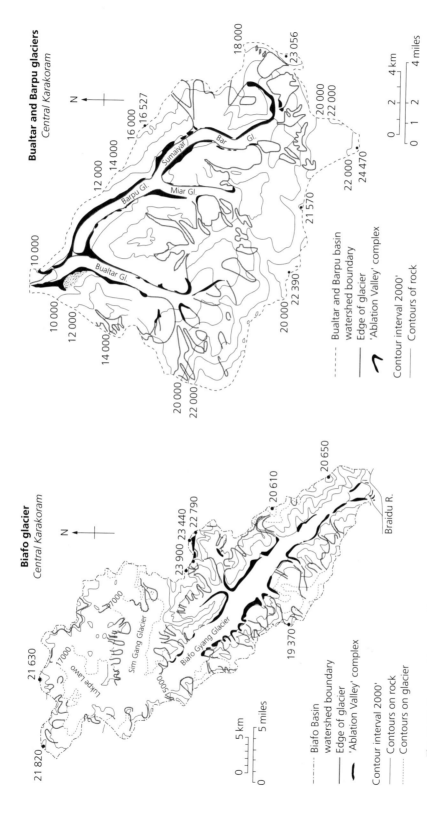

Fig. 11.18 Topographic maps of (a) the Biafo and (b) Bualtar and Barpu glaciers to show the extent of deposits in 'ablation valleys'. After Hewitt (1993) [31].

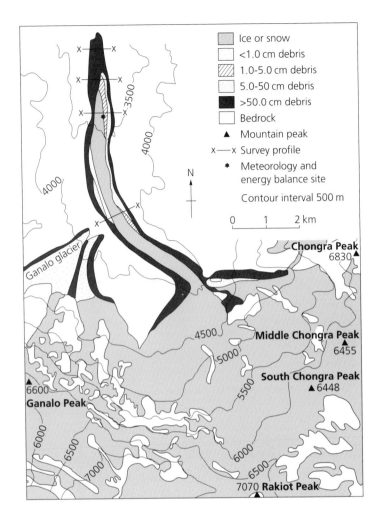

Fig. 11.19 Location of Raikot glacier, Nanga Parbat, with thickness of debris cover. After Gardner & Jones (1993) [33].

of the Nanga Parbat: above the equilibrium line this becomes englacial load; below it, it joins the supraglacial load.

• Most supraglacial material originates from collapse of lateral moraines and nearby hillslopes on to the upper part of the ablation zone.

There is little morainic material immediately below the present-day Raikot glacier, suggesting that the bulk of the sediment delivered to the terminus is disposed of efficiently by meltwater. In the past, however, the situation must have been different. Large thicknesses of glacial sediments preserved in the upper Indus valley indicate a high transfer of glacial material to the immediate proglacial area.

The transport of sediment within the glacial system can also be estimated from mass balance considerations. The ablation component of the mass balance (excluding evaporative losses) allows an estimate of meltwater yield which, when combined with meltwater sediment concentrations, allows the sediment yield to be calculated. The denudation rate thus calculated for the Raikot basin of about 2 mm y^{-1} indicates that the Raikot glacier and its meltwaters are powerful mechanisms for denudation in this part of the Himalayas and is of the same order as the estimated tectonic uplift rate of rock.

Further reading

D. Drewry (1986) *Glacial Geologic Processes.* Edward Arnold, London.

W.S.B. Paterson (1981) *The Physics of Glaciers*, 2nd edn. Pergamon Press, Oxford.

D.E. Sugden & B.S. John (1976) *Glaciers and Landscape.*
Edward Arnold, London.

References

1 E. Reclus (1881) *The History of a Mountain.* Sampson
   Low, Marston, Searle and Rivington, London, p. 121.
2 R.F. Flint (1971) *Glacial and Quaternary Geology.*
   Wiley, New York.
3 C. Embleton & C.A.M. King (1968) *Glacial and
   Periglacial Geomorphology.* Edward Arnold, London.
4 J.T. Andrews (1975) *Glacial Systems—an Approach to
   Glaciers and their Environments.* Duxbury Press, North
   Scituate.
5 D. Drewry (1986) *Glacial Geologic Processes.* Edward
   Arnold, London.
6 M. Bell & E.P. Laine (1985) Erosion of the Laurentide
   region of North America by glacial and glaciofluvial
   processes. *Quaternary Research* **23**, 154–74.
7 D.D. Braun (1989) Glacial and periglacial erosion of
   the Appalachians. *Geomorphology* **2**, 233–56.
8 W.S.B. Paterson (1981) *The Physics of Glaciers*, 2nd edn.
   Pergamon Press, Oxford.
9 K. Hutter (1983) *Theoretical Glaciology.* Reidel,
   Boston.
10 K. Hutter (1982) Dynamics of glaciers and large ice
   masses. *Annual Review of Fluid Mechanics* **14**,
   87–130.
11 R.P. Sharp (1951) Features of the firn on Upper Seward
   Glacier, St Elais Mountains, Canada. *Journal of Geology*
   59, 599–62.
12 C.C. Langway (1967) Stratigraphic analysis of a deep
   ice core from Greenland. US Army Cold Regions
   Research and Engineering Laboratory, Research Report
   77, Hanover, New Hampshire.
13 J.W. Glen (1952) Experiments on the deformation of
   ice. *Journal of Glaciology* **2**, 111–14.
14 M.F. Meier (1960) *Mode of flow of Saskatchewan
   Glacier, Alberta, Canada*, Professional Paper 351. US
   Geological Survey, Washington DC.
15 J.C. Savage & W.S.B. Paterson (1963) Borehole
   measurements in the Athabasca glacier. *Journal of
   Geophysical Research* **68**, 4521.
16 G.V. Middleton & P.R. Wilcock (1994) *Mechanics in
   the Earth and Environmental Sciences.* Cambridge
   University Press, Cambridge.
17 J.W. Glen (1955) The creep of polycrystalline ice.
   Proceedings of the Royal Society **228A**, 519–38.
18 L.E. Raraty & D. Tabor (1958) The adhesion and
   strength properties of ice. *Proceedings of the Royal
   Society* **245A**, 184–201.

19 J. Weertman (1968) Comparison between measured
   and theoretical temperature profiles of the Camp
   Century, Greenland, borehole. *Journal of Geophysical
   Research* **73**, 2691–2700.
20 J. Nye (1952) The mechanics of glacier flow. *Journal of
   Glaciology* **2**, 82–93.
21 S.S. Vialov (1958) *IASH* **47**, 266.
22 J.F. Nye The distribution of stress and velocity in
   glaciers and ice sheets. (1957) *Proceedings of the Royal
   Society* **239A**, 113.
23 M.A. Summerfield (1991) *Global Geomorphology.*
   Longman, London.
24 R.B. Alley, D.D. Blankenship, C.R. Bentley & S.T.
   Rooney (1986) Deformation of till beneath icestream
   B, West Antarctica. *Nature* **322**, 57–9.
25 G.S. Boulton (1974) Processes and patterns of glacial
   erosion. In: *Glacial Geomorphology* (eds D.R. Coates).
   State University of New York. Binghampton, pp.
   41–87.
26 C. Thorn (1988) Nivation: geomorphic chimera. In:
   Advances in Periglacial Geomorphology (ed. M.J. Clark).
   Wiley, Chichester.
27 D.E. Sugden & B.S. John (1976) *Glaciers and
   Landscape.* Edward Arnold, London.
28 R.S. Williams, Jr. (1986) Glaciers and glacial landforms.
   In: *Geomorphology from Space* (eds N.M. Short & R.W.
   Blair Jr), Special Publication 486. NASA, Washington
   DC, pp. 521–96.
29 V.R. Baker & R.C. Bunker (1985) Cataclysmic late
   Pleistocene flooding from glacial Lake Missoula: a
   review. *Quaternary Science Reviews* **4**, 1–41.
30 R.H. Meade, C.F. Nordin, W.F. Curtis, F.M. Costa
   Rodrigues, C.M. Do Vale & J.M. Edmond (1979)
   Sediment loads in the Amazon River. *Nature* **278**,
   161–3.
31 K. Hewitt (1993) Altitudinal organization of
   Karakoram geomorphic processes and depositional
   environments. In: *Himalaya to the Sea* (ed. J.F. Shroder
   Jr). Routledge, London, pp. 159–83.
32 J.F. Shroeder, Jr., M.S. Khan, R.D. Lawrence, I. Madin
   & S.M. Higgins (1989) Quaternary glacial chronology
   and neotectonics in the Himalayas of northern Pakistan.
   In: *Tectonics and Geophysics of the Western Himalaya*
   (eds L.L. Malinconico Jr. & R.J. Lillie), Special Paper
   232. Geological Society of America, Washington DC,
   275–93.
33 J.S. Gardner & N.K. Jones (1993) Sediment transport
   and yield at the Raikot glacier, Nanga, Parbat, Punjab
   Himalaya. In: *Himalaya to the Sea* (ed. J.F. Shroder Jr).
   Routledge, London, pp. 184–97.

Index